PHYSICS OF SOUND IN MARINE SEDIMENTS

MARINE SCIENCE

Coordinating Editor: Ronald J. Gibbs, Northwestern University

Volume 1—*Physics of Sound in Marine Sediments*
Edited by Loyd Hampton, 1974

An Office of Naval Research symposium
Consulting Editors: Alexander Malahoff and Donald Heinrichs
Department of the Navy

Volume 2—*Deep-Sea Sediments: Physical and Mechanical Properties*
Edited by Anton L. Inderbitzen, 1974

An Office of Naval Research symposium
Consulting Editors: Alexander Malahoff and Donald Heinrichs
Department of the Navy

In Preparation:

Natural Gases in Marine Sediments
Edited by Isaac R. Kaplan

An Office of Naval Research symposium
Consulting Editors: Alexander Malahoff and Donald Heinrichs
Department of the Navy

Suspended Solids in Water
Edited by Ronald J. Gibbs

An Office of Naval Research symposium
Consulting Editors: Alexander Malahoff and Donald Heinrichs
Department of the Navy

PHYSICS OF SOUND IN MARINE SEDIMENTS

Edited by
Loyd Hampton
Applied Research Laboratories
The University of Texas at Austin
Austin, Texas

PLENUM PRESS • NEW YORK AND LONDON

Library of Congress Cataloging in Publication Data

Symposium on Physics of Sound in Marine Sediments, Austin, Tex., 1973.
Physics of sound in marine sediments.

(Marine science, v. 1)
Conducted by the Ocean Science and Technology Division of the Office of Naval Research.
Includes bibliographies.
1. Marine sediments—Acoustic properties—Congresses. I. Hampton, Loyd, 1930- ed. II. United States. Office of Naval Research. Ocean Science and Technology Division. III. Title.
GC380.2.A25S95 1973 551.4'6083 74-8022
ISBN 0-306-35501-9

Proceedings of a symposium conducted by the Ocean Science and Technology Division of the Office of Naval Research on Physics of Sound in Marine Sediments held in Austin, Texas, May 8-10, 1973

A Division of Plenum Publishing Corporation
227 West 17th Street, New York, N.Y. 10011

United Kingdom edition published by Plenum Press, London
A Division of Plenum Publishing Company, Ltd.
4a Lower John Street, London, W1R 3PD, England

Printed in the United States of America

FOREWORD

The phenomenon of sound transmissions through marine sediments is of extreme interest to both the United States civilian and Navy research communities. Both communities have conducted research within the field of this phenomenon approaching it from different perspectives. The academic research community has approached it as a technique for studying sedimentary and crustal structures of the ocean basins. The Navy research community has approached it as an additional variable in the predictability of sound transmission through oceanic waters. In order to join these diverse talents, with the principal aim of bringing into sharp focus the state-of-the-science in the problems relating to the behavior of sound in marine sediments, the Office of Naval Research organized and sponsored an invited symposium on this subject. The papers published in this volume are the results of this symposium and mark the frontiers in the state-of-the-art.

The symposia series were based on five research areas identified by ONR as being particularly suitable for critical review and for the appraisal of future research trends. These areas include:

1. Physics of Sound in Marine Sediments,
2. Physical and Engineering Properties of Deep-Sea Sediments,
3. The Role of Bottom Currents in Sea Floor Geological Processes,
4. Nephelometry and the Optical Properties of the Ocean Waters,
5. Natural Gases in Marine Sediments and Their Mode of Distribution.

These five areas also form some of the research priorities of the ONR program in Marine Geology and Geophysics.

The Marine Geology and Geophysics Program is the section of the Navy Ocean Science Program that provides research into the ocean floor phenomena of interest to naval systems. The basic research objective of the program is to determine the acoustical, physical, chemical, and geological properties of the ocean floor. To accomplish this objective, research studies are directed toward understanding the deep ocean processes that generate oceanic sediments and crustal layers. These processes directly affect the physical and acoustic properties of the sea floor and are the origin of bathymetric, gravity, and magnetic variations in the ocean basins. Acoustic and nonacoustic submarine and antisubmarine operations; surveillance system design and construction; inertial navigation, guidance, and trajectory calculations; search, rescue, and recovery operations; and a variety of other Navy operations are directly influenced by the ocean floor.

The Marine Geology and Geophysics Program is divided into five areas, which are summarized here and detailed in the following sections.

Particular emphasis is directed toward studies of ocean floor acoustical processes, namely the physics of high frequency reflectivity of the ocean floor; the physics of low frequency acoustical propagations through the sea floor; the formation of ocean floor acoustic reflectors; and the areal variations in these acoustic phenomena.

Another part of the program is directed toward studies of the source and location of the major bathymetric, gravity, and magnetic variables, together with the application of the knowledge gained to the prediction of trends and amplitudes of these anomalies to unsurveyed areas.

Additional studies are carried out in the dynamics of sediment transport, including the effect of bottom currents on microtopography and rates of sediment deposition and erosion; the effects of sediment composition, chemical changes, and ocean floor properties on ocean floor stability; and, the relationships of sediment parameters to acoustic parameters.

The variations in acoustical and physical properties of marine sediments, the morphology of the sea floor, magnetic anomaly trends, and changes in the wavelength of marine gravity anomalies are not random phenomena, but result from discrete physical processes. Although the individual research tasks of the program are described in terms of specific program objectives, data and concepts from other classes of studies are often required for solutions of the problem. For example, ocean floor acoustic studies directed toward determining the continuity and source of prominent low frequency refracting layers of the ocean floor utilize sediment properties obtained from sediment dynamic studies and information on major structural discontinuities from physiographic and magnetic field studies. Results from the acoustic studies, in turn, provide significant constraints on gravity field models. Thus research cruises tend to be devoted to integrated marine geological and geophysical investigations and take place in geographic areas that are an integral part of the entire program. A typical cruise would involve the use of seismic reflection and refraction, precision echo sounding, gravity and magnetic sensors, sea floor coring, dredging, and bottom photographic instrumentation. These ocean-going research programs are backed by theoretical, laboratory, and specialized offshore investigations of limited duration. Requirements for specialized, expensive equipment, including a sizeable research vessel, and multidisciplinary inputs lead to a concentration of research tasks at the major oceanographic institutions.

The symposium on the physics of sound in marine sediments was formulated to enhance the scientific viability of the program in ocean floor acoustics. This field is directed towards the understanding of the physics of sound propagation through the sea floor. The prime frequencies of interest to the ONR sponsored program are 200 hertz and below (long range surveillance), 100 to 1000 hertz (short range passive surveillance and tactical sonar), and 1000 to 5000 hertz (tactical bottom bounce sonar). The specific objectives are attacked through the following outstanding scientific questions:

<u>ENERGY</u> (1) What is the distribution of energy in long range water and bottom paths? How can the bottom paths be exploited?

<u>REFRACTION</u> (2) What is the continuity and source of the prominent refracting layers of the ocean floor important in the propagation frequency range below 100 hertz? Can these refracted paths be exploited at very low, infrasonic frequencies below 10 hertz?

<u>REFLECTION</u> (3) What are the variations in reflectivity of sediments, both in the vertical sense and in horizontal continuity? In the sea floor these are important in the range below 4000 hertz. Which sediments or basement layers give rise to the prominent reflectors?

<u>COHERENCE</u> (4) How is coherence and attenuation of reflected and refracted acoustic waves affected by the topography of the ocean floor surface, by the passage of the signal through the ocean floor, and by changes in the physical properties of the sedimentary and crustal layers?

Solutions to the above problems are being pursued through three major classes of studies:

(1) <u>Underway single ship investigations</u>. These studies use 3.5 kilohertz echo sounders, sparker arrays (100-500 hertz), and airguns (20-200 hertz) as continuous underway pulsed sound sources giving an acoustic source energy in the range from 3.5 kilohertz to 20 hertz. The detectors used are multisensor towed arrays for vertical incidence reflections off the sea floor and sonobuoys for wide-angle reflections and for refraction path studies.

(2) Ocean floor instrumentation. These are used for the measurement of shallow sedimentary layers of the ocean floor, and for crustal propagation paths. Three component seismometers are used for the study of refracted P, S, and Stoneley Wave propagation in the frequency range of 1 hertz to 100 hertz through the ocean floor. The instruments are usually deployed on the sea floor for periods ranging from days to months.

(3) Physical properties of sediments. A direct relationship between the physical properties of marine sediments and their velocity gradients, acoustic reflectivities, and attenuations is known to exist. It is not known which of the physical properties or mineralogical properties are the most influential upon sound transmission through the sea floor. In-situ sound velocity measurements are carried out simultaneously with ocean coring procedures in order to relate acoustic boundaries in the sediments to mineralogical or physical boundaries. Laboratory examination of the samples is undertaken to determine their physical, mechanical, and mineralogical properties, which are compared with the in-situ measurements and to acoustic reflection and refraction horizons in the sediments as measured from the ocean surface.

The ONR objectives for this symposium were fully realized, as is shown by the content of the book. The excellent mix of both civilian and Navy laboratory authors indicates the diverse talent that was focused upon the general problem of sound propagation through the marine sediments. All aspects of the ONR program in sea floor acoustics were covered by the symposium, bringing focus upon the new research frontiers within this particular discipline.

Alexander Malahoff

PREFACE

The fact that approximately 500 references are used by the authors of the articles in this collection illustrates the problems addressed by this book. The topic of Physics of Sound in Marine Sediments is a very complex one; it is a field in which rapid changes are occurring and one which crosses many scientific disciplines. For these reasons comprehensive reporting is difficult and is very seldom timely. A need for a useful assessment of our knowledge in this area (and several others) was recognized by the Ocean Science and Technology Division of the Office of Naval Research and a series of Symposia were convened during the winter and spring of 1972-73. One of the goals of the Symposia was to obtain a useful and timely summary of our state of knowledge, and that is the purpose of this book.

The Symposium on Physics of Sound in Marine Sediments was held at Lakeway Inn near Austin, Texas, on 8, 9, and 10 May 1973. Those attending (see accompanying photograph and list) were selected to represent as many scientific interests as appropriate and, also, to represent the "user" of the information as well. Selection was also based on contributions to the field and, of those 500 references mentioned previously, approximately one fourth are citing works published by those attending the Symposium.

The papers presented here are the versions submitted by the participants after the Symposium. In many cases there were revisions to reflect the interaction and discussions which occurred at the Symposium. In general the emphasis of the papers is summary and interpretation, and thus the papers are structured differently from most journal papers. The collection of papers should therefore be useful not only to those working in this area but especially to those who have less familiarity and who desire a status report on such a diverse topic.

Special recognition must go to N. A. Ostenso and A. Malahoff of the Office of Naval Research for making this Symposium possible. One of the responsibilities of the editor was to prepare submitted papers for this volume. If there are mistakes from this process, they are the responsibility of the editor and should in no way reflect on the authors.

Loyd Hampton

Austin, Texas
February 1974

CONTENTS

A REVIEW OF THE DERIVATIONS OF THE FORMULAS FOR THE ACOUSTICAL PROPERTIES OF LIQUID-SOLID MIXTURES

AVTAR S. AHUJA

Rust College

Holly Springs, Mississippi 38635

ABSTRACT

The natural sediments run through the whole gamut from dilute suspensions of very high porosity on one end to porous solids of very low porosity on the other end. In this paper the state of the art in the derivations of formulas for the acoustical properties of deflocculated liquid-solid mixtures (of porosity varying from close to 100% to as low as 60%) is briefly reviewed. Basic assumptions, methodology, and improvements in the derivations are delineated. Derivations have been classified under three categories: scattering theory, conservation laws, and other methods. The media properties which are shown to have effect on the acoustic properties are: density, compressibility, shear viscosity, thermal conductivity, and surface tension. In the existing literature the particle has been assumed to be spherical. It has been demonstrated both theoretically and experimentally that when the suspension components differ appreciably in density, the particle shape does have effects on the acoustic parameters. Natural sediments (clay, for example) occur in shapes described as close to ellipsoidal. It is concluded in this review that the method of formulation of the wave equation for suspensions will yield acoustic parameters for ellipsoidal particles when the drag force on the ellipsoidal particle in the sound field is known.

INTRODUCTION

The naturally occurring marine sediment is a mixture of water and solid particles. Clay and silt size particles (of diameter of about 1 to 2 μ, 1 μ = 10^{-4} cm) are in suspension, and particles the size of sand (of diameter of about 500 μ) are in contact with one another. Experimental data show that natural sediments in the sea range from suspensions of particles of 1 μ to consolidated aggregates of sand grains and granules of 1000 μ; therefore, it may be said that the natural sediments occur over a broad spectrum: at one end are deflocculated dilute (high porosity) suspensions, and at the other end are the porous solids of low porosity.

It has recently been concluded [Hamilton, 1971] that with the exception of sediments existing in bays, estuaries, river deltas, or some other localities where the sediments are deposited at a fast rate and lack structural rigidity, the sediments possess enough rigidity for supporting shear waves. Mathematically speaking, the two ends of the sedimentary spectrum represent two different models: one is defined by deflocculated suspension theory and the other by porous solid theory. In the intervening area different degrees of flocculation or particulate contact are found. It is not yet clear whether one mathematical model will ever be able to cover the whole sedimentary spectrum. This paper deals with suspension theory and reviews the sound propagation in dilute and in concentrated deflocculated suspensions. Sound transmission in porous solids is not considered.

It should be remarked that the quantitative meaning of dilute suspension in terms of particle volume concentration is not definitive. Generally a dilute suspension is one of porosity of about 95% to 100%, or of particle volume concentration of about 0 to 5%; however this is not always the case in applying dilute suspension theory to experimental data. Dilute suspension theory has been shown to apply to sound propagation in kaolin-water suspension up to about 9% volume concentration of kaolin [Urick, 1948] and in blood up to about 40% volume concentration of red blood cells [Carstensen and Schwan, 1959].

Different theories for calculating acoustic parameters, such as sound velocity, viscous and thermal absorption coefficients, etc., will be reviewed. The theories require the following assumptions: (1) spherical particle, (2) particle size much less than the wavelength of the incident sound, (3) no relaxation mechanism, unless otherwise specified, and (4) no mass transfer between the particle and the surrounding fluid, i.e., no evaporation, condensation, chemical reactions, etc.

When a compressional wave falls on an obstacle, in addition to reflected and transmitted compressional waves there arise in the two media a highly damped longitudinal (thermal) wave and a highly damped transverse (viscous) wave. The energy loss is due to the complete absorption of these thermal and viscous waves in the two media. The depths of penetration for thermal and viscous waves in water, that is, the distances over which the amplitudes of waves reduce by a factor of e, are given in Table 1 for frequencies ranging from 1 kHz to 1 MHz. The depth of penetration of thermal waves in water is seen to be about the same as the size of the clay particles, i.e., 2 μ, at 10 kHz when the particle has been assumed to possess infinite viscosity. The depth of penetration of the viscous wave is equal to about the size of the particle at 100 kHz.

TABLE 1

CLAY PARTICLE OF DIAMETER 2R = 2×10^{-4} cm IN SUSPENSION IN WATER

Quantity	Frequency in Hertz			
	10^3	10^4	10^5	10^6
Wavelength of the incident compressional wave, γ (cm)	146.4	14.64	1.464	0.1464
Wavelength of the thermal wave, $2\pi\ \delta_t$ (cm)	42.38×10^{-4}	13.40×10^{-4}	42.38×10^{-5}	13.40×10^{-5}
Depth of penetration of thermal wave in suspending medium, δ_t* (cm)	6.74×10^{-4}	2.13×10^{-4}	6.74×10^{-5}	2.13×10^{-5}
Depth of penetration of thermal wave in particle, δ_t'† (cm)	3.19×10^{-3}	1.01×10^{-3}	3.19×10^{-4}	1.01×10^{-4}
Wavelength of the viscous wave, $2\pi\ \delta$ (cm)	112×10^{-4}	35.42×10^{-4}	112×10^{-5}	35.42×10^{-5}
Depth of penetration of viscous wave in suspending medium, δ* (cm)	17.84×10^{-4}	5.64×10^{-4}	17.84×10^{-5}	5.64×10^{-5}

*Defined in text.

†The value of thermal diffusivity of sandy clay has been taken to be 5.1×10^{-3} cm^2/sec (International Critical Tables).

SOUND ATTENUATION

Lord Rayleigh [1897] calculated the velocity potential Ψ of the disturbance caused by an ellipsoidal particle in the path of a plane sound wave:

$$\Psi = -\frac{4\pi V e^{-ikr}}{\lambda^2 r}\left[(1 - \beta'/\beta) + \frac{x}{r}\,\frac{4\pi(\rho'-\rho)}{4\pi\rho' + (\rho-\rho')S}\right] , \qquad (1)$$

where V is the volume of the particle, λ the wavelength of the incident sound, k the propagation constant, r the radial coordinate, β the adiabatic compressibility, ρ the density, and x the rectangular coordinate. Primed quantities indicate the suspended medium and the unprimed indicate the suspending medium. S is the shape factor and is $4\pi[1/e^2-(1-e^2)^{1/2} \sin^{-1}e/e^3]$ for an ellipsoid of flattened form (i.e., two axes equal) and is $4\pi-2\pi^2\sqrt{1-e^2}$ for a disk ($e\simeq 1$ where e is the eccentricity). Assuming that the suspending medium is frictionless, Rayleigh found that in the expansion of the potential of the disturbance in spherical harmonic terms, the zeroth order term, or the monopole coefficient, gives the contribution of the adiabatic compressibility difference and the first order term, or dipole coefficient, gives the contribution of the density difference between the particle and the suspending medium.

Realizing that the first order term in the expansion of the velocity potential of the disturbance gives the contribution of the relative motion between the particles and the medium and hence that of viscous drag, Sewell [1910] calculated the viscous absorption coefficient for suspensions containing rigid, immoveable spherical particles. Sewell's formula is valid for fog (or aerosols) at audiofrequencies, where the water particles may be assumed to be fixed due to large inertia, but it is not valid in hydrosols, where the particles move relative to the medium. As it has been pointed out by Lamb (Hydrodynamics, 1945 edition), even when the inertia of the particle (which is proportional to R^3 where R is particle radius) reduces much faster than the viscous force (which is proportional to R^2) a stage is reached when the particle begins to oscillate with air and acoustic energy attenuation is reduced. Lamb condensed and improved upon Sewell's treatment by including the moveability of the particle.

Epstein [1941] improved upon Sewell's and Lamb's treatments of viscous absorption by considering the cases of elastic solid and viscous spherical particles suspended in a viscous medium. He ignored the thermal conduction between the two phases of the suspension and reduced his general monopole and dipole coefficients to explicit form for the case of fog.

Epstein and Carhart [1953] extended Epstein's theory significantly by including the thermal conduction effects, and they calculated both the viscous and thermal absorption coefficients in explicit form for the case of fog. The thermal absorption coefficient is given by the monopole coefficient, and the viscous and thermal absorption coefficients are decoupled if the ratio of the thermal and compressional wavelengths is small. From Table 1, it is seen that this ratio is negligible. Epstein and Carhart used stress tensors for fluids and obtained the results for solids by letting the particle viscosity approach infinity. This yields the correct expression fortuitously only in the long wavelength limit. By using a stress-strain relation appropriate for isotropic elastic solids, Allegra and Hawley [1972] improved upon this situation and obtained more generally applicable results. The viscous and thermal absorption losses, when plotted against R/δ (δ is the depth of penetration of viscous wave in suspending medium) and R/δ_t' (δ_t' is the depth of penetration of thermal wave in the particle) give maxima where the viscous wavelength in the suspending medium and the thermal wavelength in the particle are approximately equal to the particle size. It is seen from Table 1 that the viscous loss for clay particles of diameter 2 μ suspended in water is predicted to be a maximum at a frequency of 1 MHz or greater, and that the thermal loss is expected to be a maximum at a frequency in the range of 0.1 MHz to 1 MHz. It should be pointed out that the critical frequencies at which the maxima in viscous and thermal losses occur depend upon the particle size, given the particle material and the suspending medium; if the particle size is increased, critical frequencies at which maxima occur will be reduced.

Epstein and Carhart [1953] assumed in their treatment that the particles remain spherical in the sound field, that the amplitude of motion is smaller than the particle radius, and that the surface tension effect is negligible. Chow [1964] improved upon the Epstein-Carhart treatment by allowing the particle large amplitude motion compared with its size and small deformation from the spherical shape, and by including the effect of surface tension between the media. He fixed the coordinate axes in the particle itself. He has shown that the effect of surface tension, which enters through the compressibility of the particle, is negligible in the mixture of liquid particles suspended in liquid or gaseous medium because of the low compressibility of the liquid particles. But for gaseous particles suspended in a liquid medium (such as air bubbles in water) the effect of surface tension is to increase the thermal dissipation several-fold (2 to 4 times). He also found that surface tension leaves the dipole coefficient unaltered.

Urick [1948] calculated the viscous absorption coefficient of a suspension in a simple manner by calculating the energy loss in the viscous boundary layer around the oscillating rigid sphere.

The equation of motion of an oscillating sphere is [Lamb, 1945]

$$F = -\tau\, \rho\, du/dt - s\, \rho\, \omega\, u \quad , \tag{2}$$

where F is the drag force experienced by the sphere, ω the angular frequency, and t the time. u is the instantaneous velocity of the particle and is $v'-v$, where v' is the velocity of the center of the particle and v the velocity of the fluid at the center of the particle if it were absent. For the rigid sphere,

$$\left.\begin{aligned} \tau &= 1/2 + (9/4)(\delta/R) \\ s &= (9/4)(\delta/R)(1 + \delta/R) \end{aligned}\right] \tag{3}$$

R is the particle radius, and $\delta[=(2\eta/\rho\omega)^{1/2}]$ is the distance over which the amplitude of the shear wave in the suspending fluid declines by a factor of e, where η is the dynamic viscosity of the suspending fluid. Replacing u by $v'-v$, Eq. (2) can be written as

$$\rho' \frac{dv'}{dt} = \rho \frac{dv}{dt} - \tau\rho \frac{d}{dt}(v'-v) - s\rho\omega(v'-v) \quad ,$$

or,

$$(\rho'/\rho + \tau)\frac{dv'}{dt} + \omega s v' = (\tau+1)\frac{dv}{dt} + \omega s v \quad . \tag{4}$$

Equation (4) can be solved to give

$$u = v' - v = -\frac{(\rho'/\rho - 1)\, e^{-i\epsilon}}{[(\rho'/\rho + \tau)^2 + s^2]^{1/2}}\, v \quad , \tag{5}$$

where

$$\epsilon = \frac{s}{\rho'/\rho + \tau} \quad . \tag{6}$$

The average energy loss in the viscous boundary layer yields the amplitude absorption coefficient:

$$X_{vis} = 1/2\, \varphi\, (\omega/c) \frac{(\rho'/\rho - 1)^2 s}{[(\rho'/\rho + \tau)^2 + s^2]} \quad , \tag{7}$$

where c is the velocity of sound in the suspending medium.

VELOCITY OF SOUND

Employing the following derivation, given by Lord Rayleigh [1899] in his interpretation of the blue of the sky, Herzfeld [1930] calculated the velocity of sound in suspension as follows.

If a plane sound wave (the primary wave) falls on a small object, a scattered wave will be emitted which attenuates completely at a large distance. If we imagine a slab of suspension of a certain thickness containing particles and extending to infinity in the plane at right angles to direction of propagation of the primary wave, then at a large distance from the slab the scattered waves from all the particles will combine to give two waves of the same frequency as of the incident or primary wave. One wave (the forward wave) goes in the same direction as the primary wave and the other wave (the reflected wave) goes in the opposite direction. The forward and primary waves give a resultant wave having a phase different from the primary wave which is proportional to the slab thickness. This phase difference has the same effect as if the resultant wave had a different velocity from the primary wave.

The formula which Herzfeld derived failed the test of the experiment, as shown by Randall [1931].

In his book A Textbook of Sound, Wood proposed a formula for velocity of sound for mixtures by regarding the mixture as homogeneous and by taking the volume-averaged values of adiabatic compressibility and density and substituting in the formula $c=(\beta\rho)^{-1/2}$. This relationship was shown by Urick [1948] to hold for suspensions containing very small particles for which sound scattering can be ignored. As the particle size is increased, scattering cannot be ignored and the assumption of homogeneity becomes untenable.

Using Rayleigh's idea mentioned above, Urick and Ament [1949] characterized the suspension by a complex propagation constant, k_s, given by

$$(k_s/k)^2 = \left[1 + \frac{4\pi N}{ik^3} a_o\right]\left[1 - \frac{4\pi N}{k^3} a_1\right] \quad , \qquad (8)$$

where N is the number of particles, the monopole coefficient is

$$a_o = - 1/3\ ik^3R^3\ (1 - \beta'/\beta) \quad , \qquad (9)$$

and the dipole coefficient is

$$a_1 = - A_1 \frac{(\rho'/\rho - 1)}{(\rho'/\rho - 1) + 3A_1/k^3R^3} \tag{10}$$

and

$$A_1 = 1/2\ k^3R^3(1+\tau+is) \quad . \tag{11}$$

For the rigid sphere, quantities τ and s are given by Eq. (3).

Urick and Ament used the above monopole (a_o) and dipole (a_1) coefficients calculated by Lamb [1945] and considered only the compressibility and moveability of the rigid spherical particles.

Using for the propagation constant,

$$k_s = \omega/c_s + i\ X_{vis} \tag{12}$$

in Eq. (8), we get the velocity of sound in suspension

$$c_s/c = 1 + 1/2\ \varphi(1 - \beta'/\beta)$$

$$- 1/2\ \varphi \frac{(\rho'/\rho - 1)[(\tau+1)\cos\epsilon + \sin\epsilon]}{[(\rho'/\rho + \tau)^2 + s^2]^{1/2}} \quad , \tag{13}$$

and the amplitude absorption coefficient X_{vis} is given by Eq. (7). φ stands for fractional volume concentration of particles.

Ahuja [1972a] included the effect of the particle viscosity in the Urick-Ament formula by assuming particle kinematic viscosity at least one order of magnitude greater than that of the suspending medium. This assumption, which is valid for the red blood cell suspension, yields unsteady flow outside the particle and steady flow inside the particle. Ahuja's formulation yields the values of c_s and X_{vis} as given in Eqs. (7) and (13), with the only exception that now

$$\tau = 1/2 + (\delta/4R) \frac{(2\eta+3\eta')^2}{[(\eta+\eta') + R\eta/3\delta]^2 + (R\eta/3\delta)^2} \tag{14}$$

and

$$s = 3/4\ (2\eta+3\eta')\ (\delta/R)^2 \times \frac{(1 + R/\delta)\ [(\eta+\eta') + R\eta/3\delta] + (\eta/3)(R/\delta)^2}{[(\eta+\eta') + R\eta/3\delta]^2 + (R\eta/3\delta)^2} \tag{15}$$

for a very viscous sphere (that is, $\eta'/\eta \geq 10$). In the case of the rigid sphere, $\eta/\eta' \to 0$, Eqs. (14) and (15) reduce to Eq. (3).

This study was begun to estimate the internal viscosity of the red cell, and was based on the assumption that the viscosity of the red cell was large compared to that of plasma, as supported by experimental evidence. It was found that since the density difference between the red cell and plasma is about 5%, the effect of red cell viscosity is negligible. However, the particle viscosity will have measurable effect on the acoustic properties of suspensions for which the density difference is significant.

Using Ament's [1959] monopole coefficient, one sees that the effect of the volume viscosity of the liquid particle on the propagation of sound in suspensions is negligible [Ahuja, 1972a]. This is because the volume viscosity of the particle enters as the product of the particle compressibility and the volume viscosity, and the liquid particle compressibility is low, of the order 10^{-12}.

Ament [1953] calculated the velocity of sound and viscous absorption coefficient by computing the effective density of the suspension in the sound field by assuming incompressible fluid flow around the particle and the equality of drag and buoyant forces on the particle. It is fortuitous that his formula for sound velocity agrees with the experiment up to 50% volume concentration in a bromoform suspension. Strictly speaking, his formula is valid only for dilute suspensions. This is because he uses Stoke's Law to describe the drag on the particle and this relation applies only in the case of a single particle in an infinite medium. In a concentrated suspension, particle interactions take place by way of hydrodynamic boundary layer of one particle interfering with that of the other (and possibly by way of particle collisions in the event of gravity settling of the particles). Ament in his treatment does not consider hydrodynamic interactions of the particles.

Chu and Chow [1965] presented a macroscopic theory for a two-phase fluid by including the drag force on the particle as a body force (in the absence of mass transfer between the suspended particle and the surrounding fluid) in the momentum equation and the heat exchange between phases as a heat source in the energy equation. They calculated the drag force and heat flux from a detailed analysis on the microscopic scale, that is, on the scale of particle dimensions. From the equation of continuity, momentum equation, and energy equation, they solved for the complex propagation constant. The real part of the propagation constant gives the sound velocity and the imaginary part gives the sum of the thermal and viscous attenuation coefficients.

Temkin and Dobbins [1966] calculated the sound velocity and attenuation coefficient by setting up the equation of continuity, momentum equation, and energy equation in each phase of the suspension and by including the drag force on the particle from Stoke's Law and the heat exchange between components from Newton's Law of Cooling. They solved these equations simultaneously and calculated the complex propagation constant. They indicated the relaxation character of the problem by pointing up the similarities between the lag of the fluctuations of temperature and velocity of the particle in comparison with those of the fluid in the acoustic field and the lag in molecular or atomic internal degrees of freedom in relaxation mechanisms. Their treatment is limited to aerosols because they used a Perfect Gas Law relation as an equation of state.

Ahuja [1972b] calculated the velocity of sound in suspension or emulsion by formulating a wave equation. He performed mass and momentum balances on a volume element of a suspension as follows.

If there are N particles of the same radius, then the fractional volume concentration of the particles is $\varphi = 4/3\ R^3N$ and the porosity is $1-\varphi$. If the particles are randomly and uniformly distributed in the suspension, then, in the absence of the acoustic field, the volume-averaged density of the suspension $\rho_s=(1-\varphi)\rho+\varphi\rho'$ and the volume-averaged adiabatic compressibility of the suspension $\beta_s=(1-\varphi)\beta+\varphi\beta'$.

The wavelength of the acoustic wave at a frequency of 1 MHz is about 0.15 cm in water (Table 1). Since there can be as many as 1000 particles (of diameter of about 1 to 2 μ) per wavelength (at 1 MHz and more at lower frequencies), the size of the suspension volume element can be so chosen that it is small enough to be much less than the incident wavelength and large enough to contain many particles in order to be treated as homogeneous in the path of the sound wave.

The cubical volume element with edges, for example, equal to 1/15 of the wavelength in water (about 0.01 cm) is about 10^{-6} cm^3. A suspension of 1% concentration will have in this volume about 2500 particles of radius 10^{-4} cm, and therefore it can be replaced, in the first approximation, by the homogeneous and continuous fluid of density ρ_s and compressibility β_s.

Let a plane acoustic wave be propagated through the suspension along the x direction. Let a suspension parallelepiped of volume dx (containing many particles) be bounded by the unit areas of the planes (perpendicular to the direction of propagation) at sections x and x+dx and the planes perpendicular to these planes. (The differential dx is assumed to be a fraction of the wavelength.)

Let the suspension be replaced by a homogeneous and continuous fluid.

In the presence of sound, field quantities such as density, pressure, temperature, etc., fluctuate. Let these fluctuations, which are functions of x and t only, be very small compared with their equilibrium values. For example, the suspension density becomes $\rho_s+\Delta_s$, where the density fluctuation $\Delta_s \ll \rho_s$. It is further assumed that, in the presence of the sound field, homogeneous and continuous fluid enters the element dx at (volume-averaged) velocity $(1-\varphi)v+\varphi v'$ and at (volume-averaged) momentum $(1-\varphi)\rho v+\varphi\rho' v'$.

Equation of Continuity

As an acoustic wave passes through a suspension (or an emulsion), its components pulsate and oscillate. The volume-averaged velocity at which the fluid enters the volume element at section x is $[(1-\varphi)v+\varphi v']$ and, therefore, the mass flux entering at x is $\{(\rho_s+\Delta_s)[(1-\varphi)v+\varphi v']\}_x$. Similarly, the mass flux leaving volume element at section x+dx is $\{(\rho_s+\Delta_s)[(1-\varphi)v+\varphi v']\}_{x+dx}$. The corresponding decrease in mass with time in the volume element is $-(\partial/\partial t)(\rho_s+\Delta_s)dx$. Hence mass balance on the suspension element (containing many particles) yields

$$-\partial/\partial t(\rho_s+\Delta_s) = \partial/\partial x\ \{(\rho_s+\Delta_s)[1-\varphi)v+\varphi v']\} \quad .$$

Since ρ_s has been assumed to be independent of space and time, and ignoring the second-order quantities such as the products $v\Delta_s$ and $v'\Delta_s$, we get

$$-(\partial\Delta_s/\partial t) = \rho_s(\partial/\partial x)[v+\varphi(v'-v)] \quad . \tag{16}$$

From thermodynamics it is known that in a single phase fluid the density fluctuation $\Delta=\rho\beta p$, where p is acoustic pressure. In the first approximation, this relation will hold in suspensions (considered homogeneous and continuous in compressibility β_s and density ρ_s); that is, $\Delta_s=\rho_s\beta_s p_s$, where p_s is the acoustic pressure in the suspension. With this value of Δ_s and Eq. (5), Eq. (16) reduces to

$$-\beta_s\ \partial p_s/\partial t = \left\{1 - \varphi\frac{(\rho'/\rho - 1)e^{-i\epsilon}}{[(\rho'/\rho + \tau)^2 + s^2]^{1/2}}\right\}\partial v/\partial x \quad . \tag{17}$$

Momentum Equation

The volume-averaged momentum density of the suspension element is $(1-\varphi)\rho v+\varphi\rho' v'$. Its total time derivative, $(d/dt)[(1-\varphi)\rho v+\varphi\rho' v']$, must be balanced by the acoustic pressure gradient, $-(\partial p_s/\partial x)$. Hence

$$(d/dt)[(1-\varphi)\rho v+\varphi\rho' v'] = -(\partial p_s/\partial x) \quad . \qquad (18)$$

The left-hand side can be written in terms of partial derivatives as

$$\partial/\partial t \ [(1-\varphi)\rho v+\varphi\rho' v'] + (1-\varphi)\rho v \ \partial v/\partial x + \varphi\rho' v \ \partial v'/\partial x \quad ,$$

from which the second and the third terms are second-order and can be ignored, and Eq. (18) can be written as

$$(\partial/\partial t)[(1-\varphi)\rho v+\varphi\rho' v'] = -(\partial p_s/\partial x) \quad .$$

By using the equation of motion Eq. (4) and Eq. (5), the momentum equation becomes

$$\rho \ \partial v/\partial t + \rho\varphi \ (\tau \ \partial v/\partial t + \omega s v) \ \frac{(\rho'/\rho - 1)e^{-i\epsilon}}{[(\rho'/\rho + \tau)^2 + s^2]^{1/2}} = - \ \partial p_s/\partial_x \quad . \qquad (19)$$

Wave Equation for Suspensions

Differentiating Eq. (19) with respect to t and Eq. (17) with respect to x and eliminating $\partial^2 p_s/\partial x\partial t$, we get, after a little algebra, the wave equation for propagation of sound in suspensions

$$\partial^2 v/\partial x^2 = (1/c^2)[1-\varphi(1 - \beta'/\beta)][1 - \varphi L \cos \epsilon + i\varphi L \sin \epsilon]^{-1}$$
$$\times [1+\varphi L(\tau \cos \epsilon + s \sin \epsilon) + i\varphi L(s \cos \epsilon - \tau \sin \epsilon)] \qquad (20)$$
$$\partial^2 v/\partial t^2 \quad ,$$

where

$$L = \frac{(\rho'/\rho - 1)}{[(\rho'/\rho + \tau)^2 + s^2]^{1/2}} \quad .$$

Comparing Eq. (20) with

$$\partial^2 v/\partial x^2 = (k_s/\omega)^2 \, \partial^2 v/\partial t^2 \quad , \tag{21}$$

and by using Eq. (12) we obtain the velocity of sound c_s (Eq. 13) and viscous absorption coefficient X_{vis} (Eq. (7)).

The effect of heat conduction has also been included in the wave equation via the complex compressibility calculated from the scattering theory [Ahuja, 1973]. All three acoustic parameters, namely, the sound velocity (c_s), viscous absorption coefficient (X_{vis}), and the thermal attenuation coefficient (X_t), have been extracted from the wave equation and they are

$$c_s/c = 1 + 1/2\ \varphi(1 - \beta'/\beta) - 3/4\ \varphi(\gamma-1)(1 - \alpha'\rho C_p/\alpha\rho' C_p')^2(\delta_t/R)$$
$$- 1/2\ \varphi \frac{(\rho'/\rho - 1)\ [(\rho'/\rho + \tau)(\tau+1) + s^2]}{[(\rho'/\rho + \tau)^2 + s^2]} \quad , \tag{22}$$

$$X_{vis} = 1/2\ (\omega/c)\varphi \frac{(\rho'/\rho - 1)^2 s}{[(\rho'/\rho + \tau)^2 + s^2]} \quad , \tag{23}$$

and

$$X_t = 3/4\ (\omega/c)\varphi[(\gamma-1)(1 - \alpha'\rho C_p/\alpha\rho' C_p')^2(\delta_t/R)(1+\delta_t/R)] \ . \tag{24}$$

In Eqs. (22-24), γ is the adiabtic constant, α the coefficient of thermal expansion, C_p the specific heat at constant pressure, and $\delta_t[=(2\sigma/\rho\omega C_p)^{1/2}]$ the distance over which the amplitude of thermal wave in suspending fluid declines by a factor e. Primed quantities stand for the particle material and the unprimed for the suspending fluid. The effect of heat conduction on sound propagation in suspensions has been included in Eqs. (22-24) under the assumption of ratio of thermal conductivity $\sigma/\sigma' \rightarrow 0$.

CONCENTRATED SUSPENSIONS

Urick's [1948] sound attenuation measurements in kaolin-water suspensions are explained by his viscous drag loss theory for dilute suspensions up to about 9% concentration. Beyond this concentration the viscous absorption coefficient is less than that predicted by the dilute suspension theory.

McCann [1969] presented a theory for calculating the acoustic attenuation coefficient of concentrated clay suspensions in terms of the long range electrical forces between the clay particles in a deflocculated suspension. Assuming that the clay lattice is cubic, with all particles equidistant from each other, McCann calculated the force exerted on a particle by its nearest neighbors in terms of particle displacements. He included this force in the equation of motion of the particle oscillating in the sound field (Eq. 4) and solved the resulting equation in the manner of Urick [1948] to yield the absorption coefficient,

$$X_{vis} = 1/2\ \varphi(\omega/c)\ \frac{[(\rho'/\rho - 1) - P\varphi^{1/3}]^2 s}{[(\tau + \rho'/\rho) - P\varphi^{1/3}]^2 + s^2} \quad . \quad (25)$$

The factor P includes the charge per unit area on each particle and is independent of the particle diameter. He obtained the constant P of his equation, 1.7, from Urick's [1948] experimental results, and the resulting equation predicts the attenuation in clay suspensions over a wide range of concentration (0 to 40%) and frequencies (10 kHz to 1 MHz).

Duykers [1970] has proposed a relaxational model for sound absorption and dispersion in kaolin-water mixture by treating it as a viscoelastic liquid. He found an excellent agreement between his theory and Urick's experiments up to particle concentration of 40%.

CONCLUDING REMARKS

Different theories for calculating acoustic parameters of liquid-solid mixtures (suspensions) have been briefly reviewed and can be classified in three categories: scattering theory, conservation laws (for each component of the suspension as well as for the suspension itself) and other (ad hoc) methods such as those of Urick [1948] and Ament [1953]. A different approach for the calculation of the acoustic parameters, by formulating a wave equation in suspensions, has been presented [Ahuja, 1972a] in somewhat greater detail.

In this review only viscous interaction between the liquid and solid components of the mixtures have been considered. One problem belonging to the sedimentary spectrum (mentioned in the Introduction), when the suspended particles rub against each other in the acoustic field, as yet has no solution. The problem of the effect of particle shape on acoustic parameters also remains unsolved.

It can be seen from Lord Rayleigh's Eq. 1 that particle shape will have an effect on the velocity potential of the disturbance of the sound wave caused by the particle, if the density difference between the particle and the suspending medium is significant. Blue and McLeroy [1968] measured the effect of particle shape on the attenuation of sound in suspensions and showed that platelet-shaped particles caused higher attenuation than needle-shaped particles.

Since the mineral contents (such as sand and clay) are not spherical and their specific gravities are also significant (varying from about 1.25 to 2.00), formulas for acoustic parameters corresponding to spherical shapes of the particle give only approximate results. Because an important component of marine sediment suspensions - clay - is plate-like, the need for the acoustic properties of suspensions containing ellipsoidal particles is obvious as a first improvement in the formulas. Once the drag force on the ellipsoidal particle in the acoustic field is known, the derivation of the formulas of acoustic properties will be simpler, using formulations of the wave equations, than by using scattering theory.

REFERENCES

Ahuja, A. S., "Effect of particle viscosity on propagation of sound in suspensions and emulsions," J. Acoust. Soc. Am. 51, 182-191, 1972a.

Ahuja, A. S., "Formulation of wave equation for calculating velocity of sound in suspensions, "J. Acoust. Soc. Am. 51, 916-919, 1972b.

Ahuja, A. S., "Wave equation in propagation parameters for sound propagation in suspensions," J. Appl. Phys. 44, 4863-4868, November 1973.

Allegra, J. R., and S. A. Hawley, "Attenuation of sound in suspensions and emulsions: theory and experiments," J. Acoust. Soc. Am. 51, 1545-1564, 1972.

Ament, W. S., "Sound propagation in gross mixtures," J. Acoust. Soc. Am. 25, 638-641, 1953.

Ament, W. S., "Wave propagation in suspensions," Armed Services Tech. Information Agency, AD 214750, 1-114, 1959.

Blue, J. E., and E. G. McLeroy, "Attenuation of sound in suspensions and gels," J. Acoust. Soc. Am. 44, 1145-1148, 1968.

Carstensen, E. L., and H. P. Schwan, "Absorption of sound arising from the presence of intact cells in blood," J. Acoust. Soc. Am. 31, 185-189, 1959.

Chow, J. C. F., "Attenuation of acoustic waves in dilute emulsions and suspensions," J. Acoust. Soc. Am. 36, 2395-2401, 1964.

Chu, B. T., and J. C. F. Chow, "On a macroscopic theory of a two phase fluid," AIAA Second Aerospace Sciences Meeting, New York, Paper No. 65-8, 1965.

Duykers, L. R. B., "Relaxation in kaolin-water mixtures," J. Acoust. Soc. Am. 47, 396-398, 1970.

Epstein, P. S., "On the absorption of sound waves in suspensions and emulsions," in Applied Mechanics (Th. von Kármán Anniversary Volume), pp. 162-188, California Institute of Technology, Pasadena, 1941.

Epstein, P. S., and R. R. Carhart, "The absorption of sound in suspensions and emulsions," J. Acoust. Soc. Am. 25, 553-565, 1953.

Hamilton, E. L., "Elastic properties of marine sediments," J. Geophys. Res., 76, 579-604, 1971.

Herzfeld, K. F., "Propagation of sound in suspensions," Phil. Mag. 9, 752-768, 1930.

Lamb, H., Hydrodynamics (Dover Publications, New York, 1945) pp. 657-661.

Lord Rayleigh, "On the incidence of aerial and electric waves upon small obstacles in the form of ellipsoids or elliptic cylinders and on the passage of electric waves through a circular aperture in a conducting screen," Phil. Mag. 44, 28-52, 1897.

Lord Rayleigh, "On the transmission of light through an atmosphere containing small particles in suspension, and on the origin of the blue of the sky," Phil. Mag. 47, 375-384, 1899.

McCann, C., "Compressional wave attenuation in concentrated clay suspensions," Acustica 22, 352-356, 1967/70.

Randall, C. R., "Ultrasonic measurements of the compressibility of solutions and of solid particles in suspensions," J. Res. Natl. Bur. Stand. 8, 79-99, 1931.

Sewell, C. J. T., "The extinction of sound in a viscous atmosphere by small obstacles of cylindrical and spherical form," Philos. Trans. R. Soc. Lond. A 210, 239-270, 1910.

Temkin, S., and R. A. Dobbins, "Attenuation and dispersion of sound by particulate-relaxation processes," J. Acoust. Soc. Am. 40, 317-324, 1966.

Urick, R. J., "The absorption of sound in suspensions of irregular particles," J. Acoust. Soc. Am. 20, 283-289, 1948.

Urick, R. J., and W. S. Ament, "The propagation of sound in composite media," J. Acoust. Soc. Am. 21, 115-119, 1949.

Wood, A. B., A Textbook of Sound (Bell, London, 1940).

ACOUSTIC WAVES IN SATURATED SEDIMENTS

ROBERT D. STOLL

School of Engineering and Applied Science

Columbia University, New York, New York 10027

ABSTRACT

This paper discusses a phenomenological model that describes the propagation of sound waves in saturated sediments. The compressibility and shearing stiffness of the skeletal frame, the compressibility of the fluid, and two major sources of attenuation are included in the model. Attenuation is attributed to two fundamentally different types of energy loss, one resulting from inelasticity of the skeletal frame and the other due to motion of the pore fluid relative to the frame, with each significant in a different frequency range.

Attenuation, dispersion, and wave velocities are found to be in favorable agreement with experimental results for both sands and fine-grained sediments over a wide range of frequencies.

INTRODUCTION

In order to accurately describe the propagation of mechanical waves in ocean sediments, it is necessary to take into account the basic mechanisms by which energy is dissipated in such a medium. Over the frequency range of interest in geophysical research, from less than 1 Hz to over 100 kHz, there are undoubtedly a number of discrete physical processes to which energy dissipation can be attributed. However, most of the available data seems to indicate that the overall response can be described, in a phenomenological sense, by grouping the losses into two fundamentally different categories--one which accounts for the effects of inelasticity of the skeletal frame in a water environment and a second which accounts for the viscosity of the water moving relative to the frame. In granular materials, such as sands and silts, the inelasticity of the frame can be traced primarily to friction losses occurring at the grain-to-grain contacts. In finer materials such as clays and silty clays, the losses that occur during small distortions of the frame have been attributed to a variety of rate-dependent processes associated with the electrochemical bonds between particles. In composite materials containing both sand and clay as well as in cemented or indurated material, it is probable that both rate-dependent processes and friction combine to produce the observed response. Fortunately (for those who attempt to formulate mathematical models for this behavior) experimental results seem to indicate that a relatively simple model is capable of reproducing the combined effects of all of these mechanisms. This model, proposed by Stoll and Bryan (1970), is based on the comprehensive theory for the behavior of saturated porous media developed by Maurice Biot. Using the generalized formulation given by Biot (1962a), including his modifications for higher frequencies, and the notion of a constant complex modulus to describe the response of the skeletal frame, most of the observed behavior can be reproduced over a wide frequency range. The theory predicts that losses in the skeletal frame dominate at low frequencies, while the viscous losses due to motion of the interstitial water become predominant at higher frequencies. The terms high and low frequency are used in a relative sense with their actual values depending on the physical properties of the particular sediment being modeled. For example in very fine clayey materials losses attributable to the skeletal frame dominate over most of the frequency range of interest while in sands the viscous losses become important over a significant portion of this range. As a result, the variation of attenuation with frequency may be quite complex, making it virtually impossible to extrapolate outside the range of experimental values without the help of a comprehensive physical theory such as the one to be described.

THEORY

Starting with a paper on consolidation of porous, elastic material in 1941 (Biot, 1941), Biot has developed a comprehensive theory for the static and dynamic response of linear, porous materials containing compressible fluid. He has considered both low and high frequency behavior (Biot, 1956a, 1956b) and has included the possibility of viscoelastic or viscodynamic response in various components of his model (Biot, 1962a, 1962b).

In the course of developing and generalizing the theory, Biot introduced several changes of notation and a number of generalizations so that some effort is necessary to extract the form most suitable for a particular application. For this reason an abbreviated derivation leading to one form of his equations is outlined below. This derivation is included to help in identifying the variables that are used and in visualizing how the response of the sediment is modeled in a mathematical way. For more rigorous and complete derivations the reader is referred to Biot's original papers, particularly (Biot, 1962a), and to a paper by Geertsma and Smit (1961).

Biot's theory predicts that in the absence of boundaries, three kinds of body waves, two dilatational and one shear, may exist in a fluid saturated porous medium. One of the dilatational waves, which is called the first kind, and the shear wave are similar to waves found in ordinary elastic media. The second kind of compressional wave is highly attenuated in the nature of a diffusion process. Compressional waves of the second kind become very important in acoustical problems involving very compressible pore fluids such as air, whereas for geophysical work in water-saturated sediments waves of the first kind are of principal interest. One exception to this may be the case of very gassy sediments where the effective compressibility of the pore fluid has been greatly reduced by the presence of dissolved or free gas.

To obtain equations governing the propagation of dilatational waves, we will first consider the case of a plane wave in a porous elastic medium filled with fluid. The model will then be generalized to include the inelasticity of the frame and the frequency dependence of viscous losses to yield a realistic model of natural sediments.

If $\overline{u}$ is a vector function giving the displacement of points in the skeletal frame and $\overline{U}$ a vector function giving the displacement of the fluid, then the volume of fluid that has flowed in or out of an element of volume attached to the frame or the "increment of fluid content" is

$$\zeta = \beta \operatorname{div} (\overline{u} - \overline{U}) \qquad (1)$$

where β is the ratio of the volume of pores to the volume of solids (porosity) of the medium. For small strains the dilatation or volumetric strain of the element attached to the frame may be written as

$$e = e_x + e_y + e_z = \operatorname{div} \bar{u} \tag{2}$$

where e_x, e_y, and e_z are the components of small compressional or extensional strain in a cartesian coordinate system. If the porous frame and the pore fluid are elastic, the strain energy, W, of the system depends on the strain components and the increment of fluid content

$$W = W(e_x, e_y, e_z, \gamma_x, \gamma_y, \gamma_z, \zeta) \tag{3}$$

where γ_x, γ_y, and γ_z are the components of shear strain. For an isotropic, linear material, W is a quadratic function of the invariants of strain, I_1 and I_2, and the increment of fluid content, ζ

$$W = C_1 I_1^2 + C_2 I_2 + C_3 I_1 \zeta + C_4 \zeta^2 \tag{4}$$

where

$$I_1 = e_x + e_y + e_z = e,$$

and

$$I_2 = e_x e_y + e_y e_z + e_z e_x - \frac{1}{4}\left(\gamma_x^{\,2} + \gamma_y^{\,2} + \gamma_z^{\,2}\right).$$

The constants C_1, C_2, C_3, and C_4 may be identified with one set used by Biot (H, C, M and μ) by writing Eq. (4) as

$$W = \frac{H}{2} e^2 - 2\mu I_2 - Ce\zeta + \frac{M}{2} \zeta^2 . \tag{5}$$

Considering total stresses on the element of volume attached to the frame, τ_{ij}, and the pressure in the pore fluid, p_f, a set of stress-strain relationships may be obtained from the strain energy, Eq. (5), by differentiation (i.e., $\tau_{xx} = \partial W/\partial e_x$, $\tau_{xy} = \partial W/\partial \gamma_z$, $p_f = \partial W/\partial \zeta$, etc). The stress-strain relations are

$$\begin{aligned}
\tau_{xx} &= He - 2\mu(e_y + e_z) - C\,\zeta \\
\tau_{yy} &= He - 2\mu(e_z + e_x) - C\,\zeta \\
\tau_{zz} &= He - 2\mu(e_x + e_y) - C\,\zeta \\
\tau_{xy} &= \mu\,\gamma_z \\
\tau_{yz} &= \mu\,\gamma_x \\
\tau_{zx} &= \mu\,\gamma_y \\
p_f &= M\zeta - Ce
\end{aligned} \tag{6}$$

From these equations it is clear that μ is the shear modulus. However, in order to understand the significance of the constants H, C, and M it is helpful to visualize two idealized quasistatic tests, involving isotropic loading. In one kind of test, termed a "jacketed" test (Biot and Willis, 1957) the saturated porous medium (shown as a granular material in Fig. 1) is placed in an impervious, flexible bag and loaded by an external pressure. The interstitial fluid in the sample is free to flow out of the bag via a tube so that the fluid pressure remains unchanged during slow loading. In the other test, called an "unjacketed" test, an uncased sample is completely immersed in fluid which is subsequently pressurized from an external source. If p' is the externally applied isotropic pressure in both cases, then

$$\tau_{xx} = \tau_{yy} = \tau_{zz} = -\, p'$$

$$\tau_{xy} = \tau_{yz} = \tau_{zx} = 0$$

and adding the first, second, and third of Eqs. (6),

$$-\, p' = (H - 4\mu/3)\; e - C\; \zeta. \tag{7}$$

For the "jacketed" test, p_f remains unchanged so that the bulk modulus of the free draining porous frame, K_b, is

$$K_b = -\, p'/e = H - 4\mu/3 - C^2/M \tag{8}$$

from the last of Eqs. (6) and Eq. (7). In a practical test the dilatation of the sample, e, could be obtained by measuring the quantity of fluid expelled from the sample provided the effect of

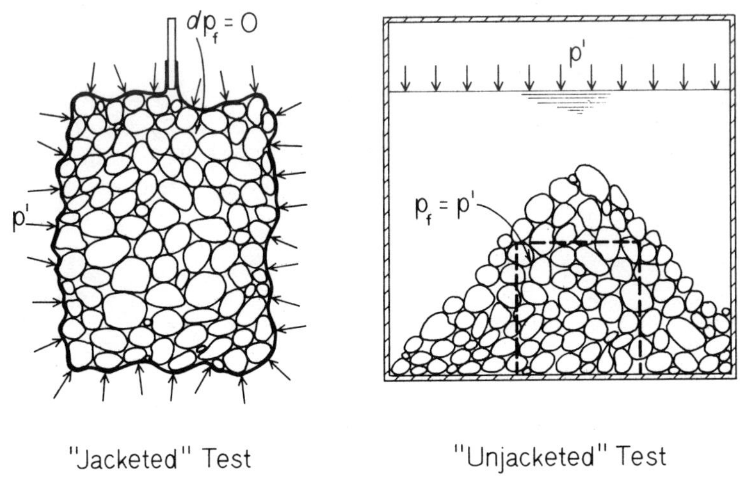

FIGURE 1

ISOTROPIC TESTS TO DETERMINE MODULI

membrane penetration on the outer surface of the specimen was negligible (which is often not so). This type of drained, isotropic loading is utilized in several standard tests used to determine the engineering properties of soil (Bishop and Henkel, 1957).

In the unjacketed test, the pressure in the pore fluid equals the applied isotropic pressure, and we may derive two measures of compliance, δ, the "unjacketed compressibility,"

$$\delta = - e/p' = \frac{1 - C/M}{H - 4\mu/3 - C^2/M} \tag{9}$$

and, γ, the "coefficient of fluid content,"

$$\gamma = \zeta/p' = \frac{H - 4\mu/3 - C}{(H - 4\mu/3 - C^2/M)M} \tag{10}$$

utilizing the last of Eqs. (6) and Eq. (7) in both cases. If the ratio of pore volume to solid volume remains constant (i.e., constant porosity) during unjacketed loading, δ equals the compressibility of the solid material composing the porous frame and γ may be expressed in terms of δ, β, and the compressibility of the pore fluid. In terms of the reciprocals of compressibility (bulk moduli)

$$\gamma = \beta(1/K_f - 1/K_r) \tag{11}$$

and

$$\delta = 1/K_r \tag{12}$$

where K_f is the bulk modulus of the fluid and K_r is the bulk modulus of the solid material composing the porous frame (the bulk modulus of the individual particles in the case of granular media). While Eqs. (11) and (12) are strictly true for materials where the porous frame is isotropic, homogeneous, and linear, they are also reasonable for cases where the frame is heterogeneous but behaves approximately like a homogeneous mass in that it undergoes the same volumetric strain as the pores.

Finally, using Eqs. (8) through (12) we may solve for Biot's coefficients in terms of the bulk moduli of the porous frame, the pore fluid, and the solid material or discrete particles composing the following frame.

$$H = \frac{(K_r - K_b)^2}{D - K_b} + K_b + 4\mu/3$$

$$C = \frac{K_r(K_r - K_b)}{D - K_b} \qquad (13)$$

$$M = \frac{K_r^2}{D - K_b}$$

where

$$D = K_r(1 + \beta(K_r/K_f - 1)) \ .$$

Having established a set of constitutive equations and the meanings of the various parameters, equations can now be written for the motion of an element of volume attached to the skeletal frame and for fluid moving into or out of the element. To simplify the derivation we consider one-dimensional motion in the x direction. The stress-equation of motion for the volume attached to the frame is

$$\begin{aligned} \frac{\partial \tau_{xx}}{\partial x} &= \frac{\partial^2}{\partial t^2}\left[\beta \rho_f U_x + (1 - \beta)\rho_r u_x\right] \\ &= \frac{\partial^2}{\partial t^2}\left[\beta \rho_f u_x + (1 - \beta)\rho_r u_x - \beta\rho_f(u_x - U_x)\right] \end{aligned} \qquad (14)$$

where ρ_f is the density of the pore fluid and ρ_r is the density of the solid material composing the frame (density of individual grains for granular material). Differentiating with respect to x, and substituting for τ_{xx} from Eq. (6) we obtain the one-dimensional form of one of the equations given by Biot (Eq. (6) of Biot, 1962a). His equation is

$$\nabla^2 \ (He - C\zeta) = \frac{\partial^2}{\partial t^2} \ (\rho e - \rho_f \zeta) \qquad (15)$$

where ρ is the total density of the saturated medium.

The second equation, which describes the motion of the fluid relative to the frame, is

$$\beta \frac{\partial p_f}{\partial x} = \frac{\partial^2}{\partial t^2}\left[\beta\rho_f U_x\right] + \frac{\beta\eta}{k}\frac{\partial}{\partial t}\left[\beta(U_x - u_x)\right]$$

or

$$\frac{\partial p_f}{\partial x} = \frac{\partial^2}{\partial t^2}\left[\rho_f u_x - \frac{\rho_f}{\beta}\left(\beta(u_x - U_x)\right)\right] - \frac{\beta\eta}{k}\frac{\partial}{\partial t}\left[\beta(u_x - U_x)\right] \ . \qquad (16)$$

The last term on the right side of this equation gives the viscous resistance to flow which depends on η, the viscosity of the pore fluid, and k, the coefficient of permeability. By differentiating Eq. (16) with respect to x and substituting for p_f from Eq. (6) we obtain the one-dimensional form of the second equation given by Biot. His equation is

$$\nabla^2 (Ce - M\zeta) = \frac{\partial^2}{\partial t^2} (\rho_f e - m\zeta) - \frac{\eta}{k} \frac{\partial \zeta}{\partial t} . \tag{17}$$

In Eq. (17), a parameter m, greater than ρ_f/β, has been substituted for ρ_f/β in the part of the inertial term corresponding to the increment of fluid flow. This has been done to account for the fact that not all of the pore fluid moves in the direction of the macroscopic pressure gradient because of the tortuous, multi-directional nature of the pores. As a result less fluid flows in or out of an element for a given acceleration than if all the pores were uniform and parallel to the gradient. The parameter m may be written as

$$m = \alpha\rho_f/\beta, \quad \alpha \geqq 1 . \tag{18}$$

For uniform pores with axes parallel to the gradient, α would equal 1, while for a random system of uniform pores with all possible orientations the theoretical value of α is 3. In real granular materials it is impossible to calculate α from theory so that it must be considered one of the variables to be determined from experiments.

Equations (15) and (17) are a pair of coupled differential equations that determine the dilatational motion of a saturated porous medium with a linear elastic frame and a constant ratio of fluid flow to pressure gradient (Poiseuille flow). Solving these equations leads to a relationship between attenuation and frequency such as shown by the broken curve of Fig. 2. It is clear from this figure that the model is not adequate to predict the behavior of real sediment at this stage of development. In fact, two major modifications are required to accomplish this. First, the viscous resistance to fluid flow must be made frequency dependent to correct for the deviation from Poiseuille flow that occurs at all but very low frequencies and, second, the inelastic nature of the frame must be taken into account.

In order to incorporate the frequency dependence of viscous resistance, Biot derived a complex correction factor to be applied to the fluid viscosity by considering the actual microvelocity field that exists within the pore channels. This problem of oscillatory motion in a closed channel is quite well known, having been solved as early as 1868 by Kirchoff. Biot's solution is

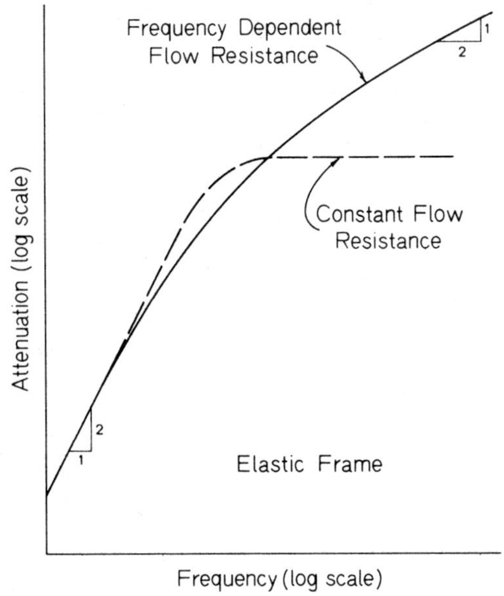

FIGURE 2

ATTENUATION VERSUS FREQUENCY FOR A LINEAR ELASTIC FRAME

written in such a way that the factor $\beta^2 F\eta/k$ gives the ratio of the friction force exerted by the fluid on the frame to the average relative velocity for oscillatory motion. Hence, η/k may be replaced by $F\eta/k$ in the frequency domain, where

$$F(\kappa) = F_r(\kappa) + i\,F_i(\kappa) = \frac{1}{4}\,\frac{\kappa T(\kappa)}{1-2\,T(\kappa)/i\kappa} \quad ,$$

$$T(\kappa) = \frac{\mathrm{ber}'(\kappa) + i\,\mathrm{bei}'(\kappa)}{\mathrm{ber}(\kappa) + i\,\mathrm{bei}(\kappa)} \quad ,$$

and

$$\kappa = a(\omega\rho_f/\eta)^{1/2}$$

The functions $\mathrm{ber}(\kappa)$ and $\mathrm{bei}(\kappa)$ are the real and imaginary parts of the Kelvin function, ω is angular frequency, and a is a parameter with the dimension of length that depends on both the size and shape of the pores. $F(\kappa)$ approaches unity for very low frequencies, thus resulting in the same equation as when Poiseuille flow is assumed. Like α, the parameter a cannot be derived theoretically for real sediments so that it must be obtained from experiments.

The solution on which Biot's complex correction factor is based is valid only for frequencies where the wavelength is large compared to the pore size. For sands, this puts the upper limit on frequencies at about 10^5 to 10^6 Hz, which is high enough to cover the frequency range of interest. When the complex correction factor is incorporated into Eq. (17), by replacing η by ηF, a relationship between frequency and attenuation such as shown by the solid curve in Fig. 2 is obtained.

Finally, to complete a physically realistic model of saturated sediment, the inelastic nature of the frame must be considered. In order to study the effects of inelasticity separate from those of viscous losses in the fluid, one may work at frequencies low enough that the inelastic effects dominate. Since low frequencies require long wavelengths, this approach often results in complications both in laboratory and in situ studies. Alternatively, tests may be performed on dry or partly saturated specimens with corrections made for the effects of wetting and full saturation. Still another possibility is to examine losses occurring during volume-constant shearing deformation and then to interpret the results in terms of dilatational motion. All three of these approaches have been used in various experimental programs. In addition many theoretical models incorporating friction losses, relaxation at grain boundaries, and other forms of dissipation have been studied. To describe all this work in detail is beyond the scope of this paper. However, we will make use of one major observation that is common to much of the work that has been reported, that is, that the logarithmic decrement is independent of frequency or amplitude for oscillations of very small amplitude. This is equivalent to saying that the material has a constant Q, or, for propagating waves, that the attenuation constant varies linearly with frequency. It is difficult to reproduce this kind of behavior with a linear visco-elastic model without resorting to an integral formulation that leads to a very cumbersome complex modulus (see Stoll and Bryan, 1970). On the other hand certain "slightly" nonlinear models are very appealing since they assume loss mechanisms that are very close to physical reality. Included in these are several models that are based on frictional losses at the grain-to-grain contacts of the particle. These models may be used to approximate the response of the skeletal frame and still result in a linear system of equations by replacing the slightly nonlinear terms by their linear equivalents according to the procedure of Kryloff and Bogoliuboff (1947). This technique, which is often used in electrical circuit analysis, leads to a set of linear equations in the frequency domain. Moreover, it bypasses the need to restrict the results to a particular physical model because many of the interesting models lead to the same form of frequency equation. In particular, several of the models for which the specific damping is constant (i.e., constant Q) may be shown to result in a constant complex modulus with small imaginary part.

In order to incorporate the inelasticity of the frame into the present model, H,C,M, and μ in Eqs. (15) and (17) are regarded as operators that may be linear viscoelastic or "slightly" nonlinear. For the conditions in most sediments it is reasonable to consider K_r and K_f, the bulk moduli of the individual particles and the fluid, respectively, as elastic constants and concentrate the inelastic effects in the operators K_b and μ which describe the response of the skeletal frame in a water environment.

To obtain a frequency equation, solutions of Eqs. (15) and (17) of the form

$$e = A_1 \exp(i(\omega t - \ell x))$$

and

$$\zeta = A_2 \exp(i(\omega t - \ell x))$$

are considered, with $\ell = \ell_r + i\,\ell_i$, the constants H,C,M and μ replaced by the appropriate operators, and η replaced by $F(\kappa)\eta$. Upon transformation to the frequency domain the following equation results

$$\begin{vmatrix} \bar{H}\,\ell^2 - \rho\omega^2 & \rho_f\omega^2 - \bar{C}\,\ell^2 \\ \bar{C}\,\ell^2 - \rho_f\omega^2 & m\omega^2 - \bar{M}\,\ell^2 - \dfrac{i\omega F\eta}{k} \end{vmatrix} = 0 \qquad (19)$$

where, in general, $\bar{H},\bar{C},\bar{M}$ and $\bar{\mu}$ are complex functions of frequency derivable by Laplace transformation in the case of linear viscoelastic operators or by the method of Kryloff and Bogoliuboff for the case of slightly nonlinear equations. Following the latter procedure, and restricting to physical models that result in constant complex moduli for the reasons mentioned above,

$$\bar{H} = \frac{(K_r - \bar{K}_b)^2}{D - \bar{K}_b} + \bar{K}_b + 4\bar{\mu}/3$$

$$\bar{C} = \frac{K_r(K_r - \bar{K}_b)}{D - \bar{K}_b}$$

$$\bar{M} = \frac{K_r^2}{D - \bar{K}_b}$$

where $\bar{K}_b = K_b + i\,K_b'$ and $\bar{\mu} = \mu + i\,\mu'$. The roots of Eq. (19) give the attenuation ℓ_i and phase velocity ω/ℓ_r as functions of frequency for the first and second kinds of dilatational wave. A similar procedure may be followed to obtain equations describing shear waves.

APPLICATIONS

When Eq. (19) is solved, utilizing physical properties typical of ocean sediments, curves such as the ones shown in Fig. 3 are obtained. For clays and other materials of very low permeability, the relationship between attenuation and frequency is linear over almost the entire range of interest. Apparently for these materials the observed dissipation is largely due to losses in the skeletal frame with viscous losses in the fluid producing a perturbing effect only at very high frequencies. On the other hand in coarser materials such as sand, the losses due to fluid motion relative to the frame appear to have a strong influence over much of the frequency range of interest. Only at very low and very high frequencies does the curve indicate a tendency for a linear relationship between frequency and attenuation. In fact the slope of the log-log curve is greater than 2 at some points and less than 1/2 at others. The position of the two curves at the low frequency end of the scale depends primarily on the damping characteristics of the skeletal frame so that either curve may be shifted up or down depending on the nature of the particular material. Also the point where deviation from a linear relationship occurs will depend on

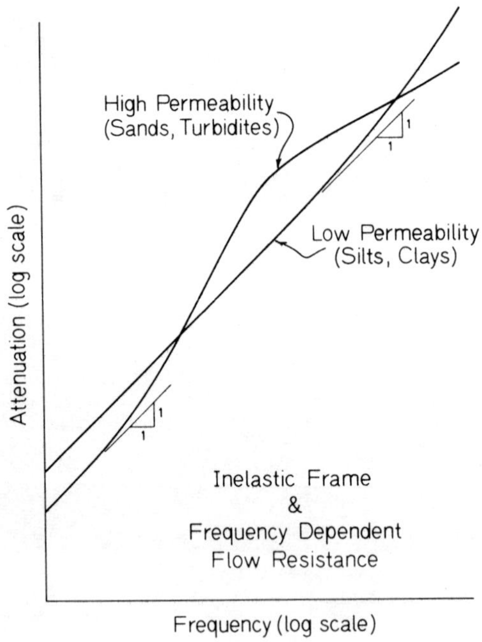

FIGURE 3

TYPICAL CURVES OF ATTENUATION VERSUS FREQUENCY
FOR A REALISTIC MODEL OF OCEAN SEDIMENT

such characteristics as porosity, permeability, damping in the frame, and the density of the fluid and solid particles.

In order to study the applicability of the proposed model to real sediments, Eq. (19) has been evaluated for several sets of data from laboratory and in situ studies. Figure 4 shows one set of curves that correspond to physical properties of sandy sediments. Figure 4 also shows most of the experimental data on sands that are available in the open literature. Some additional data from other sources may be found in papers by Hamilton (1972) and Hampton and Anderson (1973). The parameter values used to obtain the curves of

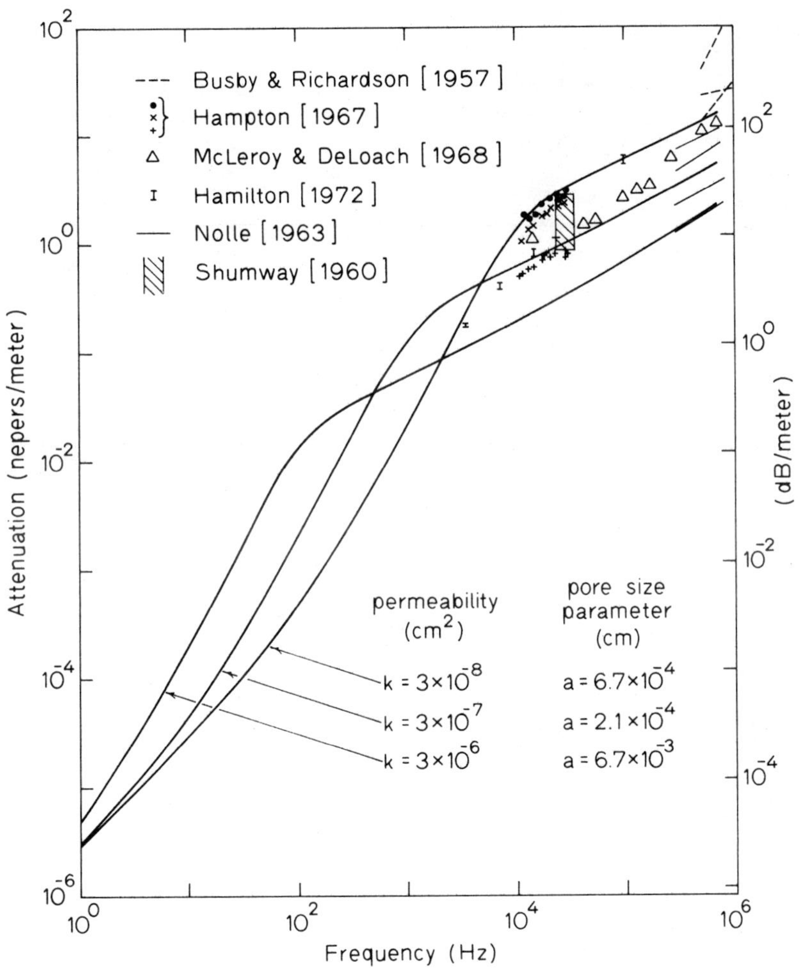

FIGURE 4

ATTENUATION VERSUS FREQUENCY FOR SANDS

Fig. 4 are given in Table 1. The complex moduli, $\overline{K}_b$ and $\overline{\mu}$, were evaluated on the basis of experiments with vibrating columns of dry and saturated sand (see Hall and Richart, 1963, and Hardin, 1965). The logarithmic decrement and the wave velocity measured in these tests were used to evaluate the required complex moduli. For example, the complex shear modulus is related to the logarithmic decrement in torsional vibrations, δ_s, by the relationship

$$\delta_s = \pi\mu'/\mu ,$$

and for small damping the real part of the modulus can be obtained from the shear wave velocity, v_s, by the relationship

$$v_s = (\mu/\rho)^{1/2} .$$

A complex modulus for longitudinal vibrations of a long column of sediment confined in a flexible membrane may be found using similar equations. Finally, the bulk modulus of the frame may be obtained from the moduli based on shear and longitudinal vibrations by using

MATERIAL PROPERTY	Values for Fig. 4	Values for Fig. 5	Values for Fig. 6
Bulk Modulus of Grains, K_r(dyn/cm^2)	3.6×10^{11}	3.6×10^{11}	3.6×10^{11}
Bulk Modulus of Fluid, K_f(dyn/cm^2)	2.0×10^{10}	2.0×10^{10}	2.0×10^{10}
Mass Density of Grains, ρ_r(g/cm^3)	2.65	2.3	2.65
Mass Density of Fluid, ρ_f(g/cm^2)	1.0	1.0	1.0
Absolute Viscosity of Fluid, η (dyn-sec/cm^2)	1.0×10^{-2}	1.0×10^{-2}	1.0×10^{-2}
Coefficient of Permeability, k(cm^2)	3.0×10^{-6} to 3.0×10^{-8}	2.6×10^{-11}	3.0×10^{-6} to 3.0×10^{-10}
Pore Size Parameter, a(cm)	6.7×10^{-3} to 6.7×10^{-4}	1.3×10^{-5}	6.7×10^{-3} to 6.7×10^{-5}
Rod Velocity of Frame, v_E(m/sec)	3.0×10^2	3.0×10^2	3.0×10^2
Shear Velocity of Frame, v_s(m/sec)	2.1×10^2	2.1×10^2	2.1×10^2
Structure Constant, α	1.25	3.0	1.25
Porosity, β	0.40	0.76	0.40 to 0.70
Log decrement for longitudinal vibrations of frame, δ_E	0.15	0.45	0.15
Log decrement for shear vibrations of frame, δ_s	0.20	0.60	0.20

TABLE 1

MATERIAL PROPERTIES USED TO OBTAIN FIGS. 4, 5 and 6

Eq. (17) of Stoll and Bryan (1970). At this point it is worth noting that there is no low frequency data available for sands other than those used in the calculations described above. Were this not the case, one could use direct field measurement of velocity (shear and dilatational) and attenuation to obtain the appropriate moduli.

The range of permeability coefficients used to obtain Fig. 4 is typical for naturally occurring sands (see Lambe and Whitman, 1969) and the constants a and α were chosen so that the theoretical curves would fit the bulk of the available experimental data. In choosing the pore size parameter a, the usual assumption that permeability is proportional to the square of pore size was made with the same proportionality constant used for all three curves. The reader is referred to Stoll and Bryan (1970) for curves showing the effects of varying each of the other parameters individually.

A second example, for parameters typical of a very fine-grained material, is shown in Fig. 5. In this case the complex moduli that determine the response of the frame and the other parameters of the model were chosen to match the data of Wood and Weston (1964). Other experimental data for similar materials may be found in Hamilton (1972) and Hampton and Anderson (1973) as well as in many other papers.

As a third example of results to be expected from the theory, the family of curves on the left side of Fig. 6 was obtained for a range of permeabilities from 3×10^{-6} to $3 \times 10^{-10} cm^2$. As before, the pore size parameter was assumed proportional to the square root of the permeability. For a fixed frequency, such as 30 kHz, a bell-shaped curve relating attenuation to pore size parameter may be derived as shown on the right side of Fig. 6. This curve is similar in all respects to experimental curves (also for 30 kHz) presented by Shumway (1960), McCann and McCann (1969), and Hamilton (1972). If the peak of the theoretical curve is brought into coincidence with the peak of the experimental curves, the pore size parameter and the mean grain diameter may be related resulting in the upper scale shown in Fig. 6. In this case the pore size parameter appears to be between 1/6 and 1/7 of the mean grain diameter. For each curve shown in Fig. 6 the porosity was chosen to match the experimental data given by Shumway (1960) and Hamilton (1972) (e.g., see Fig. 8 of Hamilton's paper). However, no attempt was made to vary the complex moduli of the frame since insufficient data is available at the required low frequencies, particularly for sands.

On the basis of the small amount of data that is available, it appears that the theory will reproduce the experimental results even better when enough data becomes available to include the variation of frame losses with size, porosity, etc.

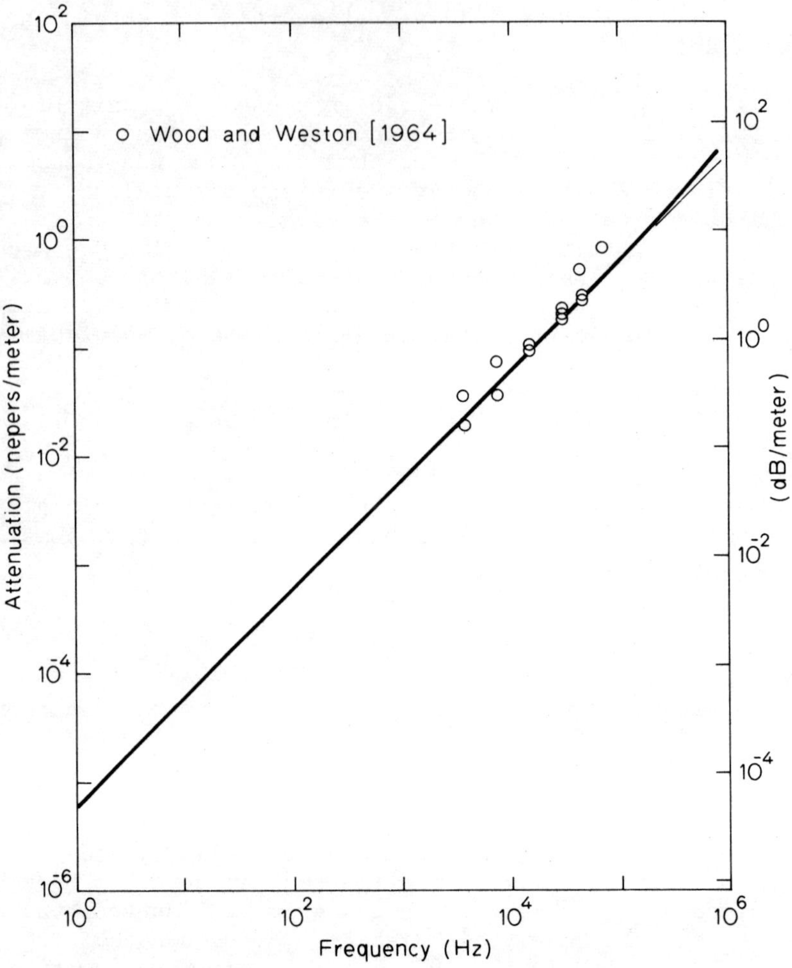

FIGURE 5

ATTENUATION VERSUS FREQUENCY FOR FINE SEDIMENT WITH LOW PERMEABILITY

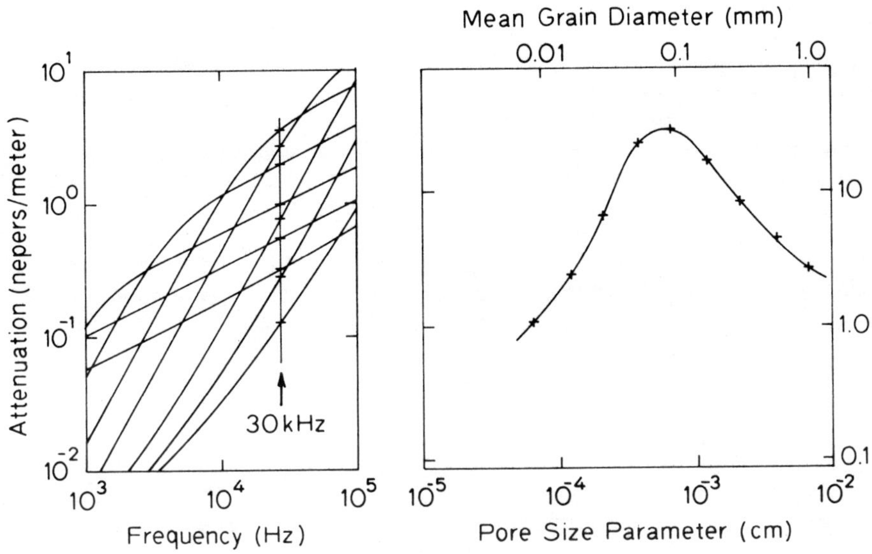

FIGURE 6

RELATIONSHIP BETWEEN ATTENUATION AND GRAIN SIZE AT FIXED FREQUENCY

SUMMARY AND CONCLUSIONS

On the basis of the foregoing discussion it appears that a relatively simple model based on Biot's theory for porous media is capable of reproducing most of the behavior that has been observed in saturated sediments. However, since there is virtually no experimental data available for sands at frequencies less than 10 kHz, it is necessary to obtain much additional information before the full range of usefulness can be evaluated. Nevertheless, several conclusions can be drawn at this time.

First, because of the complex variation of attenuation with frequency that is evident from Figs. 3, 4, 5 and 6, it is clear that extreme care should be exercised in extrapolating experimental curves out of the range where data was taken. Such extrapolation cannot be based strictly on the trend within the data interval nor should a simple first power law be used for all cases. The latter procedure has been suggested by some authors on the basis of the general trend indicated by plotting data for all different kinds of sediment on one diagram. The figures mentioned above show that large errors may result from this procedure for certain materials such as sands and other sediments with high permeability.

One of the strongest arguments for a unifying theory, such as the one discussed herein, is that it can serve as a guide in the acquisition and interpretation of meaningful experimental data. Furthermore it can be used as a starting point in the development of approximate formulas for practical use because it indicates the significant variables and those that may be neglected in certain cases. As an example, consider the problem of low frequency waves. Both the Navy and geophysical researchers in general have recently become more interested in acoustic waves at lower frequencies, so that it is probable that more work will be done in this area in the near future. For this reason it is important to examine which parameters are significant in field and laboratory experiments at low frequencies. In the higher frequency ranges, porosity and grain size have been used quite successfully as correlating parameters. On the basis of the proposed theory this result is to be expected, at least for coarser materials, because permeability is known to be primarily dependent on these two variables. On the other hand at lower frequencies, where frame losses become important, other factors such as intergranular stress and cementation become very significant. On the sea floor intergranular or effective stress depends on the overburden pressure so that it increases rapidly with depth. Thus, field studies involving long waves that pass through regions where there is a significant variation of effective stress may be difficult to interpret. In the laboratory the effective stress depends on the size of the specimen and on the amount of static confining pressure that is applied, as well as other factors. Comparisons of data from different kinds of tests can be made only after these factors have been taken into account.

Finally, it should be noted that the influence of effective stress can easily be incorporated into the proposed theory. This may be done by making the complex moduli representing the frame response dependent on this variable, when more experimental data are obtained.

REFERENCES

Biot, M. A., "General Theory of Three-Dimensional Consolidation," J. Appl. Phys., 12, 155-164, 1941.

Biot, M. A., "Theory of Elastic Waves in a Fluid-Saturated Porous Solid. I. Low-Frequency Range," J. Acoust. Soc. Am., 28, 168-178, 1956a.

Biot, M. A., "Theory of Elastic Waves in a Fluid-Saturated Porous Solid. II. Higher Frequency Range," J. Acoust. Soc. Am., 28, 179-191, 1956b.

Biot, M. A., "Mechanics of Deformation and Acoustic Propagation in Porous Media," J. Appl. Phys., 33, 1482-1498, 1962a.

Biot, M. A., "Generalized Theory of Acoustic Propagation in Porous Dissipative Media," J. Acoust. Soc. Am., 34, 1254-1264, 1962b.

Biot, M. A., and D. G. Willis, "The Elastic Coefficients of the Theory of Consolidation," J. Appl. Mech., 24, 594-601, 1957.

Bishop, A. W., and D. J. Hankel, The Measurement of Soil Properties in the Triaxial Test, Edward Arnold, Ltd., London, 1957.

Busby, J., and E. G. Richardson, "The Absorption of Sound in Sediments," Geophysics, 22, 821-828, 1957.

Geertsma, J., and D. C. Smit, "Some Aspects of Elastic Wave Propagation in Fluid-Saturated Porous Solids," Geophysics, 26, 169-181, 1961.

Hall, J. R., Jr., and F. E. Richard, Jr., "Dissipation of Elastic Wave Energy in Granular Solid," J. Soil Mech. Found. Div., A.S.C.E., 89(SM6), 27-56, 1963.

Hamilton, E. L., "Compressional-Wave Attenuation in Marine Sediments," Geophysics, 37, 620-646, 1972.

Hampton, L. D., "Acoustic Properties of Sediments," J. Acoust. Soc. Am., 42, 882-890, 1967.

Hampton, L. D., and A. L. Anderson, "Acoustics and Gas in Sediments - Applied Research Laboratories Experience," Proc. ONR Conf. Natural Gases in Marine Sediments and Their Mode of Distribution, Lake Arrowhead, Calif., 1973.

Hardin, B. O., "The Nature of Damping in Sands," J. Soil. Mech. Found. Div., A.S.C.E., 91 (SM1), 1965.

Kryloff, N., and N. Bogoliuboff, Introduction to Non-Linear Mechanics, Princeton University Press, Princeton, N. J., 1947, pp. 55-63.

Lambe, T. S., and R. V. Whitman, Soil Mechanics, John Wiley and Sons, Inc., New York, 1969, pp. 281-294.

McCann, C., and D. M. McCann, "The Attenuation of Compressional Waves in Marine Sediments," Geophysics, 34, 882-892, 1969.

McLeroy, E. C., and A. DeLoach, "Sound Speed and Attenuation, from 15 to 1500 kHz, Measured in Natural Sea-Floor Sediments," J. Acoust. Soc. Am., 44, 1148-1150, 1968.

Nolle, A. W., W. A. Hoger, J. F. Mifsud, W. R. Runyan, and M. B. Ward, "Acoustical Properties of Water-Filled Sands," J. Acoust. Soc. Am., 35, 1394-1408, 1963.

Shumway, G., "Sound Speed and Absorption Studies of Marine Sediments by a Resonance Method," Geophysics, 25, 451-467, 659-682, 1960.

Stoll, R. D., and G. M. Bryan, "Wave Attenuation in Saturated Sediments," J. Acoust. Soc. Am., 47, 1440-1447, 1970.

Wood, A. B., and D. E. Weston, "The Propagation of Sound in Mud," Acustica, 14, 156-162, 1964.

ACOUSTIC AND MECHANICAL LOADING OF MARINE SEDIMENTS

DENZIL TAYLOR SMITH

Marine Sciences Laboratories

University College of North Wales, BANGOR

ABSTRACT

Measurements of acoustic and engineering properties of sea-floor sediments show that while grain size parameters, along with porosity, dominate in any study of variations of the former, reasonably good prediction equations exist to be able to define the one knowing the other. Thus, while the static elastic moduli are several orders of magnitude less than the equivalent dynamic parameters, it is still possible to make an assessment of their magnitude from acoustic information.

Many of the measured variations can be predicted by the use of normal elastic theory, but such theory cannot account fully for the observed manner in which acoustic energy is lost in marine sediments. While inter-granular friction seems to be the main dissipating mechanism (giving a linear frequency dependence), viscous losses must also exist in any sediment.

ACOUSTIC AND MECHANICAL LOADING OF THE SEA FLOOR

The last ten years has seen an intensified interest in the physicomechanical properties of marine sediments. Some of the physical measurements are made for their own sake: thus it is to be expected that a study of the acoustic properties of sea-floor materials would be a crucial part of any examination of acoustic bottom loss. As part of such studies, it has been clearly seen that the physical properties bore interrelationships with certain mechanical properties, and many empirical equations between them have been established. From such experimental data it is possible to get a better understanding of the way sound is propagated in marine sediments.

Work at the Marine Science Laboratories of the University College of North Wales is typical of such studies now carried out at a large number of laboratories. Up to relatively recently most of the sediment material investigated has been collected from the deep North Atlantic and, as such, is predominantly clay-size with coarse fractions being generally provided either by biogenous sources (such as planktonic foraminifera in the deep ocean) or, close to the continental margin, by current action from the continent. Most of the experimental work has been carried out by laboratory studies on samples derived from cores, although more recently some measurements have been made using an in situ probe. The bulk of the studies has been devoted to the assessment of the interrelationships between acoustic properties and a large number of engineering properties, although more recently electrical properties have been measured both in the laboratory and in situ.

While the scatter-diagrams of the various parameters include both deep-sea and shelf data (collected both in the laboratory and in situ), for this paper only the deep-sea measurements will be quantified because insufficient data are available for the other environment.

ACOUSTO-MECHANICAL RELATIONSHIPS

Compressional Wave Velocity

1. Velocity increases slowly from a sub-sea-water velocity in the fine sizes to about 1.55 km/sec at 60 microns, and then increases rapidly in the coarser material (Fig. 1).

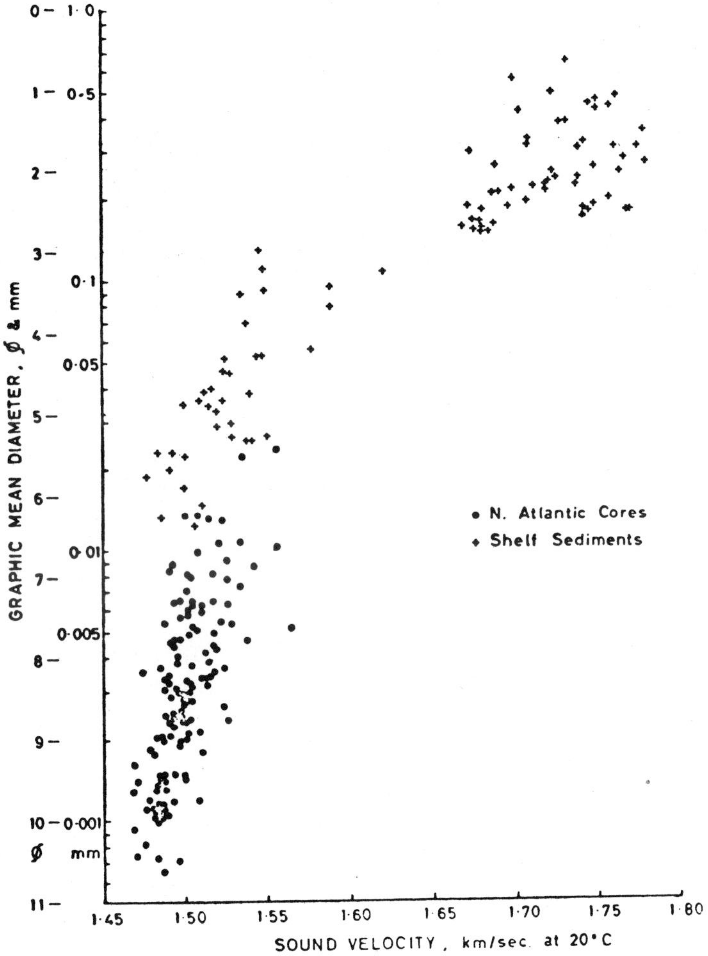

FIGURE 1

SEDIMENT SIZE PLOTTED AGAINST SOUND VELOCITY

2. In an almost negative relationship to size, porosity shows a similar relationship, the velocity increasing slowly in the high porosity sediments with decreasing porosity and then, at about 50% porosity, increasing rapidly toward the low porosity end (Fig. 2). Wood's Emulsion Equation computed separately for clay minerals and quartz is shown with the porosity-velocity variations. A similar relationship is also shown with bulk density (Fig. 3); however density and porosity have a negative linear relationship, and therefore the density data are mirror images of those for porosity.

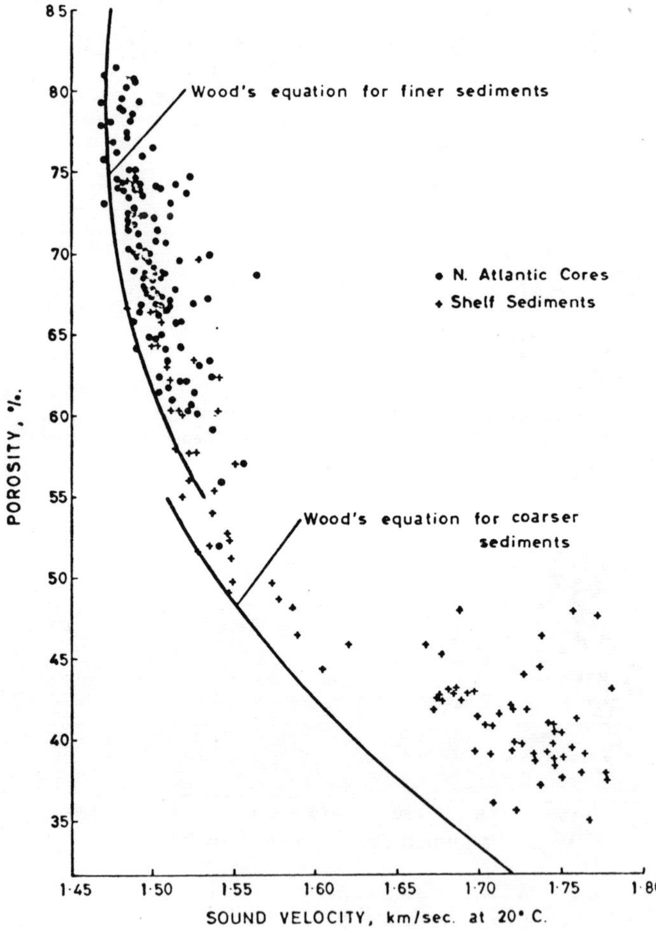

FIGURE 2

SEDIMENT POROSITY PLOTTED AGAINST SOUND VELOCITY

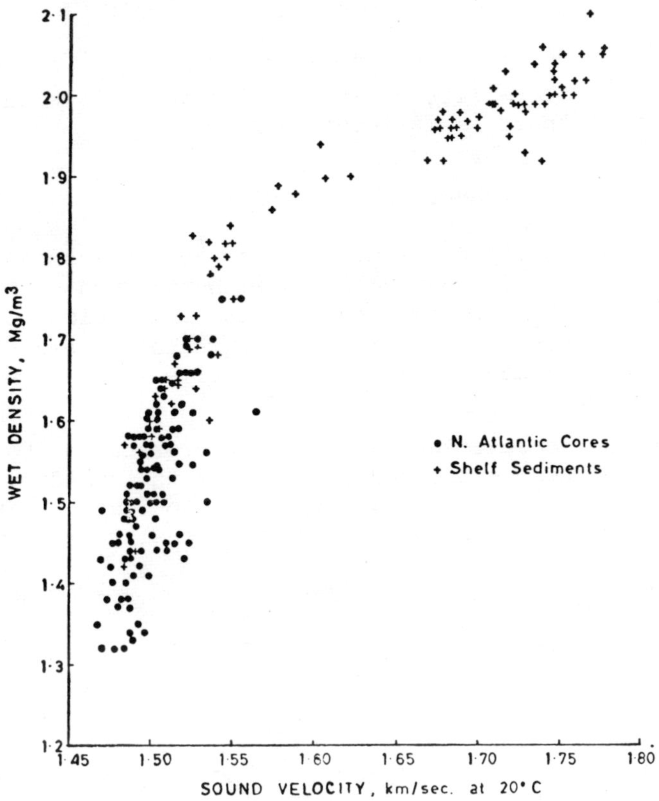

FIGURE 3

WET DENSITY PLOTTED AGAINST SOUND VELOCITY

For deep-sea sediments a multiple parameter regression analysis shows that the best equations for predicting sound velocity at 20°C are:

$$Vp = 1.336 + 0.092\ F_s + 0.101\ \rho \quad (\text{km/sec}), \quad \text{and}$$

$$Vp = 1.616 + 0.090\ F_s - 0.2\ n, \quad \text{and}$$

$$Vp = 1.416 - 0.007\ M_\phi + 0.091\ \rho, \quad \text{and}$$

$$Vp = 1.610 - 0.13\ M_d,$$

where

F_s = sand fraction
ρ = bulk density (Mg/m^3)
n = fractional volume of pores to total volume of sediment

M_ϕ = mean grain diameter expressed in phi units

M_d = median grain diameter expressed in phi units.

There appears to be little variation of velocity with frequency; or what there is is inconclusive.

Shear Wave Velocity

Owing to the paucity of data, no clear-cut assessments can be made of the variation of shear velocity with the mechanical properties. The few measurements made by the author and his colleagues agree with the pattern shown in Hamilton's recent papers (1970, 1971), although their measurements seem to give lower magnitudes in deep-sea sediments.

As far as the measurements allow an interpretation to be made, any variations are best expressed in terms of the rigidity modulus (ρV_s^2) (Figs. 4, 5, 6). This quantity seems to reach a maximum value of 200 microns, 45% porosity, and at a density of 1.95 Mg/m^3.

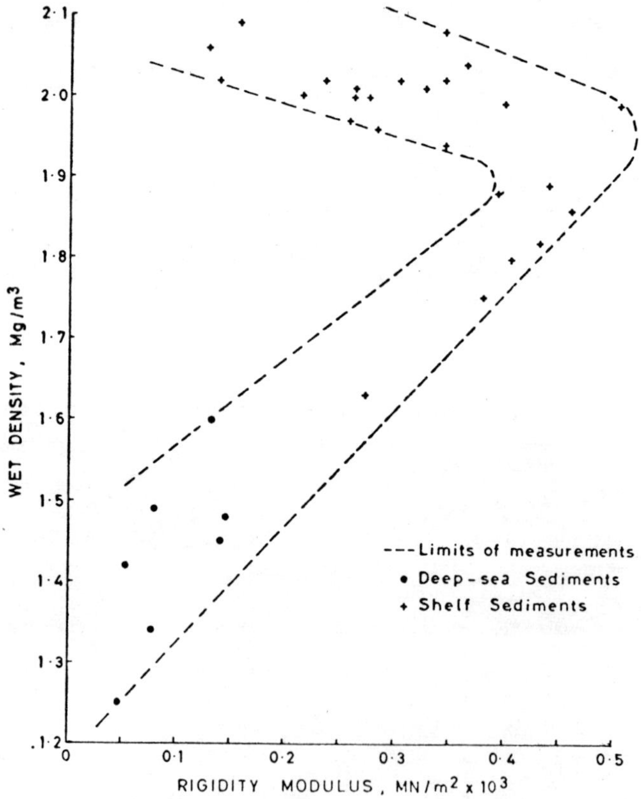

FIGURE 4

SEDIMENT WET DENSITY PLOTTED AGAINST RIGIDITY MODULUS

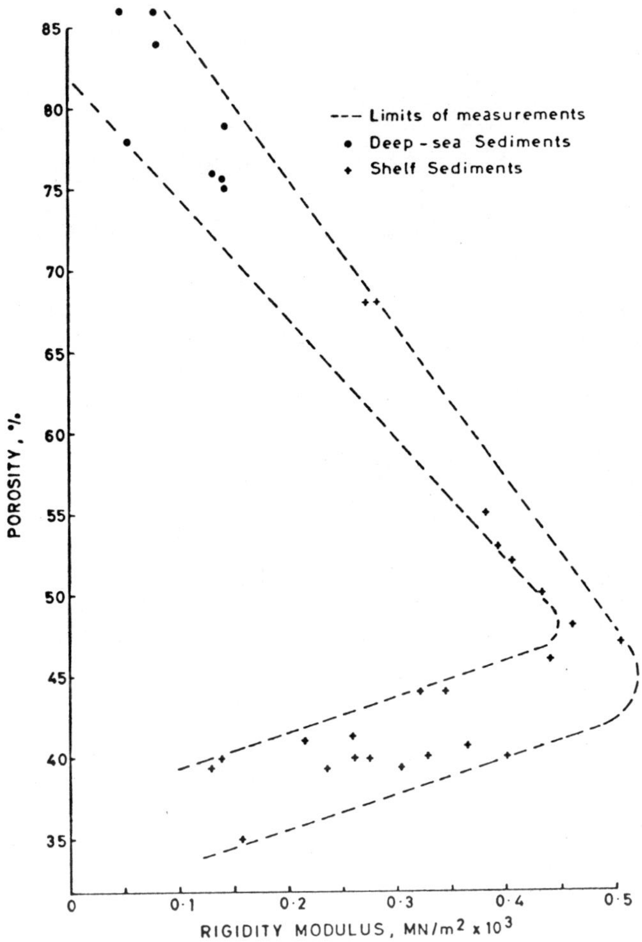

FIGURE 5

SEDIMENT POROSITY PLOTTED AGAINST RIGIDITY MODULUS

Acoustic Impedance (Z)

All of the impedance data used in a multiple regression analysis were computed from the product of compressional wave velocity and density. Following this, some in situ measurements (from both reflectivity and transducer motional admittance calculations) were carried out; since these measurements were from only a dozen sites, the good agreement between the computed and measured values could be fortuitous. It is to be expected that density and porosity would show almost perfect correlation to impedance.

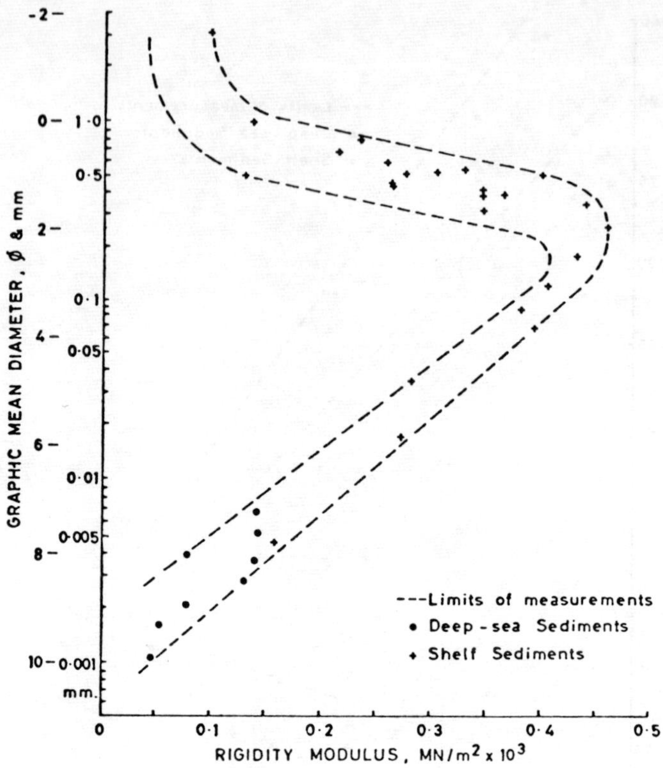

FIGURE 6

SEDIMENT SIZE PLOTTED AGAINST RIGIDITY MODULUS

Figure 7 shows the relationship in the form of normally incident reflectivity, where an impedance of 1.56 rayls has been chosen for seawater. However, other reasonably good correlations are also evident:

$$Z = 1.666\ \rho - 0.251$$

$$Z = 4.345 - 2.94\ n$$

$$Z = 1.641\ \rho - 0.008\ M_{\phi} - 0.142$$

$$Z = 1.653\ \rho + 0.121\ F_{s} - 0.243$$

$$Z = 2.733 + 0.0014\ S - 0.721\ W_{L}$$

where S = vane shear strength (g/cm^2)

W_L = liquid limit of Atterberg,

and the other parameters are as defined earlier.

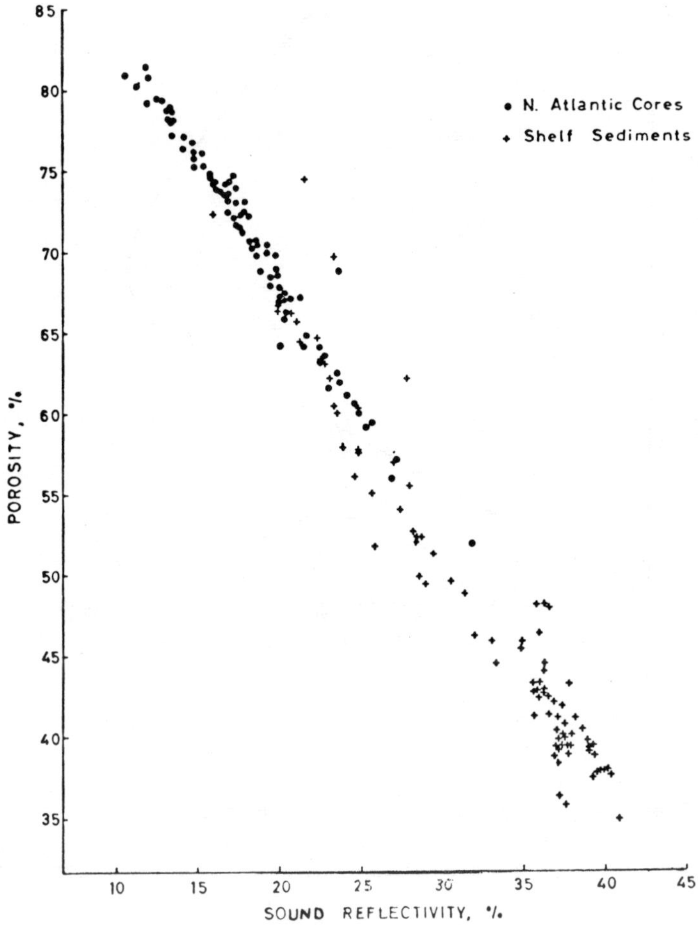

FIGURE 7

SEDIMENT POROSITY PLOTTED AGAINST SOUND REFLECTIVITY

Attenuation Coefficient

The attenuation coefficient is the pressure loss expressed in decibels per unit distance. All the data used in the regression analysis were measured on deep-sea sediments in the laboratory at 3 frequencies within the range 0.1 - 1.0 MHz. The value of the attenuation coefficient used in the analysis is that measured at 600 kHz (or, in a small number of cases, that computed for 600 kHz from frequencies either side of it). Figure 8 shows the variation of the coefficient with sediment mean grain size computed for 50 kHz to compare the variations with those measured using an in situ probe. While a perfect match is not achieved, it is seen that the computation produces the right order of magnitude.

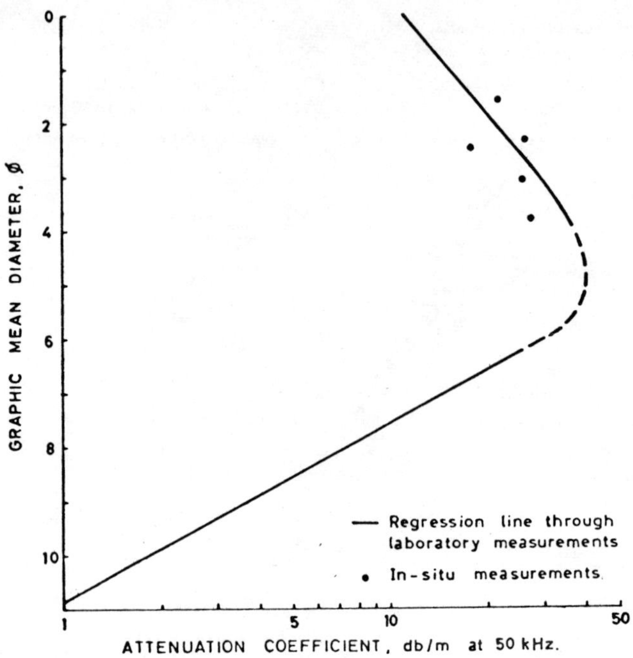

FIGURE 8

SEDIMENT SIZE PLOTTED AGAINST ATTENUATION COEFFICIENT

All the low frequency in situ measurements (30 kHz, 50 kHz, and spectral analysis of broadband seismo-acoustic pulses) show a linear frequency dependence, although these were restricted to sediments in the sand-silt range. Laboratory measurements on a complete size range gave a frequency exponent varying from 0.6 to 2.0 averaging at 1.60; an f^2 relationship almost always occurred with high porosity "red-clay" sediments of exceptionally low shear strengths.

The best equations for predicting the coefficient (in dB/cm) at 600 kHz and 20°C is either

$$A = 0.992 - 18.079\ F_s, \quad \text{or}$$

$$A = 16.002 - 1.534\ M_z$$

where $M_z = 1/3(2M_\phi + M_d)$.

Dynamic Constrained Modulus

From the data acquired in the standard soil oedometer test, which measures the change in the volume of a sample under applied pressure, it is possible to calculate a coefficient of volume change (m_v) sometimes called the unidimensional compressibility. As the sample is confined laterally in a rigid container, so that no lateral expansion can occur, then a certain amount of distortion of the sample can occur although the deformation is primarily volumetric. Applying the theory of elasticity, with lateral stress forces being zero,

$$m_v = \frac{1 - 2\sigma}{2G\ (1-\sigma)}$$

where σ and G are Poisson's ratio and rigidity modulus, respectively.

The inverse of m_v is called the constrained modulus D. It can be seen that since simple elastic theory and the wave equation define the velocity of propagation of compressional waves as

$$V^2 = (3B+4G)/3\rho$$

where B = bulk modulus = $\frac{2G(1+\sigma)}{3(1-2\sigma)}$,

then $\rho v^2 = D$.

Figure 9 shows the dynamic constrained modulus (ρv^2) plotted against the static constrained modulus ($1/m_v$), each pair of readings being either obtained on the same sample or on contiguous samples from the same core. There is thus something like 3 orders of magnitude difference between the static and dynamic results, although it is still possible to assess the one, knowing the other.

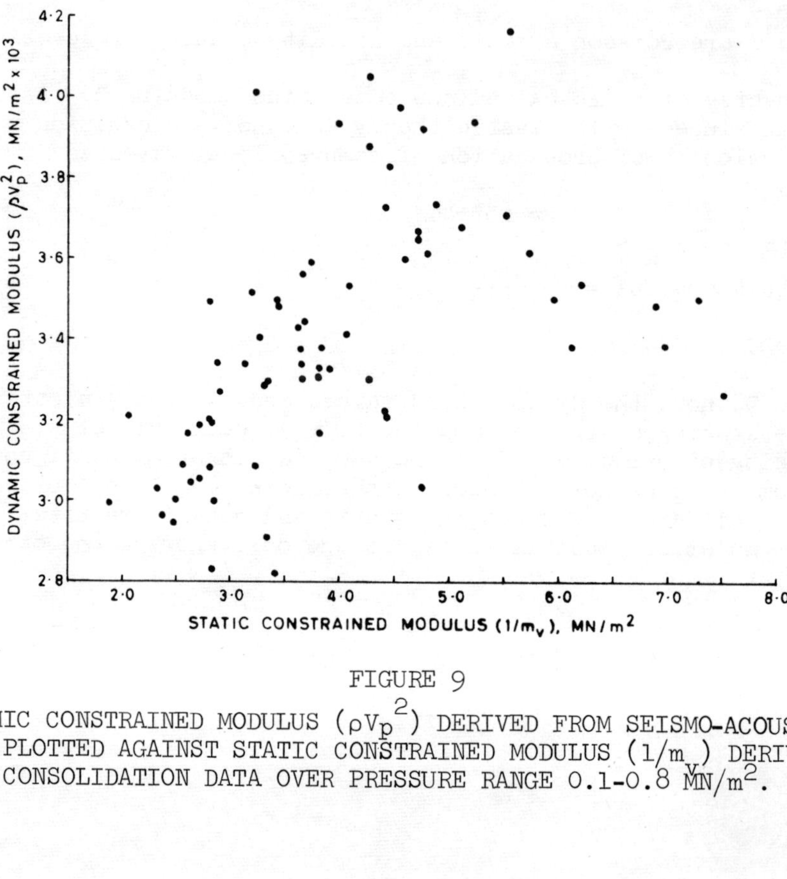

FIGURE 9

DYNAMIC CONSTRAINED MODULUS (ρV_p^2) DERIVED FROM SEISMO-ACOUSTIC DATA PLOTTED AGAINST STATIC CONSTRAINED MODULUS ($1/m_v$) DERIVED FROM CONSOLIDATION DATA OVER PRESSURE RANGE 0.1-0.8 MN/m^2.

DISCUSSION

Any theoretical discussion of the manner by which sound is propagated through a marine sediment has to take account of the interrelationships between the acoustic and mechanical properties briefly indicated above. It would be easy to explain the velocity-porosity variations by invoking simple elastic theory, as Wood did. By considering that a sediment consisted of a "suspension" of mineral grains in sea water, there would be no interaction between the grains and no rigidity; sound velocity would only depend on the relative proportions of solid and liquid and their respective compressibilities and densities. But as indicated in Fig. 2, Wood's equation simply places a lower limit on the velocity for any porosity; it does not account for the grain size variations, except insofar as there is a grain size-porosity relationship. A marine sediment, even of high porosity, has rigidity in that it can propagate a shear wave; there must therefore be some form of framework set up between the particles which has to be accounted for in any theory.

Again, as also indicated in Fig. 9, the soil and rock mechanics literature gives copious illustrations of cases where the value of Young's modulus derived from the measurement of seismic velocities is many times too high to be used in foundation design considerations. Since elastic models are usually invoked in soil mechanics to obtain the various static coefficients, does this all mean that elastic theory cannot be applied to acoustic propagation in marine sediments, or is it just a matter of stress level?

Elastic Theory Examination

Any theory of sound propagation through a marine sediment must be adequate to explain the data recorded experimentally. These are:

(1) Compressional wave velocity increases with decreasing porosity and increasing grain size.
(2) All sediments possess some rigidity (in that they transmit shear waves) and, although the evidence is not conclusive, this seems to be a maximum in the fine-sand grade. Having said this, there appears to be no correlation between shear strength and any acoustic parameter (except, in a somewhat nebulous way, with acoustic impedance).
(3) There is a slight, though generally negligible, change in velocity with frequency.
(4) Moduli, calculated from wave velocities, are many times greater than those determined statically.
(5) Compressional wave attenuation increases to a maximum in the fine-sand grade from both extremes of size.

(6) While in situ low frequency measurements on silts, sands, and rocks show that sound absorption varies linearly with frequency, precise laboratory determinations at high frequencies (0.1-1 MHz) show that the frequency exponent can be anywhere in the range 0.5-2.0. There seems to be no correlation between the exponent and any mechanical parameter of the sediment except that with very small mean grain diameters (less than 1 micron) and high porosities (greater than 80%) an f^2 relationship holds.

(7) For a fixed frequency and grain size, absorption decreases with temperature increase.

The above criteria, while fundamentally the same, differ slightly from those put forward by Hamilton in his many papers. As Hamilton points out there is no lack of ideas on the subject. Undoubtedly the most satisfactory theory would be one based on elastic concepts which have proven successful in seismology. Probably the most complete treatment of the subject is that given by Biot in a variety of papers spanning two decades from 1941 onwards (e.g., 1956). The Biot treatment allows for water in the pores to move relative to the solid particle or to remain static as in the "closed systems" suggested by other writers; its essential feature is that not only must account be taken of the properties of the individual constituents of the sediment but also due reference must be paid to intercoupling between them. Thus there is an essential structure established in sediments (via electrochemical forces in clays and grain-to-grain contact in sands) which must be allowed for in the elastic equations. Out of the analysis it is seen that in <u>sands</u>, where it must be assumed on the basis of thermodynamic considerations that a slight relative motion between the two constituents exists, two dilatational (compressional) waves and one rotational (shear) wave are possible. One dilatational wave is transmitted through the fluid and the other through the elastic structure; both waves are coupled together through the stiffnesses of the solid and fluid components of the system as well as through the coupling effect produced by motions of the solid and fluid. The dilatational wave in the fluid (the one usually measured as the sediment compressional wave) travels with the highest velocity; this velocity is higher than that in water alone because of the effect of the vibrating elastic structure. In <u>clays</u> the situation is a little confused because a skeleton is established through intermolecular-electrochemical forces which result from an intimate relationship between the solid and liquid phases. From such a situation therefore it is to be expected that little relative motion will arise. Under these conditions the measured dilatational wave velocity will be

$$V_p^2 = (P+2Q+R)/\rho$$

where P, Q, and R are Biot's coefficients and ρ is the total bulk density of the system.

$$P = \{(n-1)^2 C_p + (n-1) C_s + C_w\}/\{C_t(C_w-C_s) + C_pC_s\} + 4/3 G$$

$$Q = -n\{(n-1) C_p + C_s\}/\{C_t(C_w-C_s) + C_pC_s\}$$

$$R = nC_t/\{C_t(C_w-C_s) + C_pC_s\}$$

where C_t = bulk compressibility of the total sediment skeleton
C_s = compressibility of the solid particle
C_p = pore compressibility (coupled effect)
C_w = fluid compressibility
n = fractional porosity, and
$C_t = n C_p + C_s$.

Comparing this result with that obtained using straight elastic theory,

$$\rho V_p^2 = B + 4/3 G = P + 2Q + R$$

By manipulation of the coefficients and the compressibilities the bulk modulus,

$$B = \{C_t - (1+n) C_s + nC_w\}/\{nC_t(C_w-C_s) + C_tC_s - C_s^2\}$$

$$= [1 + \{-(1+n) C_s + nC_w\}/C_t]/\{n(C_w-C_s) + C_s - C_s^2/C_t\}$$

For a deep-sea clay at 20°C with pore water at 35 o/oo salinity

$$C_w = 4.21 \times 10^{-4} \ (m^2/MN)$$

$$C_s = 0.21 \times 10^{-4} \ (m^2/MN)$$

$$C_t = 0.3 \ (m^2/MN)$$

Thus $\{(1+n) C_s + nC_w\}/C_t \approx 10^{-3}$ and $C_s^2/C_t \approx 10^{-9}$.

The terms can therefore be ignored. Consequently the bulk modulus of a deep-sea clay can be represented by

$$B = 10^4/\{4n+0.21\} \ (MN/m^2) \quad .$$

To assess the validity of this approximation it is necessary to measure the bulk modulus and compare it to that computed by the above relationship. That is,

$$B \text{ measured} = e\, V_{\rho}^{2} - 4/3\, G$$

$$B \text{ computed} = 1/\{4n + 0.21\}$$

This comparison requires the determination of the rigidity modulus G.

Determination of G

Figures 4, 5, and 6 show the variation of rigidity modulus with density, porosity, and mean grain size derived from shear wave measurements (in situ and in the laboratory) on a small number of samples. While it would be valuable to have more determinations to define the trends completely, the results do agree with those of Hamilton. It is therefore possible to assess the rigidity modulus of any sample, provided that the mean size or porosity or density is known. Of these, undoubtedly the mean size relationship gives the most dependable assessment, since the deviations from the regression line through the data are smaller with size than with the other two parameters.

Once the rigidity modulus has been determined, the bulk modulus can be calculated and plotted against the value computed from the Biot approximation. This comparison is shown in Fig. 10. The deviations away from a perfect match can be explained in a large number of experimental ways (choice of the wrong compressibilities, inaccurate determination of porosity, erroneous assessment of G, etc.), but undoubtedly the main error is in defining the samples as perfect clays: large coarse fractions exist in some samples (usually foraminiferal skeletons) which makes the assumption of little relative motion between the phases invalid in addition to providing a different magnitude of solid compressibility. However, the overall conclusion that can be drawn is that the theory of elasticity can be used to explain the velocity phenomena observed in marine sediments. The wide disparity between static and dynamic moduli (Fig. 9), both calculated from elastic theory, can probably be accounted for by the low stress-level of the acoustic wave compared to that applied in the oedometer test: in fact, after repeated loadings and unloadings the coefficient of volume change becomes extremely small, giving a constrained modulus similar to that determined dynamically.

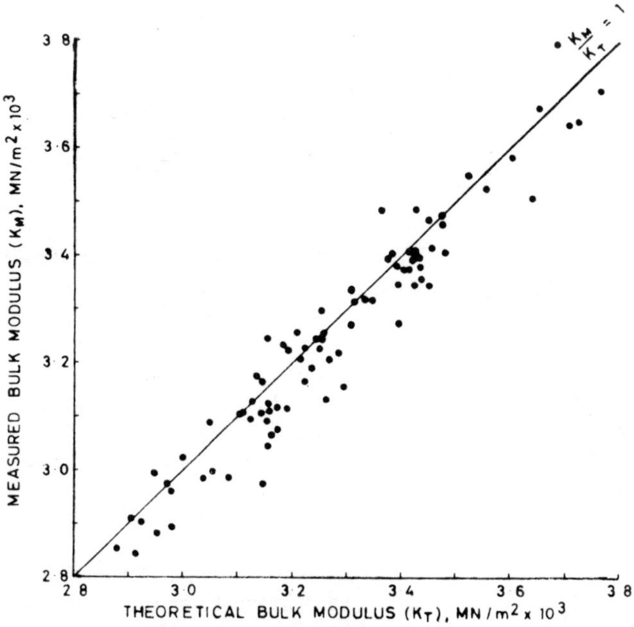

FIGURE 10

BULK MODULUS (K_M) DERIVED FROM SEISMO-ACOUSTIC DATA PLOTTED AGAINST BULK MODULUS (K_T) DERIVED FROM BIOT'S EQUATIONS

Examination of Attenuation

Although the dynamic equations developed from the theory of elasticity adequately account for the velocities measured in marine sediments, they are insufficient to explain all the details of measured attenuation. Biot's theory gives either an f^2 dependence or an $f^{1/2}$ dependence according to a frequency function related to the size of the pore-spaces: the former was effectively that for low frequencies, the latter for high frequencies. Hamilton in an excellent summary of the situation (but one dedicated to a lack of velocity dispersion and to a linear frequency dependence of attenuation), involves the Ferry model of linear visco-elasticity. However, it seems to the present writer that it is impossible to develop a single theory which can explain all the observed facts. Most certainly frequency exponents other than unity exist: the data for Fig. 8 were obtained at three frequencies centered around 600 kHz; in using the exponents of these frequencies (averaging about $f^{1.6}$ for clays) it has been possible to produce an attenuation coefficient/grain size graph which agrees with data measured in situ at 50 kHz. Certainly in real continental shelf sediments, usually in the sand-silt grade, the in situ results indicate a linear frequency dependence, but it is hard to extend this observation to pure deep-sea clays. It does seem therefore that two dissipating mechanisms are responsible for acoustic energy loss:

(a) A solid friction loss for sediments with grain-to-grain contact characterized by a linear variation of the attenuation coefficient with frequency; and
(b) A viscous loss due to interaction between solid and liquid, this being characterized by an $f^{1/2}$ dependence.

Most sediments will probably have both mechanisms operating together at the same time. For marine sediments of particle size in the sand grade which are poorly sorted and subject to some overburden pressure, grain-to-grain energy dissipation will be the dominant loss, while for well-sorted sediments which are subject to no overburden pressure the solid friction loss disappears entirely giving a square root frequency relationship; it is impossible to see how the latter situation can exist with sands outside the special conditions of the laboratory. Solid friction also disappears for sediment grains finer than sand, the loss mechanism being entirely viscous. However, as the mean diameters become smaller, the percentage of clay sizes increases, introducing the additional mechanism of interparticle attraction: this has the effect of reducing the relative motion between the particles and the fluid with a consequent decrease in viscous dissipation (McCann and McCann, 1969). In this size range the frequency dependence becomes complex, and for the very fine-grained high-porosity sediments of the deep ocean the frequency dependence approaches that experienced by propagation in seawater.

An additional argument that can be made against viscous loss as the only explanation of energy dissipation is shown by the reasonable correlation between rigidity modulus and attenuation coefficient via grain size variations (Figs. 6 and 8). That the rigidity effects can be explained in terms of intergrain contact is shown by the relationships indicated in Figs. 11 and 12: the angle of shearing friction in sands increases as grain size decreases in almost precisely the same manner as rigidity modulus over the same size range.

There is, finally, one further problem that emerges from the study: the attenuation coefficient in sands decreases as the temperature increases. Is this a function of increased fluidity of the water or a reduction of the frictional state between the grains?

Conclusion

While elastic theory can satisfactorily explain the variations of velocity with porosity and size (via the compressibilities

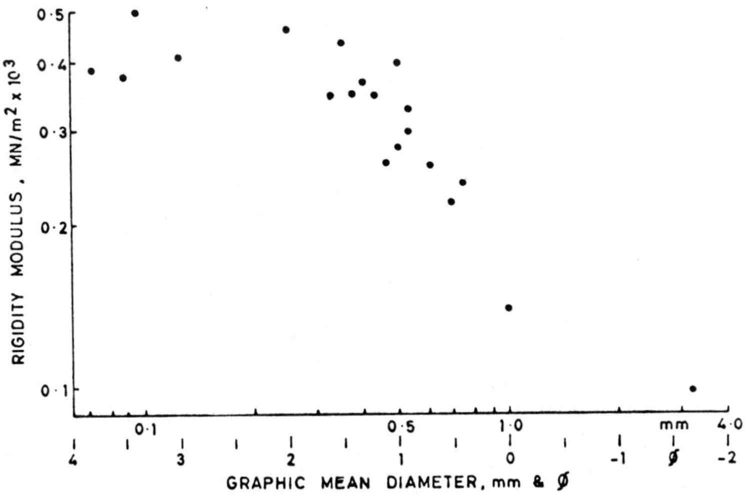

FIGURE 11

RIGIDITY MODULUS PLOTTED AGAINST SEDIMENT SIZE

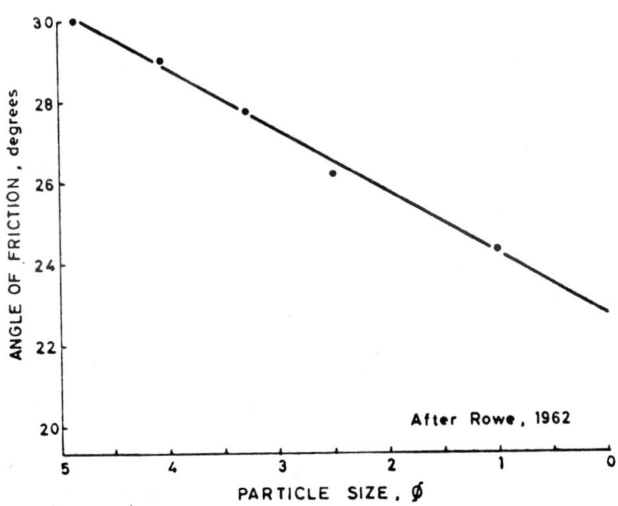

FIGURE 12

INTERNAL FRICTION PLOTTED AGAINST SEDIMENT SIZE

of water and the mineral grain in the former, and the rigidity modulus in the latter), it does not seem possible to account for energy losses in a similar way. It is possible that no all-embracing theory will emerge. However, sufficient data are now available to be able to predict either the mechanical or the acoustic condition of the sea floor, provided that certain parameters (particularly porosity and grain size) are known; these latter may be assessed by the acoustic experiment itself.

Acknowledgements

While the data indicated on the scatter-diagrams have been collected by a number of individuals in the author's laboratory, it is essential to note the special efforts of Mr. Frank Dewes in preparing the information for this paper: all the diagrams were drawn by him and he carried out much of the computation necessary for such displays. The data collection has been supported by grants from the Natural Environment Research Council, the Oceanographic Committee of NATO, and the Ministry of Defence (Navy Dept.: AUWE). The author gratefully acknowledges all this assistance.

SELECTED BIBLIOGRAPHY

Biot, M. A., Theory of propagation of elastic waves in a fluid-saturated porous solid, J. Acoust. Soc. Am., 28, 168-191, 1956.

Hamilton, E. L., H. P. Bucker, D. L. Keir, and J. A. Whitney, Velocities of compressional and shear waves in marine sediments determined in situ from a research submersible, J. Geophys. Res., 75, 4039-4049, 1970.

Hamilton, E. L., Elastic properties of marine sediments, J. Geophys. Res., 76, 579-604, 1971.

Hamilton, E. L., Compressional wave attenuation in marine sediments, Geophysics, 37, 620-646, 1972.

McCann, C., and D. M. McCann, The attenuation of compressional waves in marine sediments, Geophysics, 34, 882-892, 1969.

INFLUENCE OF ANISOTROPY AND SOIL STRUCTURE ON ELASTIC PROPERTIES OF SEDIMENTS

V. A. NACCI[1], M. C. WANG[2], AND J. GALLAGHER[3]

ABSTRACT

The theory of elasticity is used for the solution of sediment sound propagation as well as various stress distribution and soil deformation problems. This requires the determination of sediment elastic constants; yet there are many factors that make the rational and accurate prediction of field stress-strain relationship from the results of laboratory tests difficult. One of these important factors is that the soil is not an isotropic linear elastic material having a unique Young's modulus, E.

An anisotropic nonlinear elastic property, determined from an acoustic dilatational wave propagation method, a resonant column method, and consolidated undrained triaxial tests on undisturbed saturated, sedimented clay and silt samples, is studied herein. Varved silt and clay were deposited in the laboratory at various angles of stratification with respect to the horizontal plane. Clay samples were sedimented in both 0.6 Mole NaCl and 0.02 Mole quadrofos solutions. The samples obtained from sedimentation simulate the characteristics of anisotropic varved structures of various particle orientations. Data are presented to show the influence of the direction of the major principal stress with reference to the varved layers or particle orientations on both the dynamic and static stress-strain relationships.

[1]Professor of Civil and Ocean Engineering, University of Rhode Island.
[2]Assistant Professor of Civil Engineering, University of Rhode Island.
[3]Oceanographer, Naval Underwater Systems Center.

INTRODUCTION

Geological oceanographic investigations conducted jointly by the University of Rhode Island and the Naval Underwater System Center are largely directed toward the determination of significant behavioral relationships between geologic and acoustic properties of marine water-sediment aggregates in support of acoustic bottom loss studies.

To date, emphasis has been placed on the study of the variable sedimentary and acoustic structures in western North Atlantic deep ocean bottom areas having little slope and almost no roughness features, i.e., abyssal plains. Usually, these flat depositional areas contain dramatic stratigraphy, and often contain many sub-bottom acoustic reflectors. The sediments commonly found in these areas are dominated by the silt and clay grain sizes; turbidite layers also contain sand size deposits.

The analysis of sediment core samples collected by NUSC from the Tongue of the Ocean was carried out by the University of Rhode Island Civil Engineering Department in order to relate the values of mass physical and engineering properties of the samples, determined from standard terrestrial soil mechanics techniques, to values of sound velocity and elastic moduli. The results of this pilot study indicated that the stress-strain relationships for static and dynamic loading of confined and unconfined homogeneous soil samples is considerably influenced by the void ratios or confining pressures. This influence results in a general trend of increasing modulus with depth. Judging the results of this analysis, it was assumed that a short-term field load or penetration test could be a valid indicator of sound velocity. Vertical profiles of sound velocity versus depth were compared with profiles of static and dynamic elastic moduli versus depth (Gallagher and Nacci, 1968). The nonuniform character of the sound velocity profiles indicated that factors other than the void ratio, or confining pressure, appeared to be influencing the sound velocity values.

A two-year program was begun at the University of Rhode Island Civil Engineering Department in June 1966 to study improved methods for more accurately determining elastic properties of fine grained sediments. The study of elastic moduli provides for a more basic understanding of the relationships between the acoustic and dynamic-geotechnical properties of the water-sediment aggregates.

There is no adequate theory to explain the general behavior of soil subject to elastic wave propagation. One theoretical approach to this problem in the granular soil range is to regard the soil as a packing of elastic spheres in contact with each other, in which the applied pressure causes changes in the area of contact between adjacent spheres. This approach predicts that the velocity of an elastic wave through an array of spheres varies according to the 1/6th power of the confining pressures. Biot (1956, 1962) assumed that a porous material consists of a frame which is microscopically homogeneous and elastic and with the pores filled with fluid. A stress-strain relationship can be derived for the material in terms of the elastic constants of the frame and fluid, and constants relating the flow of the fluid through the pores. These constants are introduced into the equation of motion and solved for propagation characteristics.

Fine grained saturated soils are usually platy rather than spherical, often have a flocculent or 'honeycomb' structure rather than single grain contact, are highly porous and sensitive, and are somewhat anisotropic. Therefore, the approach taken in this study was to experimentally determine acoustic velocities and elastic moduli of relatively undisturbed fine-grain-particle laboratory constructed sediment samples and field core samples and to relate these parameters to porosity, particle structure, shear strength, anisotropy and confining pressure.

Longitudinal Wave Propagation in Soils

The mechanism of stress transfer of a particulate array prevails whether considering acoustic velocity, shear strength, or compressibility. Stress transfer is affected by many variables, but the most significant is porosity. Since shear strength is also governed primarily by porosity, it appears reasonable to assume that there is a correlation between strength and acoustic properties.

If a porosity-confining pressure function exists, then reliable estimates may be made of compressional wave velocity and shear wave velocity as well as shear strength. However, early in the study, it was shown that pressure-porosity relations were nonuniversal. Taylor (1967) sedimented a kaolin clay in distilled water, a 0.6 M sodium chloride solution, and a 0.02 M quadrafos solution to obtain dispersed and flocculent clay structures. While no cross-correlation of relationships between soil types is apparent for these limiting conditions, a unique relation seemingly exists among pressure, porosity, shear strength, and acoustic velocity for each soil structure.

The methods used to determine acoustic properties are based upon elastic theory. The suitability of an elastic theory to soil may be questioned; yet any system of particles may be considered elastic with the following limitations:

1. The volume of the specimen must be large in comparison to individual particle dimensions;

2. the stresses and strains are considered average and small in value; and
3. the sample must be essentially isotropic, or if composed of different layers, the sample should be isotropic within each layer.

The classical theory of wave propagation within an elastic, homogeneous, laterally unrestrained bar relates the longitudinal wave velocity, c, to the modulus of elasticity:

$$c = \sqrt{\frac{E_d g}{\rho}}$$

E_d = dilatational modulus

g = acceleration of gravity

ρ = bulk density of soil mass

This can be measured by propagating pulses through a bar. The longitudinal waves propagated may be considered dilatational waves due to the lateral restraint imposed by the chamber pressure as well as inertial effects.

The velocity of propagation may also be experimentally determined from the bar resonant frequency; the time required for a longitudinal wave to travel from one end to the other will be one fourth the period T. The velocity of propagation of this wave is

$$c = \frac{L}{T/4}$$

L = specimen length

T = period of vibration

Lawrence (1965) concluded that the most important distinction between the two tests is the amount of deformation employed and the frequencies of vibration imposed. In the resonant method, a relatively large deformation (on the order of 10^{-4} in./in.) is used whereas the pulse technique requires only small disturbances (about 10^{-6} in./in.). The frequencies used in the pulse type test (above 10,000 Hz) are generally much higher than those necessary to reach resonance (200 to 2000 Hz).

Parameters Influencing Elastic Wave Propagation and Elastic Moduli

Hardin and Black (1968) have indicated the various quantities which exert an influence on the shear modulus G as:

$$G = f(\bar{\sigma}_o, e, H, S, \tau_o, C, A, F, T, \theta, K) \quad (1)$$

in which

$\bar{\sigma}_o$ = effective octahedral normal stress (average confining pressure)

e = void ratio

H = ambient stress history and vibration history

S = degree of saturation

τ_o = octahedral shear stress

$$= 1/3 \sqrt{(\sigma_1-\sigma_2)^2 + (\sigma_2-\sigma_3)^2 + (\sigma_3-\sigma_1)^2}$$

$\sigma_i (i=1,2,3)$ = principal stresses

C = grain characteristics, i.e., grain shape, grain size, grading, and minerology

A = amplitude of vibration

F = frequency of vibration

T = secondary effects that are function of time and magnitude of load increment

θ = soil structure

K = temperature

Several of these factors such as e, C, and θ may be interrelated. For sands it has been shown (Pinson,1963; Hardin and Black, 1966; Hardin, 1965) that for a strain amplitude less than about 10^{-4} radians, and within the range of the variables studied, G is essentially independent of each of the variables listed in Eq. (1) except for $\bar{\sigma}_o$ and e. Comparisons of measured wave velocity with theoretically predicted values have shown that velocity is dependent upon the third to fifth root of intergranular stress, rather than upon the sixth root predicted by theory. This discrepancy has been resolved by Duffy and Mindlin (1957) who showed that even with carefully prepared, regular arrays of extremely high tolerance steel balls, the relationship between intergranular stress and elastic wave velocity showed the same misbehavior observed earlier in sands. However, as the ambient intergranular stress was increased (beyond about 10 psi), the data began to assume the predicted sixth root dependence. It appears, then, that wave velocity is dependent not only upon the level of intergranular pressure and void ratio but also upon the number of intergranular contacts existing at any given stress level.

Hardin and Richart (1963) showed results for crushed quartz silts and indicated the behavior which might be expected for clays, silts, and mixed soils. The velocity can vary with any power of confining pressure depending on time and the state of consolidation of the specimen.

Morgenstern and Smith (1964) observed a decrease in longitudinal wave velocity as load increments were applied, then an increase in velocity with time when load (stress) was held constant. As the load is applied, the increasing pore pressure decreases the effective stresses, and consequently the velocity. As the pore pressure dissipates, the effective stresses and the wave velocity increase.

Hardin and Black (1968) and Humphries and Wahls (1968) showed that for normally consolidated conditions, the dynamic shear modulus varies approximately with the one-half power of effective pressure for kaolin. Humphries and Wahls (1968) also showed that the shear modulus varies with two thirds power of the effective pressure for bentonite. These results indicate that theoretical techniques for a packing of granular particles cannot be used for predicting the behavior of silt and clay.

Lawrence (1965) and Hardin and Black (1966, 1968) have found that the shear modulus is a function of the mean of three principal stresses acting upon the soil and is independent of the amount of static shear stress. Hall and Richart (1963) have shown that the modulus measured in dynamic triaxial tests is influenced by the magnitude of the dynamic strain.

Hardin and Richart (1963) showed that the grain size and grading of sands retained on the No. 120 sieve had almost no effect on the shear modulus, the effect of grain shape was small, and the effect of saturation had a minor effect and occurred only at low pressure. It was also found that the frequency of vibration had no measurable effect on velocity of shear waves.

Properties of Soils Used in The Investigation

Normally consolidated laboratory sedimented silts and clays were used to study the variables affecting elastic wave propagation. The soils used were a commercially available kaolin, designated as Peerless No. 2, from the R. T. Vanderbilt Company of New York, and a Providence silt. Classification tests give the gradation curves shown in Fig. 1 and the results in Table I.

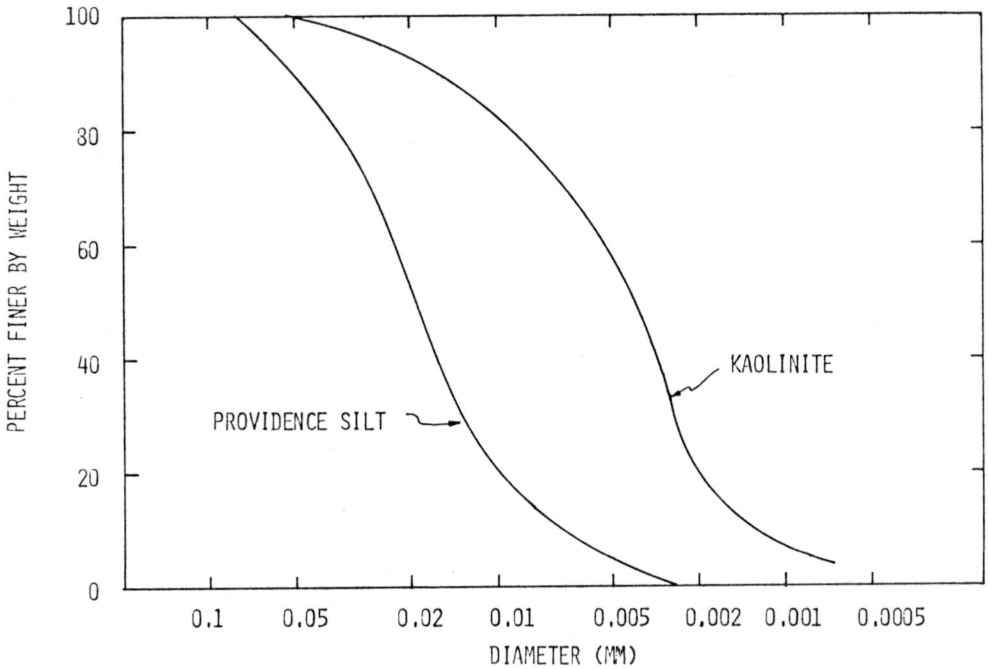

FIGURE 1

GRADATION CURVES OF TEST SOILS

TABLE I

INDEX PROPERTIES OF TEST SOILS

	Kaolin	Providence Silt
Specific Gravity	2.60	2.75
Plastic Limit	30.6%	24.0%
Liquid Limit	60.6%	28.0%

TESTING PROCEDURE AND EQUIPMENT

The longitudinal wave apparatus shown in Fig. 2 was designed and fabricated at the University of Rhode Island. The apparatus uses piezoelectric transducers which are resonant at 50 kHz. It is possible to measure pore water pressures simultaneously with sound velocity at various confining pressures. Laboratory samples of 1.4 in. (3.6 cm) diam are extruded from plastic tubes and trimmed to a length of 5 in. (12.7 cm). The samples are sheathed in filter paper, enclosed in a thin rubber membrane, and sealed to the top and bottom transducers by constricting "O"-rings. Situated at the bottom transducer is a drainage ring which leads accumulated moisture to a volume measuring burette. The top transducer is attached to weight compensating springs. The spring constants are such that isotropic consolidation is possible, provided that the

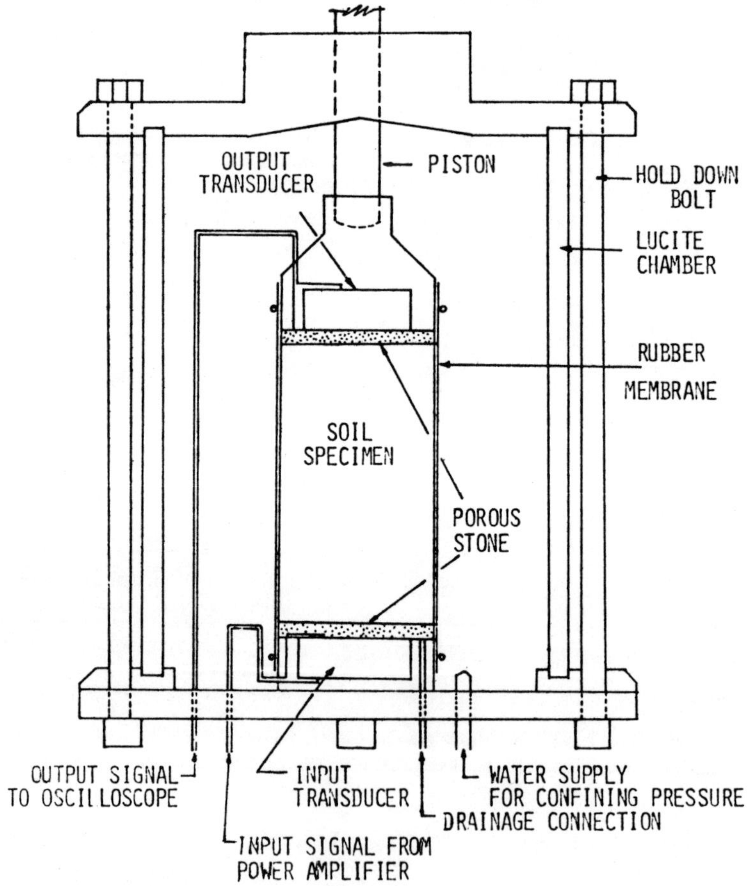

FIGURE 2

LONGITUDINAL WAVE TEST SETUP

change in sample height remains small. The acoustic velocity measuring equipment, called a V-scope, was purchased from James Electronics, Inc., Chicago, Illinois. It is composed of a digital oscilloscope, a pulse generator, and a pair of transducers. It was found that the resulting error in determining sound velocity was less than 0.5%.

The samples were allowed to consolidate for two to three days to reach various effective pressures. To completely saturate the sample at the end of the consolidation period, the cell pressure and the pore pressure were raised simultaneously to 2 kg/cm^2 and allowed to stand until the pore pressure remained constant. The pore pressure was determined by means of a mercury manometer. This process, called back pressuring, forced any entrapped air into solution and was an important step in the entire test procedure, since Hampton (1966) reported that even small quantities of entrapped gas in the test sample may dominate the acoustic properties of sediments. After the sound velocity measurement was made, the sample was removed from the testing chamber and water content was determined for the entire sample.

The torsional resonant column triaxial test device, Fig. 3, is a product of soil Dynamic Instruments, Inc., Lexington, Kentucky.

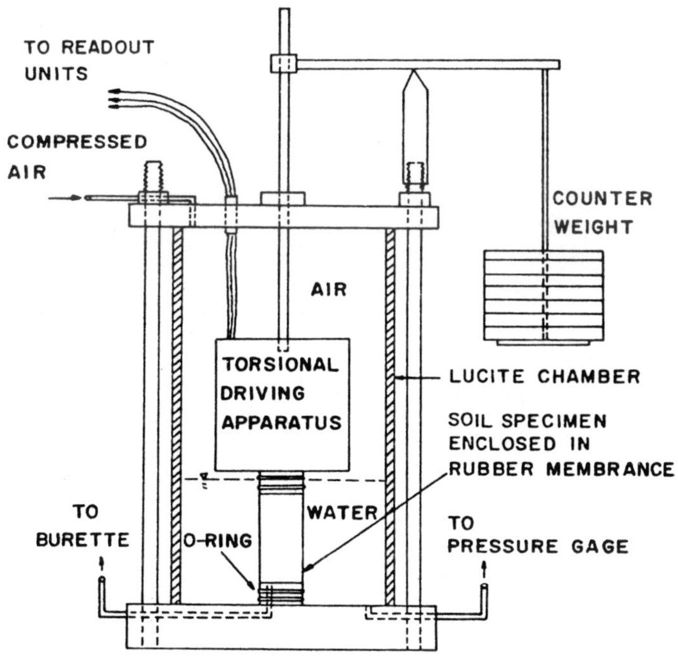

FIGURE 3

GENERAL VIEW OF SHEAR WAVE TEST SETUP

The system is composed of: (1) a cylindrical soil specimen attached to a rigid pedestal of sufficient inertia to null the motion of the attached end during vibration, (2) a specimen cap capable of applying a variable frequency sinusoidal forcing torque to the vibration end of the specimen, (3) a triaxial chamber with capability of drainage and back pressuring, and (4) the electronic components to produce the necessary sinusoidal voltage, and to display and measure acceleration, velocity, or displacement of the vibration end of the specimen.

The test procedure consists of subjecting a consolidated and back pressured column of soil to a steady state sinusoidal torsional forcing function. The frequency of the applied force is varied to obtain resonance. The resonant frequency is primarily a function of the stiffness or the modulus of the material.

The consolidated isotropic undrained (CIU) triaxial compression test was conducted by using the equipment described by Anderson and Simons (1966). The samples were consolidated for about 24 h, and a back pressure of about 3 kg/cm^2 (KSC) was used to dissolve the gas entrapped in the test sample. The tests were performed at a strain rate of 1 percent per hour.

TEST RESULTS AND DISCUSSIONS

Figure 4 illustrates the relationship between void ratio, consolidation pressure, and undrained shear strength for the laboratory sedimented kaolin. The flocculent structure was obtained by using sodium chloride solution, while the dispersed structure was produced by quadrafos solution. Test results for the laboratory sedimented Providence silt are shown in Fig. 5. Figure 5 also

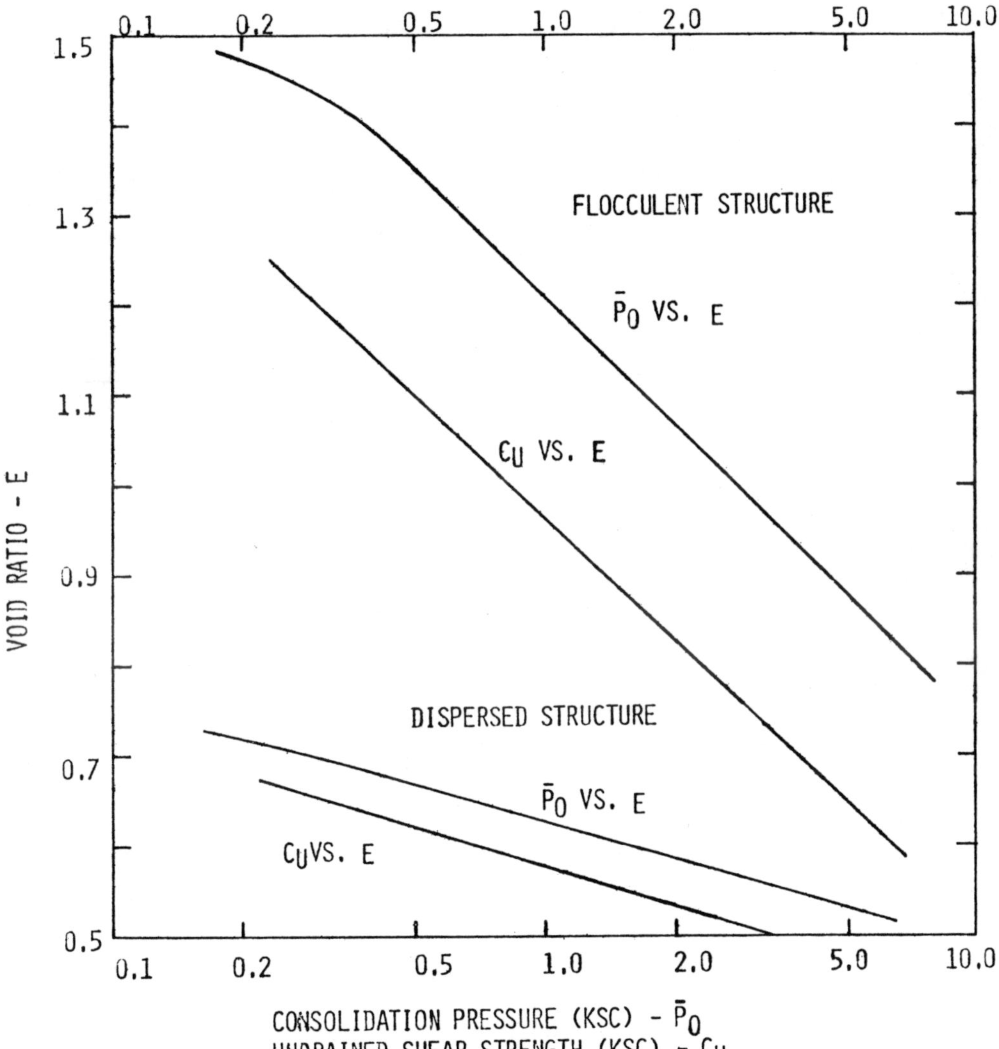

FIGURE 4

CONSOLIDATION AND SHEAR STRENGTH PROPERTIES
OF LABORATORY SEDIMENTED KAOLINITE

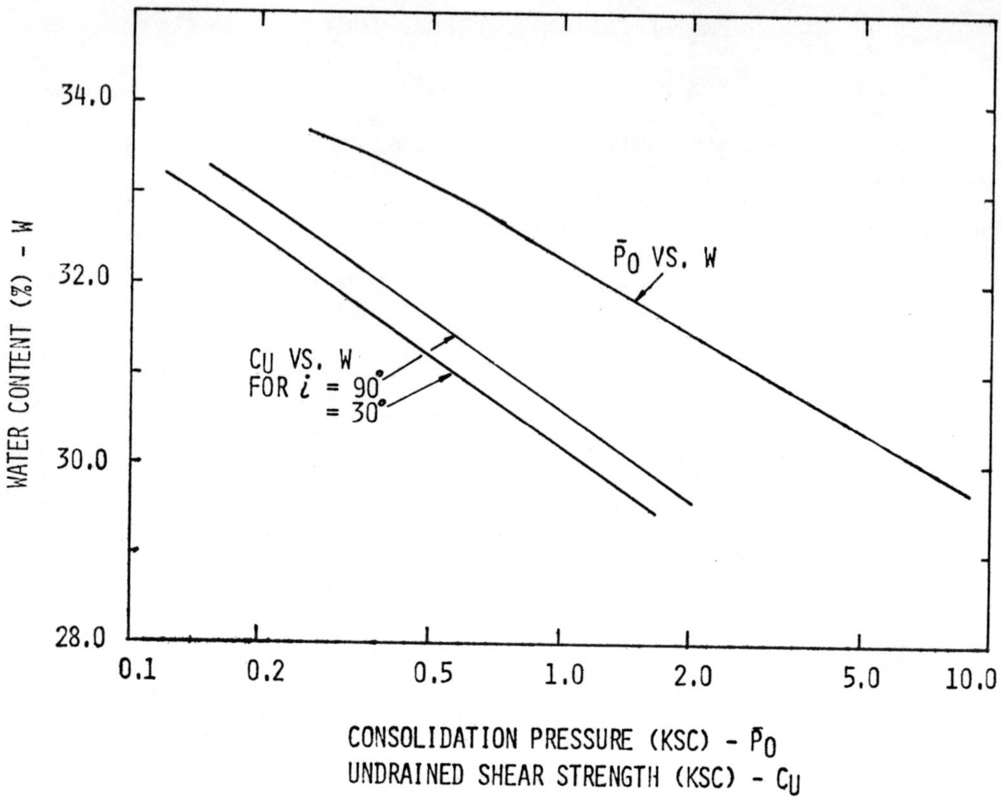

FIGURE 5

CONSOLIDATION AND SHEAR STRENGTH PROPERTIES OF LABORATORY SEDIMENTED PROVIDENCE SILT

shows the variation of the relation between shear strength and water content with various angles between loading direction and bedding plane, namely angle, i. Additional information regarding shear strength properties is summarized in Tables 2 and 3. Note that results in Table 2 are for loading normal to the bedding plane, i.e., i=0, and Table 3 presents results for various bedding plane or varve orientations.

The longitudinal acoustic wave propagation velocity was measured in laboratory sedimented kaolin under varying consolidation pressures and at an essentially constant temperature of 70°F. The observed velocity is compared with the shear strength in Fig. 6. Within the range of conditions studied, a linear relationship between wave velocity and log shear strength is seen. For a constant shear strength, the dispersed structure transmitted the longitudinal acoustic wave faster than the flocculent structure. Furthermore, the rate of wave velocity increase with an increase

TABLE 2

STRENGTH PROPERTIES OF TEST SOILS
FOR LOADING NORMAL TO BEDDING PLANE

Soil	Structure	$\bar{\phi}$(deg)	$\bar{c}$(KSC)	A_f	Cu/p
Kaolin	Flocculent	23.7	0.05	1.4 ~ 1.5	0.27
Kaolin	Dispersed	23.0	0.01	0.75 ~ 0.85	0.42
Providence silt	Honeycomb	26.7	0	1.27	0.26

where $\bar{\phi}$ = effective internal friction angle
$\bar{c}$ = effective cohesion
A_f = pore pressure factor at failure
Cu/p = undrained shear strength to consolidation pressure ratio.

TABLE 3

SHEAR STRENGTH OF PROVIDENCE SILT
FOR VARIOUS VARVE ORIENTATIONS

i(deg)	90°	67°	58.5°	40°	30°	0°
Cu/p	0.260	0.242	0.239	0.237	0.236	0.246
A_f	1.270	1.379	1.370	1.408	1.410	1.356

in shear strength was greater for the flocculent structure than the dispersed structure. The results reveal a trend for the two straight lines to approach each other at a very high shear strength corresponding to an ideal perfect solid where a unique value of wave velocity exists.

The theoretical basis for determining the static modulus of elasticity, E_s, with the CIU triaxial test follows. Assume that a normally consolidated, completely saturated soil element is initially consolidated under a system of orthogonal stress components σ_x, σ_y, and σ_z. If the stress in the x-direction is then increased by $\Delta\sigma_x$, the resulting strain in the x-direction, ϵ_x, can be expressed as follows:

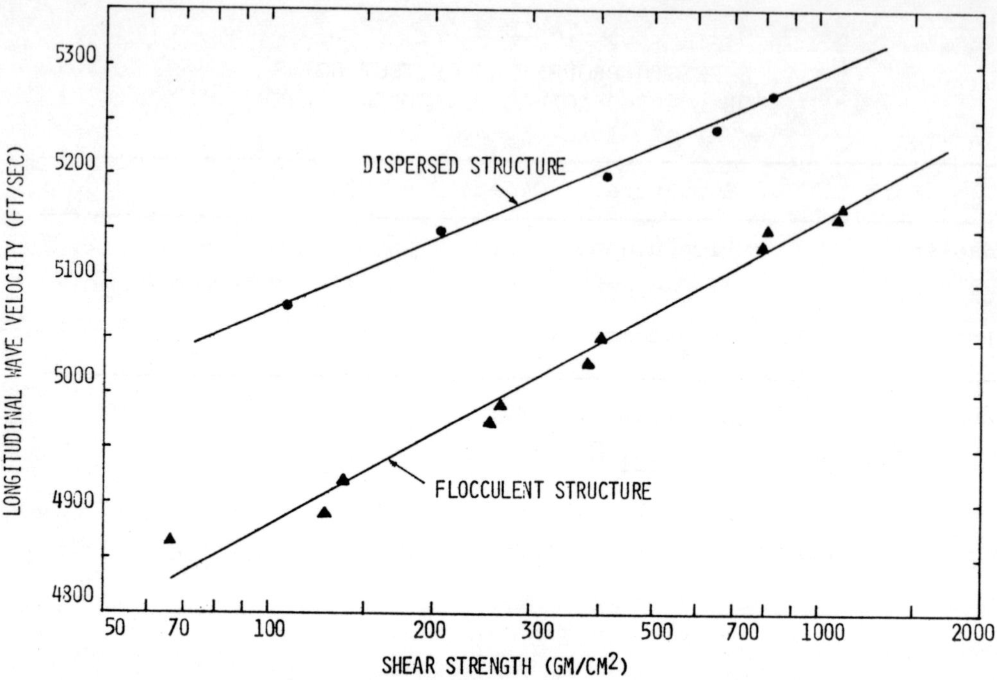

FIGURE 6

LONGITUDINAL WAVE VELOCITY VERSUS SHEAR STRENGTH
FOR LABORATORY SEDIMENTED KAOLIN

$$\epsilon_x = \frac{\Delta\bar{\sigma}_x}{E_{sx}} - \left(\frac{\Delta\bar{\sigma}_y}{E_{sz}} + \frac{\Delta\bar{\sigma}_z}{E_{sz}}\right)\mu \tag{1}$$

or

$$\epsilon_x = \frac{\Delta\sigma_x}{E_{sx}} + \Delta U_x \left(\frac{\mu}{E_{sy}} + \frac{\mu}{E_{sz}} - \frac{1}{E_{sx}}\right) . \tag{2}$$

Similarly, the strains in the y and z directions due to separately applied stress increases $\Delta\sigma_y$ and $\Delta\sigma_z$ are, respectively,

$$\epsilon_y = \frac{\Delta\sigma_y}{E_{sy}} + \Delta U_y \left(\frac{\mu}{E_{sz}} + \frac{\mu}{E_{sx}} - \frac{1}{E_{sy}}\right) \tag{3}$$

$$\epsilon_z = \frac{\Delta\sigma_z}{E_{sz}} + \Delta U_z \left(\frac{\mu}{E_{sx}} + \frac{\mu}{E_{sy}} - \frac{1}{E_{sz}}\right) \tag{4}$$

In Eqs. 1, 2, 3, and 4, E_{sx}, E_{sy}, and E_{sz} are the static moduli of elasticity in the x, y, and z directions, respectively. A Poisson's ratio, μ, of 0.5 could be used for completely saturated soils subjected to a rapid undrained loading because of little volume change due to the loading. Among the factors in Eqs. 2, 3, and 4, $\Delta\sigma$, ϵ, and ΔU can be measured; thus only three unknowns E_{sx}, E_{sy}, and E_{sz} are left for solving the equations. Furthermore, the values within the parentheses are in general small compared with the first term, and can therefore be neglected.

The stress level is one of the most important factors affecting the measured value of the static modulus of elasticity. A stress level of 25% of the ultimate stress was arbitrarily selected for the described determination of the elastic modulus. Soils generally exhibit elastic response only for very low stress levels. Therefore, not all deformations are recoverable, as illustrated in Fig. 7. The portion of the strain which is recoverable and which was used for determination of the elastic modulus approaches a constant value as the number of loading-unloading cycles increases. In this study, the elastic modulus was evaluated by using the elastic strain observed after five cycles of loading-unloading. Moore (1963) reports that the modulus he obtained at the fifth cycle was not significantly different from that obtained at the one-hundredth cycle.

The static modulus of elasticity of laboratory sedimented Providence silt was determined for various varve orientations and the results are shown in Fig. 8. The figure demonstrates that the elastic modulus increased linearly with increasing effective confining pressure. For a given confining pressure, the modulus decreased from 0° to a minimum at about 30° then increased to 90°. It is interesting to note that the minimum modulus occurred when the varve plane inclined at an angle of $45°-\phi/2$ from the direction of major principal stress (ϕ=angle of internal friction=26.7° for this material).

Figure 9 illustrates the effect of void ratio on the static modulus of elasticity of kaolin sedimented at two different varve angles. As would be expected, the modulus decreased with increasing void ratio. As for Providence silt, a greater angle of varve resulted in a larger modulus of elasticity when the void ratio was a constant. The effect of soil structure on the modulus-void ratio relation is shown in Fig. 10. For a constant void ratio, the flocculent structure exhibited a greater modulus of elasticity. Within the conditions studied, the influence of structure became less significant as the void ratio decreased.

The dynamic modulus of elasticity of laboratory sedimented soil was determined by measuring the propagation velocity of longitudinal waves. The tests were conducted at a frequency of 69 Hz

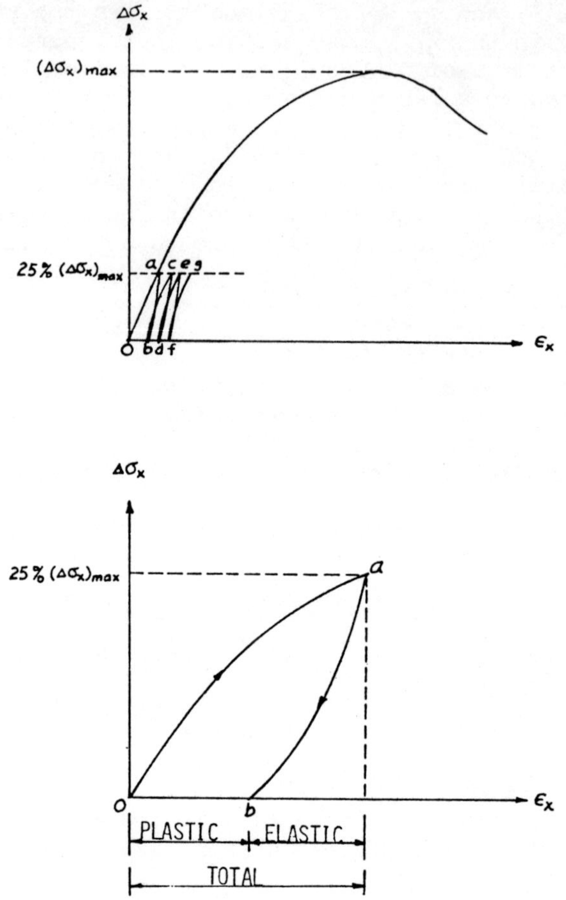

FIGURE 7

STRESS STRAIN CURVE

and a room temperature of 23° to 25°C. The velocity of wave propagation and the mass density of the test specimen were determined. The dynamic modulus was then computed from the equation:

$$E_D = \rho c^2 \tag{5}$$

in which E_D = dynamic modulus of elasticity, ρ = mass density and c = wave velocity.

Test results for laboratory sedimented Providence silt are shown in Fig. 11. As observed for the static modulus, the dynamic modulus is dependent upon the effective confining pressure and the

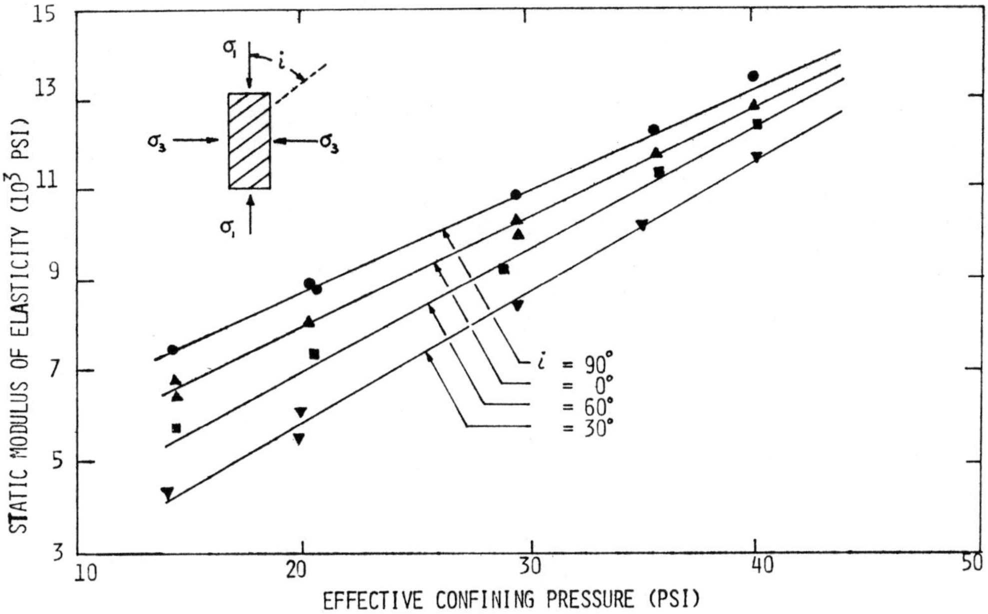

FIGURE 8

STATIC MODULUS OF LABORATORY SEDIMENTED PROVIDENCE SILT AS A FUNCTION OF EFFECTIVE CONFINING PRESSURE FOR VARIOUS VARVE ORIENTATION

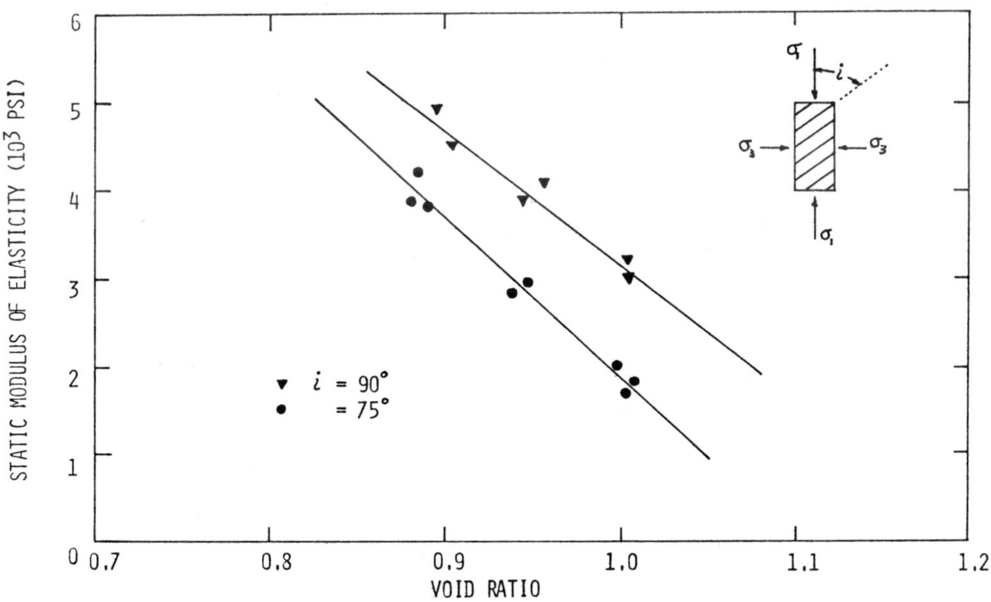

FIGURE 9

STATIC MODULUS OF ELASTICITY OF LABORATORY SEDIMENTED KAOLIN AS A FUNCTION OF VOID RATIO FOR VARIOUS VARVE ORIENTATION

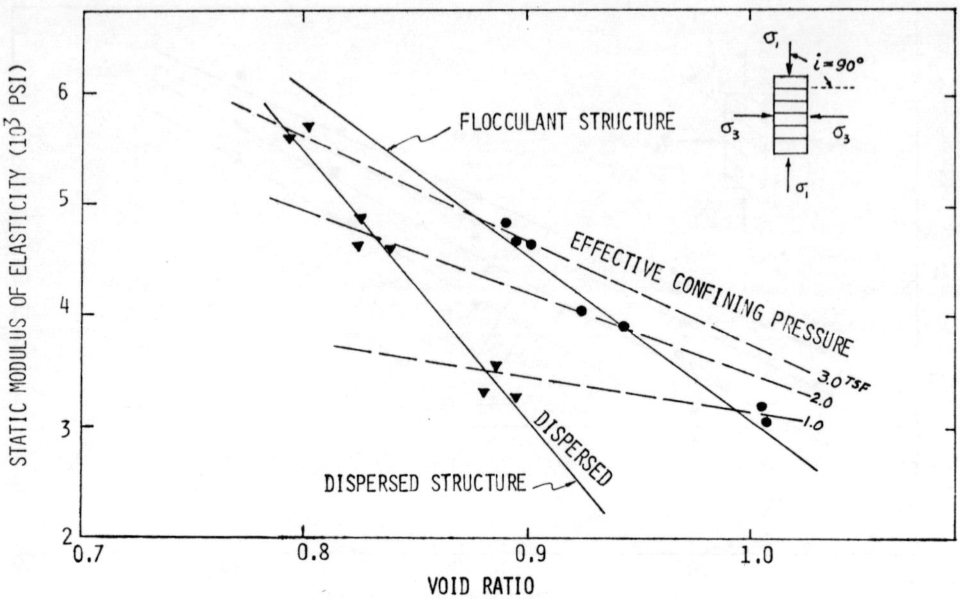

FIGURE 10

EFFECT OF VOID RATIO ON STATIC MODULUS OF ELASTICITY OF LABORATORY SEDIMENTED KAOLIN FOR DIFFERENT SOIL STRUCTURES

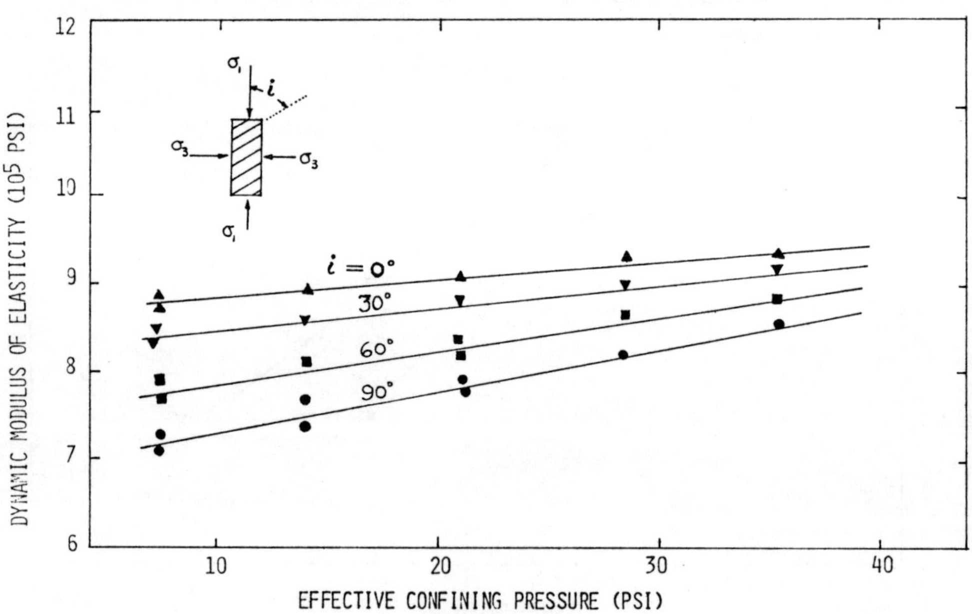

FIGURE 11

DYNAMIC MODULUS OF ELASTICITY AS A FUNCTION EFFECTIVE CONFINING PRESSURE FOR VARIOUS VARVE ORIENTATIONS - LABORATORY SEDIMENTED PROVIDENCE SILT orientation of varves. The dependency of the dynamic modulus on the effective confining pressure was less than for the static modulus. Furthermore, the effect of varve orientation on the dynamic modulus did not follow the same trend as that for the static modulus. For a constant effective confining pressure, the dynamic modulus continually increased with decreasing angle between the varve plane and the direction of deviator stress, while the static modulus first decreased then increased with increasing angle of varve orientation. This difference of behavior probably was a result of the difference in the stress levels used. In the static test, a stress level of 25% of the ultimate strength was used, whereas the stress level used in the dynamic test was an acoustic wave pressure which was only a very small percentage of the stress used in the static test. Under the higher stress level, the deformation of the test specimens involved interparticle slippage, of which a maximum occurred along the potential failure plane oriented at an angle of $45° - \phi/2$ from the deviator stress. Since varve formations possess a weak bonding between layers, a maximum deformation may be expected when the varve plane coincides with the orientation of the failure plane. For a constant stress, a greater deformation resulted in a smaller modulus. Therefore, the static modulus reached a minimum when the varves were oriented at an angle equal to the failure plane angle, namely, 30° from the deviator stress direction.

The dynamic test results for laboratory sedimented kaolin, shown in Fig. 12, indicate the same trend as that for the static test: the dynamic modulus decreased with increasing void ratio, and for a constant void ratio, a greater varve orientation exhibited smaller dynamic modulus of elasticity. Figure 13 illustrates the effect of soil structure on the relationship between the dynamic modulus and the void ratio. As observed for the static modulus, the flocculent structure again gave greater dynamic modulus than the dispersed structure when other conditions were kept equal.

The test results indicate that, under the same test conditions, the dynamic moduli were considerably greater than the static moduli. One factor which could have caused this difference is the different values of strain involved in the two different techniques for determination of the modulus. In the dynamic test, the strain was in the pure elastic range and the dynamic modulus was considered to be the Young's modulus with no effect of the transverse strain. However, in the static test, the strain extended beyond the pure elastic region resulting in a shear deformation between soil particles.

The strain rate is another important factor. In the dynamic test, the modulus was determined under a very high strain rate; thus there probably was insufficient time for either a change in the structural arrangement of the individual particles or an

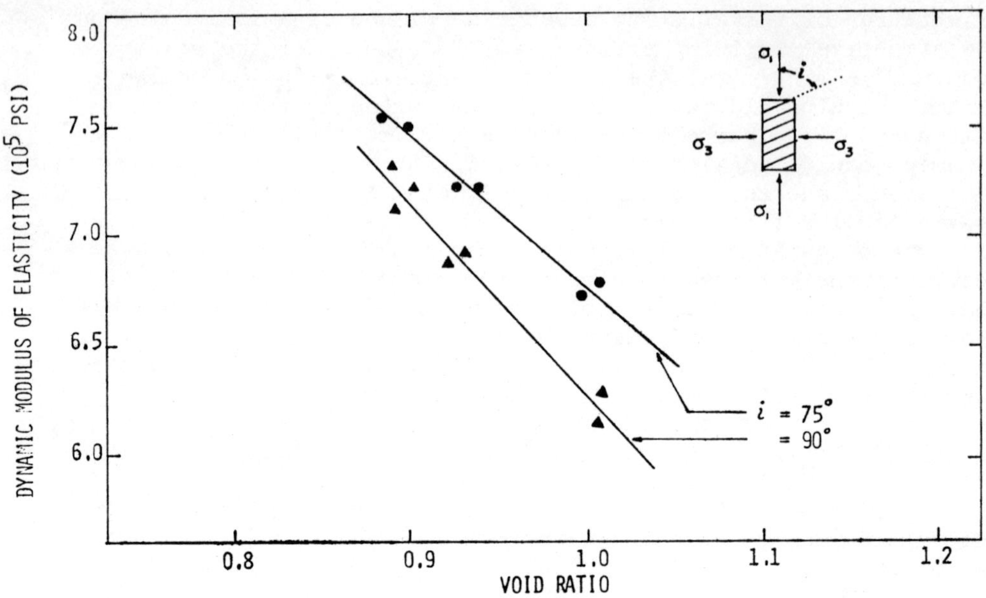

FIGURE 12

DYNAMIC MODULUS OF ELASTICITY AS A FUNCTION OF VOID RATIO FOR TWO VARVE ORIENTATIONS - LABORATORY SEDIMENTED KAOLIN

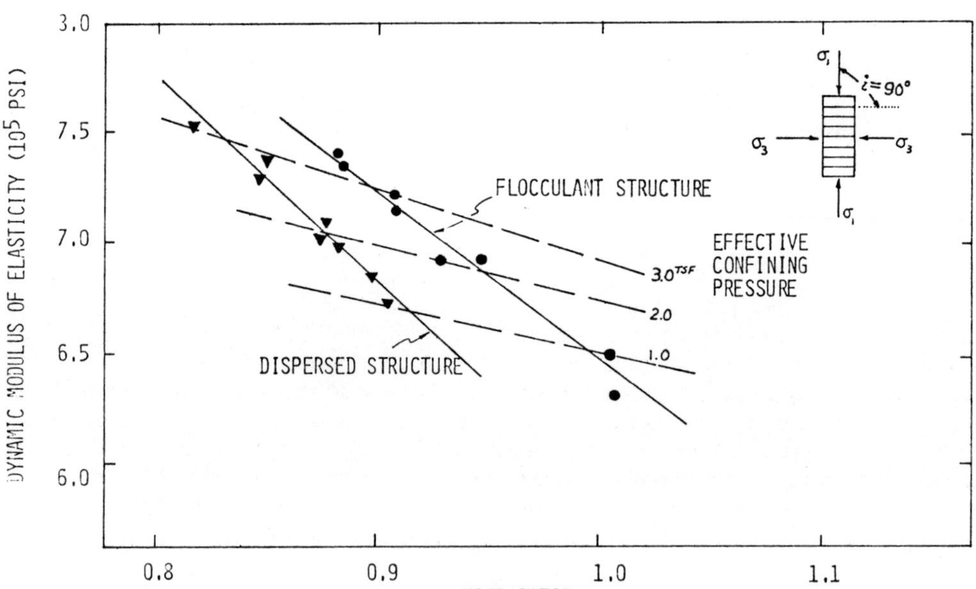

FIGURE 13

EFFECT OF VOID RATIO ON DYNAMIC MODULUS FOR DIFFERENT SOIL STRUCTURES - LABORATORY SEDIMENTED KAOLIN increase of the pore water pressure. In the static test, the strain rate was so slow that the soil had sufficient time to undergo changes in structure or to experience a pore water pressure increase. The combined effect of these two factors was a larger deformation and, therefore, a smaller modulus in the static tests.

Generally, three types of elastic waves are propagated through a saturated soil; a dilatational wave through the pore fluid, a dilatational wave through the soil skeleton, and a shear wave through the soil skeleton. Due to viscous coupling none of these waves are entirely independent; however, the shear wave is probably least influenced by pore fluid coupling. For this reason, separate studies were initiated to determine elastic moduli from shear wave propagation measurements. These tests were conducted in order to study the influence of shear strain and shear strength on the shear modulus.

In the first study, a soil specimen was restrained at its lower end. If such a body is torsionally vibrated at its lowest natural frequency, the time required for a shear wave to travel from one end to the other will be one fourth of its natural period. The propagation velocity and thence the moduli can be computed from the travel time.

In this resonant method, the test specimens were subjected to a shear strain amplitude of 10^{-4} rad. Drnevich et al. (1966) and Hall (1962) report that, for sands, the shear modulus is independent of the shear strain amplitude for shear strains less than 10^{-4} rad. For Providence silt sedimented in 0.3 M NaCl solution and subjected to a confining pressure of 20 psi, the vibration shear modulus increased as water content decreased (Fig. 14). The relation between the shear modulus and water content varied significantly with varying orientation of varves. For a given water content, the shear modulus first decreased and then increased with increasing varve orientation with respect to the direction of deviator stress. The influence of varve orientation on the shear modulus versus water content relation, however, decreased with an increase in water content. The trend of the variation of the vibration shear modulus with the varve orientation, for a constant water content, follows that for the static modulus, but was contradictory to that for the dynamic modulus determined by longitudinal wave measurements. These results, therefore, suggest that different mechanisms were involved in the propagation of compressional waves and shear waves through the layered sediment medium.

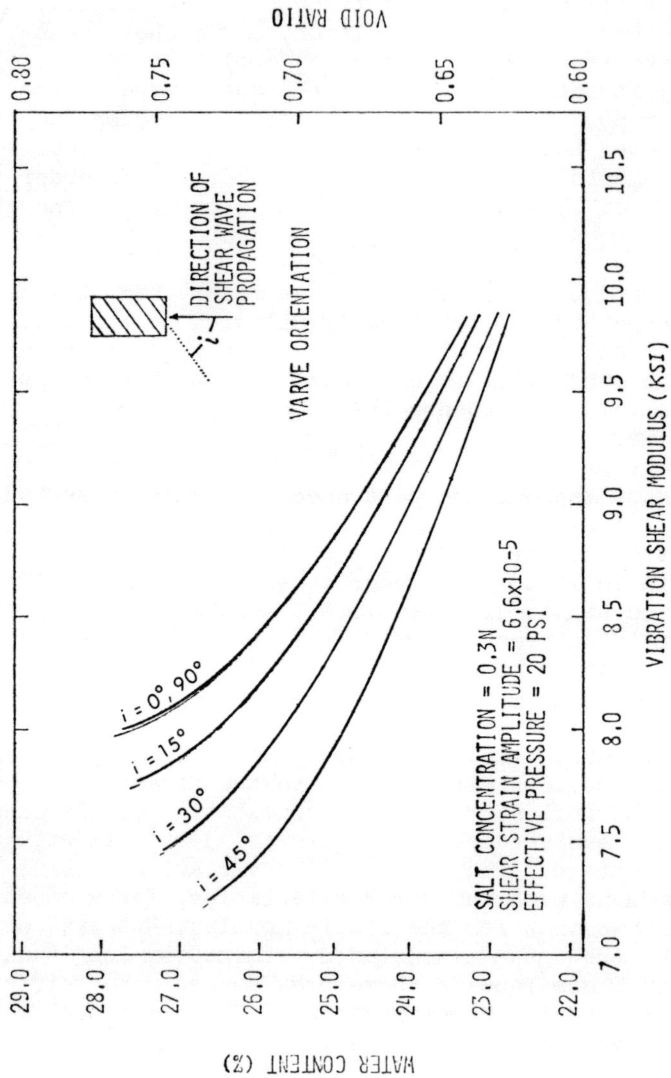

FIGURE 14

INFLUENCE OF VARVE ORIENTATION ON SHEAR MODULUS VERSUS WATER CONTENT RELATION FOR LABORATORY SEDIMENTED PROVIDENCE SILT

SUMMARY

Generally, sound velocity and strength of a material increases with increasing rigidity (dynamic moduli); yet most investigators find little relationship between sediment shear strength and sound velocity.

A laboratory study was undertaken in order to experimentally determine acoustic velocities and elastic moduli of undisturbed, fine-grained laboratory sedimented samples and to relate these parameters to porosity, particle structure, shear strength, anisotropy, and confining pressure. To date, the results of this study are:

1. The elastic modulus is useful only when boundary conditions are specified in the modulus definition. The stress-strain relationship is very sensitive to the stress system (including stress magnitude, stress rate, and stress orientation). Thus the stress system is one of the most important boundary conditions to be specified in defining the elastic modulus.

2. In varved sediments, under two different types of stress systems (static and dynamic), anisotropic nonlinear elastic moduli were exhibited.

3. The elastic modulus measured by the acoustic longitudinal wave propagation method is determined by the stress-strain relationship of the particle aggregate for compression. This dynamic modulus strongly depends on the angle between the major principal stress direction and the bedding plane.

4. The elastic modulus obtained from a static consolidated undrained triaxial test may contain information on the shear stress-strain relationship, but is not the real dilitational modulus. The static modulus of elasticity also varied with the angle between the principal stress direction and the bedding plane.

5. The dynamic shear modulus, like the dynamic compressional modulus, is greatly dependent upon confining pressure and porosity. This modulus also is anisotropic but its variation with principal stress direction differs from that of the compressional modulus.

6. Shear wave velocities are more probable to show good correlation with strength parameters than are compressional wave velocities.

BIBLIOGRAPHY

Anderson, A., and N. E. Simons, "Norwegian Triaxial Equipment and Technique," ASCE, Research Conference on Shear Strength of Cohesive Soils, Sponsored by Soil Mechanics and Foundations Division, June 1966.

Biot, M. A. (1956), "Theory of Propagation of Elastic Waves in a Fluid-Saturated Porous Solid: I. Low-Frequency Range, II. Higher Frequency Range," J. Acoust. Soc. Am. 28, 168-191.

Biot, M. A. (1962), "Mechanics of Deformation and Acoustic Propagation in Porous Media," J. Appl. Phys., 33, 1482-98.

Duffy, J., and R. D. Mindlin, "Stress-strain Relation and Vibrations of a Granular Medium," J. Appl. Mech. 24, 585 (1957).

Gallagher, J., and Nacci, V., "Investigations of Sediment Properties in Sonar Bottom Reflection Studies," NUSC Report No. 944, 6 August 1968.

Hall, J. R., Jr., "Effect of Amplitude on Damping and Wave Propagation in Granular Materials", Thesis presented to the University of Florida at Gainsville, Florida, in 1962, in Partial Fulfillment of the requirements for the degree of Doctor of Philosophy.

Hall, J. R., Jr., and F. E. Richart, Jr., "Dissipation of Elastic Wave Energy in Granular Soil," Journal of Soil Mechanics and Foundations Division, ASCE, Vol. 89, No. SM6, Proc. Paper 3698, November 1963.

Hampton, Loyd, "Acoustic Properties of Sediments," DRL-A-254 Defense Research Laboratory, The University of Texas, 1 August, 1966.

Hardin, B. O., and F. E. Richart, Jr. (1963), "Elastic Wave Velocity in Granular Soils," Journal of Soil Mechanics and Foundations Division, ASCE, Vol. 89, No. SM1, Proc. Paper 3407, February 1963, pp. 33-65.

Hardin, B. O., "Dynamic vs. Static Shear Modulus for Dry Sand," Materials Research and Standards, American Society for Testing and Materials, May 1965.

Hardin, B. O., and W. L. Black, "Sand Stiffness under Various Triaxial Stress," Journal of the Soil Mechanics and Foundations Division, ASCE, Vol. 92, No. SM2, Proc. Paper 4212, March 1966, pp. 27-42.

Hardin, B. O., and W. L. Black, "Vibration Modulus of Normally Consolidated Clay," Journal of the Soil Mechanics and Foundation Division, ASCE, Vol. 94, No. SM2, March 1968.

Humphries, W. K., and H. E. Wahls, "Stress History Effects on Dynamic Modulus of Clay," Journal of the Soil Mechanics and Foundations Division, ASCE, Vol. 94, No. SM2, Proc. Paper 5834, March 1968, pp. 371-389.

Lawrence, F. L., Jr., "Ultrasonic Shear Velocity in Sand and Clay," Report 23, MIT, Cambridge, Massachusetts, January 1965.

Moore, Peter J., "One-Dimensional Compression and Wave Propagation," MIT Dynamic Loadings, Report 21, 1963.

Morgenstern, N., and D. T. Smith, "Longitudinal Velocities in a Clay During Plastic Compression," Vol. 58, 1964/II, pp. 123-128.

Pinson, L. D., "Study of the Shear Modulus and Energy Dissipation for Statically Loaded Isolated Footings on Sand," Thesis presented to the University of Kentucky, Lexington, Kentucky, 1963.

Taylor, R. J., "Some Static and Dynamic Properties of Laboratory Deposited Kaolinite," A thesis submitted in partial fulfillment of the requirements for the Degree of Master of Science in Civil Engineering, University of Rhode Island, 1967.

THE SEISMIC WIDE ANGLE REFLECTION METHOD IN THE STUDY OF OCEAN SEDIMENT VELOCITY STRUCTURE

G. L. MAYNARD, G. H. SUTTON, D. M. HUSSONG, and L. W. KROENKE

Georgia Institute of Technology, Atlanta, Georgia 30332

ABSTRACT

Although the wide angle reflection method has been little used in the oceans, a well-developed mathematical theory and techniques for determining interval velocities within plane layers by this method have existed for 20 years. The raypaths are least-time paths. Reflection curves from interfaces below the ocean floor are generally not true hyperbolas but are higher order functions forming curved lines on a $T^2 - X^2$ plot. Certain properties and limiting cases of reflection curves allow the computation of interval velocities both by exact and by highly useful approximate methods. These methods place stringent demands upon data quality, and to provide adequate data precision the ASPER (Airgun-Sonobuoy-Precision Echo Recorder) method was developed. The high data density provided by this method allows comparison of long segments of observational travel time curves with theoretical travel time curves appropriate for various physical/geological models. The method has several limitations; for example, it requires relatively simple, uniform geologic structure, and in deep water (2 km or more) velocities are only determined to about the nearest 0.1 km/sec, or even more if vertical intervals much smaller than 200 m are used. It is sometimes useful to find a regional velocity-depth function by combining the results of several sections determined in the area.

THE SEISMIC WIDE ANGLE REFLECTION METHOD IN THE STUDY OF OCEAN SEDIMENT VELOCITY STRUCTURE

In 1921 Karcher, Perrine, Haseman, and Kite made the first successful seismic reflection measurement [Robinson, 1971]. Since then, the class of seismic methods which is loosely referred to as explorational seismology has become the best developed and most used method of indirectly determining subsurface geologic structure. Explorational seismology is generally subdivided into three fairly distinct but related categories: Normal incidence reflection, wide angle reflection, and refraction.

As implied by the term "explorational seismology", the method was first used and extensively developed by the petroleum industry to explore for potential petroleum-bearing structures; or at least the vertical reflection and refraction methods were. These two methods were adopted early and were developed rapidly in a see-saw fashion, with first one method pulling ahead in popularity and sophistication and then the other. After 25 or 30 years of this, both methods were highly developed sciences capable of yielding detailed and accurate results.

In the mid or late 1930's the scientific community adopted these methods to study the earth's crust for other purposes, and even with the simplified, and often crude, instrument systems available to it, immediately began to make indelible marks on the annals of geology. The remarkable successes that resulted when these methods went to sea in the form of the vertical reflection profiler and the crustal refraction system are now considered the norm for marine geophysical operations, especially since the development and adoption of various repetitive signal sources.

The wide angle reflection technique has had a somewhat more shadowed existence. Since the first seismic measurements were reflections, and since the experimenters needed the seismic velocities in order to establish the depth to the reflector, it is almost inconceivable that at least the elementary methods of finding velocity from wide angle reflections were not worked out almost immediately. Certainly, such methods had been worked out by the late 1920's by people in the oil exploration business. However, the basic principle for such a method was first published in 1938 by Green, who used the concept of virtual signal images and the Pythagorean Theorem.

While a sophisticated analysis of wide angle reflection seismograms is given in other papers in this volume--see Bryan, for example--it is worthwhile to review here certain aspects of analysis of an elementary but important character. The method of Green [1938] and some of the developments of Dix [1955] are sufficient to provide the ideas needed. In addition to providing an introduction to the method for the uninitiated, this will serve to fix in mind those features which must be considered in dealing with other problems.

Consider a horizontal, plane interface separating a layer of uniform thickness H and constant seismic velocity V_1 from an underlying semi-infinite medium of velocity V_2 (and assume $\rho_1 V_1 \neq \rho_2 V_2$, where ρ = density) (Fig. 1). For a source at the origin and a receiver at the horizontal range X, the interface acts as a mirror, so that the receiver is affected by the reflected seismic signal as though it came from depth 2H below the origin. By Fermat's principle, the raypath as shown must be a least-time path. For this specialized case it is also a least-distance, or straight-line, path, the length of which is given by:

$$S^2 = X^2 + 4H^2 \quad . \tag{1}$$

Dividing through this equation by V_1^2 yields

$$T^2(X) = T^2(0) + (1/V_1^2)\, X^2 \tag{2}$$

where

$T(X)$ = reflection travel time at range X,

$T(0)$ = vertical reflection-time at origin,

and a purely spatial relationship becomes a time-distance one (T-X).

Since $T^2(0)$ is a constant, the reflection curve is a hyperbola, which is asymptotic to "D", the wave traveling directly from source to receiver along the upper surface of the layer. (In marine seismology, D corresponds to the surface-channeled, direct water wave.) The slope dT/dX of this travel-time versus distance curve is an inverse apparent velocity, the numerical value of which is the seismic parameter, p, for that range. (Also, $p = (\sin\theta/V)$; Slotnick, 1959) Note that at very long ranges, where the reflected raypath becomes negligibly longer than the direct, the slope approaches that of D, a fortuitous convenience in cases where the identities of the travel-time curves are in doubt.

In order to facilitate measurement of slope, the travel-time data can be plotted as $T^2 - X^2$, (in marine geophysical work this is usually actually plotted $T^2 - D^2$, which eliminates vagaries in ship course and speed):

set

$$T^2 = \tau$$

$$X^2 = \chi$$

then equation (2) becomes

$$\tau(X) = \tau(0) + (1/V_1{}^2)\,\chi \tag{3}$$

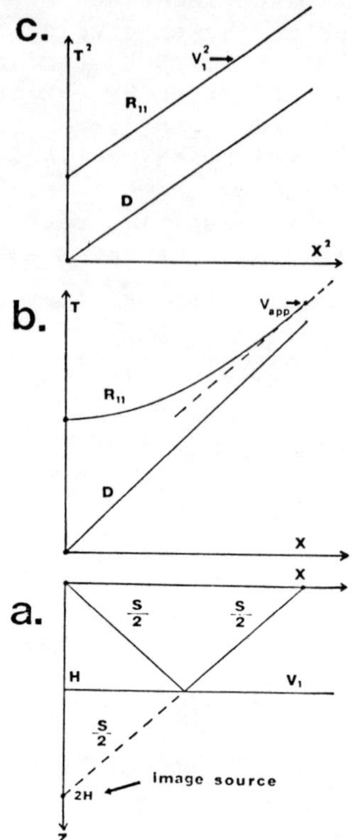

FIGURE 1

a. Reflection raypath geometry for the case of a flat-lying layer of constant thickness H having uniform seismic velocity V_1 and overlying a semi-infinite basement. S is the slant-path distance.
b. Corresponding travel time curves. Line D corresponds to the raypath running along the top of the layer (in the oceans this would be the surface-channeled direct water wave), R_{11} represents the reflection from the bottom of the layer (the ocean bottom reflection), and V_{app} indicates the inverse slope at range X of R_{11} (the "apparent velocity" of the reflection arrival).
c. Corresponding T^2 - X^2 graph. V_1^2 indicates the inverse slope of R_{11}, where V_1 is the seismic velocity of the layer.

a line with slope $1/V_1^2$, which can be measured easily and precisely; this is Green's method for finding layer velocity from wide angle reflection data. Note that, since $T^2(0)$ is a constant, when Eq. 2 is implicitly differentiated, one obtains

$$dT(X)/dX = X/(V_1^2 T) = \sin\theta/V = p \quad , \qquad (4)$$

a useful result to be used later.

Following Green's [1938] paper many other people worked at different times on various aspects of the wide angle reflection problem. However, during the 15 years following Green's pioneering paper advances in the development of the theory were limited to several significant but small steps. Dix [1955] lists many of the workers responsible for these developments.

Around 1950, Krey and Dürbaum in Europe and Dix in the United States began attempts to make more precise velocity measurements by taking into account the departure from truly hyperbolic shape of the reflection curves from interfaces below the first sub-bottom. By 1955, both Dürbaum and Dix had published solutions to the problem for the flat-layer model. Dürbaum [1954] in a mathematically formal development fit the reflection curves with a Maclaurin series and identified various coefficients in the expansion with quantities characteristic of the various layers in the model and their geometrical relationship. His results are the most general to come into widespread use thus far. In applying his theory he used terms only up to the quadratic. Clay and Rona [1965] use terms up to fourth degree and give in English a concise, straightforward derivation of Dürbaum's theory.

Dix [1955] independently arrived at equivalent results (for non-dipping layers) from physical considerations. In fact, Dix solved the problem by two approaches. On one hand, he used physically significant properties of the T - X and $T^2 - X^2$ functions to reduce the multiple-layered reflection case to a sequence of one-layer cases amenable to exact solution by Green's method; this is the reduced time-reduced distance method. On the other hand, he found useful approximations of the interval velocities of layers when the horizontal range involved is small, approximations which become exact in the limit of diminishing range. The former method, which involves a large amount of computation, was not too useful in pre-computer days, but is now quite practical; see, for example, Le Pichon et al. [1968], Houtz et al. [1968], and Bryan (this volume). The latter method has been much used ever since it was published. Dix's [1955] paper contains a comprehensive development of the theory underlying his methods. Only those portions of his development needed in this paper will be considered here. Since Dix's equation for interval velocity is a special case of Dürbaum's equation [Dürbaum, 1954; Clay and Rona, 1965], those parts of the theory which are considered here have quite general implications.

Consider two flat layers overlying a semi-infinite medium (Fig. 2). The layers have thicknesses H_1 and H_2 and seismic velocities V_1 and V_2. Let the underlying medium have velocity $V_3 (\rho_3 V_3 \neq \rho_2 V_2)$. For convenience (not necessity) let H_1 and H_2 be approximately equal; also let the upper layer have lower velocity than the second layer ($V_1 < V_2$). This sequence of velocities is used simply as a convention; regardless of the sequence in which

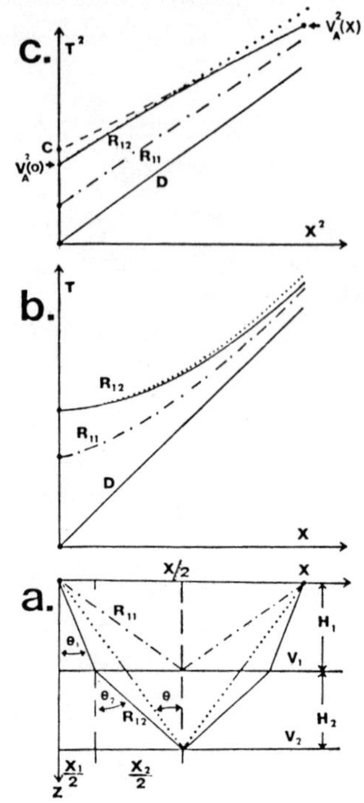

FIGURE 2

a. Reflection raypaths and a hypothetical raypath for the case of two flat-lying layers of constant thicknesses H_1 and H_2 (in general $H_1 \neq H_2$) having uniform velocities V_1 and V_2 (here $V_1 < V_2$) and overlying a semi-infinite basement. The dot-dashed line represents R_{11}, the case of Fig. 1. R_{12}, the reflection from the bottom of the second layer, is represented by the solid line; this is a least-time raypath with the Snell angles given by θ_1 and θ_2. X_1 and X_2 are the horizontal distances covered by the raypath segments of R_{12} in layer 1 and layer 2, respectively. A hypothetical least-distance raypath for R_{12} is indicated by the dotted line and its angle of incidence by θ.
b. Corresponding travel time curves. As in Fig. 1, D is the direct surface wave, and R_{11} is the reflection hyperbola for the upper layer. For R_{12}, the dotted curve is the reflection hyperbola that would correspond to the hypothetical least-distance path, and the solid curve is the true reflection quasi-hyperbola which results from the least-time nature of the raypath.
c. Corresponding T^2 - X^2 graphs. V_A^2 indicates the inverse slope of R_{12}: the slope at the origin, indicated by the dotted line, also corresponds to the constant slope of the hypothetical least-distance raypath; the slope at range X, indicated by the dashed line having T^2 intercept at C, corresponds to the true least-time raypath.

the velocities occur, the raypath is a least-time path which has shorter travel-time than would be the case along the least-distance path.

If the ray followed the least-distance path through the sequence of layers (Fig. 2-a, dotted lines) the reflection-time versus range curve (T-X) would be, as in the use of a single layer, a hyperbola (Fig. 2-b, dotted curve), and the $T^2 - X^2$ graph would again be a straight line (Fig. 2-c, dotted line), the inverse slope of which would be the square of the time weighted mean velocity through the two layers

$$\overline{V} = (H_1 + H_2)/\left((H_1/V_1) + (H_2/V_2)\right) = (V_1T_1 + V_2T_2)/(T_1 + T_2),$$

where T_i is the two-way vertical travel time through the ith layer. From similar triangles, it is clear that this squared mean velocity is the same whether the raypath is the vertical reflection at the origin or the slant-path to non-zero range X; hence the straight-line $T^2 - X^2$ relationship.

The raypath is not a straight-line path but is, by Fermat's principle, the least-time path (Fig. 2-a, solid lines). Consequently the true T-X curve (Fig. 2-b, solid curve) coincides with the hypothetical hyperbola only at the origin, falling below it to slightly smaller values of travel-time at non-zero ranges, with the discrepancy increasing with range. Since the T-X function is now a "quasi-hyperbola" at best, the $T^2 - X^2$ graph is not a straight line, but "droops" to slightly smaller values with range (Fig. 2-c, solid curve).

In the reduced time-reduced distance method mentioned above the seismic parameter, $p = \sin\theta_1/V_1$, is determined from the slope of the $T^2 - X^2$ curve at a given range. (Because of the small curvature of the $T^2 - X^2$ plot, its slope can be determined with more precision than that for the T-X curve.) The tangent line to the $T^2 - X^2$ curve can be written

$$T^2(X) = C + (1/V_A^2(X))X^2 \quad . \tag{5}$$

Then

$$p = dT(X)/dX = X/CV_A^2(X)\ T(X) = \sin\theta_1/V_1 \quad , \tag{6}$$

and

$$\sin\theta_1 = (XV_1)/(V_A^2(X)\ \ T(X)) \ . \tag{7}$$

Since all other quantities in Eq. (7) are known, $\sin\theta_1$ can be calculated. With H_1 and V_1 already determined by Green's method, it is then possible to calculate the fractions of both the total travel time and the total range which are the result of those portions of the raypath lying within the upper layer. These calculated fractions can then be subtracted from the measured total times and ranges, in effect, removing the upper layer. This reduced time-reduced distance data can then be analyzed by Green's method because the described operations effectively reduce the problem of Fig. 2 to that of Fig. 1.

The same procedure can be used for still deeper layers. For each successively deeper layer, the problem is reduced to that of one layer by compensating for the layers above. To the extent that the geologic structure can be modeled by a sequence of constant velocity layers (not necessarily increasing in velocity with depth) this method is exact. The precision of the results obtained is limited, as will be seen to be perennially the case with wide angle reflection methods, by a demand for great precision in the data.

Undoubtedly, up to the present, more wide angle reflections have been analyzed by the use of Dix's [1955] interval velocity equation than by any other method. While no claim is made that the equation gives accurate results except when horizontal range is small compared with depth to reflector and when velocity contrasts are not too great, it has nonetheless been profitably used under circumstances that were less than ideal [e.g., Meissner, 1966].

The requirement that range be small compared with depth to the reflector (and that velocity contrasts are not too large) results from the requirement that the angle of incidence at all interfaces must be small, which allows the tangents of the angles to be accurately approximated by the sines. Consider in Fig. 2-a:

$$X_1 = 2H_1 \tan\theta_1 = V_1T_1 \tan\theta_1 \tag{8a}$$

$$X_2 = 2H_2 \tan\theta_2 = V_2T_2 \tan\theta_2 \tag{8b}$$

or

$$X_1 + X_2 = V_1T_1 \tan\theta_1 + V_2T_2 \tan\theta_2 \quad . \tag{9}$$

If

$$X \ll (H_1 + H_2)$$

then

$$\tan\theta \cong \sin\theta$$

and

$$\begin{aligned} X_1 + X_2 &\cong V_1T_1 \sin\theta_1 + V_2T_2 \sin\theta_2 \\ &\cong V_1^2T_1 \sin\theta_1/V_1 + V_2^2T_2 \sin\theta_2/V_2 \end{aligned} \tag{10}$$

But by Snell's Law

$$\sin \theta_1/V_1 = \sin \theta_2/V_2$$

so that

$$X_1 + X_2 \cong (V_1^2 T_1 + V_2^2 T_2) \sin \theta_1/V_1 \quad . \tag{11}$$

Using Eq. (7) we obtain

$$X_1 + X_2 \cong (V_1^2 T_1 + V_2^2 T_2)\ (X_1 + X_2)/V_A^2(X)\ T(X) \quad . \tag{12}$$

Horizontal range cancels from the Eq. (12), and as range decreases the approximation becomes better until in the limit, for the two layer case, one can write

$$V_A^2(0)\ (T_1 + T_2) = (V_1^2 T_1 + V_2^2 T_2) \quad . \tag{13}$$

In Eq. (13), T_i is the 2-way vertical travel time through the ith layer, and $V_A^2(0)$ is the reciprocal of the slope of the $T^2 - X^2$ graph measured at X = 0. In the following we use

$$V_{RMS}^2 \equiv V_A^2(0)$$

after previous authors.

It is easily shown by mathematical induction that Eq. (13) can be extended to the n-layer case:

$$V_{RMS(n)}^2 \sum_{i=1}^{n} T_i = \sum_{i=1}^{n} V_i^2 T_i \quad . \tag{14}$$

Clearly the equation for the n-1 case is

$$V_{RMS(n-1)}^2 \sum_{i=1}^{n-1} T_i = \sum_{i=1}^{n-1} V_i^2 T_i \quad . \tag{15}$$

Subtracting (15) from (14) gives

$$V_n^2 = \left[V_{RMS(n)}^2 \sum_{i=1}^{n} T_i - V_{RMS(n-1)}^2 \sum_{i=1}^{n-1} T_i \right] \Big/ T_n \tag{16}$$

where:

V_n = interval velocity of nth layer

T_i = two-way vertical travel time through ith layer

$V^2_{RMS(n)}$ = reciprocal of slope of $T^2 - X^2$ curve for nth reflector at X = 0.

This is Dix's equation for finding the "interval velocity" of the nth layer. This equation says that if the vertical reflection time to any two interfaces producing reflection (quasi-) hyperbolas can be found and the slope at the T^2-axis of the corresponding $T^2 - X^2$ curves can be found, then the average velocity, and therefore the thickness of the interval between reflectors, can be found.

Any other techniques developed to determine interval velocities from wide angle reflection data generally require computer processing. For example, it is possible to calculate with a computer raytrace program the synthetic T-X curves for a postulated geologic structure, to then compare these synthetic reflection curves with the real ones, again by computer (typically by least squares), and to adjust parameters in the model until a satisfactory fit between the synthetic and real curves is achieved; see, for example, Le Pichon et al. [1968]. With computers, it is also possible to calculate a suite of hyperbolas corresponding to a set of RMS velocities and then sweep each hyperbola through the reflection data searching for similar curves, or to use some less simple velocity spectra method; see Taner and Koehler [1969]. Because the raypath and the geometrically shortest path coincide only in the case of a single layer, analytic representation of the time-distance relationship is complicated and is most simply given by a pair of parametric equations which are functions of the seismic raypath parameter p; Taner and Koehler [1969] derive these equations. Odegard [1972] obtained the interval velocity and thickness of a layer by iterating these equations to fit the reflection curve under the constraint that velocity and thickness are known for all the layers above the one under consideration.

Experience with computer raytracing and the construction of synthetic T-X curves, as needed for several of the above analysis techniques, quickly teaches one subsidiary lesson; a small change in the T-X curve generally represents a significant change in the model. Clear implications are that great precision is required in the reflection data if accurate results are to be obtained and that small deviations of the geologic structure from the ideal geometry may have profound results in the reflection data [Taner et al., 1970].

However, not all useful synthetic T-X curves are the extremely precise products of computers. Certainly every person engaged in the analysis of wide angle reflection and refraction data has at least a mental catalog of qualitative theoretical time-distance curves, associated with various layered models, and probably has computed and plotted many such comparisons. They are indispensable for identifying the individual reflection and refraction traces in sets of complicated data. A few such curves and their special features will be examined later when certain practical problems of analysis are considered; however, at this point it should be noted that the reverse problem, that of qualitatively identifying such curves in the data, was one of the incentives that led to the development of such seismic systems as the ASPER (Airgun-Sonobuoy-Precision Echo Recorder).

In conventional refraction operations employing two ships and explosives it has long been the practice to space small shots very closely when the ships are close to each other. In most cases this is done to accurately determine where breaks in first arrival lines occur so that the upper layers of the sea floor can be resolved by refraction methods. Attempts have occasionally been made to analyze by $T^2 - X^2$ methods the wide angle reflections obtained during such surveys, and a few studies have been made using wide angle reflections as the primary method [Knox, 1965]. Because of the quality of the data the results of most such attempts have been crude and frustrating. What is needed is some system which gives data as detailed and continuous as that long obtained routinely from precision echo sounding equipment, and since the advent of such energetic repetitive signal sources as sparkers and air cannons, from continuous vertical reflection profilers.

Clay and Rona [1965] demonstrated the solution to this problem when they substituted a free-floating radio sonobuoy for the conventional towed hydrophone array (Fig. 3), thereby achieving the data density possible with the repetitive signal source, the especially convenient and effective display characteristic of the variable density echo sounder or video facsimile recorder, and the spreading angle capabilities of the sonobuoy. Hill [1963] among others, had already described the use of sonobuoys to replace the second ship in conventional marine seismic refraction work, but the combination of instruments as assembled by Clay and Rona proved to be an especially fortuitous one for marine seismic work, as various later workers have shown [Le Pichon et al., 1968; Houtz et al., 1968; Houtz, this volume; Maynard, 1973; Sutton, et al., 1971].

The version of this combination of repetitive signal sources, sonobuoy, and precision echo recorder developed by the Hawaii Institute of Geophysics, the "ASPER" system, is a good example of the current state of the art in this type of data acquisition system. In this system (Fig. 4) the principal signal source is one or more airguns using chambers ranging in volume from 20 cubic inches up to the largest available. This is generally supplemented by, and in certain instances replaced by, a 7000-joule sparker. In some cases a 3.5 kHz pinger may be used as the sole signal source. The signal sources are fired at precise intervals in synchronization with

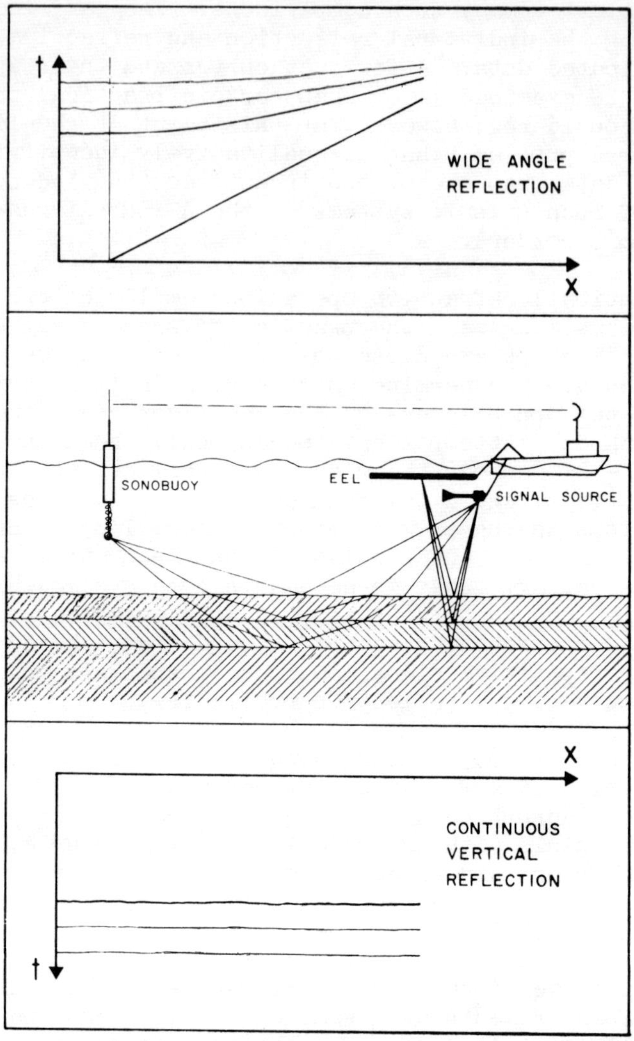

FIGURE 3

Schematic diagram illustrating various features of wide angle reflection-refraction (ASPER) and continuous vertical reflection (CVR) profiling systems and their record.

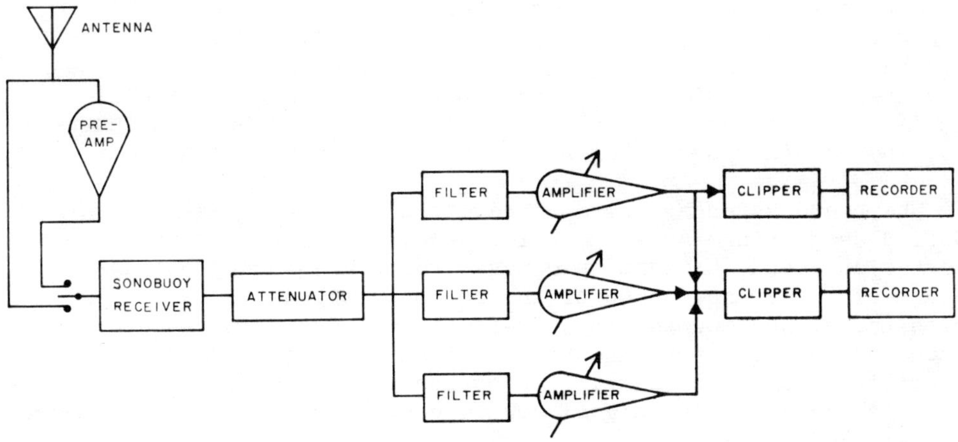

FIGURE 4

Block diagram of the ASPER system.

recorder stylus sweep. (In some systems the source interval is irregular and the sweep is triggered by the source.) Firing generally occurs at intervals of 1, 5, or 10 seconds. The 1-second interval is used in shallow water or when examining only the upper 100 m, or so, of sediment with high resolution as the objective. The 5-second interval is used in water of moderate depth or for examination of perhaps the upper kilometer of the sea floor. The 10-second interval is used for deep-water, deep-crustal studies. When more than one signal source is used at the same time, the firing times are adjusted relative to each other until the first breaks of their ocean bottom reflections coincide on the records. Signal sources are generally towed at less than 10 m depth.

While certain experiments may require the use of more specialized sonobuoys, in the majority of cases the sonobuoys used are U. S. Navy models SSQ-38A and SSQ-41A, although commercially available sonobuoys may be used. In most cases hydrophone depth is set at 18 m (60 feet) or less. For arrivals from more than a few tens of meters below the sea floor, any improvements in signal reception resulting from greater hydrophone depth are offset by the complications introduced into the record by interference between the seismic arrivals coming directly from the bottom to the hydrophone and the same arrivals reflected from the ocean surface back downward to the hydrophone.

The sonobuoy radio frequency range is between 160 and 180 MHz in the band of frequencies lying between television channels 6 and 7. Theoretically, transmission at such frequencies is line-of-sight, but in practice somewhat greater range is achieved, especially in rough water conditions. The explanation usually offered for the fact that reception is better in rough water than in calm is that in calm water conditions the direct-path signal and the signal specularly reflected from the smooth ocean-surface tend to interfere, whereas in rough water the latter is randomly scattered.

Shipboard antennas are an array of directional yagis: either modified deep-fringe-area channel 7 television antennas or intermediate-band FM radio antennas. These are turned on edge because of the polarization of the sonobuoy signals, stacked in series pairs, and mounted on a 50-foot high mast. One such array is mounted on a rotator; another is fixed and points dead astern. A sensitive mast-top preamplifier boosts the signal from the antenna and feeds the amplified signal down a coaxial cable to the sonobuoy receivers located in the laboratory. A second cable, without a preamplifier, also runs from the antenna to the sonobuoy receivers. When a transmitting sonobuoy is located near the ship, it is necessary to use such a preamplifier bypass in order to prevent overloading the system.

Shipboard sonobuoy receivers are either obtained from commercial sources or are borrowed from the Navy. The later models are of a type suitable for use in aircraft and must be used with a 60 to 400 Hz power converter. Also, the output of the Navy receiver has a maximum voltage swing of 40 volts peak-to-peak and must be passed through an adjustable attenuator to reduce the output signal to a level which the filters can handle. Commercial receivers generally are compatible with the filters.

Aboard ship the signal from the sonobuoy is paralleled into several active, maximum-flat, passband filters, each passing a frequency band one or two octaves wide with 12 dB/octave rolloff. Generally three different passbands are used: a low-frequency band with corner frequencies of 5 to 20 Hz and 25 to 40 Hz; an intermediate frequency band with corners at about 50 and 150 Hz; and a high frequency band with corners at 500 and 1500 to 2000 Hz. Experience has shown that many arrivals from deep seismic reflectors lie in the frequency range of about 20 to 25 Hz, and the low-frequency filter is always set to cover this range. The exact corner frequencies used depend upon what experimentation shows works best under the circumstances. While these low frequencies are essential for deep penetration, they lack the resolution necessary for resolving the thinner upper layers, and are below the spectrum of the sparker, which is particularly rich in the intermediate frequencies. The intermediate frequency band offers a useful compromise between good resolution and good penetration; with the signal sources used, it gives penetration of a second or more of two-way travel time into the sediments, equivalent to at least a kilometer of depth. This frequency band is generally most useful for wide-angle reflection studies of the upper kilometer of sediments. The direct water-wave is recorded well in the high-frequency range.

The output from each filter is put into a differential amplifier, the output of which is fed through a mixing network and clipper into the recorders. The recorders in the ASPER system are of the ubiquitous facsimile type used essentially universally for precise echo-sounding bathymetric work and continuous vertical seismic reflection profiling. In the ASPER system each stylus sweep in the recorder makes a segment of variable density seismogram and automatically plots a graph of seismic wave travel time versus ship travel time (Figs. 1-b, 2-b, 3), with horizontal distance implicit. The writing-point sweep is not interrupted; as the writing point gets close to the right edge of the paper another writing point starts sweeping from the left edge. The paper advances slightly during the sweep (about 38 lines per cm) producing, for example, at a 5-second signal-source firing interval, 720 correlated variable density seismograms along a 19 cm distance axis per hour.

The three frequency passbands and the signal mixing capability give the system additional flexibility. For example, if the center frequency and relative amplitude of the intermediate-frequency band are carefully chosen, seismic arrivals in this band will usually show through longer trains of low-frequency signals in the record, and produce "highlights" which delineate thin upper layers not resolved by the low frequencies. At greater penetrations, where intermediate frequencies have become attenuated, the low-frequency signal is essentially unaffected. Usually, more than one echo-sounding recorder is used (Fig. 4). As a rule, either two recorders record the same frequency bands at different relative gains, or one records the mixed passbands while the other records only the low-frequency or the intermediate-frequency passband. During reflection or refraction work requiring deep penetration, one precision echo recorder is used for low-frequencies only, while a second records low frequencies, at a different amplification, plus intermediate- and high-frequency signals. Only high- and intermediate-frequency recordings are made when shallow, detailed resolution is the objective.

The use of a silicon diode clipper between the amplifiers and recorders in the ASPER system has proved effective in conditions of low signal-to-noise ratio. When clipping level is adjusted to barely above noise level, the seismic signal, which is correlated in phase, stands out clearly against the uncorrelated background noise, thus making possible the detection and analysis of low-level signals.

The theoretical considerations predict that acceptable accuracy of results will require geometrically simple and regular geologic structure and a high degree of precision in the data. Experience bears out these predictions. For example, on the Ontong Java Plateau, reflectors in the sediments are probably as near the horizontal, parallel-plane ideal as can be found; yet attempts to find the interval velocities of vertical intervals whose thicknesses amount to about 5% or less of the total overlying water and sediment column generally gave unsatisfactory results [Maynard, 1973]. On the other hand, if the intervals are as thick as 10% of the overlying column, results were generally quite good. The limit of resolution using simple analysis methods, then, seems to be a layer whose thickness is about 7 or 8% that of the total overlying column; this is consistent with what Bryan (this volume) has found. On the Ontong Java Plateau, where water depth in the area of study averages a little more than 2 km, this means that the smallest interval in the sediments for which a reliable interval velocity can be obtained cannot be much less than 200 m thick, and it must be thicker yet near the bottom of the sediment column. This is disappointing, for in many places on the Plateau the ASPER records show 15 to 20 clear reflections from within the upper kilometer of sediments, promising a detailed velocity-depth function if all the data could be utilized.

The limitations on detail in the velocity-depth relationship implied above can often be circumvented by statistical methods. Where feasible, several ASPER profiles can be made in a limited area where geological changes are small. The results from all the stations can then be combined to give an essentially continuous local, or even regional, velocity-depth profile. Generally, such composite velocity-depth functions should be much more reliable than the results from any individual station. They also serve to indicate the accuracy to be expected at the individual stations. For example, on the Ontong Java Plateau it was found that although the velocity-depth relations determined at individual stations seemed to vary quite widely when the stations were considered separately, when these velocities were compared with the velocity that obtained at that depth in the composite, at least 90% of the time they were within 0.2 km/sec of the composite, and in more than 50% of the cases were within 0.1 km/sec. While greater precision would certainly be welcomed, accuracy in the profiles of individual stations was nonetheless sufficient to establish the existence of a velocity reversal in one area on the Plateau.

Since corresponding data points from two sets of T-D data picked from ASPER records by an experienced, careful worker, or even by two different such workers, will in most cases agree within 5 milliseconds (amounting to no more than a half cycle with the approximately 100 Hz signals required for the 1 km penetration needed to reach basement on the Plateau) it seems clear that any significant increase in the precision of the velocity values will have to come from some direction other than careful handling of the data. However, refinement of the velocity-depth relationship at a given location is not necessarily the result of increasingly complex mathematical analysis. In many cases, useful information of a qualitative--and sometimes quantitative--nature can be obtained from a record simply from examination of special features in the overall nature of the reflection curves. It is partly for this reason that a standard tool for the interpretation of wide angle reflection and refraction seismograms is a catalog of theoretical travel-time curves.

A few such curves can be found in a number of works [Ewing, 1963; Dix, 1952; Le Pichon et al., 1968; Meissner, 1966, 1967a,b; Slotnick, 1959]. However, it appears that no reasonably complete collection of them is published anywhere. One difficulty facing anyone attempting to compile such a set of curves is the great diversity of physical situations he must attempt to cover. Water depth, number of layers, layer thicknesses, and various types and magnitudes of velocity gradients are but a few of the considerations that face the modeler in even the classical horizontal flat-layer cases. In situations more complex than horizontal flat-layers, the difficulties are greater still. A set of curves constructed for a given location will be applicable at another location only if the geologic columns at both places are similar. While some information can be gained by using curves not precisely modeled after the area under investigation, the curves should be constructed with models most likely to represent the true situation.

It is instructive to examine a few curves of the type described. Some of those used in the investigation of the sediments on the Ontong Java Plateau [Maynard, 1973] provide fairly typical examples. In this investigation, several factors facilitated the construction of such curves. The predominant water depth was rather uniform from station to station. The sub-bottom structure was generally quite simple and closely approximated the flat-layer idea. Within certain limits, the thickness and interval velocity of the layers were known. Knowing these things reduced the number of models which had to be considered, made those models constructed more meaningful, and, in particular, made possible critical comparison of models giving only subtly differing travel-time curves.

From clues in the character of the reflection curves, it is sometimes possible to infer the presence of low-velocity layers and velocity gradients. Since it is often expected that such variations are most pronounced in the uppermost part of the bottom sediments, it should be instructive to examine various models which simulate a layer of sediment lying just beneath the water column and on top of a semi-infinite basement. With attention limited to the thin "sediment" layer, the velocity and other physical properties of the basement are immaterial as long as the acoustic impedance differs sufficiently from that of the overlying sediment layer to give a reflection. The velocity of the basement can thus be set equal to that of the water column (eliminating the possibility of a head wave along the sediment-basement interface). It is then informative to examine three models in which the velocity of the sediment layer is, respectively: (1) equal to that of the water column; (2) greater than that of the water column, giving the familiar case of velocity increasing with depth; and (3) less than that of the water column, simulating a low-velocity "masked" layer (Fig. 5).

In R_{1j}, let R mean "reflection", 1 indicate that we are considering only the first order bottom reflection signal (i.e., no water surface reflected multiples are considered), and j be the number of the interface which produces the reflection, as, for example, in Figs. 1 and 2. In the hypothetical case of all layers having the same uniform velocity, both R_{11} and the R_{12} reflection curves are asymptotic to D, and there is no sediment head-wave refraction (Fig. 5-a). R_{12} would have an associated head-wave refraction if velocity in the basement were greater than that in

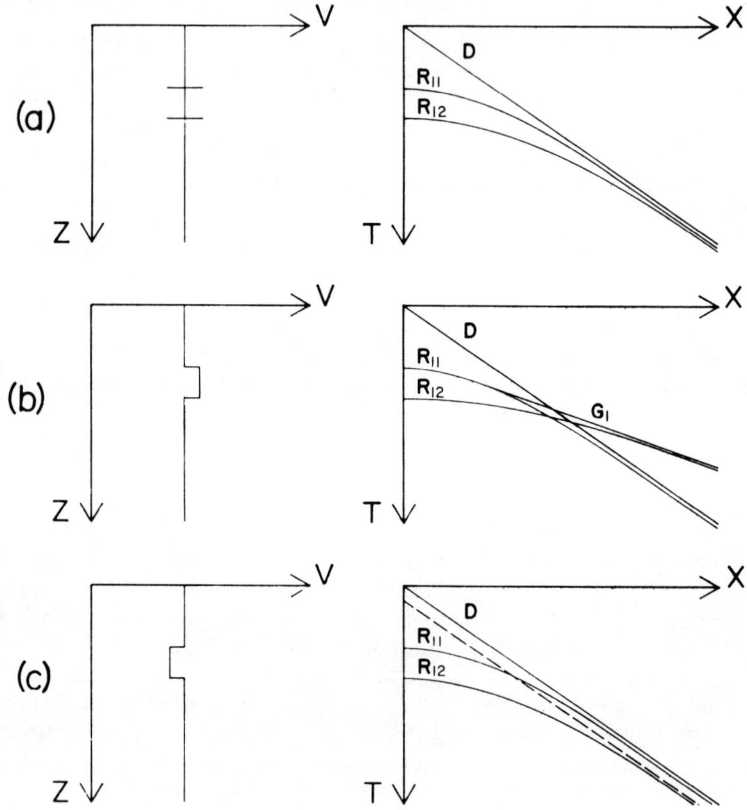

FIGURE 5

Theoretical travel-time curves for three V-D functions in homogeneous, isotropic, horizontal, plane layers. In case (a) a difference in acoustical impedance is postulated at two interfaces.

water. In the second case, the velocity of the second layer is higher than that of the upper (water) layer and the semi-infinite medium. Then R_{11} is asymptotic to D, and at critical range has tangent to it a head-wave refraction or "ground wave", G_1, to which the reflection R_{12} is asymptotic (Fig. 5-b). R_{12} would have an associated head-wave refraction if velocity in the basement were greater than that in the sediment. In the third case, the velocity of the second layer is lower than the velocity of the layers above or below: R_{11} is again asymptotic to D, but has no associated refraction. R_{12}, on the other hand, is not asymptotic to an imaginary line parallel to D and at a larger travel-time value (Fig. 5-c). The difference between travel-times for R_{11} and R_{12} approaches a constant for a masked, low-velocity second layer. As in the first case, R_{12} would have an associated head-wave refraction if velocity in the basement were greater than that in water.

Examples of all three cases can be found in the ASPER records. The topmost sediment layer most often involved is rarely thicker than a few milliseconds in two-way travel-time. Consequently, it is usually possible to tell in a qualitative way that these sediments have a seismic velocity close to that of sea water, but higher frequencies than normally employed are required for reliable quantitative measurement. For the case of the low velocity (masked) layer, both thickness and velocity can be obtained using the intercept times for R_{11} and R_{12} and the time offset between D and the asymptote to R_{12}.

Since large velocity gradients are often hypothesized to exist in the uppermost bottom sediments, it is necessary to examine the effects of velocity gradients on travel-time curves. Only linear gradients are considered here.

Consider first the simple case of a single, homogeneous layer overlying a semi-infinite medium. Let the semi-infinite medium have a linear gradient, but assume that there is no first order velocity discontinuity between the layer and the semi-infinite medium (Fig. 6-a). These are the conditions for the classical diving-wave pair, R_S and R_D [Ewing, 1963]. The raypaths of both diving waves in the semi-infinite medium are segments of circles [Slotnick, 1959] and each wave gives rise to its own branch of a travel-time curve. The branches bifurcate from a point at some minimum range A (Fig. 6-a) and form a cusp. As horizontal range increases from A, the apparent velocity of R_D increases monotonically, while the later arriving R_S gradually approaches R_{11}, and in the process becomes asymptotic to D. Depending upon the thickness of the overlying layer and velocity gradient of the semi-infinite medium, there is some angle of incidence on the interface for a ray which will emerge at the surface at range A. For a given gradient, as the layer thickness is decreased the range A decreases and the angle of incidence at A increases. Near range A, as the angle of incidence decreases, the

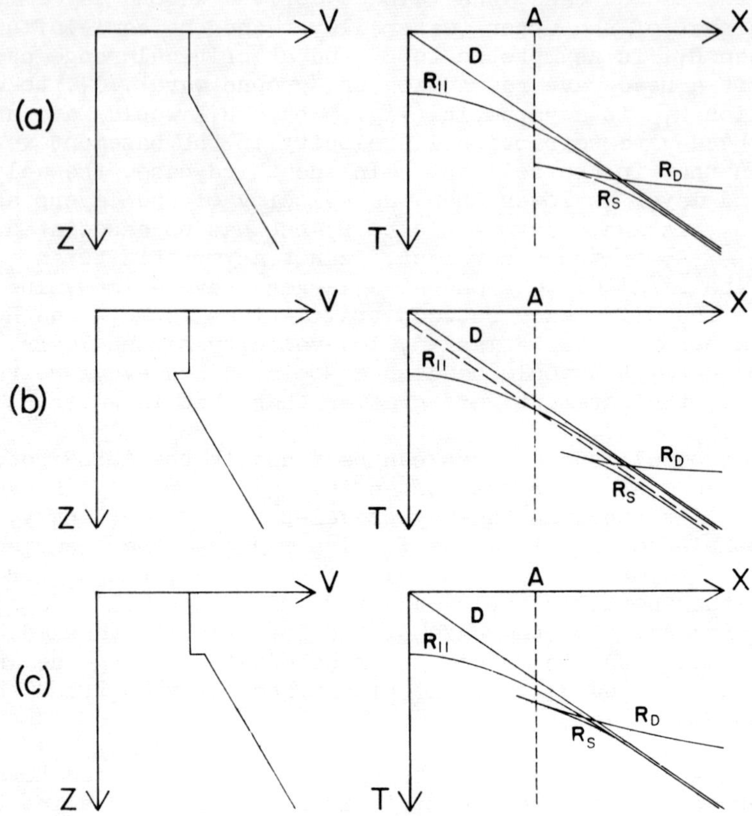

FIGURE 6

Theoretical travel-time curves for three models involving a single, constant-velocity, horizontal, plane layer over a semi-infinite medium having a positive linear velocity gradient. A difference in acoustical impedance is assumed to exist at the interface in model (a).

ray plunges more deeply, the range for emergence increases, and the high-velocity R_D branch is formed; as the angle is increased the range for emergence again increases, but the diving ray penetrates less and less of the medium and the difference between R_S and R_{11} becomes smaller.

If there is a first order discontinuity to a lower initial velocity in the underlying semi-infinite medium so that a small "low-velocity zone" is formed (Fig. 6-b), then the refracted diving-wave arrivals first emerge at the surface at a range greater than A. In this case the deep-diving refraction R_D is very similar to that of the case of no first order discontinuity, but the shallow-diving refraction, R_S, becomes asymptotic to a line parallel to, and later than, D. The analog between this case and that of a gradient-free low-velocity layer between two other layers (Fig. 5-c) is obvious. If there is a positive step in velocity at the first order discontinuity (Fig. 6-c), then the R_S and R_D arrivals first appear at a range closer than A. R_D is very similar to both preceding cases; however, R_S terminates by merging with R_{11} at the critical (head-wave) refraction range.

Consider the effects of a velocity gradient in a buried layer rather than a semi-infinite medium. In this model, assume an upper layer of constant velocity, a second layer with a linear gradient, and an underlying semi-infinite medium of constant velocity. In order to remove the added complications of head-waves arising in the semi-infinite basement, assume that the basement has the same constant velocity as the topmost layer.

Again assume that the top layer has a constant velocity which is also the initial velocity of the second layer (no first order discontinuity) and that the gradient in the second layer is positive (Fig. 7-a,b). Then R_{11} is, as usual, asymptotic to D. However, R_{12} terminates at that range at which the circular raypath of the diving waves in the second layer graze the bottom of that layer. Where R_{12} terminates, it joins one or the other of the branches of an R_S - R_D diving-wave refraction pair. The thickness of the second layer determines with which of these R_{12} merges. If Layer 2 is thick enough that range A in included in the R_S - R_D curve, then a segment of R_D is formed, R_{12} and R_D merge at some range and terminate at that point to form a cusp. As the second layer is made progressively thinner, the R_{12} curve assumes lower travel-time values throughout its length, and simultaneously the segment of R_D becomes shorter as the R_{12} and R_D maintain the cusp (Fig. 7-a). If the second layer is made sufficiently thin, the entire R_D disappears, and R_{12} then merges smoothly into R_S at range A, or some greater distance. In this last case it is difficult to determine that the R_{12} converted to an R_S curve instead of continuing onward as a reflection curve: compare Fig. 7-b with Fig. 5-a. If a good interval velocity can be obtained and it indicates that the velocity of Layer 2 is greater than that of Layer 1, then a gradient can be suspected; otherwise, it will probably go undetected. Where thicker layers are concerned, the appearance of a cusp immediately suggests a gradient.

Consider now a model which differs from the above only in that the velocity gradient in the second layer is negative instead of positive (Fig. 7-c). Then R_{11} is, as usual, asymptotic to D, but R_{12} approaches a line parallel, but delayed with respect to, D. The behavior of R_{12} is similar to that of the case of a low-velocity layer of uniform velocity; the shapes of the two reflection curves

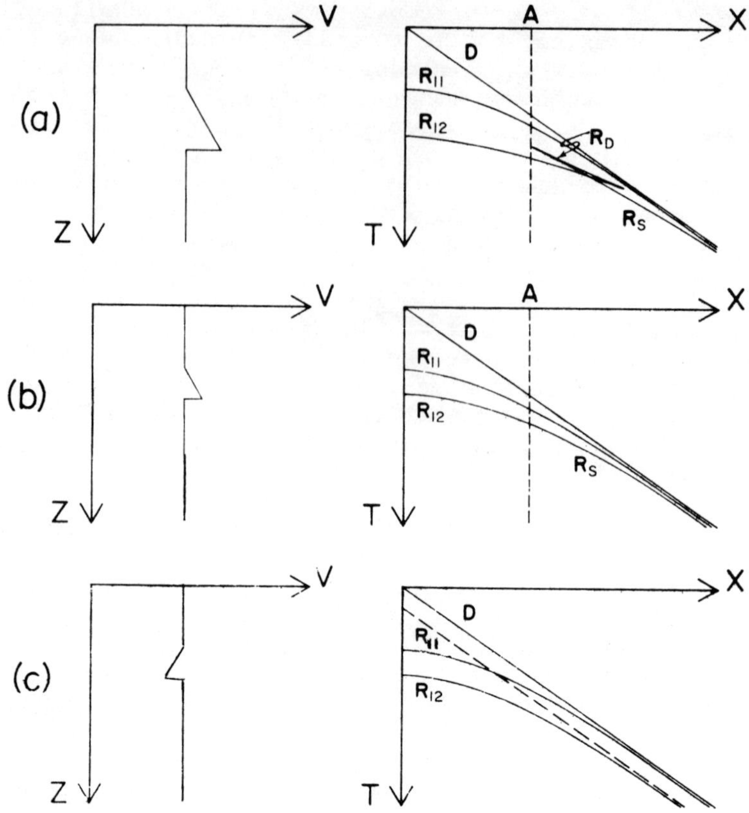

FIGURE 7

Theoretical travel-time curves for three cases of a buried horizontal plane layer having a gradient. Case (a): thick layer; case (b): thin layer; case (c): negative gradient.

may be different, but the difference is too small to be of practical use in analyzing ASPER records.

In theory, then, the occurrence of a reflection curve with an asymptote that is parallel to D but delayed in time with respect to it indicates the presence of either a masked, low-velocity layer of uniform velocity (Fig. 5-c); of a negative first order discontinuity in velocity underlaid by a positive gradient (Fig. 6-b); or of a layer with a negative gradient having no first order discontinuity at the upper interface (Fig. 7-c). It would be difficult to distinguish between the first and third cases. Furthermore, in practice, it would be difficult to identify the second case because in many reflection curves the arrivals at ranges shorter than critical (total reflection) are very weak, and it may be risky to assume their absence when they are not detected.

These are three cases of a low-velocity zone. Thus, while it may not be possible to determine the gradient, a criterion does exist for identifying the low-velocity zone: if a reflection curve is asymptotic to a line which is parallel to and later than a head-wave refraction line, then a masked layer of lower velocity than the overlying material is indicated.

The preceding paragraphs apply primarily to thin layers. However, this should not imply that things are always much simpler when thicker layers are involved. There are conditions under which it would be possible to confuse ordinary reflection and head-wave refraction arrivals with those from diving waves. The problem would most likely arise because of the difficulty of distinguishing between a cusp due to a velocity gradient and a "pseudo-cusp" resulting from the amplitude increase at critical reflecting and refracting range in layers having uniform velocities. In models consisting of a sequence of uniform, horizontal layers of constant velocity, each reflection usually has associated with it two head-wave refraction lines: its asymptote, and its own tangent refraction line (R_{12}, G_1, and G_2 in Fig. 8-a). Hence, if the arrivals from ranges shorter than critical reflection range are weak (as they frequently are), then it is possible that R_{12} and G_2 could be mistaken for an R_S - R_D pair (note inserts in Fig. 8-a), falsely indicating a gradient. The converse also holds true. Each head-wave refraction has two reflection curves associated with it: the curve to which the refraction is tangent, and the curve which is asymptotic to the refraction (Fig. 8-b). Thus, the combination of R_{12}, G_2, and R_{13} (or similarly, R_{11}, G_1, and R_{12}) in a horizontal, constant-velocity layer situation could be confused with an R_{13} which had associated with it an R_S - R_D pair arising from a gradient, especially when the R_{13} reflection is weak. Also, the reflection curves and refraction lines tend to "bunch" at certain ranges, so that the identity of the individual curves is difficult to maintain throughout the record. Therefore, it frequently is not possible to distinguish between the reflections and refractions. However, for practical purposes, if reflections near the critical angle are mistaken for refractions and are analyzed as such, not much error results because both tend to fade out quickly with range in thick sediments, insuring that the reflection is actually very little different from the refraction in arrival time and slope.

When combined with the several-cycle-long seismic pulse typically obtained from airguns, the tendency for the curves and lines on ASPER records to cross over each other and to form bunches can introduce an added complexity into the records: spurious arrivals. While the following should apply strictly only to refraction lines, it might also be applicable to reflection curves at long ranges where curvature is small.

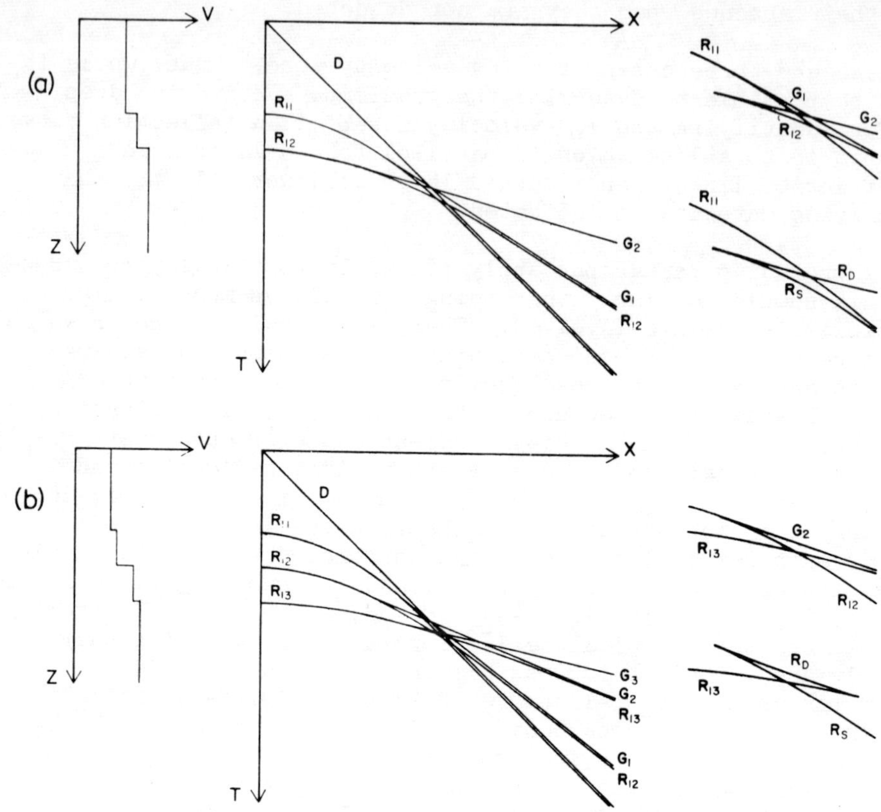

FIGURE 8

Travel-time curves illustrating some difficulties encountered in distinguishing between reflections, diving-wave refractions, and head-wave refractions.

Interference between two head-wave refraction lines from layers with velocities V_1 and V_2 can produce a spurious "masked layer", with apparent velocity $V_m = 2V_1V_2/(V_1 + V_2)$, between the two real layers. This results from the super-position of two propagating sine waves of the same frequency but of differing velocity:

$$S = \sin 2\pi f \left[\frac{1}{V_1} X - T\right] + \sin 2\pi f \left[\frac{1}{V_2} X - T\right]$$

$$= 2 \cos 2\pi f \left[\frac{1}{2}\left(\frac{1}{V_1} - \frac{1}{V_2}\right) X\right]$$

$$\cdot \sin 2\pi f \left[\frac{1}{2}\left(\frac{1}{V_1} + \frac{1}{V_2}\right) X - T\right] .$$

The first factor is an amplitude term, and the second factor is a propagating sine wave of the same frequency as the two which were added but having a slope of $1/V_m$. It has not been possible to positively identify such spurious arrivals in ASPER records, but the possibility that they may exist occasionally makes the use of some arrivals doubtful. Interference phenomena of this type from the associated past-critical-range reflection probably accounts for the irregularities in slope and amplitude observed in some refraction lines.

When Figs. 5-a and 7-b are compared, one notes that it would be difficult to distinguish between two other cases: a reflection coming from interfaces between a pair of layers having the same constant velocity (but different acoustical impedance) and a reflection that converts to an R_s diving-wave refraction at some range beyond A because of the effects of a positive gradient in a thin intervening layer. If the data were exceptionally good or the layer were thick enough, it might be possible to detect the higher interval velocity by wide angle reflection methods.

A number of seismologists have observed that when used simultaneously, continuous vertical reflection, wide angle reflection, and refraction do not necessarily measure the same parts of a layer or even the same layer. A common observation is that horizons which produce very strong reflections in the continuous vertical reflection profiles sometimes produce very weak reflection lines in a wide angle reflection profile; conversely, strong wide angle reflections may come from interfaces which are hardly detectable in otherwise excellent continuous vertical reflection profiles. It frequently appears that refractions come from greater depths than do the associated wide angle reflections. In addition, refraction methods measure a horizontal velocity, whereas wide angle reflection methods determine an average vertical velocity over an interval of depth.

When different signal frequencies are used in the different methods, unlike results are not too surprising because in many cases the reflecting "interface" is a transition zone between layers rather than a first order discontinuity. These zones are often complicated ones with reflection and refraction characteristics related in complicated ways to wavelength [Nakamura and Howell, 1964] and angle of incidence [Meissner, 1966]. In all cases, it seems fairly certain that interference phenomena enter in. In continuous vertical reflection profiles the reflection from a horizon is often seen to vary in strength with location. Occasionally, reflections from a horizon fade from a strong arrival to one too weak to be discernible, while above or below this horizon another reflection may intensify. Similar variations occur in wide angle reflections and refractions. Thus, it is not surprising that analysis is sometimes difficult and that comparisons between methods are not always straightforward.

Obviously the wide angle reflection method has several shortcomings, some severe, some merely inconveniences. In its present state of development, the method requires that the geologic structure be quite simple and have regular reflection surfaces. Even when the geologic structure approaches the ideal, demands on data precision are stringent. Under the best of circumstances velocities determined are reliable to no better than perhaps 0.1 km/sec in individual seismic sections. Despite sophisticated and imaginative analysis techniques and the use of computers, determination of the simplest of gradients in most layers is still largely guesswork, and the study of more complex zones of transition is even more speculative. In short, the time-distance relations are subtle and complex, often too subtle and complex to be unraveled.

On the other hand, except for those rare cases where a sonic well log is available or can be made, the best, most detailed estimate of the velocity-depth relationship in marine sediments can be made by the wide angle reflection method. Furthermore, the relationships established by the method will often agree well with those found by refraction methods, especially when many determinations are made in a homogeneous area and all results are statistically combined. While it is often discouraging to correlate between seismic sections determined by the wide angle reflection or the refraction method and a normal incidence reflection profile, if the results of all three explorational seismic methods are integrated, the geologic structure is generally accurately defined, often with surprising detail.

Acknowledgments

We would like to thank R. Meissner, G. B. Morris, and D. A. Walker for their many helpful consultations on matters concerning analysis and interpretation of seismograms. We appreciate the help of M. E. Odegard and G. Fryer, who provided several computer programs and valuable assistance in their use. N. J. Thompson, W. N. Ichinose, and R. C. Mitiguy kindly provided the answers to many questions of an engineering or technical nature. The assistance and suggestions of E. McAfee with the manuscript, and of C. McCreery, S. Dang, and R. Rhodes with the figures are gratefully acknowledged.

This research was supported under Office of Naval Research contracts Nonr 3748(05) NR 083-603 (1967-1968, 1968-1969) and N00014-70-A-0016-0001 (1969-1970) and under National Science Foundation contracts NSF GA 1639 and GA 15792.

This paper constitutes Hawaii Institute of Geophysics contribution number 587.

REFERENCES

Clay, C. S., and P. A. Rona, Studies of seismic reflections from thin layers on the ocean bottom in the western North Atlantic, J. Geophys. Res., 70 (4), 855-969, 1965.

Dix, C. H., Seismic Prospecting for Oil, 245 pp., Harper and Brothers, New York, 1952.

Dix, C. H., Seismic velocities from surface measurements, Geophysics, 20 (1), 68-86, 1955.

Dürbaum, H., Zur Bestimmung von Wellengeschwindigkeiten aus reflexionsseismischen Messungen, Geophys. Prospecting, 2, 151-167, 1954.

Ewing, J. I., Elementary theory of seismic refraction and reflection measurements, in The Sea, Vol. 3, edited by M. N. Hill, pp. 3-19, Interscience, New York, 1963.

Green, C. H., Velocity determinations by means of reflection profiles, Geophysics, 3 (4), 295-305, 1938.

Hill, M. N., Single-ship seismic refraction shooting, in The Sea, Vol. 3, edited by M. N. Hill, pp. 39-46, Interscience, New York, 1963.

Houtz, R., J. Ewing, and X. Le Pichon, Velocity of deep-sea sediments from sonobuoy data, J. Geophys. Res., 73 (8), 2615-2641, 1968.

Knox, W. A., A deep-ocean sedimentary velocity function, J. Geophys. Res., 70 (8), 1999-2001, 1965.

Le Pichon, X., J. Ewing, and R. E. Houtz, Deep-sea sediment velocity determinations made while reflection profiling, J. Geophys. Res., 73 (8), 2597-2614, 1968.

Maynard, G. L., Crustal layer of seismic velocity 6.9 to 7.4 kilometers per second under the deep oceans, Science, 168, 120-121, 1970.

Maynard, G. L., Seismic wide angle reflection and refraction investigation of the sediments on the Ontong Java Plateau, Ph.D. dissertation, Hawaii Institute of Geophysics, University of Hawaii, Honolulu, 1973.

Meissner, R., Seismic wide angle measurements in the Bavarian Molasse basin, Geophys. Prospecting, 14 (1), 1-6, 1966.

Meissner, R., Exploring deep interfaces by seismic wide angle measurements, Geophys. Prospecting, 15 (4), 598-617, 1967a.

Meissner, R., Engebnisse der Weitwinkelmessungen im bayerischen Molassebecken, 1 und 2, Gerlands Beiträge zur Geophysik, 76 (3), & Portig K.-G., Leipzig, 1967b.

Nakamura, Y., and B. F. Howell, Marine seismic experiment: Frequency spectra of refraction arrivals and the nature of the Mohorovicic discontinuity, Bull. Seismol. Soc. Amer., 54 (1), 9-18, 1964.

Odegard, Mark E., Interpretation of wide angle reflection data using the parametric form of the time-distance curve, a paper presented at the 42nd Annual Meeting of the Society of Explorational Geophysicists, abstract in Geophysics, 38 (1), 187, 1973.

Robinson, W. B., in Society and Industry Briefs, Geophysics, 36 (3), A-30, 1971.

Slotnick, M. M., Lessons in Seismic Computing, 268 pp., Society of Exploration Geophysicists, 1959.

Sutton, G. H., G. L. Maynard, and D. M. Hussong, Widespread occurrence of a high-velocity basal layer in the Pacific crust found with repetitive sources and sonobuoys, in The Structure and Physical Properties of the Earth's Crust, Geophys. Monogr. Ser., Vol. 14, edited by J. G. Heacock, 193-209, 1971.

Taner, M. T., and F. Koehler, Velocity spectra-digital computer derivation and applications of velocity functions, Geophysics, 34 (6), 859-881, 1969.

Taner, M. T., E. E. Cook, and N. S. Neidell, Limitations of the reflection seismic method: Lessons from computer simulations, Geophysics, 35 (4), 551-573, 1970.

SONOBUOY MEASUREMENTS IN THIN LAYERS

GEORGE M. BRYAN

Lamont-Doherty Geological Observatory
of Columbia University
Palisades, New York 10964

ABSTRACT

The application of standard sonobuoy techniques to thin subbottom layers at 3.5 kHz requires modification of the usual method of analysis. For a thin layer in deep water (thickness ratios greater than 15), velocity computed by the standard $t^2(x^2)$ method is sensitive to the detailed velocity structure in the water. An alternative approach, which compares paths of the same ray parameter rather than the same shot position, gives the solution for a subbottom layer independently of the velocity structure above the layer. The method has been tested on simulated layer structures and on actual sonobuoy data. The results indicate that the technique will be useful for thickness ratios as high as 100. Systematic variations in the results from real data suggest the possibility of lateral inhomogeneities in the layer structure.

INTRODUCTION

Variable-angle reflection data from an air-gun source and radio-sonobuoy receiver are used routinely to determine sound velocities in marine sediments (Houtz, this volume). The standard analysis (Le Pichon et al., 1968), which assumes straight line paths in the water column and in the sediment layers, is adequate for layered structures commonly resolved by the air-gun. Typical layer thicknesses involved are as small as 10% of the water depth. Recent attempts to apply the technique to much thinner layers at 3.5 kHz suggested that the standard analysis is inadequate when the water depth is more than about 15 times the layer thickness. In what follows, this empirical result is interpreted analytically and an alternative approach is presented.

THE STANDARD METHOD

The standard method of analysis (Le Pichon et al., 1968) is a modified $t^2(x^2)$ method which assumes straight-line path segments in each layer, including the water column. Referring to Fig. 1, an average velocity and depth for the water column (layer 1) is determined from the slope and intercept of a linear fit to $t_1{}^2$ as a function of x^2. With these parameters and the derivative of $t_2(x)$ we can compute that part of path t_2 which is in layer 1. This part is removed from path t_2 to give reduced time-distance data in layer 2. The reduced data are then fitted to a $t^2(x^2)$ line to yield velocity and thickness of layer 2. The procedure can be extended to any number of layers, including those with arbitrary dip, the results for each layer depending on the results for the overlying layers.

The assumption of straight-line paths leads to errors which may be significant in certain cases. In particular, errors in the water column which are negligible in comparison with the times and distances involved in the water column can be appreciable compared with the reduced data involved in an underlying layer, which is very much thinner than the water column. Yet these errors are introduced into the reduced data when the water column is subtracted out.

If, for example, the straight-line assumption introduced an absolute error of Δt in the travel time in the water, then the relative error, in the water, is of order $\Delta t/t_1$. The relative error of the reduced data in the underlying layer is of order

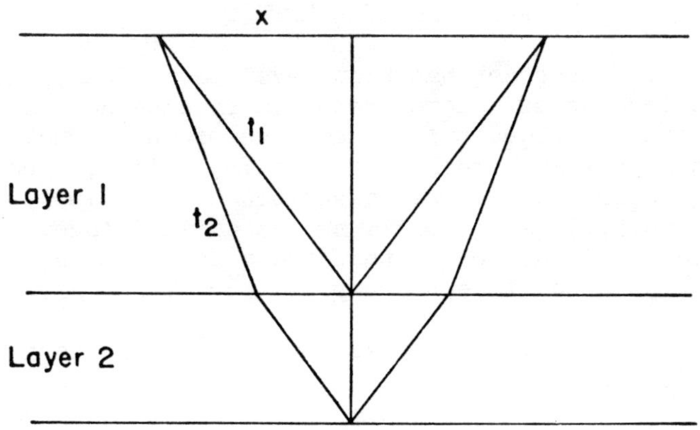

FIGURE 1

Geometry for the standard method of sonobuoy data reduction.

$$\frac{\Delta t}{t_2-t_1} = \frac{\Delta t}{t_1} \cdot \frac{t_1}{t_2-t_1}$$

That is, the relative error is magnified approximately by the ratio of water depth to layer thickness. The relative error in the velocity of the thin layer is therefore similarly magnified.

The straight-line assumption is reasonably successful for thickness ratios up to about 15. For larger ratios the computed velocity in the thin layer is sensitive to the detailed velocity structure in the water. If this structure were known in sufficient detail, ray-tracing techniques in the water column could be expected to give considerable improvement. However, for the ratios of interest here (50 to 100), the analysis requires more detailed information on the water structure than is ordinarily available.

THE RAY PARAMETER METHOD

An alternative approach has been developed which yields the velocity and thickness of a homogeneous layer embedded in an arbitrary structure without prior knowledge of the structure above the layer. An approximate method to do this, valid in the limit of small angles of incidence, was described by Dix (1955). The present scheme is applicable to all angles of incidence. It assumes, as does Dix's approach, that the velocity structure above the layer is a function of depth only.

A ray path is then characterized by a fixed ray parameter p given by Snell's law:

$$p = \frac{\sin \theta \ (z;p)}{c(z)} \tag{1}$$

where z is depth, c(z) is the velocity structure, and θ (z;p) is the angle which the path p makes with the vertical at depth z. For incremental travel time dt along the path we have

$$dz = cdt \cos \theta = (1-p^2c^2)^{1/2} \ cdt \tag{2a}$$

$$dx = cdt \sin \theta = pc^2dt = pc(1-p^2c^2)^{-1/2} \ dz \tag{2b}$$

Rearranging and integrating from the surface to depth z_1 we have

$$t(p) = \int_0^{z_1} c^{-1} \ (1-p^2c^2)^{-1/2} \ dz \tag{3a}$$

$$x(p) = \int_0^{z_1} pc(1-p^2c^2)^{-1/2} \ dz \tag{3b}$$

These equations give the travel time and x-distance along path p in terms of integrals along the path. In practice, t and x are known over a range of p, and the equations must be solved for the velocity structure c(z).

Now consider a homogeneous layer of velocity c_1 bounded by reflectors at depths z_1 and $z_1 + \Delta z$ (Fig. 2). A given ray parameter p defines a reflection path to each reflector. Unlike the usual approach, these two paths are not associated with the same value of x but rather with the same value of p. In fact, above z_1 the two paths are everywhere parallel. The differences in travel time and x-distance for these two paths are then, by Eqs. (3),

$$\Delta t(p) = 2 \int_{z_1}^{z_1+\Delta z} c^{-1}(1-p^2c^2)^{-1/2} \ dz \tag{4a}$$

$$\Delta x(p) = 2 \int_{z_1}^{z_1+\Delta z} pc(1-p^2c^2)^{-1/2}\, dz \tag{4b}$$

where the factor of 2 takes account of the return path. For a uniform velocity c_1 in the layer, Eqs. (4) can be integrated

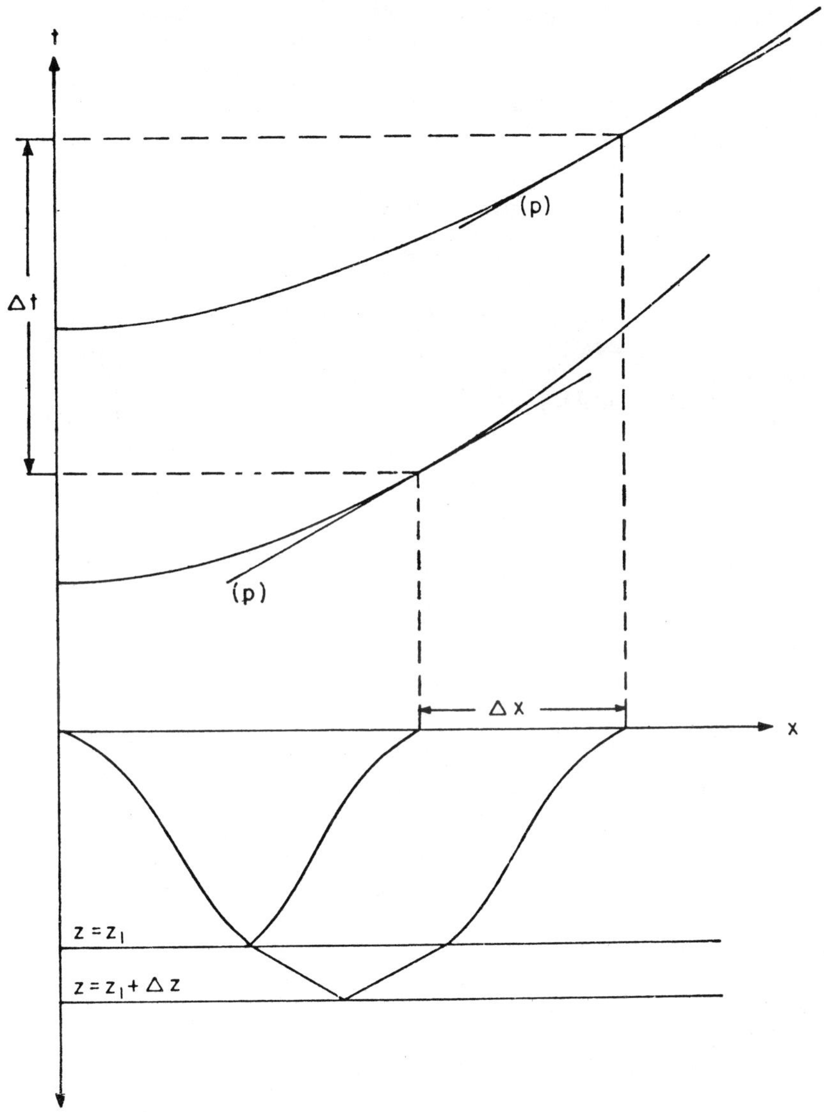

FIGURE 2

Geometry for the ray parameter method of sonobuoy reduction.

to give

$$\Delta t = 2c_1^{-1}\,(1-p^2c_1^{\,2})^{-1/2}\,\Delta z \tag{5a}$$

$$\Delta x = 2pc_1\,(1-p^2c_1^{\,2})^{-1/2}\,\Delta z \tag{5b}$$

These are readily solved for the velocity and thickness of the layer.

$$c_1 = \sqrt{\frac{1}{p}\,\frac{\Delta x}{\Delta t}} \qquad \Delta z = \frac{1}{2}\,c_1\Delta t\,\sqrt{1-p^2c_1^{\,2}} \tag{6a,b}$$

The contribution from the structure above z_1 has canceled out because of the choice of paths.

In order to apply this approach to the sonobuoy experiment it is necessary to match paths from the two reflectors for the same value of p. The upper part of Fig. 2 shows travel time as a function of x for each of the reflectors. These travel time curves are the basic data from the sonobuoy experiment. Each point on such a curve corresponds to a particular value of p which is just the derivative of the curve at that point, since from Fig. 3 we have

$$\frac{c_o\,\delta t}{\delta x} = \sin\theta_o = pc_o \tag{7}$$

where c_o and θ_o are the velocity and path angle at z=0. The symbol for differentials has been changed here to emphasize the fact that we are differentiating along the t(x) curve at fixed z (i.e., z=0) rather than along a ray path (fixed p) as in Eqs. (2).

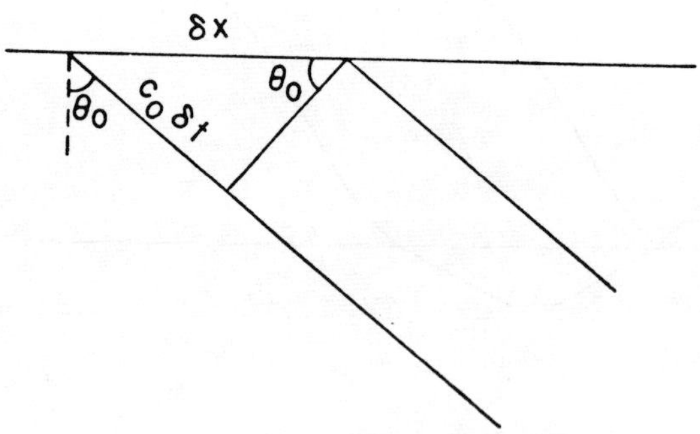

FIGURE 3

Relationship between the derivative of the time-distance curve and the ray parameter.

The procedure then is to differentiate the t(x) curves of the sonobuoy data, solve for t(p) and x(p), and compute the differences Δt and Δx for each value of p. Equations (6) then give a series of velocities and thicknesses which can be averaged to give best estimates. A complete program to carry out this scheme has been written and tested on mathematical models and on real sonobuoy data.

The model studies were based on a homogeneous layer of velocity 2.0 km/sec beneath a homogeneous water column of velocity 1.5 km/sec. The water depth was fixed at 5.0 km and the layer thickness was varied to give thickness ratios from 10 to 500. For each model, time-distance curves were plotted by computer, measured by the technique normally used for sonobuoy records, and fed into the program outlined above for computing the layer parameters. The results are summarized in Fig. 4a, which shows averages and standard deviations for the layer velocity at several thickness ratios. It can be seen that the results are quite good out to a ratio of 50, begin to deteriorate at 100, and are very poor at 500. When the method is applied to real sonobuoy data the results parallel the model results but with considerably greater variability at any given thickness ratio. Thus, while the method seems to be useful on real data out to ratios of about 25, serious difficulties arise around 50.

The systematic deterioration of the computations in the direction of thinner layers is a consequence of increasing relative error in Δx as Δx itself decreases with layer thickness. When the uncertainty in the value of x associated with a given value of p (see Fig. 2) is of the order of Δx, the computation breaks down. Fortunately in this limit of very small Δx it is possible to make an approximation which avoids the need for computing Δx.

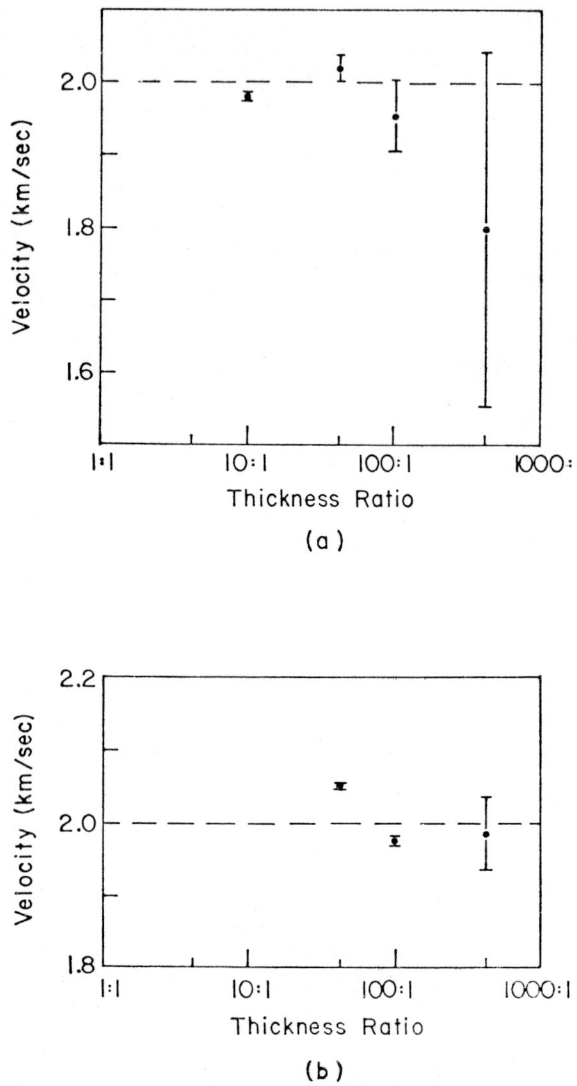

FIGURE 4

Results of model calculations using (a) the ray parameter method and (b) the thin layer approximation for various thickness ratios. Model velocities were 1.5 and 2.0 km/sec for upper and lower layers respectively.

THE THIN LAYER APPROXIMATION

In Eqs. (6) we split Δt into two parts: Δt_1, the time interval between the two reflection curves at a fixed x; and Δt_2, the time increment corresponding to moving out a distance Δx along one of the curves. Thus from Eq. (6a) we have

$$\frac{1}{pc_1^2} = \frac{\Delta t}{\Delta x} = \frac{\Delta t_1}{\Delta x} + \frac{\Delta t_2}{\Delta x} \tag{8}$$

For sufficiently large thickness ratios the two t(x) curves of Fig. 2 are practically straight and parallel across the interval Δx, and the term $\Delta t_2/\Delta x$ approaches p, the slope of both curves at x in this limit. Making this substitution and using Eq. (5b) to eliminate Δx we obtain

$$p^2 = \frac{1}{c_1^2} - \frac{(\Delta t_1)^2}{4(\Delta z)^2} \tag{9}$$

This equation is independent of the two quantities, Δx and Δt_2, which become unstable in the limit of very thin layers. Fitting a straight line to p^2 as a function of $(\Delta t_1)^2$, we obtain the velocity from the intercept and the thickness from the slope.

Application of this scheme to the model used above yields the results shown in Fig. 4b for ratios of 50:1, 100:1, and 500:1. At 10:1 the approximation involved is inadequate and no solution is obtained, but beyond 50:1, where the exact method begins to fail, the approximate method gives good results out to 500:1.

Preliminary tests on real sonobuoy data at 3.5 kHz indicate that the thin layer approximation will be useful in very thin layers where the exact ray parameter method is poor and the standard method fails entirely. Figure 5 shows a portion of a sonobuoy record taken in 5 km water depth. The vertical time scale is 50 ms/division and the horizontal scale represents a total of about 2 1/2 km. The lower set of relatively sharp traces is the primary data; the upper, diffuse set is a surface-reflected echo of the lower set.

We assumed a homogeneous layer between the sea floor and the first prominent subbottom reflector at about 60 ms into the sediment (indicated by an arrow in Fig. 5). This corresponds to a thickness ratio of about 100:1. The plot of p^2 against $(\Delta t_1)^2$ is shown in Fig. 6. The data tend to group along three distinct line segments corresponding to the following sets of solutions for average velocity and thickness.

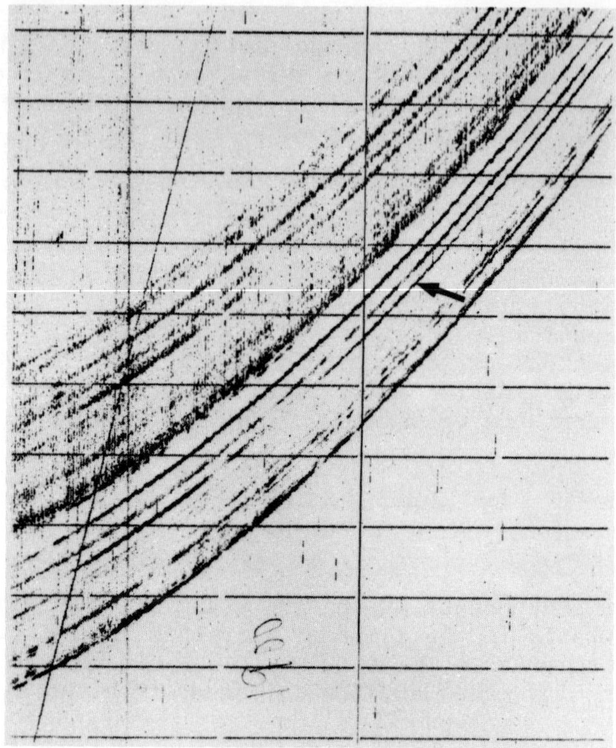

FIGURE 5

Sonobuoy record at 3.5 kHz in 5 km of water. Vertical scale is 50 ms/division; horizontal scale represents a total of about 2 1/2 km. Arrow points to horizon yielding the data of Fig. 6.

1.70 km/sec and 55 m
1.66 km/sec and 52 m
1.60 km/sec and 49 m

These values are quite reasonable for deep sea sediments and the spread is not excessive even if it must be attributed to experimental error. On the other hand, since the point of reflection moves along the bottom as the ship moves away from the buoy, the variation may be real. This would explain the systematic shift of the data from one line to another as p is varied. The depth recorder shows gentle undulations of the sea floor with amplitude of 3 to 4 m and wavelength of about 2 km. The range of p in Fig. 6 corresponds to a distance across the bottom of about 1.5 km.

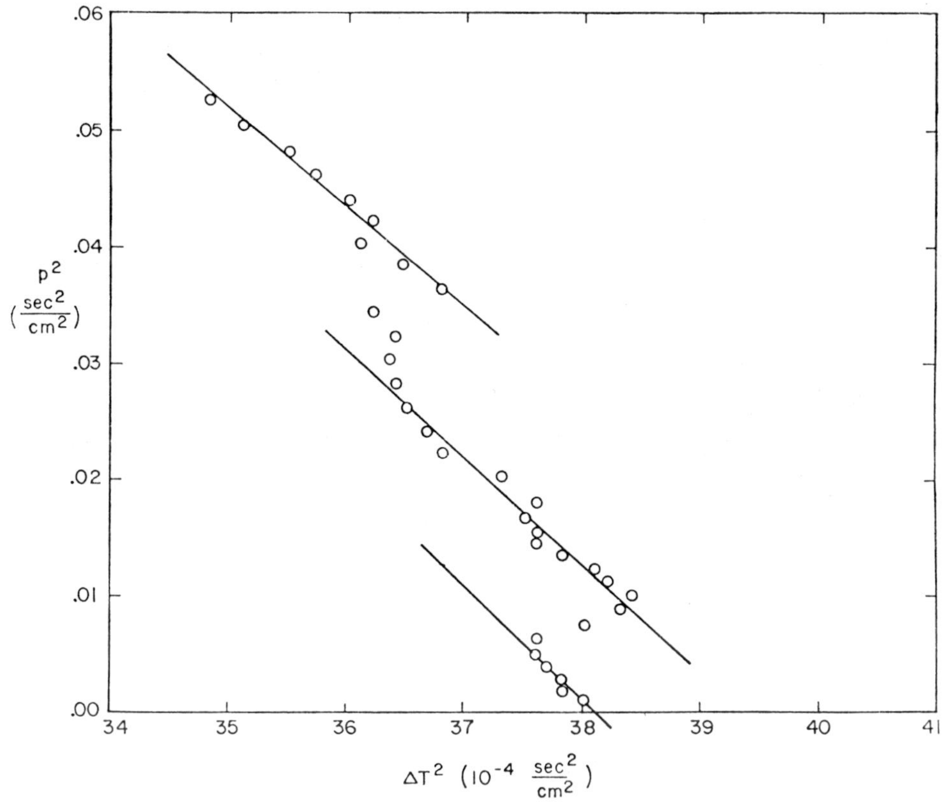

FIGURE 6

Plot of p^2 against Δt^2 for the record of Fig. 5. The straight lines correspond to the layer parameters given in text.

Further testing with real data is required to demonstrate convincingly that such variations in the sediment layer can be resolved. Lateral variations in the water column, for example, would give similar results. However, it should be pointed out that, unlike the standard method, the ray parameter method can in principle follow the variations in a layer whose velocity and thickness change gradually over the lateral distance swept by the sonobuoy run. Similarly, the method does not depend on an accurate determination of the origin of the t versus x curves; any portion of the sonobuoy record which spans a sufficiently large range of p^2 and $(\Delta t)^2$ to define a straight line will serve. The chief advantage of the method, of course, is that the solution of a layer at depth does not require detailed knowledge of the structure above the layer.

Acknowledgments

This work was supported by the Office of Naval Research under Contract No. N00014-67-A-0108-0004, and by National Science Foundation (I.D.O.E.) Grant No. GX 34410.

REFERENCES

Dix, C. Hewitt, Seismic Velocities from Surface Measurements, Geophysics, Vol. XX, No. 1, pp. 67-86, 1955.

Houtz, Robert E., Preliminary Study of Global Sediment Sound Velocities from Sonobuoy Data, Physics of Sound in Marine Sediments, Plenum Press, Loyd Hampton, ed., (this volume).

Le Pichon, Xavier, John Ewing, and Robert E. Houtz, Deep-Sea Sediment Velocity Determination made while Reflection Profiling, J. Geophys. Res., Vol. 73, No. 8, pp. 2597-2614, 1968.

THE INTERACTION OF UNDERWATER ACOUSTICS AND MARINE GEOPHYSICS

A. C. KIBBLEWHITE

Physics Department, University of Auckland

Auckland, New Zealand

ABSTRACT

Acousticians have always been concerned with the way in which the ocean and its boundaries influence the transmission of underwater sound. Through the study of such effects, underwater sound has also become a valuable tool for the investigation of the environment itself. New Zealand experiments at low frequencies have provided numerous cases of such duality. Some of the more unusual examples are presented to emphasize the interaction of acoustics and marine geophysics at frequencies below 1000 Hz.

INTRODUCTION

This conference is concerned with the acoustic properties of the sea floor. For reasons which are different in the two cases, geophysicists and acousticians alike will be discussing the topic, in the main, from the viewpoint of frequencies in the range of a few kilohertz. On the one hand, in the study of the upper marine sediments, a compromise between penetration and resolution forces the geophysicist into this part of the spectrum; on the other, operational issues predetermine the frequency range for the naval acoustician concerned with active sonars.

With this emphasis it would be easy to overlook the interaction of the marine floor and acoustics in other areas of sonar technology. By drawing primarily on New Zealand experience, this review is intended not only to draw attention to other aspects of this general subject, but also to give examples of the broader interaction of ocean acoustics and marine geophysics. The examples cited include studies of underwater volcanism, geodesy, topographic mapping, marine seismics, hydrological boundaries, and the attenuation of low frequency sound in the sea.

Over the past twenty years New Zealand research in underwater acoustics has been confined to three institutions. For some years the Department of Physics at the University of Auckland has been studying the relationship between the fluctuation in acoustic signals and thermal microstructure in the sea [Sagar, 1955, 1957, 1960; Brownlee et al., 1974]. The Department of Electrical Engineering at the University of Canterbury has recently entered the field in a project aimed at employing the principles of a sonic blind-aid in a fish-finding sonar [Smith, 1972]. But, predictably, the greatest involvement in underwater acoustics has been at the Defence Scientific Establishment (until 1970 known as the Naval Research Laboratory), a section of the Ministry of Defence. It is primarily the activities of this group with which this review is concerned, although reference is also made to some aspects of the current marine acoustic program at the University of Auckland.

While obviously concerned with many aspects of underwater acoustics, New Zealand's geographic position in the Southern Hemisphere has given the Defence Scientific Establishment a special interest in the transmission of low frequency sound in the ocean. In the early 1960's accounts of numerous shallow water experiments carried out on the New Zealand Continental Shelf, together with associated theoretical publications [Kibblewhite and Denham, 1965, 1966, 1968; Denham, 1966, 1969] point to the general objective of acoustic modelling.

During this phase, the significance of short and long term environmental changes on sound transmission was studied in the waters east of a permanent field station operated by the laboratory on Great Barrier Island. It was from the continuous program based on this station that the first New Zealand examples of the application of acoustics to marine geophysics arose. Others came to light as the acoustics program expanded to include long range transmission studies in deep water.

I. VOLCANISM, SEISMOLOGY, AND UNDERWATER ACOUSTICS

With the large low frequency acoustic source levels involved, underwater volcanism and seismicity can contribute significantly to the ambient noise level in the oceans, and must be of concern to the acoustician concerned with low frequency sound transmission. There is thus a direct interaction, through the underwater ambient noise field, between acoustics and these aspects of marine geophysics. Three examples will emphasize this relationship.

Underwater Volcanism

From the viewpoint of global tectonics New Zealand represents an unusual geophysical province. It lies between the southern end of the Pacific island arc system in the north and an offshoot of the midoceanic ridge system in the south. The New Zealand Alpine Fault, which lies between these features, is one of four famous transcurrent faults (the others being the Philippine Fault, the Atacania Fault, and the San Andreas Fault) in which major horizontal movements between sections of the crust have taken place. Further, a massive negative isostatic anomaly (-150 mgals) is found in the center of the North Island. Deep earthquakes occur in this region and appear to lie in a Benioff Zone which dips west of the anomaly at an angle of approximately 60°. West of the anomaly and parallel to this general feature lies a line of active (or recently active) volcanoes, running from Mt. Egmont in the west to the Bay of Plenty in the northeast (Fig. 1). Prior to 1963 it appeared that this line of volcanism stopped at White Island in the Bay of Plenty, but evidence from underwater acoustics modified this situation and revealed that the volcanic chain extends in a north-easterly line to Raoul Island in the Kermadec Island Group.

Early acoustic experiments based upon the Great Barrier Island hydrophones were sometimes marred by what was initially believed to be hum interference brought about by variation in earth currents associated with the station's power plant. In due course transistor technology permitted upgrading of the sea-installation. With this change it became apparent that the interference was acoustic in nature and geophysical rather than biological in origin. Its intermittent nature made identification a slow process, but eventually it was established that an underwater volcano was active in the area. During a period of more intense activity the position of the volcano was determined by acoustic measurements, and the general characteristics of the signal were established [Kibblewhite, 1966]. The source level varies over wide limits and depending upon the intensity of the activity the spectra can extend up to a hundred or up to a thousand hertz.

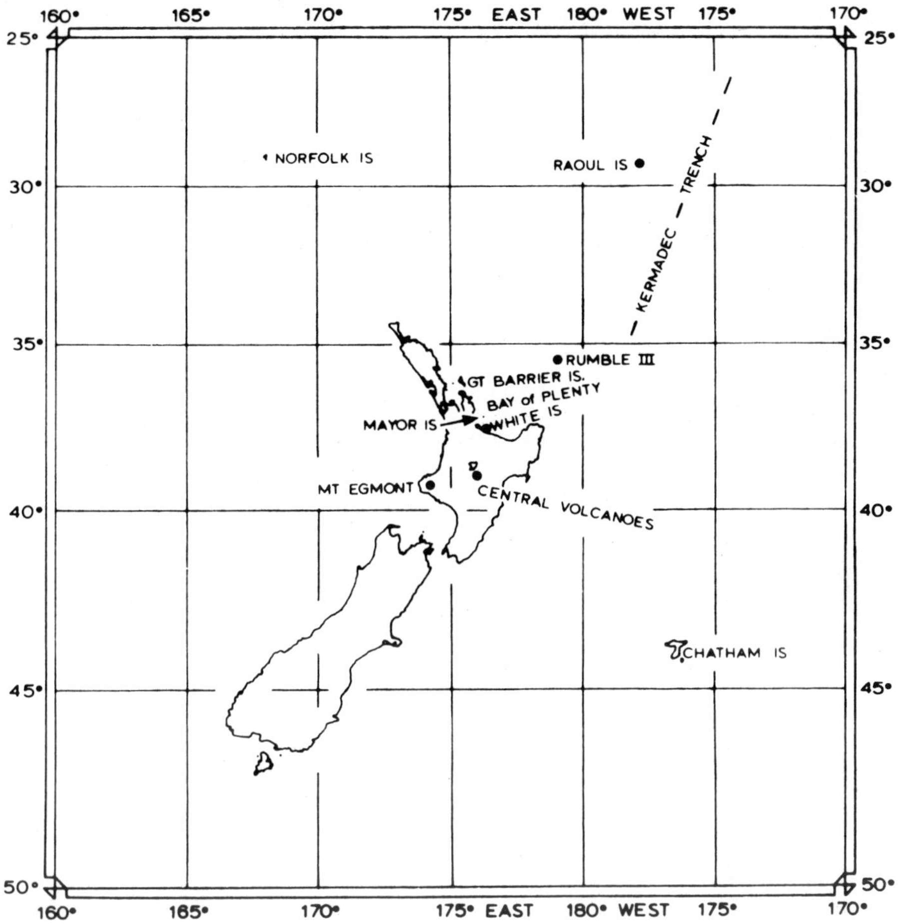

FIGURE 1

THE NEW ZEALAND VOLCANIC CHAIN RUNNING FROM MT. EGMONT ON THE WEST COAST OF THE NORTH ISLAND THROUGH THE CENTRAL VOLCANIC AREA TO WHITE ISLAND IN THE BAY OF PLENTY. RUMBLE III AND RAOUL ISLAND LIE ON AN EXTENSION OF THIS LINE.

Often the spectra are characterized by strong harmonics, which could be confusing to sonar systems operated in the area.

Location was established by correlation techniques and cross fixing from two receiving sites. The position established placed the source on the line joining the volcanic island Raoul in the Kermadec Group with the extension of the line through the active volcanoes in the North Island of New Zealand (Fig. 1). At the time this area of the sea floor was considered to be featureless and of no great morphological interest.

Subsequent surveying revealed a group of topographical features, all of which were shown to be magnetic and thus of probable volcanic origin. A series of measurements involving acoustics, magnetics, dredging and chemical analysis finally established at least two of the peaks as being currently active, the most active being Rumble III. [Kibblewhite and Denham, 1967a; Kibblewhite, 1967].

While underwater volcanism had been observed before [Dietz and Sheehy, 1959; Snodgrass and Richards, 1956; Richards, 1963] this study probably represents the first example of the identification and location of a previously unknown center of underwater volcanism by means of underwater sound. Experience from this study has aided the identification of other underwater volcanism. For instance observations made at a position south of Kandavu Island (Fiji) pointed to a volcano of considerable acoustic output on the Norfolk Island Ridge [Kibblewhite, 1966]. Later, on the basis of the acoustic characteristics of Rumble III, [Norris and Johnson, 1969] reinterpreted signals received at Hawaii and originally thought to be T-phases, as originating from volcanic sources spread throughout the Pacific Ocean. The most distant source, predicted to be on the Tuamotu Archipelago, in the vicinity of Rapa Island, was subsequently confirmed by a survey in the area [Johnson, 1970]. In this case the propagation path was some 4800 km.

T-Phase Signals

While T-phase signals have been recorded on hydrophones in New Zealand waters, nothing comparable to the studies carried out in the North Pacific has ever been undertaken. Any paper, however, which deals with the interaction of underwater acoustics and marine geophysics would be incomplete without reference to work on T-phase recording. In this regard the study by Johnson and Norris [1968] deserves particular mention. Besides providing data on SOFAR channel sound-velocity structure in the Pacific Ocean, their study of T-phases has contributed significantly to the general field of seismology.

Microseisms

Besides the intermittent high intensity seismic and volcanic signals which occur from time to time, the general microseismic activity of the earth, much of which has its origin in sea-surface oscillations, must contribute to the general ambient noise in the oceans. Urick [1967] estimates, in fact, that, in the dominant frequency range in which microseisms occur, the pressure amplitude is probably of the order of 120 dB above 1 μPa. Storm forecasting on the basis of microseismic activity induced in the ocean floor by sea-swell can reasonably be regarded as another area in which underwater acoustics impinges on other aspects of marine geophysics.

Underwater Acoustics and Geothermal Activity

One novel intended application of sub-aqueous acoustics in New Zealand will involve the monitoring of changes of phase. There is a hot lake in the Inferno Crater at Waimangu (one of the New Zealand thermal areas), which undergoes comparatively rapid fluctuations in level, at times reaching 40 feet. It is believed that this fluctuation is caused by the development and condensation of steam in the pipe beneath the lake (Pers. comm.: R. F. Keam and E. F. Loyd). At one stage in the fluctuation rapid condensation is believed to occur and it is proposed to test this hypothesis by setting up hydrophones to detect the postulated implosions of condensing steam.

II. GEODESY, SEA-FLOOR MAPPING, AND UNDERWATER ACOUSTICS

The fifth ship in the OPERATION CHASE program was detonated on 24 May 1966 off Cape Mendocino on the west coast of the United States. The New Zealand Defence Scientific Establishment was among the organisations advised of the explosion in advance, and the underwater signals were recorded in New Zealand (Fig. 2) [Kibblewhite and Denham, 1969].

Two separate hydrophone systems were used to record the explosion--one at Great Barrier Island and the second a temporary installation laid off Mayor Island. In spite of the extensive topography in the mid-Pacific Ocean between Cape Mendocino and New Zealand, signals were received at both stations (Fig. 3). Two aspects of the received signals are of interest in the present context: (1) the geodetic implications of the total travel time of the signal, and (2) the topographic mapping made possible by the reverberation produced by reflectors throughout the Pacific Ocean.

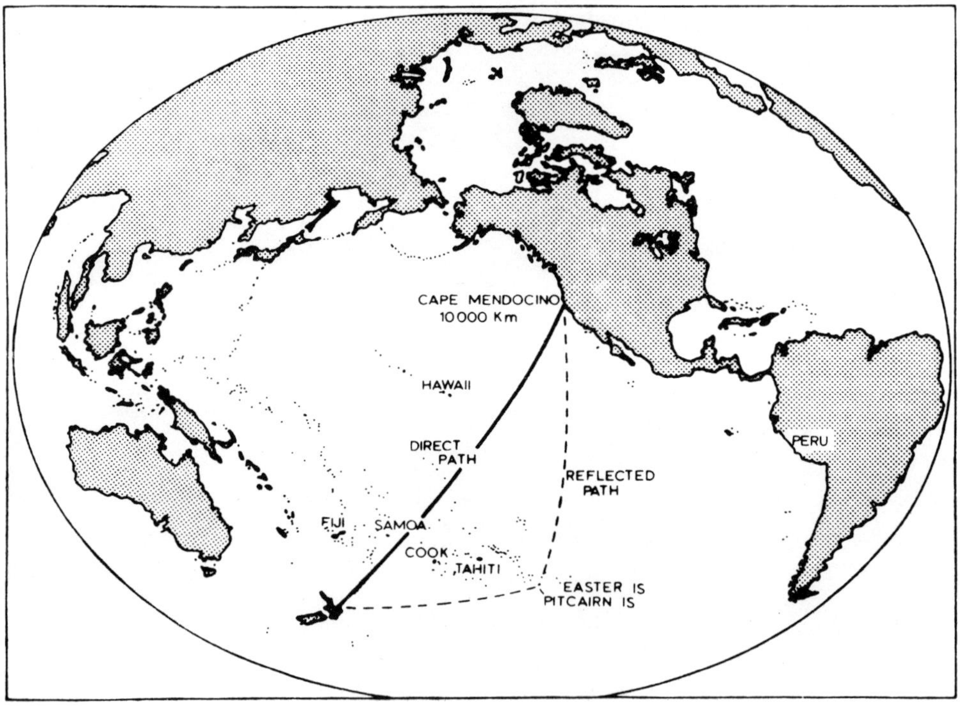

FIGURE 2

THE CHASE V TRANSMISSION PATHS
FOR THE NEW ZEALAND RECORDING STATIONS

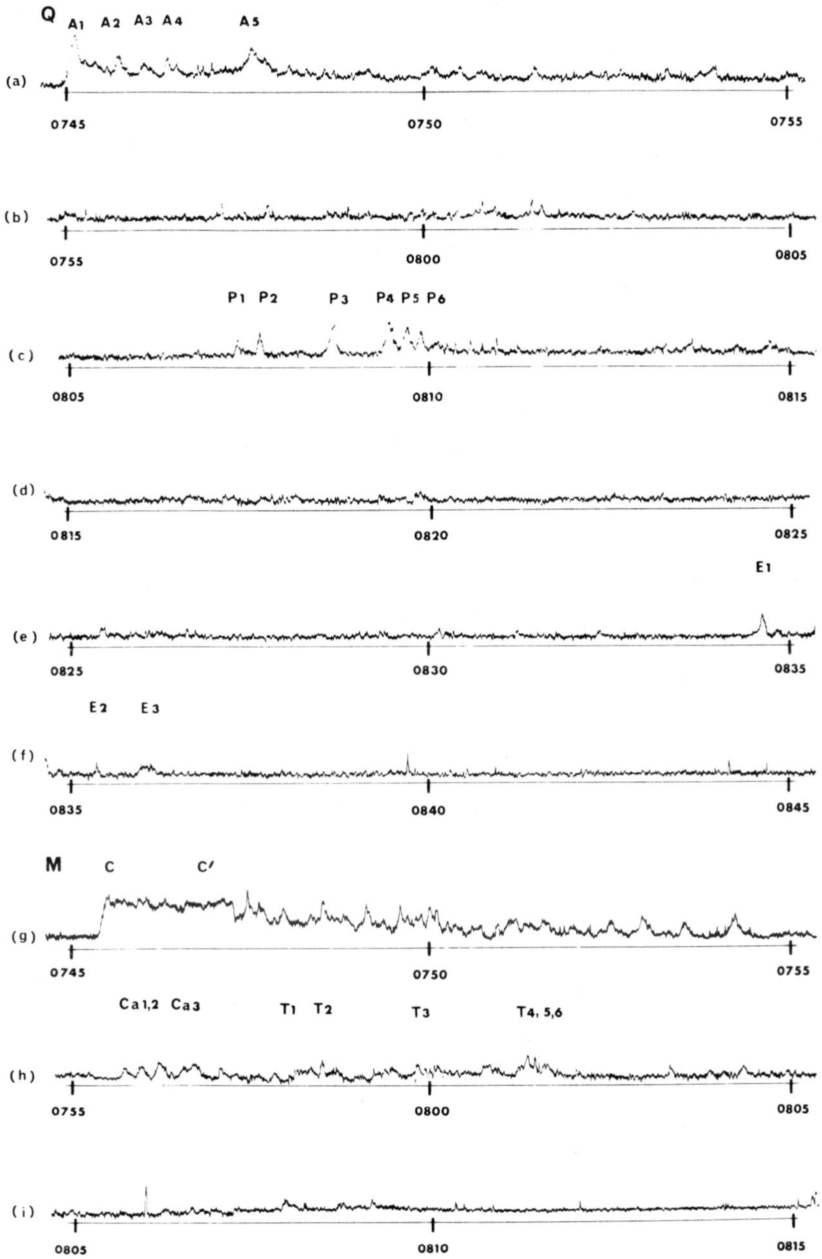

FIGURE 3

PRESSURE LEVEL RECORDS FOR THE 25 Hz OCTAVE BAND OF THE ACOUSTIC SIGNALS FROM THE CHASE V EXPLOSION, RECEIVED BY THE HYDROPHONES AT GREAT BARRIER ISLAND (Q) AND MAYOR ISLAND (M).

III. TRAVEL-TIME MEASUREMENT AND ITS GEODETIC IMPLICATIONS

A salient feature of the experimental arrangement for the CHASE V explosion was the accuracy with which the detonation and arrival times and detonation and receiver positions could be determined. With these data likely to be available, the possibility existed of examining the nature of the propagation path and the propagation mechanism, over a distance of some 10,000 km. To this end therefore the positions of the New Zealand recording stations were determined as precisely as possible in terms of the national grid. In the final analysis, it was concluded that the geographical positions of the individual hydrophones were known to an accuracy of ±10 m in the case of the Great Barrier Island installation and ±100 m in the case of the Mayor Island station. With these positions known, the true geodetic distance from each hydrophone to the detonation point off Cape Mendocino was calculated.

The nature of the signal received in New Zealand introduced some uncertainty as to arrival times, but taking all ambiguities into account, the arrival times measured were considered accurate to ±1 sec at Great Barrier Island and ±0.5 sec at Mayor Island.

With the length of the geodesic and signal travel time available, it is possible to evaluate a mean sound-velocity for this path. It was considered more instructive, however, to assume rectilinear propagation and use available sound-velocity data to calculate a travel-time over this path by numerical integration along the geodesic. Such a procedure was followed in an attempt to test the proposal originally put forward by Ewing, Worzel, and Talwani [1959] that SOFAR signals could be used to interconnect the geodetic networks of land masses separated by large oceanic distances.

Neglecting the uncertainty in the source and hydrophone positions, the calculated values of the geodetic arc lengths for the Great Barrier Island and Mayor Island distances were estimated to have a computed accuracy of ±1 m. These distances led to travel times of 6950.0 and 6978.5 sec respectively, which are in both cases less than the measured values of 6961.0 and 6983.8 sec. The difference is of the order of 11 sec for Great Barrier Island and 5 sec for Mayor Island.

There were several pieces of evidence to indicate that the arrival at Great Barrier Island was not direct, but resulted from a reflecting surface close to the direct path. The presence of a small difference between the calculated and observed travel times is thus not surprising for this station. The Mayor Island discrepancy on the other hand was not due to either of these causes, and it was proposed that the 5.3 sec difference in travel time (or 8 km difference in path length) could have arisen from a combination of effects, viz:

i. an error in the detonation position
ii. an error in the measured arrival time
iii. a systematic error in the sound-velocity profiles
iv. horizontal refraction, due to variations in the channel velocity along the path, adding to the path length.

With the data available it was not possible to assess accurately the relative significance of these effects. The angle of arrival of the signal at Mayor Island confirmed however that sound transmission over the path was very close to rectilinear and that little horizontal refraction had occurred.

As the result of recent work it is now believed that the principal error arises from the influence of the sound-velocity profile on the shape of any SOFAR signal. In many cases this effect would set a limit, not foreseen by Ewing et al., to the precision of any geodetic measurement made over long distances. Reference to this effect will be made later. While an error is clearly present, perhaps it is more surprising that it is as small as it turned out to be.

Topographic Mapping

Topographic reflections were recorded for nearly 1 hour after the main arrival (Fig. 3). Many sections of the reverberation signal could be identified with known topographic features in the South Pacific; others gave evidence of features which are as yet uncharted. In discussing this aspect of the interaction of acoustics and marine geophysics attention will be restricted to the spectacular series of discrete arrivals, P1 to P6 (the largest of which rose some 25 dB above the background reverberation level), the group E1 to E3, and the section between these.

There was no difficulty in establishing that the reflections P1 to P6 arose from the general area of Pitcairn Island. A detailed analysis, in which travel-time and azimuth data were examined, identified P1, P3, and P6 with Oeno, Pitcairn, and Henderson Islands respectively, and P2 and P4 with two charted seamounts [Kibblewhite and Denham, 1971(a)]. Available charts do not indicate any feature which could account for reflection P5, and the implication is that one as yet uncharted seamount, comparable in size to Pitcairn Island, exists at position Lat. 24° 21'S, Long. 128° 34'W (Fig. 4).

Reflections E1 and E3 arose from Easter Island and Sala-y-Gomez Island, respectively, E2 being connected with a minor feature between the two. The reverberation level between Ducie Island and Easter Island indicates that the reflecting zones in

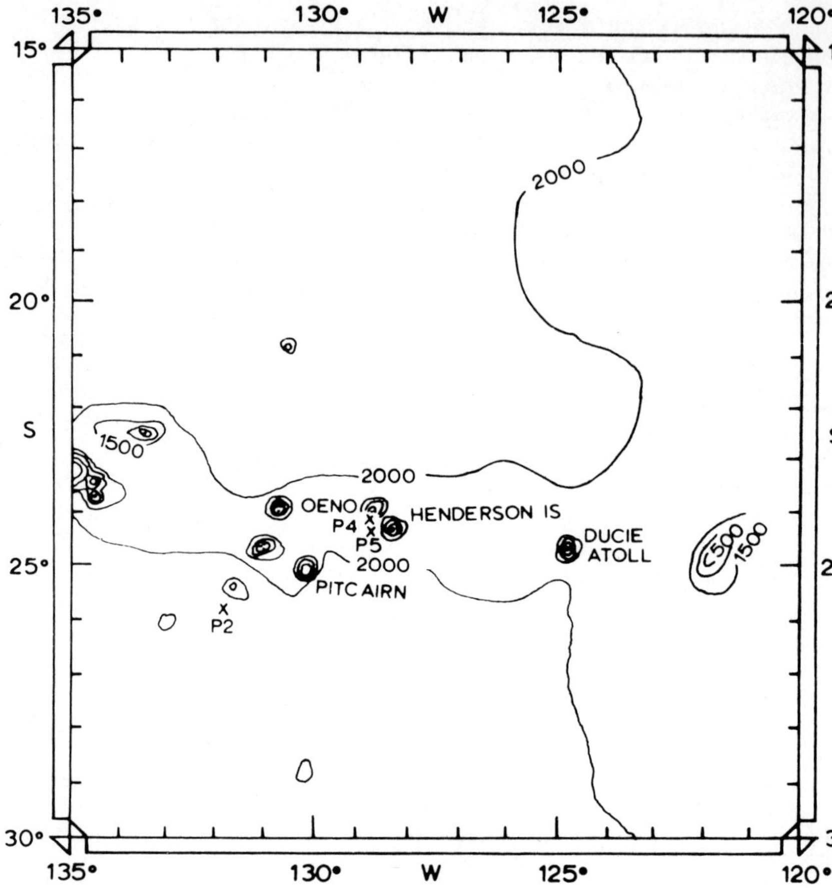

FIGURE 4

THE TUAMOTU RIDGE TOPOGRAPHY IN THE VICINITY OF PITCAIRN ISLAND

this area are comparatively minor topographic features. Use was made of this information in selecting the transmission path for a recent long range propagation experiment between New Zealand and South America, PROJECT KIWI [Bannister et al., 1973].

Various authors [Luskin et al., 1952; Northrop, 1968] have drawn attention to the possibility of mapping underwater topography by means of large underwater explosions. The CHASE V experiment confirmed this possibility, but also demonstrated the remarkable effectiveness of the technique given the right geometry for the experiment. From a distance of nearly 5500 km from the Tuamotu Ridge it was possible to resolve discrete features which were separated by distances of only 40 km, and establish their absolute geographic positions to an accuracy of better than ±20 km.

IV. UNDERWATER ACOUSTICS AND PHYSICAL OCEANOGRAPHY

Because of the influence of the physical properties of sea water on underwater sound, the acoustician relies heavily on the physical oceanographer for the appropriate environmental data. There are, however, situations in which the physical oceanographer can be aided in turn by the acoustician, the measurement of sea-floor bathymetry, surface wave height, internal waves and deep ocean currents providing well known examples. Another application revealed by studies in the Southern Hemisphere concerns the identification of the position of major hydrological boundaries.

In the course of the South Tasman Sea experiment referred to earlier [Kibblewhite and Denham, 1967b], the propagation path crossed the Subtropical Convergence. It was found that a marked change in the transmission characteristics occurred at a range which coincided with one defined position of this hydrological boundary (Fig. 5). Convergence zones which were indistinct south of this feature became much more prominent north of it and were spaced at the 32-33 mile range interval expected. Furthermore, the absolute levels of the received signals decreased by a small though discernible amount, and the attenuations experienced on the two sides of the "boundary" were noticeably different, those in the water mass north of the "boundary" being smaller.

Similar results were observed later in transmission experiments to the east of New Zealand [Kibblewhite and Denham, 1971(b)]. Once again, changes in signal level and in attenuation were observed at the approximate position of the Subtropical Convergence. The attenuations south of the boundary were comparable to those observed south of the boundary in the Tasman Sea, while those north of it agreed reasonably with values obtained by others in the Pacific and Atlantic Oceans.

Another more recent experiment, PROJECT KIWI, has confirmed this effect very dramatically [Bannister et al., 1973]. Along a great circle transmission path between the East Cape of New Zealand and the Galápagos Islands distinct regions were identifiable in terms of the attenuation values observed (Fig. 6). On this occasion the signals were recorded at both ends of the transmission path.

These results suggest that a suitably designed propagation experiment can be used as a sensitive method of identification of significant hydrological boundaries, and emphasise, moreover, that the attenuation of sound at very low frequencies is dependent on the oceanic water in which the measurements are carried out.

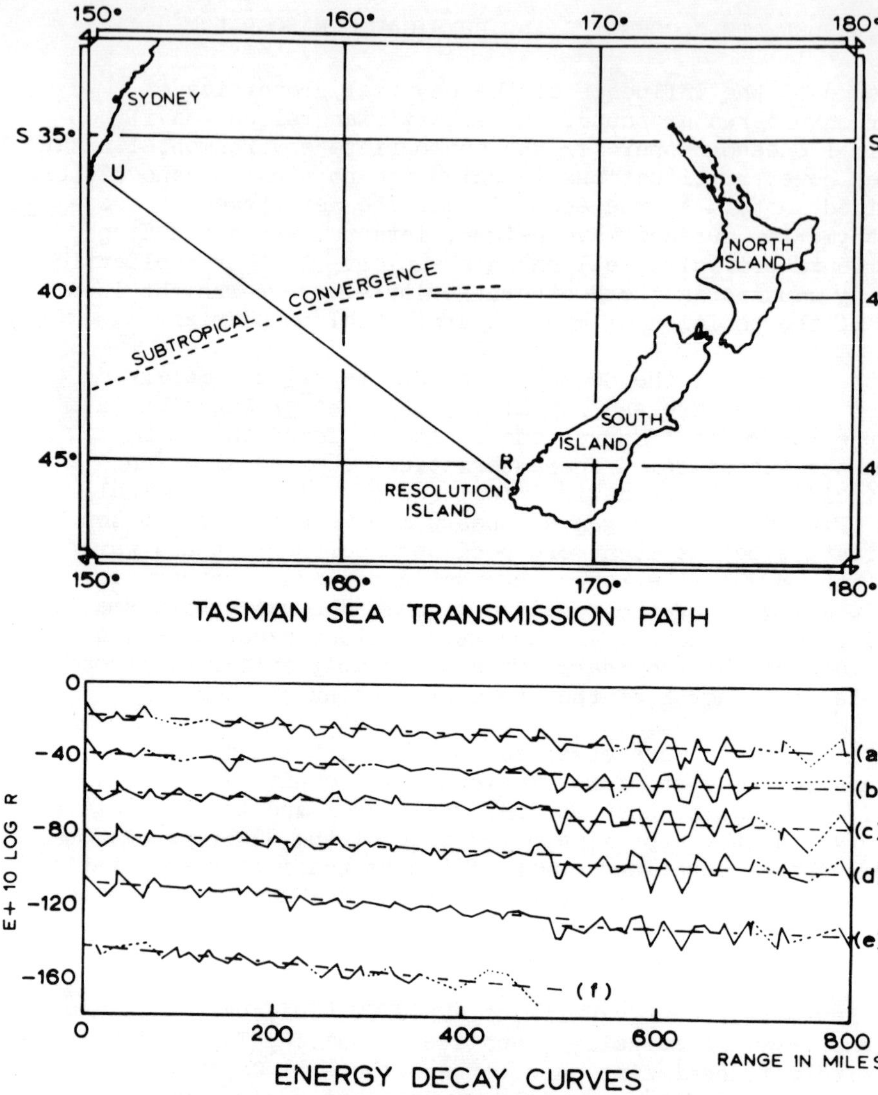

FIGURE 5

THE TRANSMISSION PATH AND ENERGY DECAY CURVES OF THE TASMAN SEA EXPERIMENT. THE ENERGY DECAY CURVES (CORRECTED FOR CYLINDRICAL SPREADING) FOR THE OCTAVES (a) 16-35 Hz, (b) 35-75 Hz, (c) 75-150 Hz, (d) 150-300 Hz, (e) 300-600 Hz, (f) 600-1200 Hz, SHOW A CHANGE IN TRANSMISSION OCCURS AT THE APPROXIMATE POSITION OF THE SUBTROPICAL CONVERGENCE.

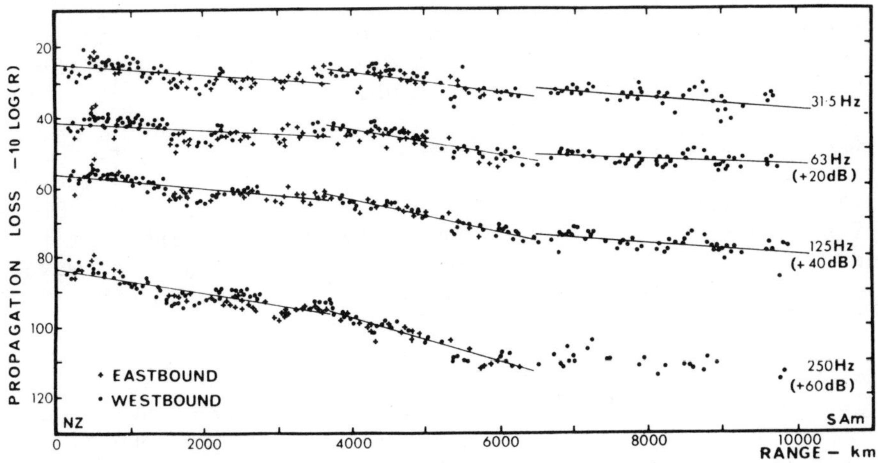

FIGURE 6

THE TRANSMISSION LOSS CURVES RECORDED IN THE EXPERIMENT, PROJECT KIWI, IN WHICH RECORDING TOOK PLACE AT BOTH ENDS OF THE PATH. THREE DISTINCT HYDROLOGICAL REGIONS ARE APPARENT.

Marine Seismics

Any paper dealing with the interaction of acoustics and marine physics would be incomplete without a reference to marine seismics. However, as the emphasis in this paper is intentionally directed towards other features of New Zealand work, only brief reference will be made to this area of activity.

The first marine refraction study in New Zealand was carried out by C. B. Officer [1959]. Since this time commercial oil interests have surveyed extensively on the Continental Shelf. Probably the most extensive scientific activity has centered on the ELTANIN with a series of reflection profiles surveyed around the New Zealand Continental Shelf by R. Houtz and others of Lamont-Doherty Geological Observatory.

Hochstein [1967] made some refraction measurements in the southwest Pacific. These were carried out in the vicinity of the Cook Islands (as part of the New Zealand Eclipse Expedition), to investigate the structure of some of the volcanic islands. A preliminary interpretation of gravity measurements had shown that the steep gravity gradients observed near the reef edge on most of these islands cannot be explained by the assumption that the islands have a uniform density, and it had been postulated that a core of denser material exists. It was the aim of the seismic studies to check the core hypothesis, and also to investigate the structure of the crust under these islands.

In the course of this cruise Hochstein [1970] also carried out seismic refraction measurements in Suva Harbour (Fiji). These revealed a mud layer between 50 and 150 m thick covering the bottom of the harbour entrance and the inner harbour. The most significant feature of this survey was that the compressional wave velocity in the mud was shown to be unusually low, approaching 1 1/2 times the sound velocity in air in some locations.

More recently the Geophysics Division of the Department of Scientific and Industrial Research and the University of Auckland have begun a series of investigations on the Continental Shelf. Deep sea coastal studies are also planned.

At the other end of the scale, the University of Auckland has also recently carried out a reflection survey for the Auckland Harbour Board, using a coherent source. This investigation involved about 200 miles of surveying. It was unusual in that the area of interest was only submerged at high tide and that an extensive series of boreholes was available for correlation of seismic echoes and subsurface strata. Useful results were obtained, even though the water depth rarely exceeded 3 m.

V. SOFAR SIGNALS--FACTORS CONTROLLING THEIR SHAPE

The Classical SOFAR Signal

It is well known that the deep ocean phenomenon of SOFAR propagation occurs as a consequence of a minimum in oceanic sound velocity profiles. This minimum, which defines the depth of the "SOFAR axis", permits the transmission of acoustic signals to great distances. Signals received at long ranges from an explosive source detonated in this "SOFAR channel" generally consist of many successive pulse arrivals culminating in a high peak pressure followed by a sharp cutoff. This composite arrival has become known as the classic SOFAR signal.

SOFAR signals recorded in New Zealand waters to date have not exhibited this classic shape, but are characterised rather by a "triangular" and "inverted SOFAR" shape (Fig. 7). In most

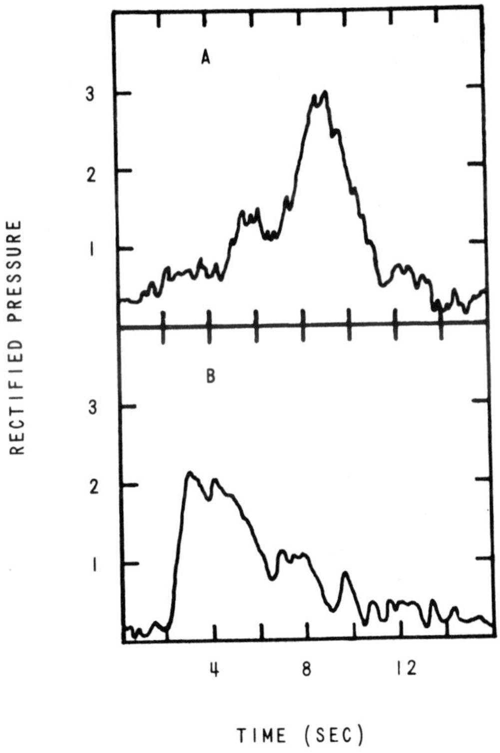

FIGURE 7

SOFAR SIGNALS RECORDED ON THE EAST COAST OF NEW ZEALAND.
(A) TRIANGULAR SHAPE; SOURCE, NORTH PACIFIC; RANGE 7650 km.
(B) INVERTED SHAPE; SOURCE, SOUTH PACIFIC; RANGE 1530 km.

experiments the receiving hydrophones have been positioned on the edge of the Continental Shelf and not in the SOFAR channel, so that the "tail" evident in each signal is almost certainly due to reverberation produced through interaction with the sea-floor near the hydrophone. The source of the trianglular signal depicted in Fig. 7 was located in the North Pacific and the propagation path was characterized by velocity-profiles similar to (A) in Fig. 8. Therefore, the signal is interpreted as being of classic shape but with reverberation obscuring the sharp cutoff. The propagation path of the inverted signal lay to the east of the hydrophone at a latitude of about 40°S and was characterised by profiles similar to (B) and (C). Although reverberation again obscures the latter part of this signal, the abrupt start in this instance indicates that, in contrast to the triangular signal, the main pressure peak has arrived first. It appears therefore that these two signals would also have been of quite different shape had they been recorded in their respective deep channels, where the propagation is governed solely by the velocity profiles.

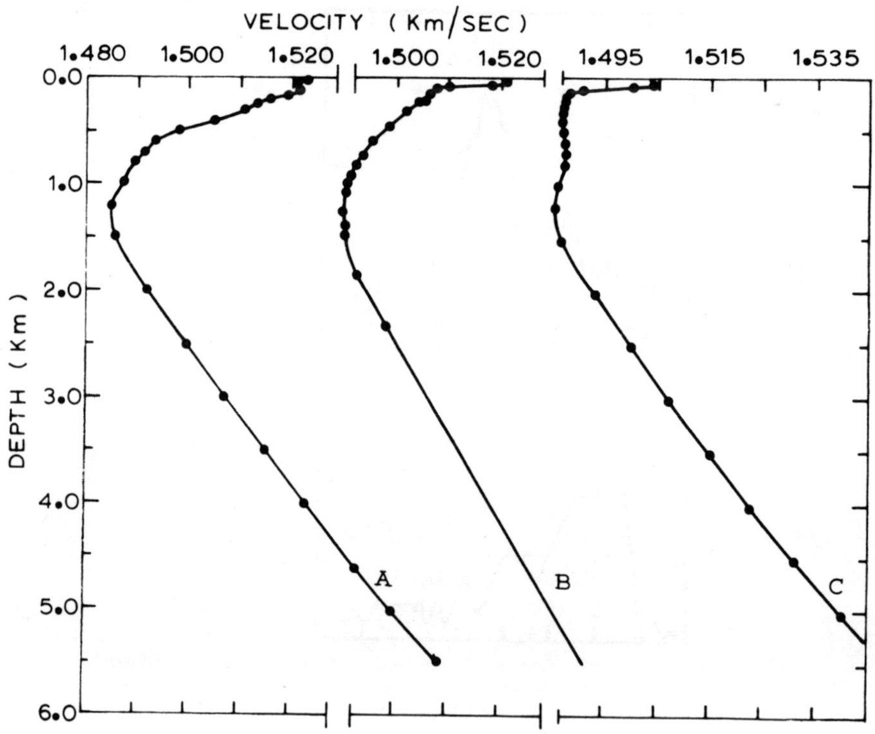

FIGURE 8

CHARACTERISTIC VELOCITY PROFILES CONSIDERED BY GUTHRIE.
(A) THE "BILINEAR" PROFILE; (B) THE "COSH" PROFILE;
(C) THE "RECTANGULAR" PROFILE.

A Ph.D. student at the University of Auckland is examining the influence of the sound velocity and bathymetric profile on SOFAR signal shape as a research project. He has used wave theory to generate theoretical SOFAR signals from the normal modes of a given channel, and has shown that the shape of these signals is indeed sensitive to the velocity profile [Guthrie, 1974]. Guthrie obtains the modes without resort to analytic approximation by solution of the wave equation using experimental velocity profiles which are interpolated in a smooth fashion. He also generates signals closely resembling those of wave theory by using the ray approximation.

In his examination Guthrie uses the three typical profiles shown in Fig. 8. The first (A), recorded at latitude 28°S in winter, is fairly representative of the class of warm water profiles which can be roughly classified as bilinear. Closer to the equator such profiles have a sharper minimum. The second (B) typifies those with a broader minimum. It was recorded at latitude 39°S in summer and, to emphasise its major difference from profile A, has been classified as a "cosh" type. The third (C) has been classified "rectangular" in view of the close approximation to isovelocity conditions above the axis. This particular profile was recorded in the autumn at latitude 43°S. The locations of profiles B and C lie near to the propagation path over which the unusual SOFAR signal discussed earlier was observed.

Figures 9 a, b, and c are a comparative study of SOFAR signal shapes as a function of hydrophone depth, calculated for the three profiles A, B, and C respectively. In each case the range is 1500 km, the source is on the axis, the center frequency is 25 Hz and the pulse length, τ, is taken as 0.1 sec. Mode arrivals are marked along the time axis of the lowest signal.

It is apparent that for a wide range of hydrophone depths about the axis, the classic SOFAR signal is obtained for profile A. Both signal-peak and duration grow and decay as the hydrophone moves from near the surface, through the axis (where the slow, low order modes contribute) into deeper water.

In the case of profile B, the classic shape is observed over a more restricted depth range about the axis, and the triangular shape for the deepest and shallowest positions is quite unlike those for profile A. The signal peak again grows near the axis, but more striking is the 50% drop in duration (compared with profile A), caused by the lower profile gradients above and below the axis.

The most obvious feature of the signals produced by profile C is that they are double peaked and display no classic SOFAR shape whatever. The first peak is the true SOFAR peak and arises due to

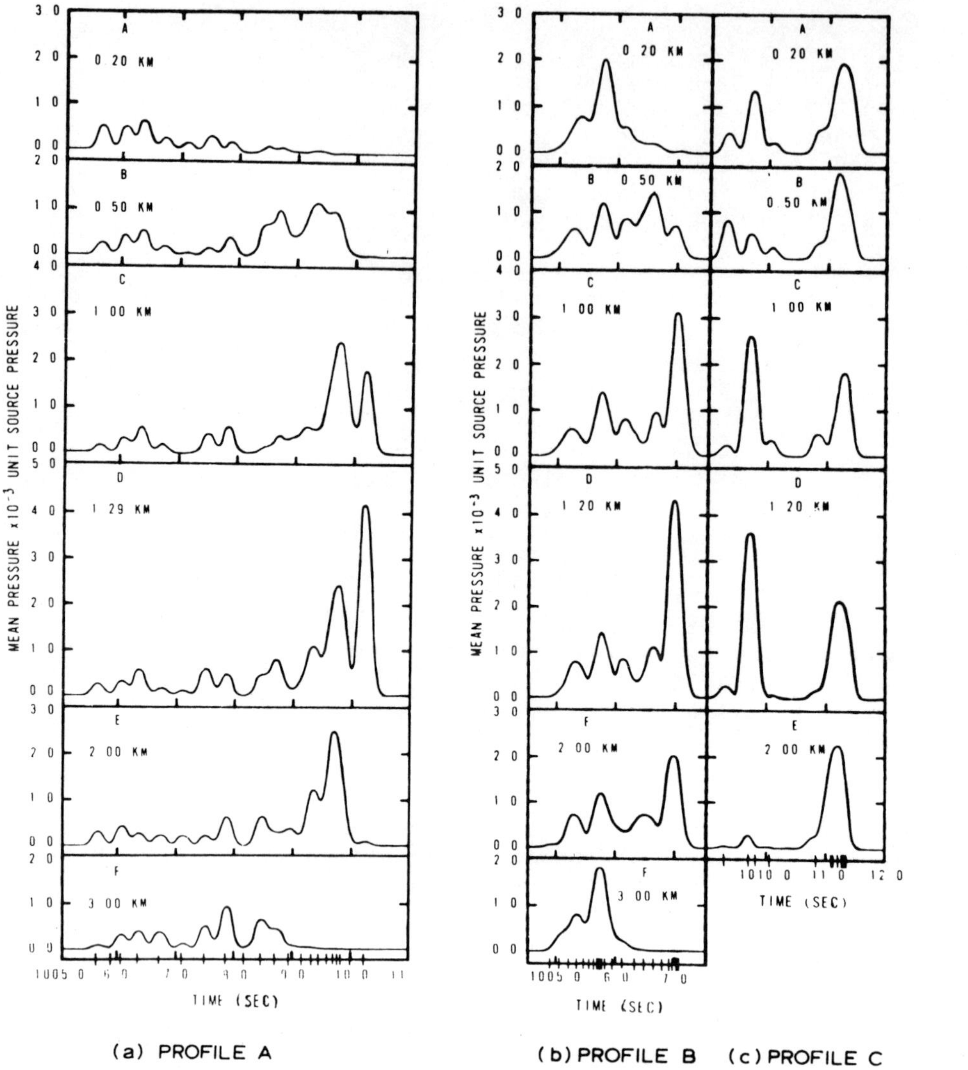

FIGURE 9

SYNTHETIC SOFAR SIGNALS DETERMINED FOR THE SOUND VELOCITY PROFILES A, B, AND C, SHOWN IN FIG. 8. IN EACH CASE THE FREQUENCY CONSIDERED IS 100 Hz, THE SOURCE AND HYDROPHONE ARE AT AXIS DEPTH, AND THE RANGE IS 1500 km.

energy trapped close to the axis, while the second peak is caused by a collection of high order modes travelling at similar group velocity. Again the growth of the SOFAR peak in the vicinity of the axis is very evident, as is the short duration of the signals when compared with those of profile A.

In a general ray picture these synthetic SOFAR signal-shapes can be qualitatively understood. We expect most of the energy to be trapped close to the channel-axis and since the axial rays are slowest in profiles with overall curvature greater than a cosh profile (of which type bilinear profiles are the extreme example), such profiles will produce the classic SOFAR signal. In profiles more rectangular than a cosh profile, the axial rays arrive first, having a shorter distance to travel at approximately the same velocity, and an inverted SOFAR signal results. This effect is apparent in profile C where the early peak is due to the low order modes (which correspond to near axial rays) and, furthermore, with increasing mode number (corresponding to initial ray angle) group velocities decrease. However, for sufficiently large mode number the linear section of the profile below the axis dominates the rectangular section above, and group velocities begin to increase for each subsequent mode. Finally, it can be seen that an exact cosh profile neatly divides these types of profile; all rays travel at the same group velocity with exact focusing at regular points along the axis. A signal received on the axis would therefore be a spike at discrete ranges.

Signal Shape and Sound Ranging

The variation of SOFAR signal shape with velocity-profile has a number of implications for long range sound transmission.

The most obvious is the implication for sound ranging and geodetic surveying. Estimates of range based on the maximum of the received signal and the "axis" sound-velocity will depend upon the shape of the profile. In the examples quoted, for effectively the same axial velocity profiles B and C produced travel times for the peak signal of 1007.070 and 1009.292 sec, respectively. In the extreme case of the double peak in profile C, the 1.5 sec between peaks at 1500 km give a range uncertainty of 2.4 km unless the peak arising from energy confined near the axis can be identified by an analysis similar to that presented by Guthrie. In the event of the axial rays being intercepted by some bottom feature extending into the SOFAR channel a significant error would result.

In the light of this evidence one must ask whether the most significant factor involved in producing the difference between the measured and calculated travel-times for the CHASE V signal (approximately 5 sec) was not the effect described by Guthrie.

The order of magnitude of the effect could well have produced this discrepancy over a path length of 10,000 km. With the actual sound-path involving all three profiles, it is a complicated matter to answer this question with certainty, but Guthrie is planning to look at the problem of combining the effects of different profiles over long oceanic paths. However, even if such a theoretical analysis proves feasible, the dynamics of the SOFAR channel and the problem of sampling the velocity-profile with sufficient density would appear to discount earlier claims [Ewing et al., 1959] that geodetic measurements based on SOFAR signals could be made to an accuracy of 50 m over a range of 1600 km.

Calculations of SOFAR Axis Velocities

In a directly equivalent way, estimates of mean SOFAR velocity based on SOFAR arrivals must be treated with caution. Many estimates of Pacific SOFAR velocities produced by Johnson and Norris [1968], based on T-phase arrivals, will not have taken such effects into account. In view of the varying depths involved with seismic sources the errors could be significant.

Classification of Signals

Any deep water hydrophone will detect many large transient signals in the course of a 24 hour period. When working at low frequencies, events originating in the North Pacific or indeed the Atlantic Ocean can register in the New Zealand area, and vice versa. Such events can confuse experiments in progress in either area. Because of the frequency of occurrence of such extraneous signals, precise timing control of "true" signals is required. When using aircraft to drop explosive sources from 10,000 ft, the time of detonation is not always easy to establish with the precision necessary to differentiate the true event when two signals are received close together. In such situations, and others, classification can presumably be aided by a study of SOFAR signal shapes.

VI. THE INFLUENCE OF THE SEA BED ON THE TRANSMISSION OF SOUND IN THE DEEP OCEAN

The Attenuation Curve for Seawater

It is generally accepted that the attenuation of sound in the ocean at frequencies greater than 100 Hz follows an expression of the form

$$\alpha = \frac{0.1f^2}{1+f^2} + \frac{40f^2}{4100+f^2} + 2.8 \times 10^{-4} f^2 \quad \text{dB/kyd}$$

where f is in kHz.

Recent work [Fisher et al., 1973] suggests that the first term is due to a boric acid-borate system relaxation; the second arises from the magnesium sulphate relaxation absorption; and the third from viscosity [Schulkin and Marsh, 1962].

Below 100 Hz the situation is less certain. Various experiments [Urick, 1966; Kibblewhite and Denham, 1967(b); 1971(b)] have indicated a tendency for the α vs f curve to level off at $\alpha \approx 0.8 \times 10^{-3}$ dB/kyd and increase for frequencies below 30 Hz, but the existence of excess attenuation and the reality of such a minimum has been discounted by others [Berman and Guthrie, 1972]. Whatever one's conviction, it must be agreed by all that the experimental difficulties at these frequencies, the assumption of simple cylindrical spreading in deriving the "attenuation" values, and the scatter in the experimental data obtained produce uncertainty in this part of the spectrum.

As we have seen, Guthrie [1973] has shown that the time dependence of propagation within the SOFAR channel is very dependent upon the form of the velocity-profile which defines the channel. It is pertinent to ask what effect, if any, the velocity-profile has on the attenuation in the channel. Specifically, does a channel with a well defined minimum give less attenuation than one with a broad minimum.

A related question concerns how and at what frequencies do the channel boundaries (sea-surface and sea-bed) manifest themselves. Throughout this Conference we have been concerned in one way or another with the interaction of the sea-floor and acoustic signals, with the emphasis arising from active sonar usage. It is well known that the sea-floor is also important in shallow-water low-frequency transmission but in deep water its influence on low-frequency transmission is less well understood.

Theoretical Attenuation Coefficients

Guthrie (Pers. comm.) is extending his work to examine the preceding questions. He has been able to calculate the attenuation coefficients for the various modes involved using an approach based on quantum mechanical perturbation theory. He has examined, specifically, a typical oceanic velocity-profile, where the pertinent parameters have the following values:

Channel depth: $H = 5.0$ km

Attenuation coefficients:

$$\alpha_{water} = 1.1 \times 10^{-7} f^2 \text{ dB/km (f in Hz)}$$

$$\alpha_{bottom} = \text{constant}$$

Density contrast: $\rho_b/\rho_w = 1.5$

Velocity ratio: $c_b/c_w = 1.05$

Some of his results are summarised in Fig. 10.

At a sufficiently high frequency any mode is trapped by the velocity-profile alone and therefore is essentially contained within the water column. Guthrie's theoretical analysis shows that in this case, as expected, the mode attenuation is just the water attenuation, $\alpha_w = 1.1 \times 10^{-7} f^2$ dB/km.

As the frequency decreases the attenuation coefficient follows α_w until the velocity-profile is no longer capable of trapping the mode completely. The mode now penetrates into the channel bottom and the attenuation coefficient rises rapidly as the proportion of acoustic energy travelling in the bottom increases.

Guthrie has shown that the lowest modes carry the bulk of the SOFAR energy for sources near the axis. Hence any excess attenuation process will manifest itself in a typical SOFAR transmission experiment only if these lowest modes are affected. In the case of bottom leakage this can only occur by direct interaction of the low order modes with the bottom or by indirect coupling (through horizontal velocity gradients) to higher order modes which are leaking energy into the bottom. However, as Guthrie points out, the experimental observation of the classic SOFAR signal, in which most of the energy arrives in a single peak, indicates that there is little coupling of this sort and that modes do tend to retain their identity as they propagate.

Guthrie's analysis of the theoretical attenuation coefficient for the lowest mode, with $\alpha_b = 10^{-5}$ dB/km (although low, this attenuation is 100 times the extrapolated water column attenuation at 1 Hz), shows that there is no evidence of bottom loss above

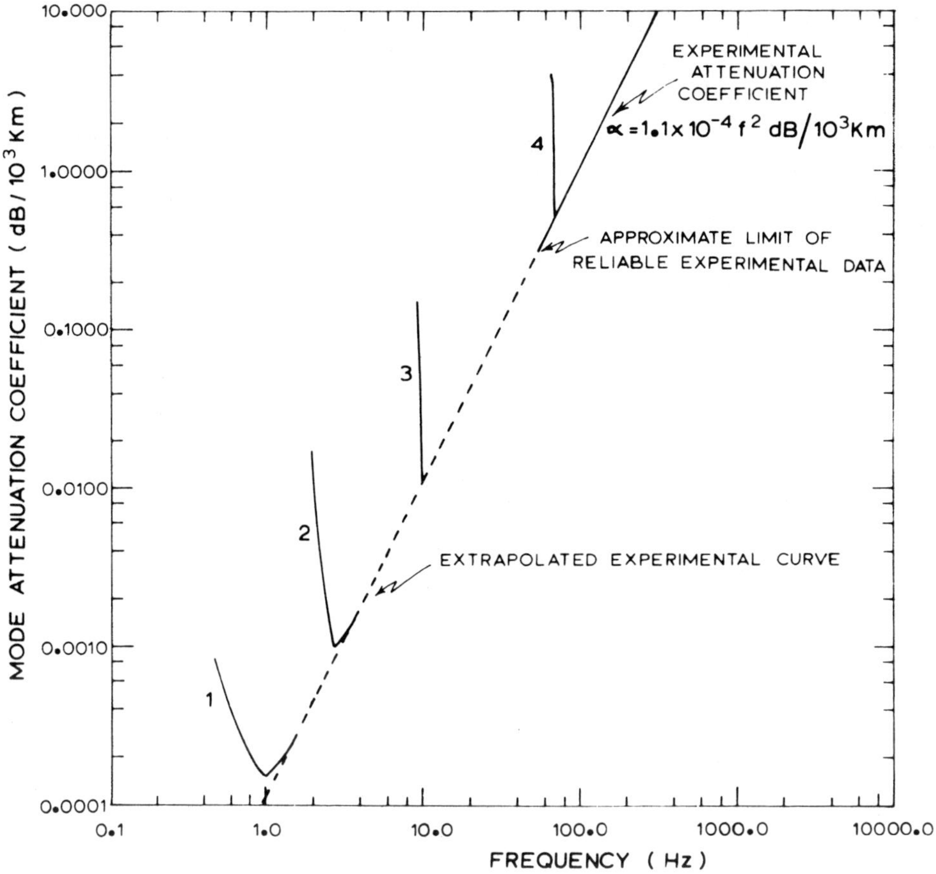

FIGURE 10

THEORETICAL ATTENUATION COEFFICIENTS SHOWING THE EFFECT OF ATTENUATION IN THE CHANNEL BOTTOM. α_w IS TAKEN TO BE 10^{-4} f^2 dB/10^3 km and α_b AS INDICATED.

CURVE 1: LOWEST MODE; $\alpha_b = 10^{-2}$ dB/10^3 km

CURVE 2: LOWEST MODE; $\alpha_b = 10^3$ dB/10^3 km

CURVE 3: 10th MODE ; $\alpha_b = 10^3$ dB/10^3 km

CURVE 4: 80th MODE ; $\alpha_b = 10^3$ dB/10^3 km

THE CHANNEL DEPTH = 5.0 km. A TYPICAL OCEANIC SOUND VELOCITY PROFILE IS USED BUT THE FORM OF THE CURVES IS ESSENTIALLY INDEPENDENT OF THE PROFILE CONSIDERED.

1.5 Hz and that only below 1 Hz does the attenuation increase rapidly (Fig. 10). The resulting value of α at 1 Hz is well below α_b because the fraction of the mode propagating in the bottom is very small. Even for an extremely strong α_b (curve 2), there is no bottom loss for the lowest mode above about 3.0 Hz in such a 5.0 km channel. This is the lowest frequency at which the mode is trapped by the profile alone.

Higher modes extend over greater depths and therefore suffer bottom loss at higher frequencies. However above 10 Hz there are sufficient modes (all propagating with $\alpha_n \approx \alpha_w$, where n is the mode number) to give the time dispersion typical of SOFAR propagation. The analysis shows further that at least 10 modes are present at 10 Hz (curve 3) and that well in excess of 100 modes are present at 100 kHz (curve 4).

Finally, we observe that these results are essentially independent of the velocity-profile used since the ocean structure is very uniform below the axis. It thus transpires that the important parameter governing the onset of bottom absorption is the channel depth, which, at any given frequency, must be sufficiently shallow for the low modes to sense its presence.

To summarise, Guthrie has shown:

i. In acoustic propagation over a typical deep (5 km) ocean path, bottom loss becomes significant only in the region of 1 to 3 Hz, i.e., at an order of magnitude below frequencies currently of interest.

ii. Only when the channel depth is reduced to about 2.0 km do losses become significant in the 10 to 100 Hz region.

iii. The velocity-profile is not critical, except that for a shallow axis the bottom depth would have to be correspondingly shallower before losses become significant. A tentative "rule of thumb" is given as: if the axis to bottom depth exceeds 1.0 km, bottom loss will not be significant above 10 Hz.

The Excess Attenuation below 100 Hz

We are thus left with the following questions, both of which are very significant from an operational point of view:

i. Are the high "attenuations" observed below 100 Hz real or a consequence of the experimental procedure?

ii. If real, to what effect in the water column can the attenuation be attributed?

The apparent lack of influence of the marine floor on the

transmission of low frequency sound over long ranges in deep water seems to pose a new problem for the acoustician.

Attenuation and the Velocity-Profile

While Guthrie's analysis has shown that the velocity-profile has very little effect on bottom loss, it can be expected that profile shape, which determines how near the surface the low-modes lie and how well they are confined to the axis region, will play an important role in determining the effect of a surface attenuating layer. For example, profiles with a shallow or very broad minimum will enable even the lowest modes to have relatively large amplitudes even near the surface. There is experimental evidence [Kibblewhite and Denham, 1967(b), 1971(b); Bannister et al., 1973] to suggest that the attenuation in the Southern Ocean, where the SOFAR axis approaches the surface, is higher than in other oceanic waters. The developments in the next few years will be instructive.

REFERENCES

Bannister, R. W., R. N. Denham, D. G. Browning and W. H. Thorp, "A study of the effect of bottom topology on long range sound propagation in the South Pacific," paper presented to the 85th Meeting of the Acoustical Society of America (Paper B1), 1973, Boston, Massachusetts.

Berman, A., and A. N. Guthrie, "On the medium from the point of view of underwater acoustics," J. Acoust. Soc. Am. 51, 994-1009 (1972).

Brownlee, L. R., F. H. Sagar, and D. E. Ash, "Measurement of wave propagation in a weakly homogeneous medium - the sea," J. Acoust. Soc. Am. (in press), 1974.

Denham, R. N., "Intensity-decay laws for sound propagation in shallow water of variable depth," J. Acoust. Soc. Am. 39, 1170-1173 (1966).

Denham, R. N., "Asymptotic solution for the sound field in shallow water with a negative velocity gradient," J. Acoust. Soc. Am. 45, 365-375 (1969).

Dietz, R. S., and M. J. Sheehy, "Transpacific detection of Myojin volcanic explosions by underwater sound," Bull. Geol. Soc. Am. 65, 941-956 (1954).

Ewing, M., J. L. Worzel, and M. Talwani, "Some aspects of physical geodesy," in Contemporary Geodesy, C. A. Whitten and K. H. Drummond (eds.), Geophys. Monograph 4, (American Geophysical Union, Washington, D. C., 1959), pp. 7-21.

Fisher, F. H., E. Yeager, R. Bressel, and J. Miceli, "Origin of the relaxational sound absorption in sea-water below 1 kHz," J. Acoust. Soc. Am. 53, 364(A)(1973).

Guthrie, K. M., "A wave theory of SOFAR signal shape," J. Acoust. Soc. Am. (in press), 1974.

Hochstein, M. P., "Seismic measurements in the Cook Islands, South-West Pacific Ocean," New Zealand J. Geol. Geophys. 10, 1499-1526 (1967).

Hochstein, M. P., "Seismic measurements in Suva Harbour (Fiji)," New Zealand J. Geol. Geophys. 13, 269-281 (1970).

Johnson, R. H., "Active submarine volcanism in the Austral Islands," Science 167, 977-979 (1970).

Johnson, R. H., and R. A. Norris, "Geographic variation of SOFAR speed and axis depth in the Pacific Ocean," J. Geophys. Res., 73, 4695-4700, 1968.

Kibblewhite, A. C., "The acoustic detection and location of an underwater volcano," New Zealand J. Sci. 9, 178-199 (1966).

Kibblewhite, A. C., "Note on another active seamount in the South Kermadec Ridge Group," New Zealand J. Sci. 10, 68-69 (1967).

Kibblewhite, A. C., and R. N. Denham, "Experiment on propagation in surface sound channels," J. Acoust. Soc. Am. 38, 63-71 (1965).

Kibblewhite, A. C., and R. N. Denham, "Experiment on sound propagation in shallow water under isovelocity conditions," J. Acoust. Soc. Am. 40, 1337-1344 (1966).

Kibblewhite, A. C., and R. N. Denham, "The bathymetry and total magnetic field of the South Kermadec Ridge seamounts," New Zealand J. Sci. 10, 52-67 (1967(a)).

Kibblewhite, A. C., and R. N. Denham, "Long range sound propagation in the South Tasman Sea," J. Acoust. Soc. Am. 41, 401-411 (1967(b)).

Kibblewhite, A. C., and R. N. Denham, "Experiment on sound propagation in shallow water with velocity structure," J. Acoust Soc. Am. 44, 104-112 (1968).

Kibblewhite, A. C., and R. N. Denham, "Hydroacoustic signals from the CHASE V explosion," J. Acoust. Soc. Am. 45, 944-956 (1969).

Kibblewhite, A. C., and R. N. Denham, "The CHASE V explosion - submarine topographic reflections from the vicinity of Pitcairn Island," Deep Sea Res. 18, 905-911 (1971(a)).

Kibblewhite, A. C., and R. N. Denham, "Low-frequency acoustic attenuation in the South Pacific Ocean," J. Acoust. Soc. Am. 49, 810-815 (1971(b)).

Luskin, B., M. Landisman, G. B. Tirey, and G. R. Hamilton, "Submarine topographic echoes from explosive sound," Bull. Geol. Soc. Am. 63, 1063-1068 (1952).

Norris, R. A., and R. H. Johnson, "Submarine volcanic eruptions recently located in the Pacific by SOFAR hydrophones," J. Geophys. Res., 74, 650-664, (1969).

Northrop, J., "Submarine topographic echoes from CHASE V," J. Geophys. Res., 73, 3909-3915, (1968).

Officer, C. B., "On some off-shore seismic refraction profiles in the Cook Strait, Tasman Bay and Golden Bay areas of New Zealand," New Zealand J. Geol. Geophys. 2, 350-354 (1959).

Richards, A. F., "Volcanic Sounds: Investigation and analysis," J. Geophys. Res., 68(3), 919-928, (1963).

Sagar, F. H., "Fluctuations in intensity of short pulses of 14.5 kc sound received from a source in the sea," J. Acoust. Soc. Am. 27, 1092-1106 (1955).

Sagar, F. H., "Comparison of experimental underwater acoustic intensities of frequency 14.5 kHz with values computed for selected thermal conditions in the sea," J. Acoust. Soc. Am. 29, 948-965 (1957).

Sagar, F. H., "Acoustic intensity fluctuations and temperature micro structure in the sea," J. Acoust. Soc. Am. 32, 112-121 (1960).

Schulkin, H. W., and M. Marsh, "Sound absorption in sea water," J. Acoust. Soc. Am. 34, 864 (1962).

Snodgrass, J. M., and A. F. Richards, "Observations of underwater volcanic acoustics at Barcena Volcano, San Benedicto Island, Mexico, and in Shelikof Strait, Alaska," Trans. Amer. Geophys. Union 37, 97-104 (1956).

Smith, R. P., Transduction and audible displays for broadband sonar systems, Ph.D. dissertation, University of Canterbury, New Zealand (1972).

Urick, R. J., "Long-range deep-sea attenuation measurement," J. Acoust. Soc. Am. 39, 904-906 (1966).

Urick, R. J., Principles of underwater sound for engineers, (McGraw Hill Book Co., Inc., New York, 1967).

UNDERWATER SOUND TRANSMISSION THROUGH THE OCEAN FLOOR

R. J. URICK

Naval Ordnance Laboratory

Silver Spring, Maryland 20910

ABSTRACT

Predictions of sound transmission in the deep ocean ordinarily rely on ray diagrams and ray trace calculations, wherein the ocean floor is assumed to be an impenetrable boundary that does nothing more than reflect sound back up into the sea. Yet, there is a growing body of experimental evidence to indicate that refracted sound traveling through the ocean floor makes a significant contribution to the sound in the sea. This evidence includes recent work with a geophone as a sensor of underwater sound, a measurement of sound transmission up the continental slope, and some recent findings on low-frequency, low-angle bottom loss. In this paper, this experimental data will be reviewed in terms of the role that the sea bed may play in long-range sound transmission in the sea.

INTRODUCTION

In the years since World War II, the detection ranges of sonar targets -- whether they be submarines, fish, or the ocean bottom of echo sounders -- have grown longer, often by an order of magnitude. New propagation paths have come to be utilized, such as the bottom-reflected path in bottom bounce sonars, and the magnitude and properties of the reflected sound from the deep sea bed have been extensively studied. The great improvement in sonar performance has been brought about by using increasingly lower frequencies as a way to avoid the dissipative effects occurring at higher frequencies. The useful frequencies of underwater sound now extend downward to a few Hertz, and the ranges of interest are measured in tens and even hundreds of miles.

At such ranges and at such low frequencies, the ocean bottom plays an important role in sound transmission in the sea. This role is more than just acting as a lossy container for the sound in the overlying fluid. It is often a transmitter of sound in the form of the longitudinal, shear, and boundary waves familiar to seismologists. By doing so, the sea bed materially affects the propagation of sound in the sea above it.

The importance of this role is not always realized. Evidence for it exists in acoustic measurements that are hard to interpret other than in terms of propagation via sub-bottom paths. In this paper some examples of such observations -- some of which resulted from NOL field programs -- will be described.

SHALLOW WATER

In shallow water, the dominant effects of the ocean floor on the propagation of sound in the liquid medium above have long been recognized. Sound at a frequency too low to be trapped in the fluid layer is found to be well-propagated in shallow water; at such frequencies the water serves merely as a coupling fluid between the source of sound and the earth.

The fact of sub-bottom propagation was recognized during World War II in research motivated by the needs of the newly emergent acoustic mine. In a now-classic series of papers, Worzel et al. (1948) described measurements in a variety of shallow water areas using both hydrophones and geophones to record shots fired at a distance. A sample of one of their oscillograph recordings is reproduced in Fig. 1. Here, the traces show a shot fired 4000 yards away in 53 feet of water as received by a geophone and a hydrophone filtered to pass high and low frequencies. The high frequency water arrival occurring at 2.4 seconds is preceded by a low frequency forerunner called the "ground wave". This precursor was recognized at the time to be the refracted sound, well-known to seismologists, that travels a higher velocity through the earth. The ground wave forms the major part of the low-frequency signal picked up by the hydrophone.*

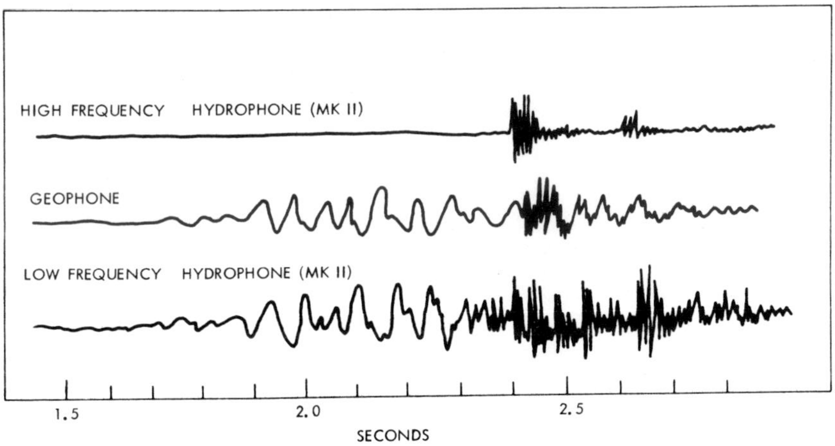

FIGURE 1

THE "GROUND WAVE" AS RECORDED IN SHALLOW WATER IN NOVEMBER 1943. SHOT 48, CHESAPEAKE BAY (SOLOMONS, MD.), RANGE 4000 YARDS (WORZEL ET AL., 1948)

*Some hesitancy should be felt in attributing signals picked up by bottom-mounted hydrophones to acoustic pressures in the water. Hydrophones are often vibration-sensitive and may respond more as seismometers than pressure-sensitive hydrophones.

When the frequency is sufficiently low, such earthborne underwater sound can travel long distances through the bottom and be influenced by refraction at great depths. That this is the case is suggested by the diverse and anomalous results of a short acoustic survey of U. S. Gulf and East coast waters made by NOL (Urick, 1971) entirely from a Navy P-3 Orion aircraft. Sonobuoys were used for reception of Navy Mk 61 explosive sound signals dropped from the aircraft in a radial pattern out to a distance of 30 miles from the sonobuoy at the center. The results were presented as transmission loss contours centered about the receiving location for various one-octave frequency bands.

Examples for two widely separated locations are shown in Fig. 2. The site of the contour map at the left was in the Gulf of Mexico; that of the map at the right was off the U. S. East coast. Notice the great difference in the transmission at 63 Hz between these two locations. At a distance of 30 miles the transmission loss differs by 40 dB, or a ratio of 10,000 to 1, in sound energy density.

It seems impossible to account for this vast acoustic difference by the nature of the water column and the general nature of the bottom. For example, Fig. 3 gives the bottom contour and BT traces at the spots in the pattern shown by the arrows where the air-dropped BT's were taken. There is nothing markedly different between the two areas; both are underlaid by Tertiary sediments down to a depth of 20,000 feet.

However, the velocity profiles in the sub-bottom sediments at the two sites are somewhat different, as can be seen in Fig. 4. It is possible (though not likely) that the difference in velocity profile is the answer. In any event, anomalies of this sort appear to be not uncommon in shallow water at long ranges, where penetration of the earth by low frequency sound to great depths can be expected to occur.

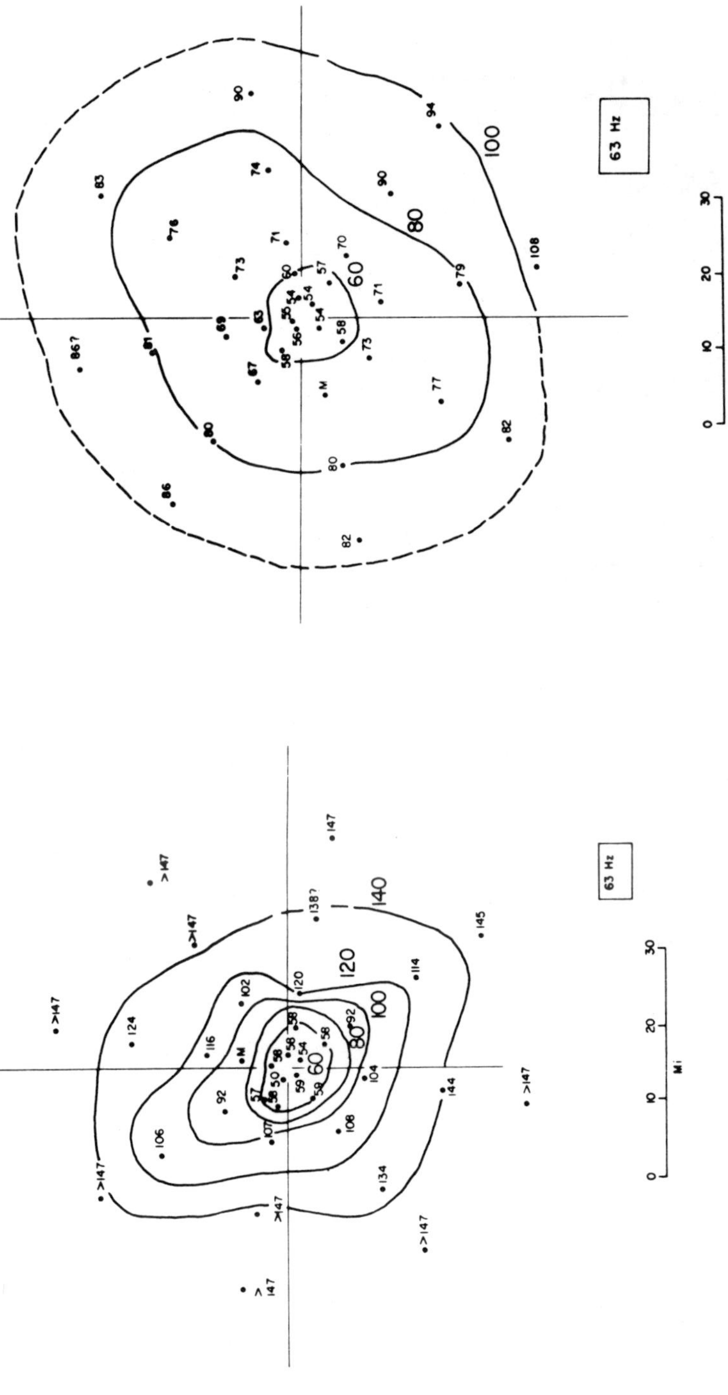

FIGURE 2

LOSS CONTOURS IN AN OCTAVE BAND CENTERED AT 63 Hz FOR TWO SHALLOW WATER LOCATIONS. POINTS ARE DROP LOCATIONS. THE NUMBER AT EACH POINT IS THE MEASURED TRANSMISSION LOSS AT THAT POINT. THE DROP PATTERN EXTENDS OUT TO 30 MILES FROM THE RECEIVING LOCATION AT THE CENTER.

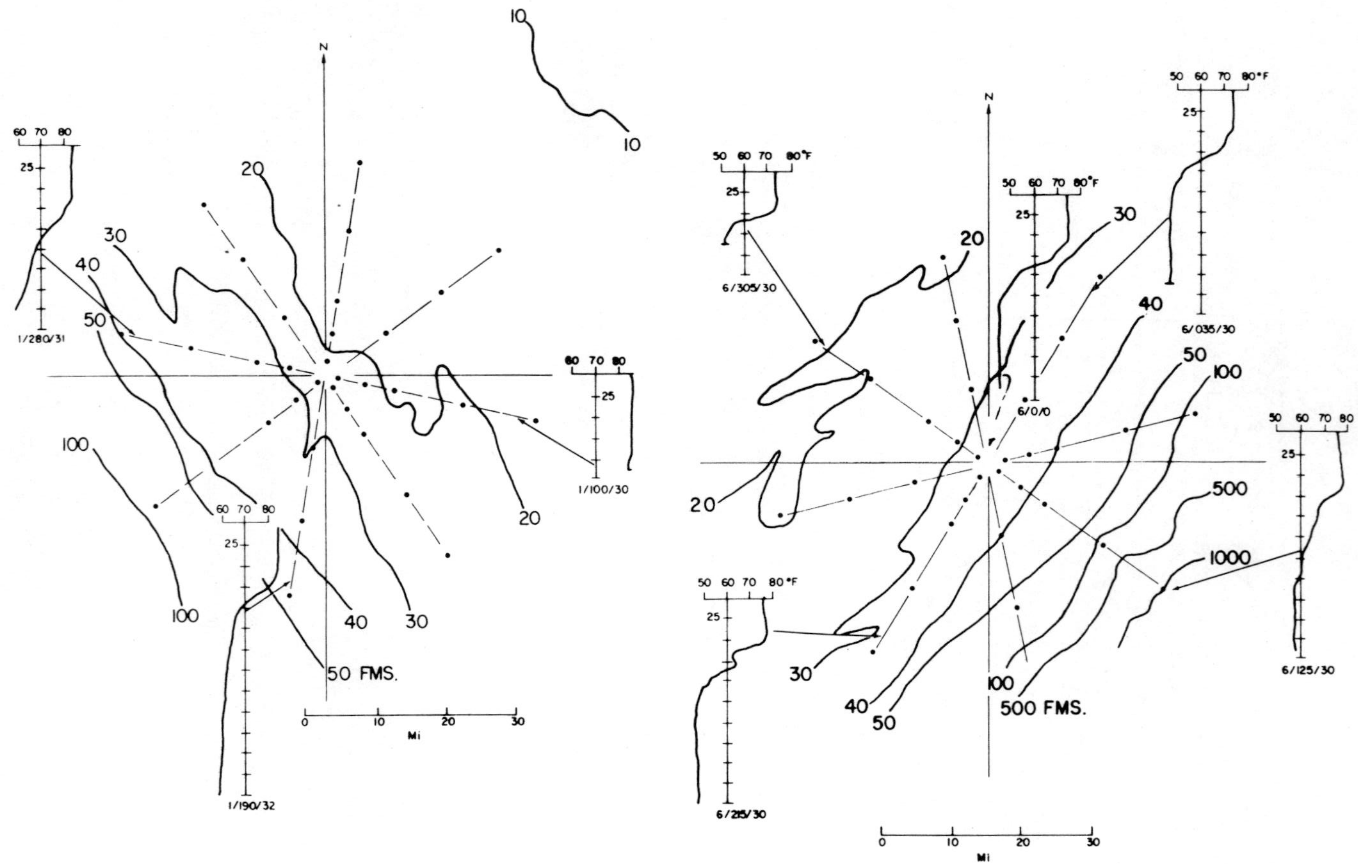

FIGURE 3

ENVIRONMENTAL DATA FOR THE TWO LOCATIONS FOR WHICH ACOUSTIC DATA IS GIVEN IN FIG. 2. SOLID LINES ARE BOTTOM CONTOURS; VERTICAL TRACES SHOW THE BT OBTAINED FROM AN/SSQ-36 BT-BUOYS DROPPED AT THE LOCATIONS IN THE PATTERN SHOWN BY THE ARROWS.

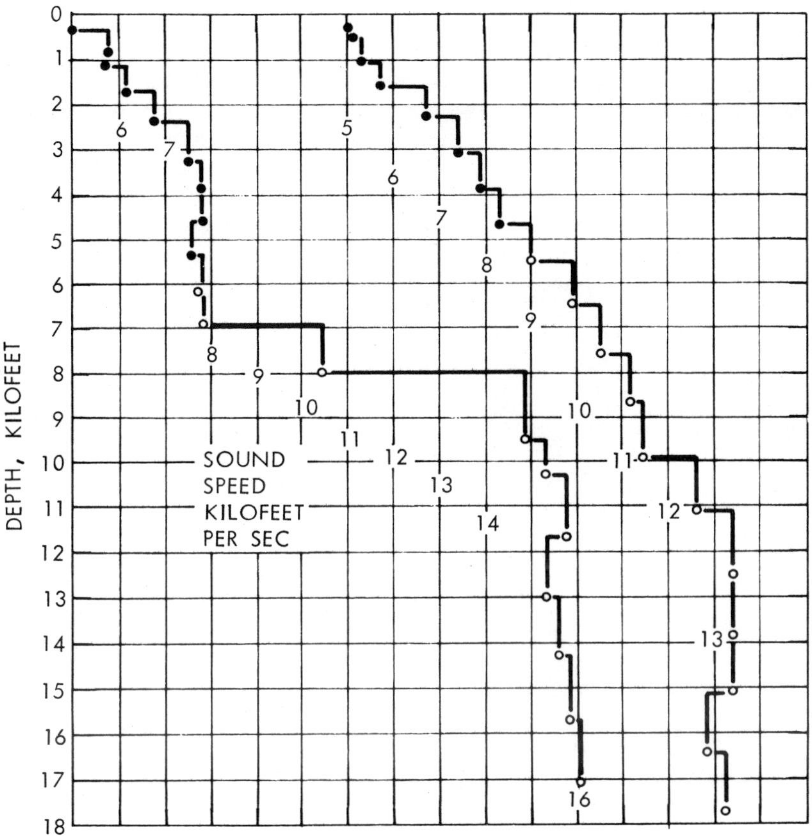

FIGURE 4

VELOCITY PROFILES AT THE TWO LOCATIONS.
DATA COURTESY OF SHELL OIL COMPANY.

DEEP TO SHALLOW WATER

Propagation of sound in the transition water between that commonly referred to as being "deep" to that called "shallow" has received comparatively little attention. Whatever we may mean by "shallow" and "deep", it is clear that this must be a more complicated condition, as far as propagation is concerned, than for propagation in water of uniform depth. For, in propagating up a slope, sound encounters the bottom at successively shorter intervals because of the increasing steepening of sound rays on successive reflections from the sloping bottom. Moreover, somewhere along the slope the angle with the bottom will become optimum for the conversion of sound into the various propagating waves (e.g., longitudinal, transverse, surface) capable of being transmitted through the bottom. We would therefore expect bottom transmission to play an especially important role in this kind of propagation.

Two examples from NOL work may be cited. One experiment was done in 60 feet of water off Ft. Lauderdale, Florida, where a three-component geophone was firmly imbedded by divers in the coral-sand bottom (Urick and Colvin, 1968) in addition to a bottom-mounted hydrophone placed nearby. The hydrophone was designed to be insensitive to acceleration; it was found from measurements on a shake-table, wherein the unit was subjected to known accelerations, that the acceleration-response of the unit was negligibly small compared to its response to pressure, and that it was indeed a pressure-sensor rather than an accelerometer.

Figure 5 is a cross-section in the direction of bottom slope (nearly east-west) from the location of the geophone-hydrophone combination shown at the left. Freighters and tankers traveling southbound, and therefore at right angles to the plane of the cross-section, served as sources of sound. A typical passing distance was 1500 yards from the geophone-hydrophone combination. Under the existing conditions of a strong negative temperature gradient, it is evident that sound from the source at S could not reach the receivers without at least one encounter with the bottom.

Figure 6 shows the result of analyses of recordings from two freighters, one at a range of 3300 yards to the southeast of the instrument site, the other at a range of 6400 yards. These show the signal-to-noise ratio at frequencies between 20 and 300 Hz, as received by the four sensors. The hydrophone, or pressure sensor, had a S/N ratio of about 5 dB that is roughly the same at all frequencies. The vertical geophone signals were generally buried in noise at nearly all frequencies. The two horizontal geophone components had a S/N ratio of 10-15 dB at low frequencies; indeed, the signal from these sound sources persisted long after all trace of it had vanished on the two other units. We conclude that sound from these sources must have traveled principally in the bottom as

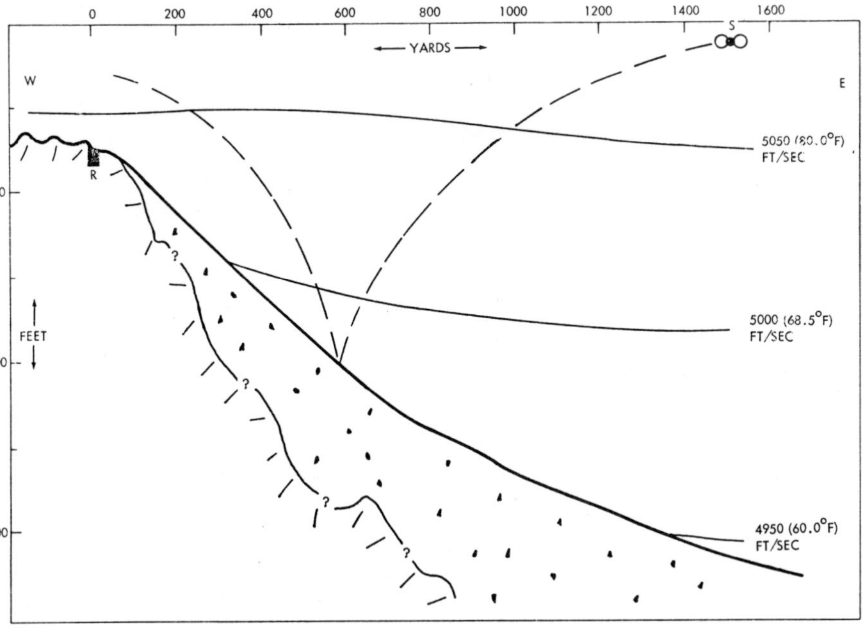

FIGURE 5

WEST-TO-EAST CROSS-SECTION OFF FT. LAUDERDALE, FLA., SHOWING THE BOTTOM PROFILE AND SOUND VELOCITY CONTOURS. THE INSTRUMENT PLANT IS AT R AT THE LEFT, WHILE S AT THE RIGHT DENOTES THE LOCATION OF A MERCHANT VESSEL PASSING FROM NORTH TO SOUTH NORMAL TO THE CROSS-SECTION. THE DASHED CURVE IS A SOUND RAY LEAVING S HORIZONTALLY AND STRIKING THE BOTTOM.

a wave or waves having principally horizontal particle motion and therefore not coupled to the lossy water medium. What is likely to have occurred was that the waterborne sound, on striking the bottom, became converted to earthborne sound traveling up the slope with lower loss than the reflected, poorly trapped, sound remaining in the water layer.

The same explanation is suggested by the results of another series of measurements over much greater distances (Urick, 1972).

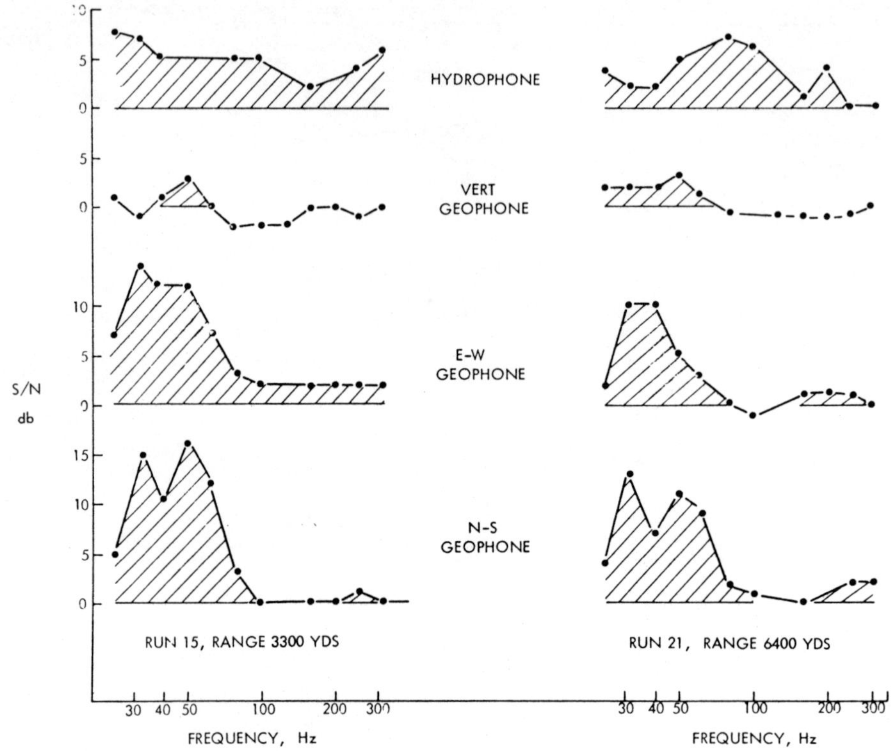

FIGURE 6

SIGNAL-TO-NOISE RATIO SPECTRA FOR THE FOUR SENSORS, FOR TWO VESSELS AT THE INDICATED RANGES.

Figure 7 is a portion of a chart of the area off the coast of Delaware. In this experiment, sonobuoys were dropped by an aircraft at two locations on the continental shelf. One, called the "inshore" buoy, was placed in 25 fathoms of water; the other, called the "offshore" buoy, was placed in 20 miles to seaward at the 100-fathom line where the break in the bottom profile -- between the gentle slope of the shelf and the much steeper slope of the continental rise -- begins. While one aircraft monitored and recorded the output of these sonobuoys, another aircraft dropped standard Navy explosive sound signals at different distances out to 100 miles in the direction of the bottom slope (Run 1 in Fig. 7). This same aircraft then flew north and began an inbound run with a similar series of explosive drops (Run 2 in Fig. 7). At each range, two sound signals were dropped, one detonating at 60 feet and the other at 2000 feet, the axial depth of the deep sound channel in the area. The tape recordings made in the field were later analyzed in the laboratory, and were converted to transmission loss by means of a known acoustic

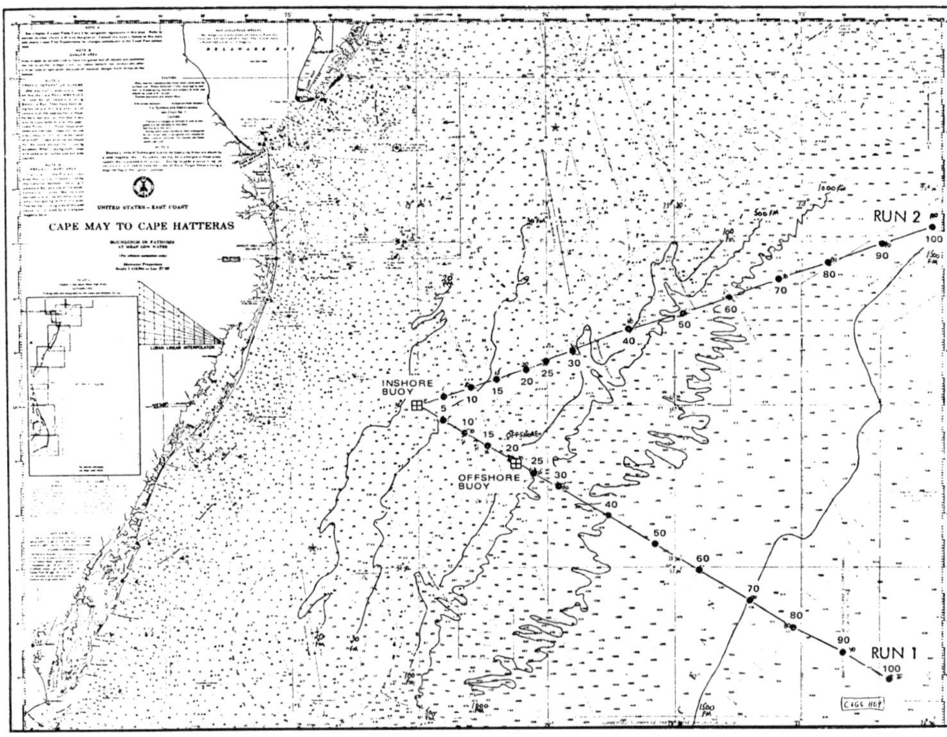

FIGURE 7

CHART SHOWING AREA OF AIRCRAFT RUNS AND SONOBUOYS. EXPLOSIVE CHARGES WERE AIR-DROPPED AT THE LOCATIONS SHOWN BY THE DOTS. NUMBERS AT EACH DROP LOCATION ARE MILES FROM THE INSHORE BUOY. DEPTH CONTOURS ARE 20, 30, 100, 500, 1000, and 1500 FATHOMS. USC&GS CHART 1109.

output of the sound signals. The results are shown in Figs. 8 and 9 for the 2000-foot shots and the 60-foot shots, respectively. Each figure shows plotted data for two frequencies for the inshore and offshore buoys. Turning to the 2000-foot shots first, we observe that the slope of the plots is compatible with cylindrical spreading plus the attenuation of deep water propagation. Not only is the slope comparatively small, but the loss to the offshore buoy is less than it would be to a point on the axis of the sound channel in deep water in the absence of the sloping bottom. Sound appears to travel down the channel axis until it encounters the sloping bottom, from which it is reflected once or twice up into the shallow water of the shelf. The ray diagram for this geometry is seen in the upper part of Fig. 10. There is no need to resort to bottom propagation to account for this kind of transmission.

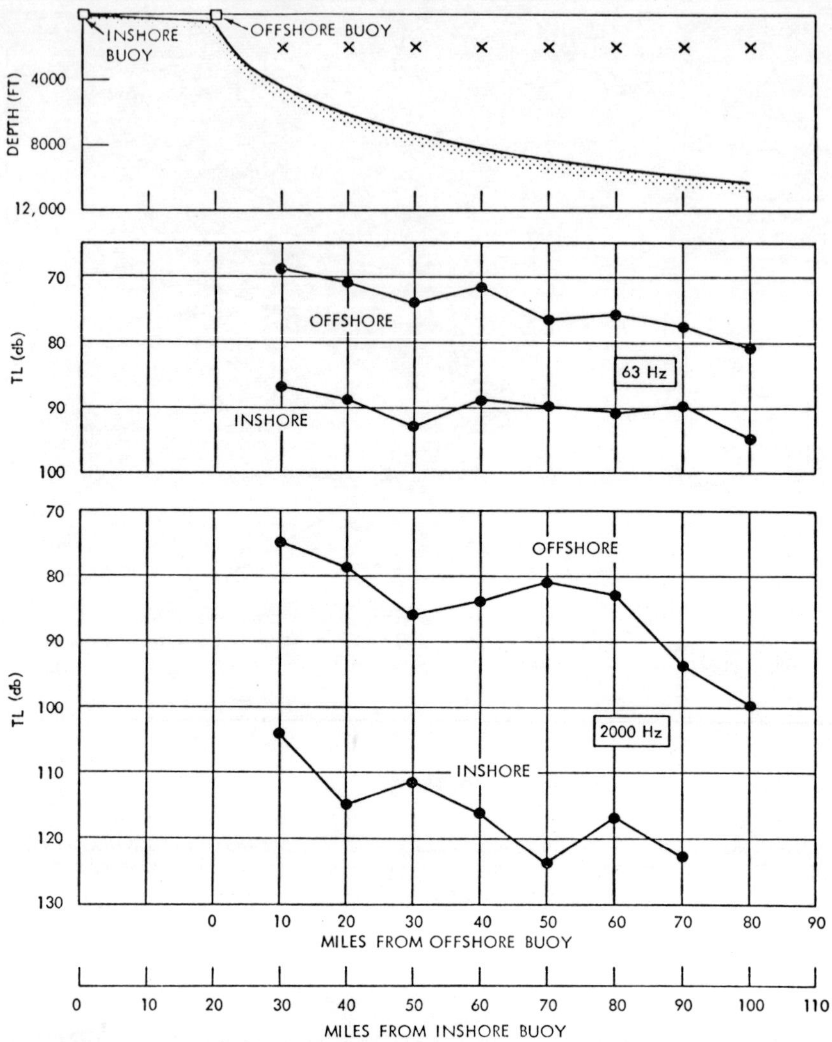

FIGURE 8

RESULTS FROM THE 2000-FOOT SHOTS, RUN 1. A CROSS-SECTION IS SHOWN AT THE TOP. PLOTTED POINTS GIVE THE MEASURED TRANSMISSION LOSS IN OCTAVE BANDS CENTERED AT 63 Hz AND 2000 Hz FOR THE TWO BUOYS.

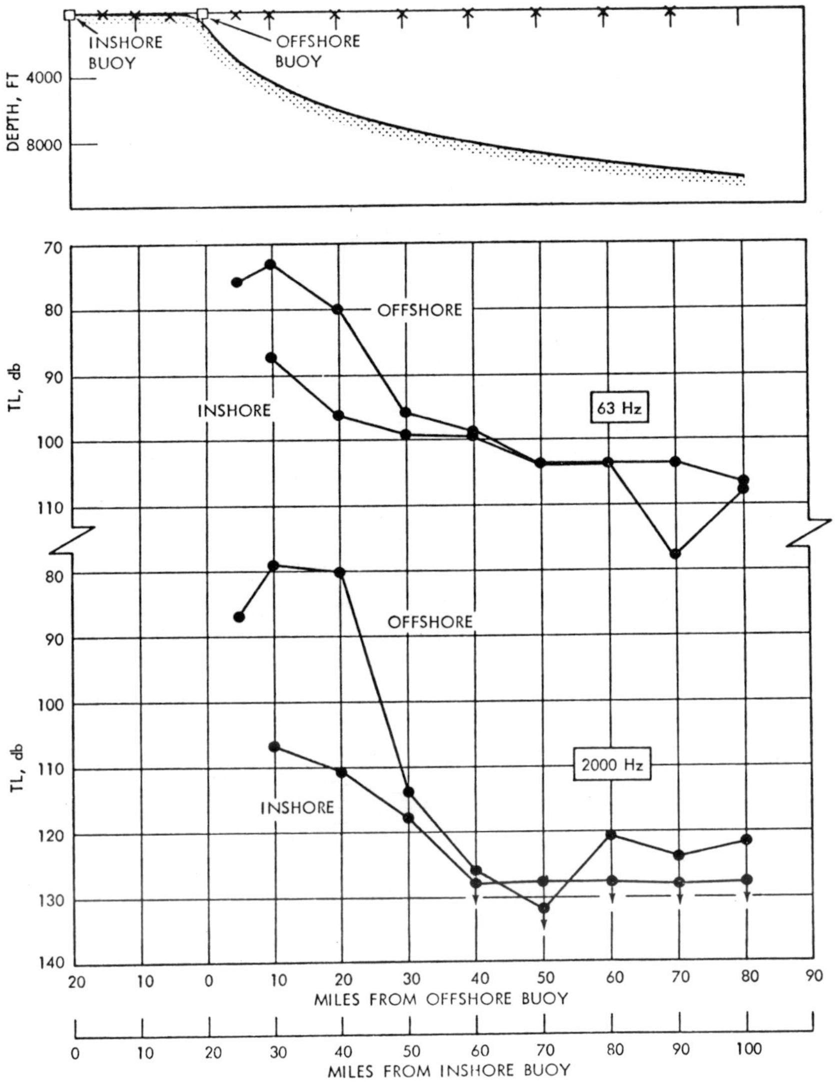

FIGURE 9

SAME AS FIG. 8, BUT FOR THE 60-FOOT SHOTS.

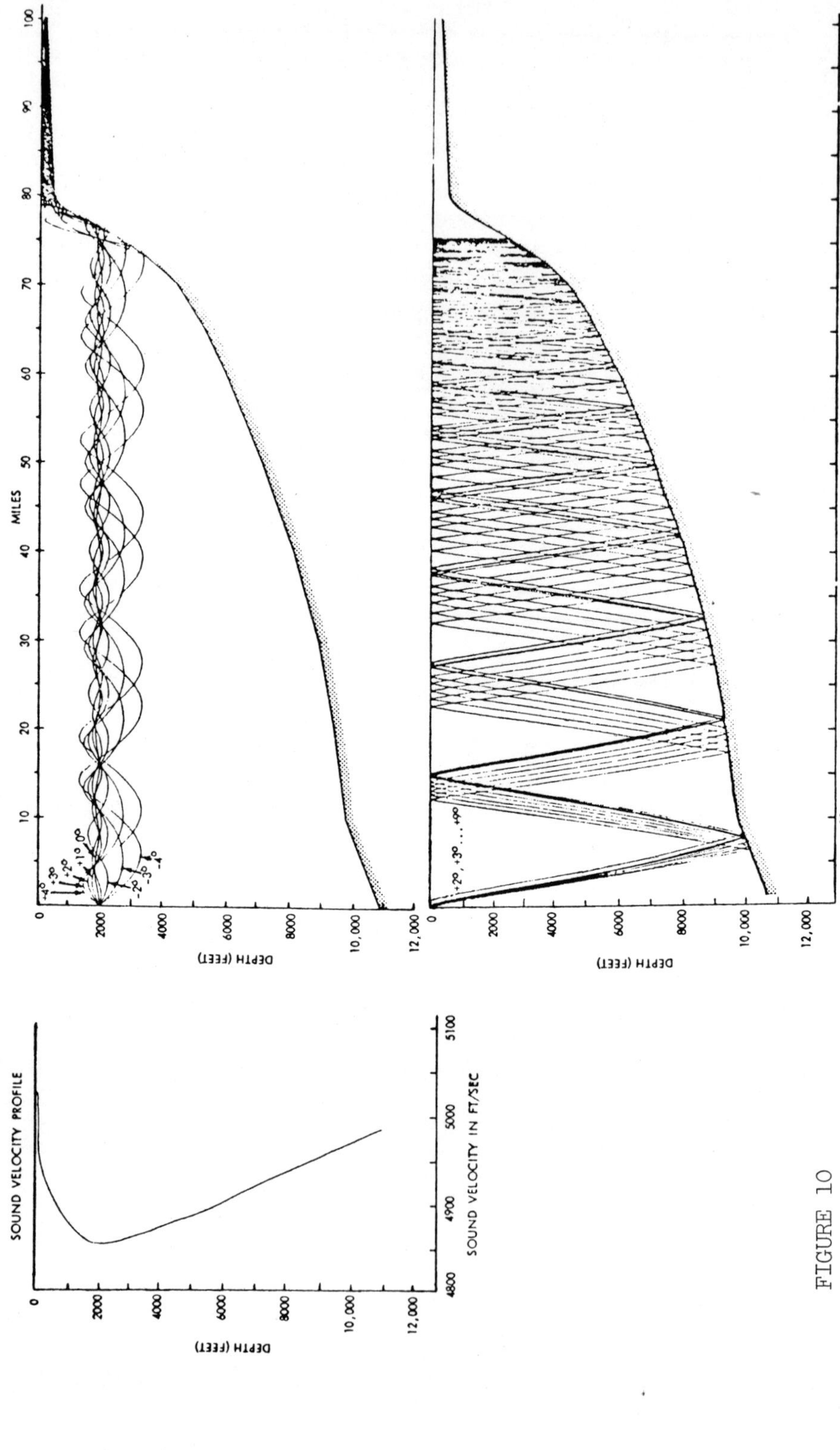

FIGURE 10

RAY DIAGRAMS FOR A SHOT AT 100 MILES FROM THE INSHORE BUOY. UPPER DIAGRAM IS FOR A 2000-FOOT SHOT, LOWER FOR A 60-FOOT SHOT. RAY INTERVAL IS 1°. VELOCITY PROFILE USED FOR THE DIAGRAMS IS GIVEN AT THE UPPER LEFT.

On the other hand, the transmission of the 60-foot shots requires additional explanation. As seen in Fig. 9, the received signals fall off rapidly in the region of the steep slope. Some sound is received at all ranges out to the limit of 100 miles. According to the ray diagram in the lower part of Fig. 10, no sound can reach long distances via all-water paths. The low level residual sound at long ranges, unless it represents forward scattering of waterborne sound, is likely to have been bottom-propagated. However, neither the purposes of the experiment nor the airborne method of conducting it permitted the gathering of evidence to support or deny this explanation.

DEEP WATER

A suggestion of a bottom-refracted contribution to underwater sound in uniformly deep water is provided by measurements summarized in Fig. 11. The curves show reflection loss from the deep sea bed as a function of grazing angle at frequencies of a few hundred cycles and below. They are based on data collected in a variety of deep ocean areas and were compiled by the U. S. Naval Oceanographic Office. Reflection loss is defined to be the ratio, in decibels, between the known amount of acoustic energy of an explosive shot at a unit short distance, to that received at a longer distance, after reflection from the deep sea bed at a known angle, when the transmission loss in the water (using ray trace calculations and an assumption as to the attenuation coefficient) has been accounted for. That is to say, reflection loss is what is left over after the loss in the water medium is subtracted. We see from the figure that the loss, as so defined, tends to become negative at small grazing angles. More signal is being returned from the sea bed than is allowed by transmission in the sea alone.

One possible source of the additional sound is reflection from thick layers or strong interfaces deep within the bottom. Another is refraction within a homogeneous bottom, brought about by the relatively strong gradient of sound velocity produced by compaction in sedimentary materials. This can be seen by the aid of a ray diagram, in which the rays are continued past the bottom interface and into the lower medium. Figure 12 shows two of the rays from such a ray diagram; the bottom has a constant velocity gradient of 1 sec^{-1} (Faust, 1951). Two rays reach a shallow depth at a range of 18 miles after encountering the ocean floor. These are the 10°

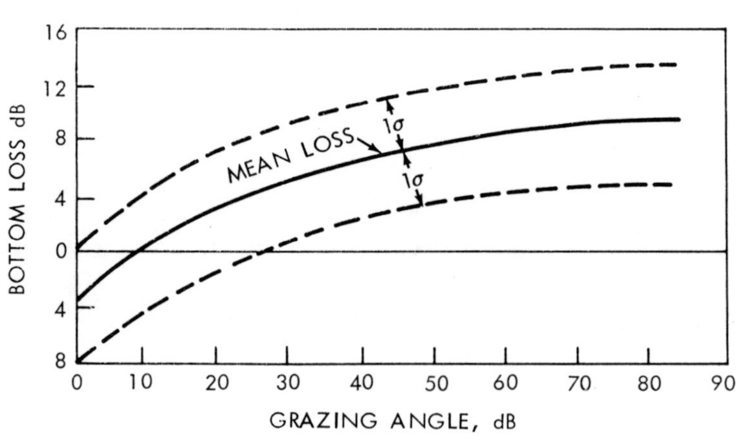

FIGURE 11

RESULTS OF A COMPILATION OF LOW FREQUENCY MEASUREMENTS OF BOTTOM LOSS. DASHED CURVES SHOW LIMITS OF ONE STANDARD DEVIATION.

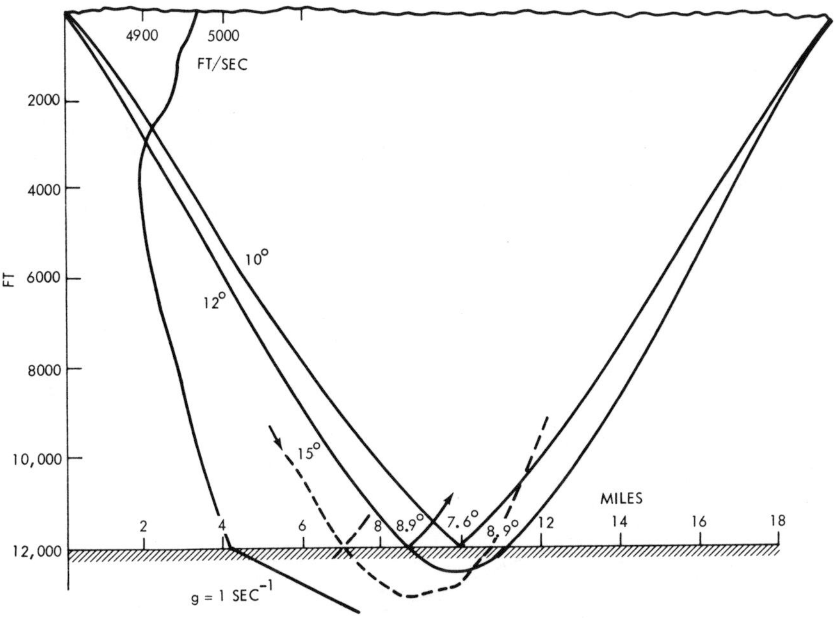

FIGURE 12

TWO RAYS (THE 10° AND 12° RAY) FROM A COMPUTER-DRAWN RAY DIAGRAM USING THE VELOCITY PROFILE GIVEN AT THE LEFT. A PORTION OF THE 15° RAY IS ALSO SHOWN. THE TWO RAYS MEET NEAR THE SURFACE AT A RANGE OF 18 MILES, WHERE THE 10° RAY IS REFLECTED AND THE 12° RAY HAS PENETRATED INTO THE BOTTOM.

ray reflected at an angle of 7.6°, and the 12° ray refracted into the sedimentary bottom and returned back to the sea by the strong velocity gradient in the bottom. Calculations show that the travel time along these paths is nearly the same; the longer path length along the refracted path is compensated for by the higher velocity within the bottom. The contributions along these ray paths are therefore indistinguishable and the results from the bottom are larger than can be accounted for by reflection from the water-bottom interface alone.

A SOURCE OF NOISE

A completely different way in which the sea bed influences underwater sound is that it may be a source of the ambient noise background in the sea. This possibility, that the seismic motion of the ocean floor contributes to the low frequency noise background of the sea, was suggested some years ago by Wenz (1962) on the basis of measurements of the seismic background made on land. More recent data obtained with seismographs located on the ocean floor lend support to this hypothesis. Figure 13 shows the results of converting sea bed motion measurements (displacement, velocity) taken from the literature into acoustic pressures in the overlying

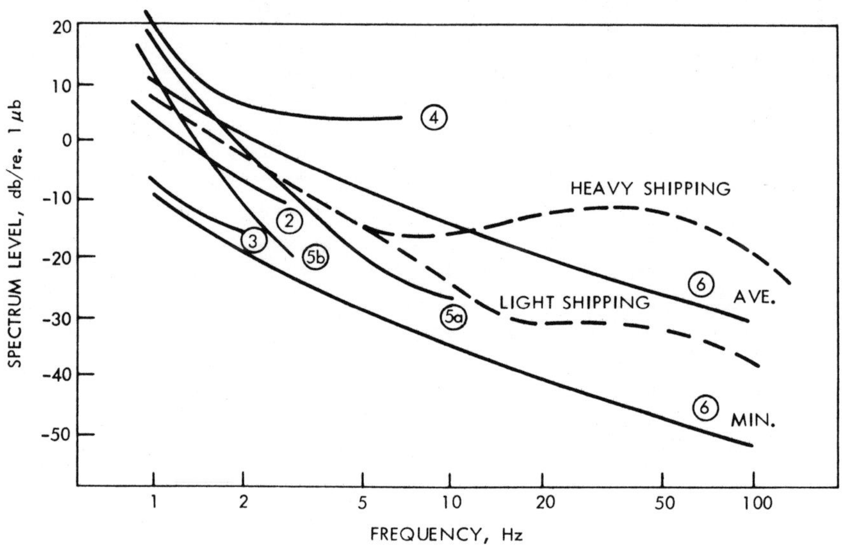

FIGURE 13

SEISMIC UNREST AS A SOURCE OF THE AMBIENT ACOUSTIC BACKGROUND OF THE SEA. THE DASHED CURVES SHOW THE AVERAGE MEASURED BACKGROUND OF SEA NOISE IN DEEP WATER. THE SOLID CURVES SHOW PREDICTED SPECTRA USING PUBLISHED DATA ON SEA BED MOTION. THE CURVES ARE NUMBERED BY LITERATURE SOURCES, AS FOLLOWS:

2, Latham, G. V., and Sutton, G. H., "Seismic Measurements on the Ocean Floor," J. Geophys. Res., 71, 2545, 1966.

3, Schneider, W. A., and Backus, M. M., "Ocean Bottom Seismic Measurements Off the California Coast," J. Geophys. Res., 69, 1135, 1964.

4, Bradner, H., and Dobbs, J. G., "Comparative Seismic Noise on the Ocean Bottom and on Land," J. Geophys. Res., 69, 4339, 1964.

5, Prentiss, D. D., and Ewing, J. I., "Seismic Motion of the Deep Ocean Floor," Bull. Seism. Soc. Am., 53, 765, 1973.

6, Brune, J. N., and Oliver, J., "Seismic Noise of the Earth's Surface," Bull. Seism. Soc. Am., 49, 349, 1959.

water, on the assumption of plane-wave radiation from the vibrating boundary. The results are compatible with the existing, though admittedly scanty, measurements of ambient sea noise at low frequencies. They suggest that the seismic unrest of the ocean floor creates significant noise in the sea, at least in those regions of the world's oceans where ship traffic is light.

CONCLUSION

We have, in the foregoing, examined a number of bits and pieces of experimental evidence pointing toward an active, dynamic role of the ocean floor in affecting sound in the sea. None of them is conclusive or unequivocal. But the suggestion is clear that the deep ocean floor is more than just a reflecting and scattering boundary for underwater sound. At the low frequencies of current interest for long-range sound transmission, the sea bed sometimes appears to be a refractive lossy medium sustaining various kinds of propagating waves that contribute to the sound field in the sea at long ranges. In addition, it is likely to be the dominant source of noise in the sea in regions of high seismic unrest. This complex behavior must be included in the sophisticated models now being developed for more accurate low-frequency predictions of sound in the sea.

REFERENCES

Faust, L. Y., 1951, Seismic Velocity as a Function of Depth and Geologic Time: Geophysics, v. 16, p. 192.

Urick, R. J., Some Perspectives and Recent Findings in Shallow Water Acoustics, Naval Ordnance Laboratory Technical Report 71-204, 1971.

Urick, R. J., Sound Transmission from Deep to Shallow Water, Naval Ordnance Laboratory Technical Report 72-1, 1972.

Urick, R. J., and Colvin, G. M., Detection of Underwater Sounds by a Geophone Planted in the Bottom, Naval Ordnance Laboratory Technical Report 68-102, 1968 (Unpublished).

Wenz, G. M., Acoustic Ambient Noise in the Ocean: Spectra and Sources, J. Acoust. Soc. Am. 34, 1936 (1962).

Worzel, J. L., and Ewing, M., Explosion Sounds in Shallow Water, article in Propagation of Sound in the Ocean, Geological Society of America Memoir 27, 1948.

GEOACOUSTIC MODELS OF THE SEA FLOOR

EDWIN L. HAMILTON

Naval Undersea Center

San Diego, California 92132

ABSTRACT

A "geoacoustic model" is defined as a model of the real sea floor with emphasis on measured, extrapolated, and predicted values of those properties important in underwater acoustics and those aspects of geophysics involving sound transmission. Such models are also important in other aspects of geology and geophysics. The real sea floor cannot be defined by any single geoacoustic model; therefore, it is important that acoustic and geophysical experiments at sea involving the sea floor be supported by a particular model of the area. These models can then be used to reconcile experiment and theory. However, it is possible to use geologic and geophysical judgment to extrapolate a general model over wider areas. A sufficient collection of models from diverse environments will allow predictions of bottom models in similar areas of the world's oceans.

A geoacoustic model should detail the real sea floor. It can then be used in studies of reflection and refraction of compressional and shear waves over wide frequencies, in geologic studies of stratigraphy, sedimentology, and geologic history, and in various studies in the field of geophysics such as the reflection and refraction of sound, and gravity computations.

The production of a geoacoustic model of the sea floor requires assembly of data from a wide variety of sources in the fields of oceanography, geology, and geophysics. A model thus brings into focus and utility data from many scientific disciplines and operations at sea and in the laboratory. In general, a model details the true thicknesses and properties of various sediment and rock layers overlying the earth's crust. The gross layering may be all that is required in some geologic and geophysical studies, but the acoustician must be supplied sufficient detail to study insonified areas at various sound frequencies. The information required for a complete geoacoustic model should include the following for each layer; in some cases, the state of the art allows only rough estimates, or information may be non-existent.

1. Properties of the overlying water mass from Nansen casts and velocimeter lowerings.
2. Sediment information (from cores, drilling, or geologic extrapolation): sediment types, grain-size distributions, densities, porosities, compressional and shear wave attenuations and velocities, and other elastic properties. Gradients of these properties with depth; for example, velocity gradients and interval velocities from sonobuoy measurements.
3. Thicknesses of sediment layers (in time) determined at various frequencies by continuous reflection profiling.
4. Locations, thicknesses, and properties of reflectors within the sediment body as seen at various frequencies.
5. Properties of rock layers. Those at or near the sea floor are of special importance to the underwater acoustician.
6. Details of bottom topography, roughness, relief, and slope, for example, as seen by underwater cameras, and deep-towed equipment.

Recent studies have provided restrictive parameters for any elastic or viscoelastic model for water-saturated sediments (e.g., velocity dispersion is negligible or absent, and the dependence of attenuation on frequency is close to f to the first power, at frequencies of most interest). These parameters and elastic and viscoelastic models which can be applied to marine sediments are reviewed, and a particular viscoelastic model (with concomitant equations) is recommended.

I. INTRODUCTION

During the past fifteen years the Naval Undersea Center, formerly part of the Navy Electronics Laboratory, has fielded a number of major and minor expeditions whose basic purposes were research in underwater acoustics and reconnaissance into far areas which included the Western Pacific and Indian Oceans. A typical station was occupied for 2 to 3 days.

It was soon discovered that the best results were obtained when the scientific staff included acousticians and engineers supported on board by physical oceanographers, biologists, geologists, and geophysicists.

The main responsibility of the geologists and geophysicists on this type of expedition is to produce a geoacoustic model of the sea floor for areas where the bottom was insonified by the acoustic experiments. These models are then furnished the acousticians, and the geologic and geophysical data are later published in Navy documents and in outside scientific journals. We have found it essential, in this work, that basic research support the applied research. This is best done by the same individuals. As a result, we have gradually improved our models, frequently in accordance with requests from the acousticians.

A "geoacoustic model" is defined as a model of the real sea floor with emphasis on measured, extrapolated, and predicted values of those properties important in underwater acoustics and those aspects of geophysics involving sound transmission. Such models are also important in other aspects of geology and geophysics. In general, a geoacoustic model details the true thicknesses and properties of various sediment and rock layers overlying the earth's crust.

One of the important uses of geoacoustic models is in reconciling experimental results with theory when sound is reflected or refracted from the sea floor. This requires a full range of information about the sea floor, from basic physics of sound in marine sediments, through empirical relationships between properties, to outright estimates based on geological probabilities. Various aspects of these subjects will be included in this report.

II. ELASTIC AND VISCOELASTIC MODELS FOR MARINE SEDIMENTS

Introduction

In order to understand and predict sound propagation in the sea floor, it is necessary to use, as a minimum, values of sediment density, sound velocity, and sound attenuation. Some models require other elastic constants, or velocity gradients. All of these properties are interrelated in various theories of elasticity and viscoelasticity.

The subject of elastic and viscoelastic models and equations which can be applied to marine sediments was discussed in 4 recent reports (Hamilton et al., 1970; Hamilton, 1971a,b, 1972.) The following is mostly excerpted from these reports, which can be consulted for fuller discussions.

In the following discussion (unless otherwise noted) it is assumed that the medium is a porous, gas-free, uncemented, mineral structure, fully saturated with water, and is macroscopically isotropic. The stress is that of a compressional or shear wave of low amplitude, and strains are of the order of 10^{-6} or less. The frequency range is from a few Hertz to at least several hundred kHz, or into the MHz range. Wave lengths are much greater than grain size; otherwise, Rayleigh scattering can occur, and attenuation is related to the fourth power of frequency (e.g., Busby and Richardson, 1957).

The study of elastic and viscoelastic models which can be applied to dry and saturated porous rock and sediments has concerned a large number of scientists and engineers. Many of these studies have considered the extent to which these media can be described by the equations of Hookean elasticity, or those of the Kelvin-Voigt, Maxwell, or other viscoelastic models. It is somewhat surprising, therefore, to discover such a wide diversity of models, equations, and opinions on such an important subject. In the case of water-saturated, natural sediments, the reason for this diversity appears to be that, until recently, experimental evidence capable of restricting model parameters has been scarce.

In the absence of sufficient experimental evidence it has been possible to construct rather elegant theoretical approaches, altered if necessary by constants to fit available data. To derive such theoretical models one must start with assumptions. In the case of water-saturated sediments, some of the less tenable of these assumptions have been that (1) all water-saturated sediments are analogous to suspensions of mineral particles in fluids, (2) all the mineral particles are spheres, (3) Poiseuille flow operates in natural sediments, (4) pore water necessarily moves relative to the mineral frame or sediment structure, (5) sediments lack rigidity, in which case the shear modulus is zero, and Poisson's ratio is 0.5, (6) attenuation is related to the square or square root of frequency, and (7) appreciable velocity dispersion is permissible.

All of the above assumptions are invalid in part or in whole. In recent papers, Hamilton (1970a, 1971a) discussed several aspects of sediment structure and elasticity. Some conclusions are pertinent: almost all saturated sediments have mineral particles which are not spheres, near suspensions are unusual, and almost all sediments have sufficient rigidity to allow transmission of shear waves. Yong and Warkentin (1966) indicate Poiseuille flow (through small tubes) does not hold for natural sediments. Two critically important parameters, discussed below, concern velocity dispersion and the dependence of wave-energy damping on frequency.

The equations of Hookean elasticity do not account for energy damping. Consequently, an adequate model must be anelastic if energy damping is considered. In the selection of an appropriate anelastic model, a critical factor is the extent of relative movement of pore water and mineral particles under the light stress of a compressional or shear wave. If the pore water moves significantly relative to mineral structure, then viscous damping and velocity dispersion must be considered. If the pore water does not move significantly with respect to the solids, the effective density of the medium is the sum of the mass of pore water and solids in a unit volume, there is negligible or no velocity dispersion, and energy damping is negligibly dependent on, or independent of, frequency and is not dependent on viscosity of pore water and permeability of the mineral structure. The dependence of energy damping on frequency is different in the two cases. Consequently, two critical parameters for anelastic models are velocity dispersion, if any, and the dependence of energy damping on frequency.

Velocity Dispersion

The subject of the variation of velocity with frequency ('velocity dispersion') in marine sediments was recently discussed at some length (Hamilton, 1972). The conclusions were as follows.

In sands, a number of studies (cited in Hamilton, 1972) have reported no dispersion over restricted frequency ranges. Those studies and comparison of the values of velocity in similar sands at different frequencies indicate that velocity dispersion, if present, is negligibly small from a few kHz to the MHz range. The evidence indicates that velocity dispersion in higher porosity silt-clays, if present, is negligible over a frequency range from less than 1 kHz to 2 MHz. However, it must be stated that most tests of velocity dispersion were made over only an order of magnitude of frequencies or less, which is not necessarily enough to show dispersion. In other words, it cannot be stated on the basis of present experimental evidence that velocity dispersion is nonexistent (it could be within experimental errors in velocity measurements), especially over very wide frequency ranges (from a few Hz to several MHz).

Energy Damping

The relationship between frequency and wave-energy loss, or damping, is a critical parameter in selection of an appropriate anelastic model for any medium. Recent reviews have summarized a large number of laboratory and field studies of wave-energy losses in rocks in which the specific attenuation factor, $1/Q$, and the logarithmic decrement have been shown to be approximately independent of frequency over a range of at least 10^8 Hz (Knopoff and MacDonald, 1958; Knopoff, 1965; White, 1965; Bradley and Fort, 1966; Attwell and Ramana, 1966). Attwell and Ramana included some sediment data.

Evidence that the specific attenuation factor, $1/Q$, is independent of frequency implies that attenuation in dB/unit length, α, increases linearly with frequency, f (e.g., White, 1965, p. 98). Recent summaries of work in this field in the case of compressional waves (White, 1965; Knopoff, 1965; Attwell and Ramana, 1966) indicate that, for most rocks, there is a small variation around linearity in the range of frequencies of most interest in underwater acoustics and marine geophysics; that is, in the relationship, $\alpha = kf^n$, the exponent "n" is approximately one. These studies included dry rocks (usual in the laboratory) and in situ measurements in rock strata which, below groundwater levels, are saturated.

Recent measurements of attenuation in water-saturated, natural sediments over a frequency range from 3.5 to 1500 kHz can be summarized as follows (see Hamilton, 1972, for references). Tables (in Hamilton, 1972) list 25 values of the exponent of frequency, n, (in $\alpha = kf^n$) between 0.94 and 1.26; however, all but two of the values fall between 0.94 and 1.11. The writer believes the best of his experimental data are in fine sand off San Diego ($n = 1.007 \pm 0.060$). When these data are combined with Shumway's averaged data for fine sand from the same area, $n = 0.992 \pm 0.065$.

Data listed in tables in Hamilton (1972) are plotted in Fig. 1 (frequency vs attenuation).

Several measurements of energy damping have been added to Fig. 1 since publication in 1972. These are:

Material	Frequency	Attenuation dB/m	Reference
Sandy Shale (borehole)	12.5 Hz	0.001	Mack, 1966
Sand and boulder clay, Cardigan Bay, Wales	150	0.03	Li and Smith, 1969
		0.01	
		0.04	
		0.10	
	600 Hz	0.08	
		0.08	
		0.16	
		0.32	
Fine sand and shell	30 kHz	18.7	Muir and Adair, 1972
	100 kHz	61.7	
Mud (clay-silt), Baltic Sea	1.2 kHz	0.29	Schirmer, 1971
	1 MHz	383	
Mediterranean and Atlantic:			
Abyssal plains (avg.)	12 kHz	0.70	Bennett, 1966
Non-abyssal plains (avg.)	12 kHz	0.36	
Silty sand, Santa Barbara shelf (avgs.)	15 kHz	3.3	Igarashi, 1973
	30 kHz	9.5	
	60 kHz	20.8	

The references in the above table report the exponent of frequency, n, as follows: Li and Smith, 1969 - 0.71, 1.28, 1.07, and 0.81; Muir and Adair, 1972 - 1.00, Schirmer, 1971 - 1.07; Igarashi, 1973 - 1.13 to 1.26.

It can be seen in Fig. 1 that most of the data are consistent with an approximate first-power dependency of attenuation on frequency over a wide frequency range. The upper and lower bounds of the data-plot probably define the area in which most natural marine sediments and sedimentary rocks will lie. With regard to sediment type, the silt-clays, or 'mud' (squares) lie in a narrow band along the lower side of the data plot, and the sands (circles), and mixtures--silty sands, sandy silts, etc. (triangles)--along the top. Extrapolation of the silt-clay and mixed-type data to frequencies below 1 kHz, using a first-power dependency, results in attenuation values in accord with the silt-clay data of Bennett (1967; 40 to 900 Hz), the clay-sand data of Tullos and Reid (1969; 50 to 400 Hz), and the sand-boulder-clay data of Li and Smith (1969; 150 and 600 Hz).

It should be emphasized that in Fig. 1, three sediment types are differentiated: the sands, silt-clays, and mixtures of these two end types; they are shown on the same plot for convenience. As far as significant differences between sediment types in the dependence of attenuation on frequency are concerned, the totality of the experimental evidence does not support Mizikos (1973) who calls for f^2 for muds, nor Stoll and Bryan (1970) or Stoll (1973) who follow Biot (1956) in suggesting $f^{1/2}$ or f^2 for sands (depending on frequency).

In summary, the experimental evidence indicates that the dependence of attenuation on frequency for both mud and sand is close to f^1, and does not support any theory calling for a dependence of attenuation on $f^{1/2}$ or f^2 for either (or both) sediment types or mixtures. However, as in the discussion of velocity dispersion, the case should not be overstated. There is no single data set covering more than two orders of magnitude in frequency, and data is scarce below 1 kHz. These data are enough to show that dependence of attenuation on frequency is more nearly f^1 than $f^{1/2}$ of f^2 (but is not enough to verify an exact dependence) for the following: silt-clays or muds from a few Hz to at least 1 MHz, and from 1 kHz to at least 1 MHz for sands, and from 150 Hz to 1 MHz for mixed types. More information is especially needed for attenuation in sands at frequencies below 1 kHz.

Review of Elastic and Viscoelastic Models

In the field of soil mechanics, large static or dynamic stresses have to be considered, and over the full range of stresses, sediments may be elastic, viscoelastic, or plastic. Yong and Warkentin (1966, pp. 80-94) have a good discussion of various models and elements within the models which describe this behavior.

In the fields of soil mechanics and foundation engineering, the Hookean model and equations are commonly used for derivations of dynamic elastic constants and studies of vibrating loads (e.g., Barkan, 1962; Heukelom, 1961; Jones, 1958; Evison, 1956; Hardin and Richart, 1963; Hall and Richart, 1963; Richart and Whitman, 1967). However, the dynamic moduli from most velocity data are for very small strains on the order of about 10^{-6}, and corrections to moduli should be made for greater strains (Whitman et al., 1969, have a correction curve).

In the fields of physics and geophysics, studies of the elasticity of minerals and rocks have demonstrated that the elastic equations of the Hookean system adequately define the velocities of compressional and shear waves; these equations are conveniently interrelated in a table by Birch (1961, p. 2206). This field has been summarized by Birch (1966) and by Anderson and Lieberman (1968); papers of special interest are by Christensen (1966a,b), Brace (1965a,b) and Simons and Brace (1965).

Although the elastic equations of the Hookean model adequately account for wave velocities in most earth materials, they do not provide for wave-energy losses in these media. To account for both wave velocities and energy losses, various anelastic (viscoelastic and 'near-elastic') models and equations have been proposed. Viscoelastic models frequently favored are the Kelvin-Voigt, Maxwell, or some other combination of Hookean elastic springs and Newtonian dashpots (see Yong and Warkentin, 1966, for a concise resume) or some variation of Biot's models (1956) in which a basic assumption involves movement of pore water of the Poiseuille type (at lower frequencies).

In his various papers, Biot (e.g., 1956, 1962) discussed the full range of systems in which water within pore spaces does or does not move with the solids upon imposition of a small stress such as that of a sound wave. In some of these acoustic models this movement or flow of water through the sediment mineral structure was considered to be of the Poiseuille type. In the last several decades it has been determined that the simple flow equations of the Poiseuille type (derived from flow of water through tubes) do not hold for real, in situ sediments. These equations have to be considerably altered, even for clean sands, and are not applicable to relatively impermeable clays (Yong and Warkentin, 1966). In other words, models based on Poiseuille-type flow of pore water are probably not applicable to natural sediments.

One model which has been especially studied in connection with rocks and sediments is the Kelvin-Voigt model, in which, as originally defined, compressional-wave velocity varies with frequency; and attenuation, at frequencies of most interest in underwater acoustics and geophysics, increases with the square of frequency. White (1965, pp. 110-112) has a thorough discussion of theory and experimental evidence on this subject, and concludes (p. 112) that neither velocity nor attenuation shows this frequency dependence, and the Voigt solid cannot be considered an adequate model of earth materials. The evidence of this report and earlier ones (Hamilton et al., 1970; Hamilton, 1971a, 1972) are in accord with this conclusion.

A Viscoelastic Model for Water-Saturated Sediments

Given macroscopic isotropy, small sinusoidal stresses, wave lengths much greater than grain size, and frequencies from a few Hertz to at least several hundred kHz (and probably in the MHz range for most natural sediments), the restrictive parameters for any elastic, 'nearly elastic', or viscoelastic model for marine sediments can be summarized as follows: (1) almost no marine sediments can be considered suspensions, (2) almost all have non-spherical mineral particles which form structures which have sufficient regidity to transmit shear waves, (3) Poiseuille flow (through small tubes) probably does not hold for relatively impermeable silt-clays nor for natural sands, (4) velocity dispersion is absent or negligibly present, and (5) the dependence of attenuation on frequency is close to f^1. Some relative movement of pore water and mineral frame cannot be excluded on the basis of present evidence, although the above parameters indicate that, if present, it should be small.

The model proposed below is within, or accounts for, the above restrictions, and has several advantages. It is a good working model which does not specify the mechanics of attenuation. It is an anelastic model which includes provision for velocity dispersion and non-linear dependence of attenuation on frequency; the user is thus not committed, a priori, to no-velocity-dispersion or to any particular f^n relationship. The model also indicates clearly those factors involving velocity dispersion and non-linear attenuation which, if negligible, can be dropped. It indicates clearly under what conditions Hookean elastic equations can be used to interrelate wave velocities and other elastic moduli. And, interestingly, this model has been widely used in studies of rocks and the earth's crust, as well as in the properties of polymers, and in some soil mechanics studies.

Most publications of the equations for the recommended model start with a priori assumptions that 1/Q is independent of frequency, that linear attenuation (e.g., dB/m) is related to the first power of frequency, and that there is no velocity dispersion. As a consequence, these publications do not include the equations which show just what has been dropped by accepting these assumptions. The equations below will include those factors usually dropped upon a priori acceptance of the preceding assumptions.

It should be emphasized that other models are not excluded if they are within the above stated parameters. The whole subject merits much more experimental and theoretical study.

A model and concomitant equations within the parameters noted above are a case of linear viscoelasticity. The basic equations of linear viscoelasticity have been summarized in an excellent treatise by Ferry (1961). For the model recommended in this paper, the basic equations (Adler, Sawyer, and Ferry, 1949) have been discussed in different form, including neglect of negligible factors, by Nolle and Sieck (1952), Ferry (1961, pp. 93-94), Krizek (1964), White (1965), Krizek and Franklin (1968), Hamilton et al. (1970), and others.

In the above model, Lamé elastic moduli μ and λ are replaced by complex moduli, $(\mu + i\mu')$ and $(\lambda + i\lambda')$, in which μ,λ, and density govern wave velocity and the imaginary moduli, $i\mu'$ and $i\lambda'$, govern energy damping. The following (Ferry, 1961, pp. 11-13) illustrates the stress-strain relations in this model. For a sinusoidal wave, if the viscoelastic behavior is linear, the strain will be out of phase with stress. The stress can be vectorially decomposed into two components: one in phase with strain and one 90° out of phase. For a shear wave, the complex stress/strain ratio is $\mu^* = \mu + i\mu'$. The phase angle, φ, which expresses energy damping is, in this case: $\tan \varphi = \mu'/\mu$.

The basic derivations of the above model are in Ferry (1961) and White (1965) and will not be repeated here. Without assumptions as to negligible factors, the equations of the model advanced by Bucker (in Hamilton et al., 1970, p. 4046), or in Ferry (1961, p. 94, p. 419), reduce to the following for both compressional and shear waves (with some changes in notation).

$$\frac{1}{Q} = \frac{aV}{\pi f - \frac{a^2V^2}{4\pi f}} \tag{1}$$

where

1/Q is the specific attenuation factor, or specific dissipation function

a is the attenuation coefficient

V is wave velocity

f is frequency (circular frequency, $\omega = 2\pi f$).

Subscripts (p or s) can be inserted in Eq. (1) when referring to compressional or shear waves.

When energy damping is small (i.e., $\lambda' \ll \lambda$ and $\mu' \ll \mu$; White, 1965, p. 95; Ferry, 1961, p. 123; $r \ll 1$, where $r = aV/2\pi f$), the term in the denominator of Eq. 1, $a^2V^2/4\pi f$, is negligible and can be dropped. This leaves the more familiar expression (e.g., Knopoff and MacDonald, 1958; White, 1965; Bradley and Fort, 1966; Attwell and Ramana, 1966):

$$\frac{1}{Q} = \frac{aV}{\pi f} \tag{2}$$

$$\frac{1}{Q} = \frac{2aV}{\omega} = \frac{\Delta}{\pi} = \tan\varphi \quad . \tag{3}$$

Additionally

$$\frac{1}{Qp} = \tan\varphi P = \frac{\lambda' + 2\mu'}{\lambda + 2\mu} \tag{4}$$

$$\frac{1}{Qs} = \tan\varphi s = \frac{\mu'}{\mu} \tag{5}$$

$$\frac{\Delta E}{E} = \frac{2\pi}{Q} \tag{6}$$

$$\alpha = 8.686a \tag{7}$$

where (in addition to those symbols already defined)

Δ is the logarithmic decrement (log of the ratio of two successive amplitudes in an exponentially decaying sinusoidal wave)

$\tan\varphi$ is the loss angle

$\Delta E/E$ is fraction of strain energy lost per stress cycle

α is attenuation in dB/linear measure (e.g., dB/cm).

Equations involving compressional- and shear-wave velocities in Hamilton et al. (1970), or in Ferry (1961), are (in Ferry's notation)

$$(\lambda + 2\mu) = \rho V_p^{\,2}(1 - r^2)/(1 + r^2)^2 \tag{8}$$

$$\mu = \rho V_s^{\,2}(1 - r^2)/(1 + r^2)^2 \tag{9}$$

where

$r = aV/2\pi f$

λ = Lamé's constant

μ = rigidity

ρ = density.

In Eqs. (8) and (9), the term, $(1 - r^2)/(1 + r^2)^2$, indicates the degree of velocity dispersion for linear viscoelastic media. When damping is small (defined above), this term is negligible, and can be dropped, as implied by Ferry (1961, p. 94). This leaves the more familiar Hookean equations

$$(\lambda + 2\mu) = \rho V_p^{\,2} \tag{10}$$

$$\mu = \rho V_s^{\,2} \quad . \tag{11}$$

This means that if the factor $(1 - r^2)/(1 + r^2)^2$ in Eqs. (8) and (9) and the term in the denominator of Eq. (1), $a^2V^2/4\pi f$, are considered negligible and dropped, that wave velocity, 1/Q, and the log decrement are independent of frequency, and linear attenuation is proportional to the first power of frequency.

Computations using the data of Hamilton (1972), and from the literature, indicate that most water-saturated rocks and sediments qualify under the above definitions as media with 'small damping'. For example computations from Hamilton (1972, Table 1) indicate that the factor $(1 - r^2)/(1 + r^2)^2$ for compressional waves at 14 kHz is 0.9992 in fine sand, and an average of 0.9997 for four silty clays; in Pierre shale (McDonal et al., 1958), the factor for shear waves is about 0.992. Equations (2) through (7), (10), and (11) should apply to both water-saturated sediments and rocks.

Those investigators who wish to include velocity dispersion and 1/Q or a log decrement dependent on frequency, and linear attenuation not proportional to the first power of frequency, can consider Eqs. (1), (8), and (9). The results of computations involving wave velocities, densities, and associated elastic constants will be negligibly different if one uses viscoelastic Eqs. (1), (8), and (9), or the classic Hookean elastic equations (e.g., Eqs. 10 and 11).

Computations of Elastic Constants

The computation of elastic constants for saturated sediments was discussed at length in Hamilton (1971a), and was reviewed in Hamilton (1971b, 1973). The general subject will only be briefly noted in this section.

To compute elastic constants in saturated sediments using Hookean elastic equations, as justified in the preceding section, requires values of density, and any two other constants. Density and compressional velocity are easily measured or can be reasonably predicted for most common sediment types (Hamilton, 1971a,b, 1973). One more elastic constant is required to compute the others. The third constant selected (Hamilton, 1971a) to use in computations was the bulk modulus (incompressibility). The theoretical basis of this computation follows Gassmann (1951).

Gassmann (1951) formulated a 'closed system' in which pore water does not move significantly, relative to the mineral frame (no movement of water in or out of a unit volume), the effective density of the medium is the sum of the mass of water and solids in a unit volume, wave velocity and energy damping (e.g., 1/Q) are independent of frequency, and Hookean elastic equations can be used in studying wave velocities unless energy damping is to be considered. The closed system as a special case in the elasticity or viscoelasticity of saturated, porous media has been noted in many studies (references in Hamilton, 1972).

The bulk modulus was selected as the third constant to use in computing the other elastic constants because it appears possible to compute a valid bulk modulus from its components. The equation used in this computation (Gassmann, 1951) is

$$\kappa = \kappa_s \frac{\kappa_f + Q}{\kappa_s + Q} \qquad Q = \frac{\kappa_w(\kappa_s - \kappa_f)}{n(\kappa_s - \kappa_w)} \tag{12}$$

where

κ_s = aggregate bulk modulus of mineral solids

κ_f = frame bulk modulus ("skeletal" bulk modulus of Gassmann, 1951)

κ_w = bulk modulus of pore water

n = decimal-fractional porosity of sediment.

Good values for the bulk modulus of distilled and sea water, κ_w, and most of the common minerals of sediments, κ_s, have been established in recent years. This leaves only a value for the frame bulk modulus, κ_f, needed to compute a bulk modulus for the water-mineral system.

A contribution of Hamilton (1971a) was in derivation of a relationship between sediment porosity and the dynamic frame bulk modulus (Fig. 2). Using this relationship, the frame bulk modulus was derived for each sample, and used with the bulk moduli of pore water and minerals to compute the system bulk modulus with Eq. (12).

The computed bulk modulus, and measured density and compressional-wave velocity, were then used to compute the other elastic constants. Those equations using these three constants were favored. The equations are

$$\text{Compressibility, } \beta = \frac{1}{\kappa} \tag{13}$$

$$\text{Lamé's constant, } \lambda = \frac{3\kappa - \rho V_p^2}{2} \tag{14}$$

$$\text{Poisson's ratio, } \sigma = \frac{3\kappa - \rho V_p^2}{3\kappa + \rho V_p^2} \tag{15}$$

$$\text{Rigidity (Shear) Modulus, } \mu = \frac{3}{4}\left(\rho V_p^2 - \kappa\right) \tag{16}$$

$$\text{Velocity of shear wave, } V_s = (\mu/\rho)^{1/2} \quad . \tag{17}$$

Tables of measured and computed elastic constants for various sediment types are in Hamilton (1971a); up-dated tables are in Hamilton (1973).

Buchan et al. (1972, p. 271) attribute to Gassmann (1951) an incorrect equation (in the form of compressibilities) showing the relation between the frame, or structure, bulk modulus, porosity, and the other component moduli of the system, or aggregate, bulk

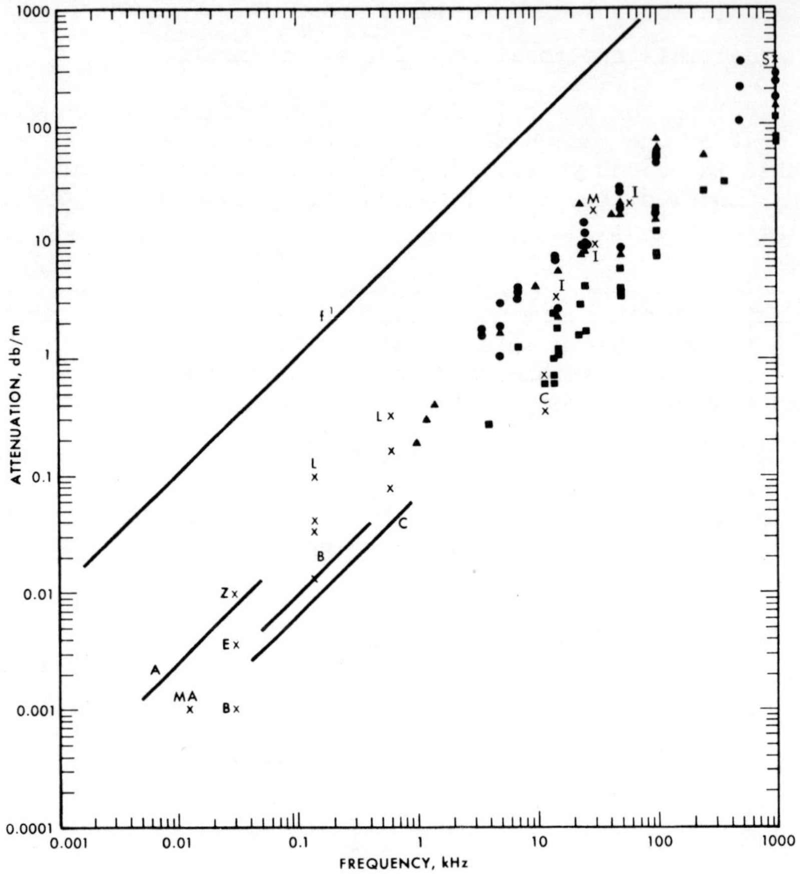

FIGURE 1

Attenuation versus frequency in natural, saturated sediments and sedimentary strata; revised figure from Hamilton (1972). Symbols: circles:sands (all grades); squares:clayey silt, silty clay; triangles:mixed sizes (e.g., silty sand, sandy silt, sand-silt-clay); sand data at 500 and 1000 kHz from Busby and Richardson (1957). Low-frequency data: line "A"-Zemstov, 1969 (land, sedimentary strata); "B"-Tullos and Reid, 1969 (Gulf of Mexico coastal clay-sand), "C"-Bennett, 1967 (sea floor, reflection technique); Z-Zhadin (in Vasil'ev and Gurevich, 1962); E and B-Epinatyeva et al., and Berzon (in Zemstov, 1969). Line labeled "f^1" indicates slope of any line having a dependence of attenuation on the first power of frequency. Data added to this revised figure: MA - Mack, 1969 (sand shale): M - Muir and Adair, 1972 (fine sand and shell); S - Schirmer, 1971 (mud, Baltic Sea); C (at 12 kHz) - Bennett, 1966 (avg. abyssal plains, avg. non-abyssal plains: I-Igarashi, 1973 (silty sand).

modulus. This equation is the one used by Laughton (1957). The difference between Gassmann's equation (Eq. 12, above) and the one used by Laughton and noted by Buchan et al. was discussed in Hamilton (1971a, pp. 583-584).

Corrections to density, compressional velocity, and the bulk modulus from laboratory to in situ conditions allow prediction of in situ elastic constants. Such in situ computations were discussed with a numerical example in Hamilton (1971b).

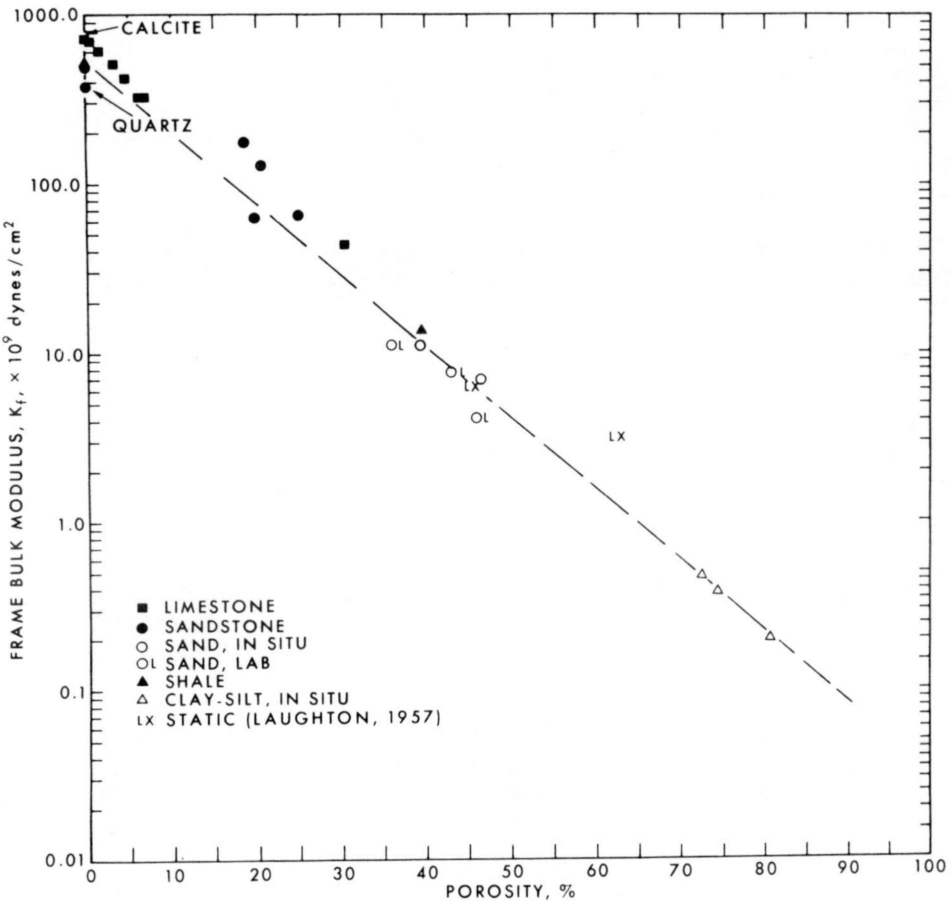

FIGURE 2

Porosity versus dynamic frame bulk modulus, K_f; see Hamilton (1971a) for discussion and regression equations.

III. DATA REQUIRED TO CONSTRUCT GEOACOUSTIC MODELS

Introduction

The real sea floor cannot be defined by any single geoacoustic model; therefore, it is important that acoustic and geophysical experiments at sea involving the sea floor be supported by a particular model of the area. However, it is possible to use geologic and geophysical judgment to extrapolate a general model over wider areas. A sufficient collection of models from diverse environments will allow predictions of bottom models in similar areas of the world's oceans.

A geoacoustic model should detail the real sea floor. It can then be used in studies of reflection and refraction of compressional and shear waves over a wide range of frequencies, in geologic studies of stratigraphy, sedimentology, and geologic history, and in various other studies in the field of geophysics (e.g., gravity computations).

The production of a geoacoustic model of the sea floor requires assembly of data from a wide variety of sources in the fields of oceanography, geology, and geophysics. A model thus brings into focus and utility data from many scientific disciplines and operations at sea and in the laboratory. The gross layering may be all that is required in some geologic and geophysical studies, but the acoustician must be supplied sufficient detail to study insonified areas at various sound frequencies.

Data Required and Methods

In an ideal production of a geoacoustic model, the following data should be derived at sea and in the laboratory. In addition, associated information from all available sources, published and unpublished, should be sought and selectively used.

Data for a bathymetric chart. The first requisite of a geoacoustic model is a good bathymetric chart of the insonified (and adjacent) area. Data required includes: (1) all available sounding data from government sources and oceanographic institutions (published and unpublished), (2) a careful record of all ship's movements on station, (3) continuous echo sounding records, (4) a Nansen cast or other data which allows corrections from echo-sounder to true depths, (5) location by satellite navigation methods. In the laboratory, the smoothed ship's track is plotted with soundings, and all available data is used to produce a good contoured chart of the insonified and adjacent areas. It can rarely be assumed that any given published chart of an area is valid. Very little of the sea floor has been charted properly, in detail.

Data to determine layer thicknesses and locations of reflectors. Continuous seismic reflection profiling determines travel time between impedance mismatches, or reflectors. Air-gun power sources can obtain data at low frequencies on the order of 20 to 50 Hz. Electric 'sparker' sources usually are operated between about 80 and 250 Hz. Layering to depths of 5 to 20 m in silt-clay sediments can be seen at 12 kHz by the normal echo sounder operating on a short ping. The 3.5 kHz system frequently shows reflectors in silt-clays to depths of 40 to 60 m.

Given travel time in a sediment layer, the true thickness can be derived if the interval velocity, or velocity gradient, is known. At present, these data are usually acquired from wide-angle reflection measurements using expendable sonobuoys (LePichon et al., 1968; Houtz et al., 1968).

Water-mass data. To predict in situ sediment surface properties, it is necessary to have information on the sound velocity, density, and salinity of the sea water at the water-sediment interface. These data can be derived from a normal Nansen cast; a curve showing sound velocity versus water depth is particularly useful.

Data on sea-floor relief. Details of bottom topography, roughness, relief, and slope are required for some acoustic studies. These can be determined by surface echo sounders (especially those with narrow beams), underwater cameras, and deep-towed equipment.

Data on rock layers. Rock layers at or near the sea floor are important to the underwater acoustician or geophysicist. At low frequencies all of the sediment column and deeply buried rock layers can be important. Information is required on, at least, the density, compressional-wave velocity, and attenuation in these rock layers.

Data for sediment properties. Sediment samples from gravity and piston corers, box corers, or other samplers are required to obtain sediment physical properties. Sound velocities can be measured aboard ship, or the samples can be preserved under sea water in a refrigerator for velocity and other measurements in the laboratory.

In the laboratory, the minimum physical property measurements should include grain size analyses (mean grain size, and percentages of sand, silt, and clay), bulk grain density, saturated density, porosity, and additional sound velocity measurements. Other properties can be computed or predicted through these measured properties.

In shallow water, the best data can be obtained by in situ

measurements for some properties (e.g., as in Hamilton et al., 1970; Hamilton, 1972).

If all of the above data are not available, or if there is no data at all, certain in situ predictions can be made following Hamilton (1971b). Predictions of layer thicknesses and attenuations (not included in the 1971 report) will be briefly noted below. An up-to-date resumé of the writer's data is in Hamilton (1973).

IV. COMMON GEOACOUSTIC MODELS

Among an almost infinite variety of geoacoustic models there are four common types. Two of these are of the continental shelf and two are of deep-sea areas. Actual gross models will illustrate these types.

Shallow-Water Geoacoustic Models

A common stratigraphy in continental shelves is a top layer of soft mud, or clay-silts, overlying harder silts and sands. This is common because during lowered sea levels of the Pleistocene sand was deposited over wide areas of the shelf, and then covered with silt-clays as sea level rose. Figure 3 is a record from the shallow Asiatic continental shelf, which illustrates this model. The record was made with a 12 kHz echo sounder using a short ping. The old sea floor, composed of sands, silts, and clays, was uneven and gullies were eroded in lower areas. Subsequent deposition of silty clay has covered most of the area. Sediment samples of the lower sand-silt were taken where the material crops out on the present sea floor. The gross model for the maximum sediment thickness is illustrated in Fig. 4.

FIGURE 3

Record from 12 kHz echo-sounder on the Asiatic continental shelf. The old, gullied sea floor has been covered by more recent muds, but crops out at, or near, the present sea floor (left side of figure). Line below labels is a multiple of the sea-floor reflection.

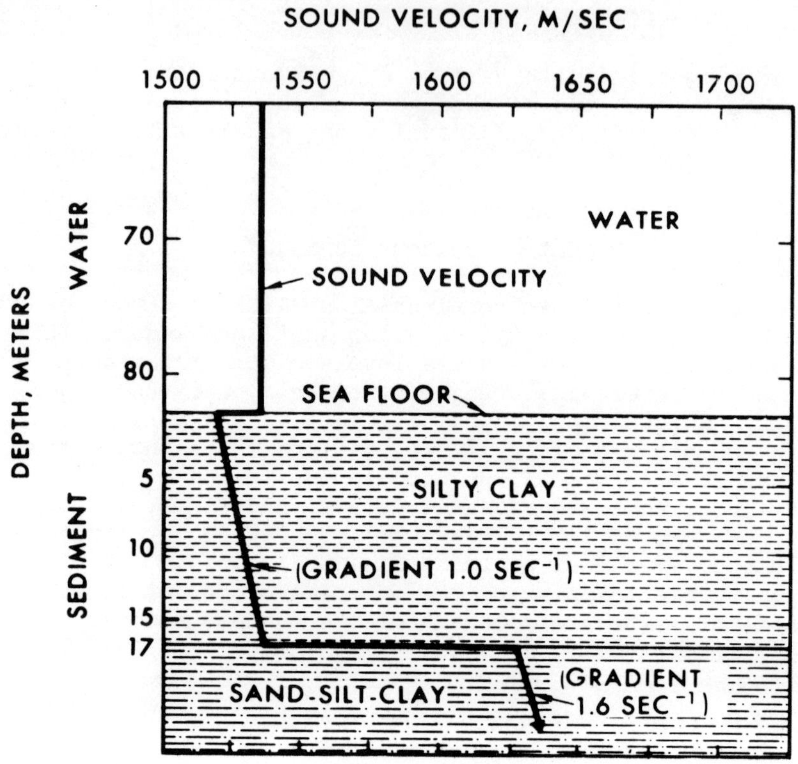

FIGURE 4

Diagram of gross geoacoustic model for thickest section of the sea floor shown in Fig. 3. The velocity gradient is assumed; velocities in sediments from cores. This is one of two common shallow-water geoacoustic models (mud over sand or rock).

The other common shallow-water model is a layer of thick sand, usually overlying rock, illustrated by Fig. 5. The sand usually forms a high-density, high-velocity (hence high-impedance) layer in which sound attenuation is also high. Subbottom layers are not usually acoustically detected in these areas.

Deep-Sea Geoacoustic Models

In the deep-sea there are two common models: one of abyssal plains and one of abyssal hills.

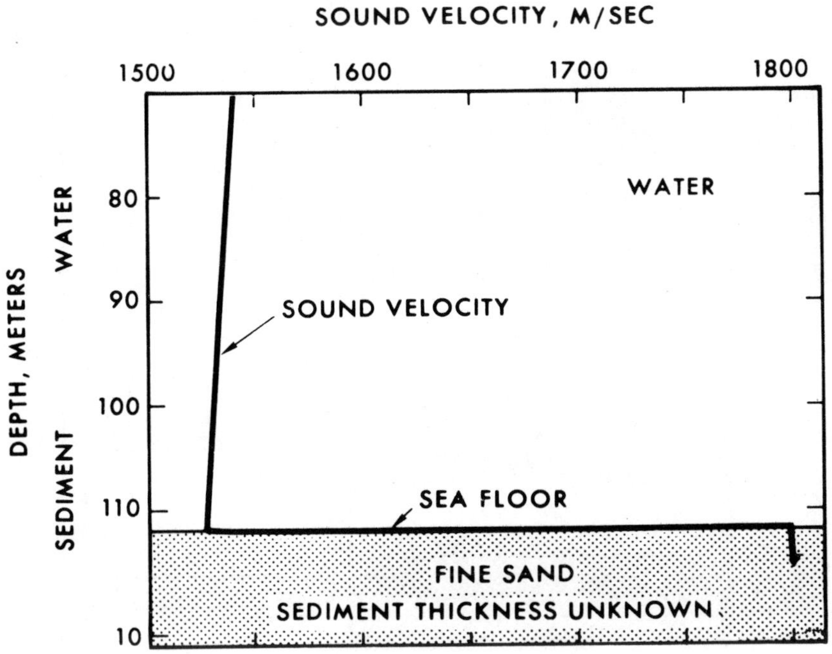

FIGURE 5

One of two common shallow-water geoacoustic models (a thick layer of sand over rock).

In the abyssal hills, there is usually a single layer of pelagic silt-clays, with or without volcanic ash layers, over volcanic or sedimentary rock. The sediment layer may be quite thin as demonstrated in the Pacific by the Deep Sea Drilling Project, and by reflection surveys (e.g., Ewing et al., 1968). This type of geoacoustic model is illustrated in Fig. 6. The general area is in the north central Pacific, west of the Aleutian Abyssal Plain. The area is in the volcanic ash zone as described by Horn et al. (1969).

As previously noted, when sedimentary layers are thin, or when frequencies are very low, the model must include properties of the rock layers. Much of the Pacific has a silt-clay layer (overlying basalt) in which one-way sound travel time is 0.05 sec or less (Ewing et al., 1968); these layers will usually be 50 to 100 m thick.

Rock velocities can usually be obtained from refraction surveys in or near the area of interest. For example, in the area

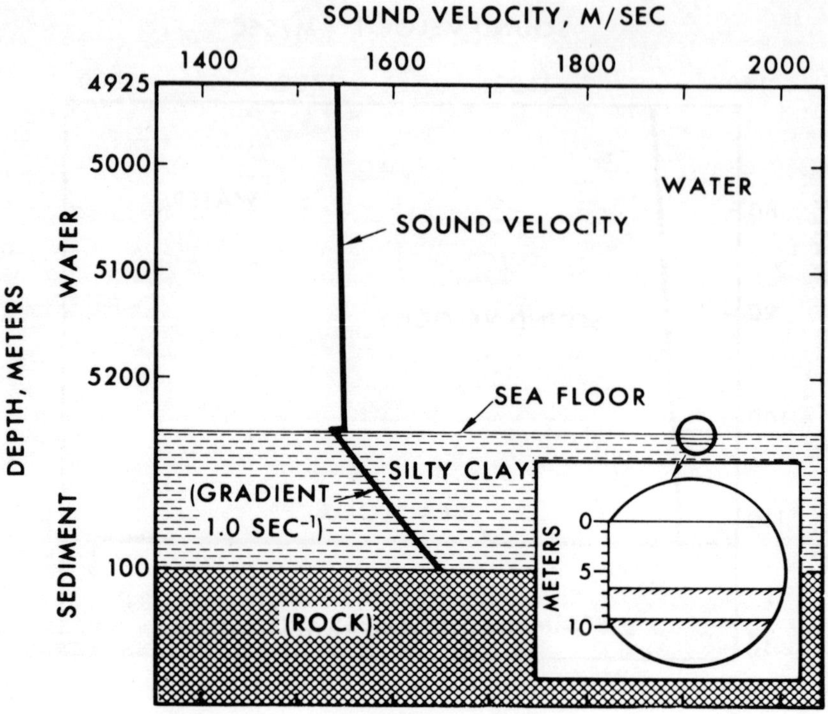

FIGURE 6

One of two common deep-sea geoacoustic models (abyssal hills). This model, from North Pacific abyssal hills, represents a thin (100 m) layer of pelagic deep-sea clay overlying basalt. The inset figure shows reflectors (seen on 12 kHz records) which are probably formed by volcanic ash. Tables 1 through 5 giving numerical details are appended.

represented by Fig. 6, the rock is basalt as determined by the Deep Sea Drilling Project, and an average velocity in the top of the basalt is 5.7 km/sec (Houtz et al., 1970). Given a rock velocity, the best procedure to get density, at present, is to enter diagrams relating density and velocity (e.g., Nafe and Drake, 1967; Dortman and Magid, 1969). Approximate attenuation values for different rock types can usually be derived from the literature (e.g., Balakrishna and Ramana, 1968; Levykin, 1965).

The less common deep-sea geoacoustic model is of abyssal plains where turbidity currents have laid down alternating sequences of silt-clays and silt-sands (turbidites). These sediments over rock can vary in thickness from a few meters to over 2000 m. The section might comprise hundreds of layers. Most of these layers have been deposited by flows which top the natural levees of great undersea channels, and individual layers vary in thickness and cannot usually be correlated over appreciable distances. Only a few of these layers are usually cored, and some deeper layers are seen as reflectors by echo sounders and reflection equipment. For the acoustician who requires a fully-layered model, the geologist-geophysicist should detail the layering (reflectors) as deep as he can from cores and reflection records, and then accept an alternation of probable 'average layers' to the full thickness of the sediment layer.

Figure 7 represents the deep-water turbidite model; it is from the southern Japan Sea Abyssal Plain. The total thickness of the turbidite layer was determined by reflection profiling (to get sound-travel time in the sediment layer), and an interval velocity measurement using the sonobuoy technique developed at Lamont-Doherty Geological Observatory by Houtz and his colleagues (Houtz et al., 1968; Le Pichon et al., 1968). In the inset is a diagram from the 12 kHz echo sounder which shows detailed layering to about 10 m. The first line (at 0 m) is the sea floor. About 1 m below the sea floor is a strong reflector. The corer dropped through a silty clay layer and stopped in a sand layer. Measurements in the cored sediment indicated that the first meter had a velocity a little more than one percent less than velocity in the bottom water. In situ layer velocities and densities were: first layer, 1467 m/sec and 1.23 g/cm^3; second layer, 1819 m/sec and 2.02 g/cm^3. The linear velocity gradient is from the sediment-surface velocity through the sonobuoy interval velocity.

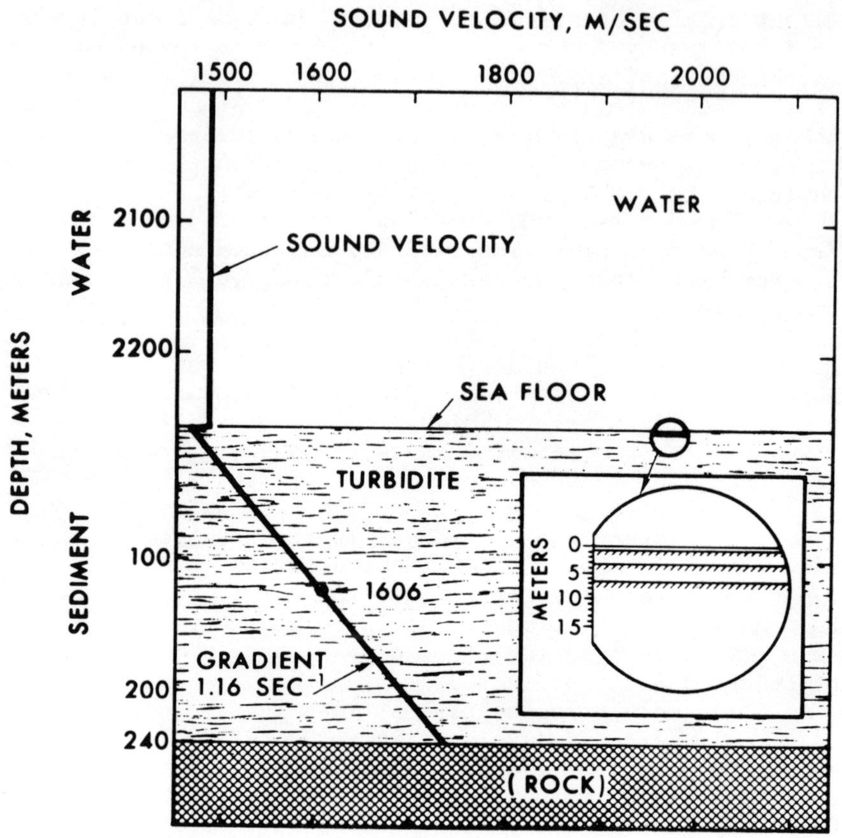

FIGURE 7

One of two common deep-sea geoacoustic models (abyssal plains). This model, from the Japan Sea Abyssal Plain, represents 240 m of multi-layered turbidites overlying rock. The interval velocity (1606 m/sec) was measured by the sonobuoy technique. The linear gradient is from the sediment surface velocity through the interval velocity. The inset figure shows reflectors as seen in 12 kHz records. The first layer was silty clay and the second was sand.

Acknowledgments. This work was supported by the Office of Naval Research (Code 483), and the Naval Ship Systems Command (Code PMS 302).

APPENDIX

DETAILED GEOACOUSTIC MODEL

General

The large, gross models illustrated and discussed are of interest and utility, but for most studies of reflection and bottom loss, it has been determined that a fully-layered model must be used to reconcile experiment with theory; recent examples of this are in reports by Hastrup (1970) in the Mediterranean, and by Morris (1970) in the Pacific. For the acoustician who needs fully-layered models, all available data, estimates, and predictions are used to indicate probable layering and sediment properties down to and including the 'acoustic basement' (usually sedimentary or volcanic rock). In these estimates, available data from other sources are used: the general geologic and geophysical literature, Navy reports, unpublished data, data on similar sediments, and a certain amount of geologic 'intuition and judgment.'

At the Naval Undersea Center, our models include a gross figure such as described, plus a topographic chart of the area of the experiment, and slope and relief as seen by the echo sounder and lowered camera. The acoustician requires quantitative information. He should not have to try to measure it off figures or curves. Consequently, each model is accompanied by 5 tables and a general information sheet.

In the following section, the 5 tables that accompany the model represented by Fig. 6 are explained. The numbers in the tables are examples; values not in parentheses were measured. The subject of predictive methods and corrections from laboratory to in situ properties was discussed in a recent report (Hamilton, 1971b).

A general information form sheet is furnished for each station; it accompanies the 5 tables. The general information sheet includes station location (latitude and longitude), maximum and minimum water depths in fathoms and meters (both echo-sounder and corrected depths), the general geographic area (e.g., the north-central Pacific), the geomorphic province (e.g., abyssal hills), and a brief description of the sea floor (topography, sediment distributions, stratigraphy, structure).

TABLE 1

In situ properties of bottom water.*

True Depth, m	T, °C	S, ppt	P, kg/cm^2	Sound speed, m/sec	Density, g/cm^3	Impedance, g/cm^2sec $\times 10^5$
5251	1.58	34.69	547.4	1546.6	1.05174	1.62662

* At location of model: coring site

TRUE DEPTH was determined by correcting the echo-sounder depth to true depth. This additive correction is based on the velocity profile in the water mass derived from station Nansen casts. A small table supplied by the Acoustic Propagation Division (Code 503, at NUC), is inset in the bathymetric chart to indicate the value of these corrections and to permit correction of the echo-sounder depths on the contour charts to true depths. Such corrections can also be obtained from NAVOCEANO (1966).

TEMPERATURE, SALINITY, PRESSURE, and SOUND SPEED were derived from Nansen-cast data at the indicated true depth.

IMPEDANCE was computed by the formula: density × sound speed.

TABLE 2

In situ physical properties of sediments (other than acoustic).*

Layer No.	h, m	Sediment type	Mean diam., mm	Sand, %	Silt, %	Clay, %	n, %	ρ_s, g/cm^3
1	6.2	Silty clay	0.004	9.8	39.0	51.2	76.3	2.65
2	0.1	(Volcanic ash)	—	—	—	—	(65.0)	(2.7)
3	3.0	(Silty clay)	—	—	—	—	(75.0)	(2.65)
4	0.2	(Volcanic ash)	—	—	—	—	(65.0)	(2.7)
Rock	—	Basalt	—	—	—	—	—	—

Total thickness of sediment over rock: 100 m
*At location of model: coring site.

Notes:
(1) Column headings: h is thickness; n is porosity; ρ_s is density of solids.
(2) Values in parentheses are predicted.
(3) For a complete, estimated geoacoustic model, assume alteration of Layers 3 and 4 to full thickness of sediment (100 m).
(4) Rock type (basalt) determined from velocity, acoustic reflection records (traced into seamounts), and Deep Sea Drilling Project in general area.

The THICKNESSES of sediment layers were determined from cores, 12 kHz, and 3.5 kHz echo-sounder records, acoustic reflections (sparker), sonobuoys, and probabilities of layering (determined in similar sediments or taken from reports of other institutions for the area).

Below coring depths in the sediment, thicknesses of relatively thin layers between reflectors can be computed by measuring sound travel time between reflectors from an echo sounder record (12 or 3.5 kHz) and multiplying by a velocity extrapolated from layers above (using a velocity gradient, as discussed below). The true thicknesses of the reflecting layers (reflectors) can be estimated from the thicknesses of cored layers in probably similar sediments above or elsewhere. For example, Horn et al. (1969) reported that white volcanic ash in the northeast Pacific ranged in thickness from 1 to 29 cm (avg. 6.5 cm); the thicker sections would be those closest to sources of the ash; the mean grain size of these ashes was 5.42 phi. Similar information is usually available in the literature (e.g., Horn et al., 1971).

There are three usual alternatives when computing true sediment thicknesses for relatively thick layers in an area where no interval velocities have been measured: (1) use an equation or curve for mean velocity versus travel time from a similar area, (2) use a predicted linear gradient and a predicted V_0 (discussed below), or (3) assume an interval velocity. Of these three alternatives, the last one is most commonly used, and is the least accurate. These alternatives are briefly discussed, with examples, in Hamilton (1973) and Hamilton et al. (1973).

There are now sufficient published data to show that more areas of turbidites have reasonably close velocity gradients in the upper, unlithified layers. For example, at a one-way travel time of t = 0.2 sec, the computed thickness of a layer using the Atlantic and Gulf of Mexico equations of Houtz et al. (1968), and those for the Central Bengal Fan and Kamchatka Basin (Hamilton et al., 1973) are, respectively, 347 m, 341 m, 351 m, and 343 m: a variation of less than 3 percent. Thus, if one is computing sediment layer thicknesses for an area of turbidites where no measurements have been made, the use of equations for the most similar area will probably yield reasonable results. If the sediment type is calcareous ooze, the equations for the Pacific Equatorial Zone (Houtz et al., 1970) are recommended.

Given a linear gradient, a, the sediment surface velocity, V_0, and one-way travel time, t, the thickness of a layer can be computed (Houtz and Ewing, 1963) by

$$h = V_0(e^{at} - 1)/a$$

where e is the base of natural logarithms.

This is a very useful equation because V_0 can be closely estimated (Hamilton, 1971b) and one-way travel time in a layer can be measured from a reflection record; and, the velocity gradient can usually be reasonably estimated (see Hamilton, 1973, or Hamilton et al., 1973, for fuller discussion).

SEDIMENT TYPE, MEAN DIAMETER OF MINERAL GRAINS, and PERCENTAGES OF SAND, SILT, AND CLAY were determined from grain size analyses. These follow the nomenclature scheme discussed by Shepard, 1954; however, the Wentworth scale was used for sand sizes. The Wentworth scale is based on median diameter of mineral grains; very fine sand (0.062 to 0.125 mm); fine sand (0.125 to 0.250 mm); medium sand (0.250 to 0.500 mm); coarse sand (0.500 to 1.000 mm); silt (0.062 to 0.004 mm); and clay (less than 0.004 mm).

POROSITY is the volume of voids or pore space divided by the volume of the sample; this was determined from evaporation of pore water and corrected for dried salts (see Hamilton, 1971b, for methods).

DENSITY OF SOLIDS, the bulk density of the mass of dried mineral grains (without salts evaporated from the pore water), was determined by the pycnometer method.

Total thickness of sediment over rock (100 m) was determined from one-way reflection time in the sediment and sediment interval velocity as discussed above.

TABLE 3

In situ acoustic properties of sediments.*

Layer No.	Sound velocity, m/sec	Velocity gradient, sec^{-1}	Velocity ratio	Density, g/cm^3	Impedance, $g/cm^2 sec \times 10^5$
1	1539	(1.0)	0.994	1.44	2.216
2	(1595)	—	(1.03)	(1.63)	(2.600)
3	(1545)	—	—	(1.45)	(2.240)
4	(1598)	—	—	(1.63)	(2.605)
Rock	(5700)	—	—	(2.6)	(14.82)

*At location of model: coring site

Notes:

(1) Values in parentheses are predicted. Velocity increased in lower layers by the amount indicated by the velocity gradient.

(2) For a complete, estimated geoacoustic model, assume alternation of Layers 3 and 4 to full thickness of sediment (100 m).

(3) Velocity in basalt from an average for the general area from Houtz et al., 1970. Density in basalt from velocity-density relationship of Nafe and Drake (1967).

SOUND VELOCITY was determined in the laboratory at approximately 200 kHz, and corrected to in situ values (Hamilton, 1971b); the values in parentheses are predictions. (See the following paragraphs for velocities in the lower layers.)

VELOCITY GRADIENT. The linear velocity gradient (in meters per second per meter, or sec^{-1}) shows the increase in velocity with depth in the sediments. At those stations where interval velocities were measured with sonobuoys, the overall linear gradient was established by using the sediment-surface velocity and the layer-interval velocity. The interval velocity is the actual velocity at a depth in the sediment of one-half the layer thickness. These gradients are smoothed, average gradients through the whole layer or layers, and do not reflect the various velocities in individual layers; consequently, only one gradient is usually given or estimated. Lacking actual measurements, a gradient is predicted (in parentheses); see Hamilton (1973) or Hamilton et al. (1973) for a discussion. These linear gradients usually vary between 0.5 and 2.0 sec^{-1}; a reasonable, over-all prediction is 1.0 sec^{-1}.

For many stations, the tables should indicate estimated properties for lower layers. The estimate is made by relating the properties to those in a higher layer or by predicting sediment type and properties from similar sediments.

The total thickness of sediments in models required by acousticians varies with frequency, energy, grazing angle, etc. Consequently, the known layering can be alternated, as previously indicated, to the sediment-rock interface to furnish a rational basis for geoacoustic models in areas where layered turbidites are present. The velocity gradient should then be used to correct (increase) velocity in each lower layer, a procedure which requires that the impedance also be corrected. For example, assume a velocity of 1500 m/sec and a gradient of 1.0 sec^{-1} in Layer 1. A similar layer at a depth of 100 m in the sediment body will have a velocity of 1600 m/sec. For these purposes it can be assumed that the density is the same, although small increases are probably present in thicker sections.

VELOCITY RATIO is computed by dividing the sediment velocity by the bottom-water velocity; it is the same in the laboratory as in situ (Hamilton, 1971b). At greater sediment depths, the ratio is not given or estimated. If computations are made for deeper layers, the velocity gradient in the sediment must be considered (see previous paragraphs).

DENSITY is the saturated bulk density of a unit volume of sediment, in situ, as corrected from laboratory measurements.

IMPEDANCE is the product of the in situ values of density and velocity.

TABLE 4

In situ, computed elastic constants of sediments.*

Layer No.	β	κ	σ	μ	λ	V_s
1	0.3055	3.2736	0.484	0.1028	3.2051	267
2	0.2623	3.8128	0.468	0.2505	3.6457	392
3	0.3003	3.3298	0.485	0.0986	3.2641	261
4	0.2623	3.8128	0.466	0.2622	3.6379	401

*β, compressibility, dynes/cm^2 $\times$ 10^{-10}; κ, bulk modulus, dynes/ dynes/cm^2 $\times$ 10^{10}; σ, Poisson's Ratio; μ, rigidity modulus, dynes/cm^2 $\times$ 10^{10}; λ, Lamé's Constant, dynes/cm^2 $\times$ 10^{10}; V_s, shear-wave velocity, m/sec.

Data and the method used to compute elastic constants of sediments are in

Hamilton, E. L., "Elastic Properties of Marine Sediments," J. Geophys. Res., 76, 579-604, 1971

Hamilton, E. L., 1971b, "Prediction of In-situ Acoustic and Elastic Properties of Marine Sediments": Geophysics, v. 36, p. 266-284

All of these values were computed by using the measured density, measured compressional (sound) velocity, and a computed value for the bulk modulus (corrected to in situ conditions) with Eqs. (13) to (17), discussed earlier and in Hamilton (1971a,b).

TABLE 5

Predicted attenuation of compressional (sound) waves*

Layer No.	Recommended	k Probable Max.	Probable Min.
1	0.08	0.18	0.04
2	0.12	0.38	0.09
3	0.08	0.19	0.04
4	0.12	0.38	0.09

*To determine an equation which can be used at any frequency, substitute k into the equation

$$\alpha = kf^1$$

where

a is attenuation of compressional (sound) waves in dB/m

k is a constant

f is frequency in kHz.

For deeper layers, alternate values shown for Layers 3 and 4

Data and the method used to predict attenuation are in

Hamilton, E. L., 1972, "Compressional-wave attenuation in marine sediments": Geophysics, v. 37, p. 620-646.

A study by Hamilton (1972) indicated that attenuation of sound in marine sediments was approximately dependent on the first power of frequency. In the equation $\alpha = kf^n$ (where attenuation, α, is in dB/m and frequency, f, in kHz), if the exponent "n" is taken as one, the only variable is the constant "k". Relationships between k and mean grain size and porosity (Hamilton, 1972, Figs. 3-5) in the sediment layers have been used to derive a value of k. This value (and the probable maximum and minimum values) can be substituted into the equation above to derive an equation which can be used at any frequency.

REFERENCES

Adler, F. T., W. M. Sawyer, and J. D. Ferry, "Propagation of transverse waves in viscoelastic media," J. Appl. Phys. 20, 1036-1041 (1949).

Anderson, O. L., and R. C. Liebermann, "Sound velocities in rocks and minerals: experimental methods, extrapolations to very high pressures, and results," in Physical Acoustics, Principles and Methods, W. P. Mason (ed.) (Academic Press, New York, 1968), pp. 330-466.

Attwell, P. B., and Y. V. Ramana, 1966, "Wave attenuation and internal friction as functions of frequency in rocks": Geophysics, v. 31, p. 1049-1056.

Balakrishna, S., and Y. V. Ramana, "Integrated studies of the elastic properties of some Indian rocks," in The Crust and Upper Mantle of the Pacific Area, L. Knopoff, C. L. Drake, and P. J. Hart (eds.), Geophys. Mon. No. 12 (Amer. Geophys. Union, Washington, D.C., 1968), pp. 489-500.

Barkan, D. D., Dynamics of Bases and Foundations (McGraw-Hill Book Co., Inc., New York, 1962).

Bennett, L. C., Jr., In situ measurements of acoustic absorption in unconsolidated marine sediments, Ph.D. Thesis, Bryn Mawr College, 1966, p. 107.

Bennett, L. C., Jr., "In situ measurements of acoustic absorption in unconsolidated sediments" (abstract), Trans. Am. Geophys. Union 48, 144 (1967).

Biot, M. A., "Theory of propagation of elastic waves in a fluid-saturated porous solid. I. Low-frequency range; II. Higher-frequency range," J. Acoust. Soc. Am. 28, 168-178; 179-191 (1956).

Biot, M. A., "Mechanics of deformation and acoustic propagation in porous media," J. Appl. Phys. 33, 1482-1498 (1962).

Birch, F., "The velocity of compressional waves in rocks to 10 kilobars," Part 2, J. Geophys. Res., 66, 2199-2224, 1961.

Birch, F., "Compressibility: Elastic constants," in Handbook of Physical Constants, S. P. Clark, Jr. (ed.), Geol. Soc. Amer. Memoir 97, 98-174 (1966).

Brace, W. F., "Some new measurements of linear compressibility of rocks," J. Geophys. Res., 70, 391-398, 1965a.

Brace, W. F., "Relation of elastic properties of rocks to fabric," J. Geophys. Res., 70, 5657-5667, 1965b.

Bradley, J. J., and A. N. Fort, Jr., "Internal friction in rocks," in Handbook of Physical Constants, S. P. Clark, Jr. (ed.), Geol. Soc. Amer. Memoir 97, pp. 175-194 (1966).

Buchan, S., D. M. McCann, and D. T. Smith, "Relations between the acoustic and geotechnical properties of marine sediments," Quarterly J. Engin. Geology 5, 265-284 (1972).

Busby, J., and E. G. Richardson, 1957, "The absorption of sound in sediments": Geophysics, v. 22, p. 821-828.

Christensen, N. I., "Compressional wave velocities in single crystals of alkali feldspar at pressures to 10 kilobars," J. Geophys. Res., 71, 3113-3116, 1966a.

Christensen, N. I., "Shear wave velocities in metamorphic rocks at pressures to 10 kilobars," J. Geophys. Res., 71, 3549-3556, 1966b.

Dortman, N. B., and M. Sh. Magid, "New data on velocity of elastic waves in crystalline rocks as a function of moisture," Soviet Geol. 5, 123-129 (1968); translation: Intern. Geology Rev. 11, 517-523 (1969).

Evison, F. F., "The seismic determination of Young's Modulus and Poisson's Ratio for rocks in situ," Geotechnique 6, 118-123 (1956).

Ewing, J., M. Ewing, T. Aitken, and W. J. Ludwig, "North Pacific sediment layers measured by seismic profiling," in Geophys. Mon. No. 12, The Crust and Upper Mantle of the Pacific Area, L. Knopoff, C. L. Drake, and P. J. Hart (eds.), pp. 147-173 (Amer. Geophys. Union, Washington, D.C., 1968).

Ferry, J. D., Viscoelastic Properties of Polymers (John Wiley and Sons, Inc., New York, 1961).

Gassmann, F., "Über die Elastizität Poröser Medien," Vierteljahrsschrift Naturforschenden Gesellschaft in Zürich 96, 1-23 (1951).

Hall, J. R., Jr., and F. E. Richart, Jr., "Dissipation of elastic wave energy in granular soils," J. Soil Mech. & Foundations Div., Amer. Soc. Civil Engin. SM 6, 27-56 (1963).

Hamilton, E. L., "Sound velocity and related properties of marine sediments, North Pacific," J. Geophys. Res., 75, 4423-4446, 1970a.

Hamilton, E. L., "Sound channels in surficial marine sediments," J. Acoust. Soc. Am. 48, 1296-1298 (1970b).

Hamilton, E. L., 1970c, "Reflection coefficients and bottom losses at normal incidence computed from Pacific sediment properties": Geophysics, v. 35, p. 995-1004.

Hamilton, E. L., "Elastic properties of marine sediments," J. Geophys. Res., 76, 579-604, 1971a.

Hamilton, E. L., 1971b, "Prediction of in situ acoustic and elastic properties of marine sediments": Geophysics, v. 36, p. 266-284.

Hamilton, E. L., 1972, "Compressional-wave attenuation in marine sediments": Geophysics, v. 37, p. 620-646.

Hamilton, E. L., "Prediction of deep-sea sediment properties: state of the art," in Proceedings of a Symposium, Physical and Engineering Properties of Deep-Sea Sediments, A. L. Inderbitzen (ed.), in press, 1974.

Hamilton, E. L., H. P. Bucker, D. L. Keir, and J. A. Whitney, "Velocities of compressional and shear waves in marine sediments determined in situ from a research submersible," J. Geophys. Res., 75, 4039-4049, 1970.

Hamilton, E. L., D. G. Moore, E. C. Buffington, J. R. Curray, and P. H. Sherrer, "Sediment velocities from sonobuoys: Bay of Bengal, Bering Sea, Japan Sea, and Northeast Pacific," in preparation, 1974.

Hardin, B. O., and F. E. Richart, Jr., "Elastic wave velocities in granular soils," J. Soil Mech. & Foundations Div., Amer. Soc. Civil Engin. SM 1, 33-65; and discussions SM 5, 103-118 (1963).

Hastrup, O. F., "Digital analysis of acoustic reflectivity in the Tyrrhenian Abyssal Plain," J. Acoust. Soc. Am. 70, 181-190 (1970).

Heukelom, W., "Analysis of dynamic deflections on soils and pavements," Geotechnique 11, 224-243 (1961).

Horn, D. R., M. N. Delach, and B. M. Horn," Distribution of ash layers and turbidites in the North Pacific," Bull. Geol. Soc. Amer. 80, 1715-1724 (1969).

Horn, D. R., M. Ewing, B. M. Horn, and M. N. Delach," Turbidites of the Hatteras and Sohm Abyssal Plains, western North Atlantic," Marine Geology 11, 287-323 (1971).

Houtz, R. E., and J. I. Ewing, "Detailed sedimentary velocities from seismic refraction profiles in the western North Atlantic," J. Geophys. Res., 68, 5233-5258, 1963.

Houtz, R. E., J. I. Ewing, and X. Le Pichon, "Velocity of deep-sea sediments from sonobuoy data," J. Geophys. Res., 73, 2615-2641, 1968.

Houtz, R., J. Ewing, and P. Buhl, "Seismic data from sonobuoy stations in the northern and equatorial Pacific," J. Geophys. Res., 75, 5093-5111, 1970.

Igarashi, Y., "In situ high-frequency acoustic propagation measurements in marine sediments in the Santa Barbara shelf, California," Naval Undersea Center Tech. Publ. 334 (NUC TP 334), Naval Undersea Center, San Diego, California (1973), p. 40.

Jones, R., "In situ measurement of the dynamic properties of soil by vibration methods," Geotechnique 8, 1-21 (1958).

Knopoff, L., "Attenuation of elastic waves in the earth," Physical Acoustics, W. P. Mason (ed.) (Academic Press, New York, 1965), Vol. III-B, pp. 287-324.

Knopoff, L., and G. J. F. MacDonald, "Attenuation of small amplitude stress waves in solids," Rev. Modern Phys. 30, 1178-1192 (1958).

Krizek, R. J., "Application of the one-sided Fourier transform to determine soil storage and dissipation characteristics," Proc. Symp. Soil-Structure Interaction, University of Arizona, Tucson, Arizona (1964).

Krizek, R. J., and A. G. Franklin, "Energy dissipation in a soft clay," Proc. Int. Symp. Wave Propagation and Dynamic Properties of Earth Materials (University of New Mexico Press, Albuquerque, New Mexico, 1968), pp. 797-807.

Laughton, A. S., 1957, "Sound propagation in compacted ocean sediments": Geophysics, v. 22, p. 233-260.

Le Pichon, X., J. Ewing, and R. E. Houtz, "Deep-sea sediment velocity determination made while reflection profiling," J. Geophys. Res., 73, 2597-2614, 1968.

Levykin, A. I., "Longitudinal and transverse wave absorption and velocity in rock specimens at multilateral pressures up to 4000 kg/cm^2," (English ed.) Izvestiya, Acad. Sci., USSR, Physics of the Solid Earth 2, pp. 94-98 (1965).

Li, W. N., and D. T. Smith, "Identification of sea-floor sediments using underway acoustics," Geophysical Prospecting 17, 231-247.

McDonal, F. J., F. A. Angona, R. L. Mills, R. L. Sengbush, R. G. Van Nostrand, and J. E. White, 1958, "Attenuation of shear and compressional waves in Pierre shale": Geophysics, v. 23, p. 421-439.

Mack, H., 1966, "Attenuation of controlled wave seismograph signals observed in cased boreholes": Geophysics, v. 31, p. 243-252.

Mizikos, J. P., "Use of acoustic (or seismic) waves for deep sea bottom layers identification and computation of their physical and engineering properties," in Proceedings of a Symposium, Physical and Engineering Properties of Deep-Sea Sediments, A. L. Inderbitzen (ed.), in press, 1974.

Morris, H. E., "Bottom-reflection-loss model with a velocity gradient," J. Acoust. Soc. Am. 48, 1198-1202 (1970).

Muir, T. G., and R. S. Adair, "Potential use of parametric sonar in marine archeology," Ms., Applied Research Laboratories, The University of Texas at Austin, Austin, Texas.

Nafe, J. E., and C. L. Drake, "Physical properties of rocks of basaltic composition," in Basalts, H. H. Hess and H. Poldervaart (eds.), (Interscience Publ., New York, 1967), pp. 483-502.

NAVOCEANO, Handbook of Oceanographic Tables (U.S. Naval Ocean. Office SP-68, 1966).

Nolle, A. W., and P. W. Sieck, "Longitudinal and transverse ultrasonic waves in synthetic rubber," J. Appl. Phys. 23, 888-894 (1952).

Richart, F. E., Jr., and R. V. Whitman, "Comparison of footing vibration tests with theory," J. Soil Mech. & Foundation Div. Amer. Soc. Civil Engin. SM 6, 143-193 (1967).

Schirmer, F., Eine Untersuchung akustischer Eigenschaften von Sedimenten der Nord-und Ostsee, Ph.D. Thesis, University of Hamburg (1971).

Shepard, F. P., "Nomenclature based on sand-silt-clay ratios," J. Sedim. Petrol 24, 151-158 (1954).

Simons, G., and W. F. Brace, "Comparison of static and dynamic measurements of compressibility of rocks," J. Geophys. Res., 70, 5649-5656, 1965.

Stoll, R. D., "Acoustic waves in saturated sediments," in Proceedings of a Symposium, Physics of Sound in Marine Sediments, L. D. Hampton (ed.), in press, 1974.

Stoll, R. D., and G. M. Bryan, "Wave attenuation in saturated sediments," J. Acoust. Soc. Am. 47, 1440-1447 (1970).

Tullos, F. N., and C. Reid, 1969, "Seismic attenuation of Gulf Coast sediments": Geophysics, v. 34, pp. 516-528.

Vasil'ev, Y. I., and G. I. Gurevich, "On the ratio between attenuation decrements and propagation velocities of longitudinal and transverse waves," (English ed.) Izvestiya, Acad. Sci., USSR, Geophys. Series 12, pp. 1061-1074 (1962).

White, J. E., Seismic Waves: Radiation Transmission, and Attenuation (McGraw-Hill Book Co., Inc., New York, 1965).

Whitman, R. V., R. J. Holt, and V. J. Murphy, "Discussion of: vibration modulus of normally consolidated clay," by B. O. Hardin and W. L. Black, J. Soil Mech. & Foundation Div., Amer. Soc. Civil Engin. 95, SM2, 656-659, 1969.

Yong, R. N., and B. P. Warkentin, Introduction to Soil Behavior (The Macmillan Co., Inc., New York, 1966).

Zemtsov, E. E., "Effect of oil and gas deposits on dynamic characteristics of reflected waves," Int. Geol. Rev. 11, 504-509 (1969).

SOUND PROPAGATION CALCULATIONS USING BOTTOM REFLECTION FUNCTIONS

H. P. BUCKER

Naval Undersea Center

San Diego, California 92132

ABSTRACT

In many modes of sonar operation there is appreciable reflection of the sound waves from the marine sediments. Depending upon the method of signal processing, calculations of sonar performance require various information relating the reflected sound wave to incident sound wave. The first part of this paper discusses various sound propagation situations, such as bottom bounce, shallow water, surveillance, etc., and what information about the bottom reflection function is required for useful propagation calculations.

The second part of the paper describes mathematical models that have been used at NUC to convert sediment parameters into plane wave reflection coefficients. These include a multi-layer viscoelastic model that has worked well at active sonar frequencies and a multi-layer absorbing liquid model with sound speed gradients in the sediment layers that is being used in the 10 to 500 Hz frequency range. The extension of bottom reflection models to cases where both the rigidity in the sediment and sound speed gradients must be considered will be briefly discussed.

At the Naval Undersea Center (NUC) there are a number of computer programs that have been developed over the years to calculate the sound fields generated by various sound sources such as an active sonar transducer array or a transiting submarine. In many cases a significant part of the sound field will reflect or interact in some way with the ocean bottom. If the bottom is reasonably smooth we have developed methods that account for the effect of the bottom in terms of a bottom reflection coefficient. First, a program is used that generates the bottom reflection coefficients as a function of grazing angle in tabular form. Then the reflection loss table serves as one of the inputs for the propagation program. This short paper describes how the reflection loss tables are used and how they are calculated for typical ocean sediment models.

The use of the bottom reflection function in ray theory calculations is illustrated by Fig. 1. First, the significant rays that can travel from the source to the receiver are identified.

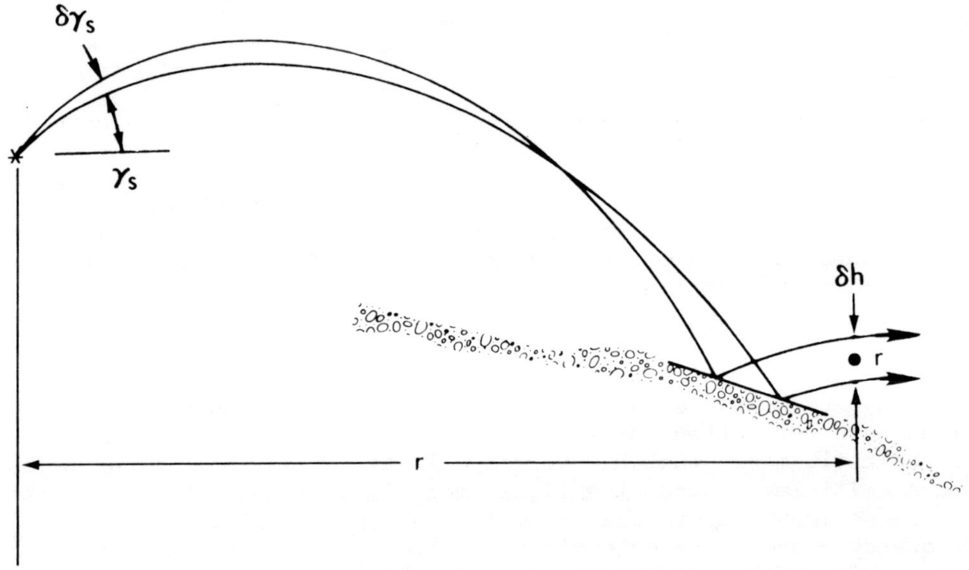

FIGURE 1

RAY THEORY DEPICTION OF SOUND FIELD

These are called eigen-rays. While there are, in general, an infinite number of eigen-rays, only a few are usually required at each receiver since the higher angle rays carry a negligible amount of sound energy. Finding an eigen-ray is usually done by locating a pair of neighboring rays that bracket the receiver. This is shown in Fig. 1 where a ray with source angle γ_s passes on one side of the receiver at point r while a second ray with source angle $\gamma_s + \delta\gamma_s$ passes on the opposite side. It follows that there is an eigen-ray similar in path to the rays shown in Fig. 1 with source angle γ^*_s such that $\gamma_s < \gamma^*_s < \gamma_s + \delta\gamma_s$. To calculate the sound intensity associated with the eigen-ray we note that the sound energy radiated by the source per unit time into an annular ring bounded by the rays shown in Fig. 1 and called E_1 is given by (assuming axial symmetry about the source)

$$E_1 = 2\pi \; I_1 \cos \gamma_s \; \delta \gamma_s \quad . \tag{1}$$

Here I_1 is the sound intensity at one unit length from the source point. The sound energy E_r, arriving at the annular ring containing the receiver point r per unit time, is equal to

$$E_r = 2\pi \; r \; I_r \; \delta h \tag{2}$$

where r is the horizontal range to the receiver, I_r is the sound intensity at r, and δh is the separation of the rays at r. Except for losses at the bottom reflection E_r is equal to E_1. This can be expressed as

$$E_r = |R|^2 \; E_1 \quad . \tag{3}$$

In the above equation R is the bottom reflection coefficient (in general a complex number) that relates the sound pressure immediately before the reflection to the sound pressure immediately after, i.e.,

$$p\ (\text{after}) = R \cdot p\ (\text{before}) \quad . \tag{4}$$

The modulus of R is related to the loss of energy at the bottom

$$\text{Bottom Loss (in dB)} = -20 \log_{10} |R| \quad . \tag{5}$$

The argument of R is equal to the phase shift of the ray upon bottom reflection.

The intensity of the eigen-ray at r can be calculated by substituting Eqs. 1 and 2 into Eq. 3 and solving for I_r.

$$I_r/I_1 = |R|^2 \cos \gamma_s \ \delta \gamma_s/(r\ \delta h) \quad . \tag{6}$$

Obviously this equation will not be correct in the neighborhood of a caustic line (a caustic line consists of points where neighboring rays cross and $\delta h \to 0$). In this case a more sophisticated formulation must be used.

The random sum total acoustic field at r is calculated by finding all significant eigen-rays and summing their intensity contributions. The alternative (called the coherent phase intensity) is to sum the pressures of each eigen-ray. The acoustic pressure, P, at point r of the eigen-ray shown in Fig. 1 can be written as

$$P_r = (\rho\ c_r\ I_r)^{1/2} \left[\int_o^r (\omega/c) d\ell + \arg(R) - \pi/2 - \omega t \right] \quad . \tag{7}$$

In the above equation, ρ is the density of water, c is the sound speed, ω is the angular frequency, $d\ell$ is an element of length along the ray path, and arg(R) is the argument of R. The π/2 phase shift results because the eigen-ray touches a caustic associated with the crossing ray paths in Fig. 1. The individual pressures are summed to give a total sound pressure. Ray theory provides a convenient and reasonably accurate picture of a sound field. However, as is well known, [1] the ray solution is a solution of the wave equation only in the limit as the frequency increases without limit. In many cases the more accurate wave solutions can be used to refine the ray theory picture.

The wave equation which governs the propagation of small amplitude waves follows directly from the second law of motion and the equation of continuity. Most wave solutions are derived from a line integral that represents a sum of cylindrical waves traveling at different angles. One convenient solution [2], called the natural mode solution, results when the basic line integral is transformed into a contour integral. The solution can then be written as a sum of residues (called the natural mode sum) plus possible branch line contours that remove any branch points from the interior of the contour. The standard form of the natural mode solution is written in terms of a potential function θ that is related to the sound pressure by the relation $\theta = p/(\rho)^{1/2}$. The natural mode sum for θ is given by

$$\theta = (2\pi/r)^{1/2} \sum_n \Phi(z_o, k_n)\, \Phi(z, k_n)\, N_n \exp(i\, k_n\, r) \quad . \tag{8}$$

Here, z is the depth of the receiver, z_o is the depth of the source, N is a generalized normalization function for each mode, and k is the horizontal wave number. The subscript n indicates that k is equal to k_n which is one of the values of k for which Φ satisfies the boundary conditions. An example of this will be given later. The function Φ is a solution of the z-separated part of the wave equation, i.e.,

$$\begin{aligned} d^2\Phi/dz^2 + (\omega^2/c^2) + (d^2\rho/dz^2)/(2\rho) \\ - (3/4)(d\rho/dz)^2 - k^2\Phi = 0 \end{aligned} \quad . \tag{9}$$

To illustrate the above, consider the two layer shallow water problems first solved by Pekeris [3] as shown in Fig. 2. The water layer has sound speed c_1 and density ρ_1 and the bottom half-space, representing the bottom sediments, has sound speed c_2 and density ρ_2.

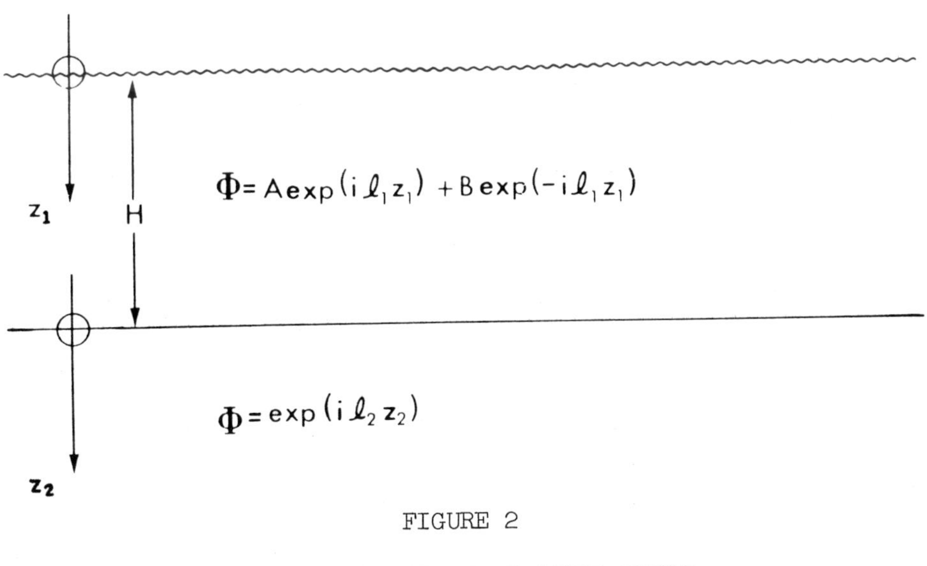

FIGURE 2

WAVE THEORY DEPICTION OF SOUND FIELD;
LIQUID BOTTOM

In each layer c and ρ are constant so that Eq. 9 becomes

$$d^2 \Phi/dz^2 + \ell^2\Phi = 0, \qquad \ell^2 = \omega^2/c^2 - k^2 \quad . \tag{10}$$

Thus, Φ can be written as

$$\Phi = A \exp(i\, \ell_1 z_1) + B \exp(-i\, \ell_1 z_1) \tag{11}$$

$$\text{in layer 1, } \ell_1^2 = \omega^2/c_1^2 - k^2,$$

$$\Phi = C \exp(i\, \ell_2 z_2) + D \exp(-i\, \ell_2 z_2) \tag{12}$$

$$\text{in layer 2, } \ell_2^2 = \omega^2/c_2^2 - k^2 \quad .$$

To satisfy the Sommerfeld radiation conditions D is set identically to zero and for convenience C is set equal to one. To satisfy the surface boundary condition Φ must be zero at $z_1 = 0$. Finally we have two interface conditions, namely continuity of pressure, $\sqrt{\rho}\,\Phi$, and continuity of the vertical component of particle velocity, $(1/\rho)\, \partial/\partial z\, (\sqrt{\rho}\,\Phi)$, at the interface between the two liquids, i.e., at $z_1 = H$ and $z_2 = 0$. The interface conditions for this case can be written as

$$\sqrt{\rho_1}\, [A \exp(i\, \ell_1 H) - \exp(-i\, \ell_1 H)] = \sqrt{\rho_2} \tag{13}$$

and

$$(\ell_1/\sqrt{\rho_1})\, [A \exp(i\, \ell_1 H) + \exp(-i\, \ell_1 H)] = (\ell_2/\sqrt{\rho_2}) \quad .$$

These equations require that

$$\exp(-2i\, \ell_1 H) = (\rho_1\ell_2 - \rho_2\ell_1)/(\rho_1\ell_2 + \rho_2\ell_1) \quad . \tag{14}$$

Equation 14, called the dispersion equation, will be satisfied only for certain values of $k = k_n$, $n = 1,2,3, \ldots$, which correspond to natural propagation modes of the channel.

The natural mode problem can be posed in a more general, but also more simple, form by the use of a bottom reflection coefficient. This is shown in Fig. 3. Here $\exp(i\, \ell_1 z_1)$ represents a downward traveling plane wave and $\exp(-i\, \ell_1 z_1)$ the reflected wave. R is the bottom reflection coefficient. The origin of the z coordinate system has been set at the bottom of the channel because it is customary to refer R to this system. The surface boundary condition is that Φ (at $z = -H$) must be zero and this leads directly to the dispersion equation,

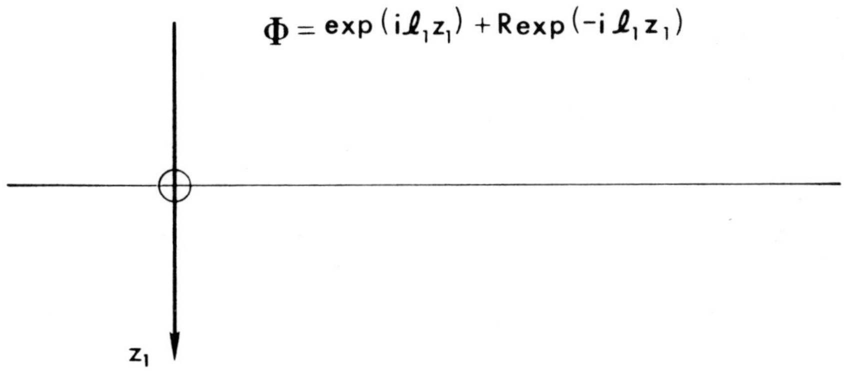

FIGURE 3

WAVE THEORY DEPICTION OF SOUND FIELD; GENERALIZED BOTTOM

$$R \exp(2i\ \ell_1 H) = -1 \quad . \tag{15}$$

If the bottom is modeled as a liquid half-space then R is the usual Rayleigh reflection coefficient, $R = (\rho_2\ell_1 - \rho_1\ell_2)/(\rho_2\ell_1 + \rho_1\ell_2)$, and Eq. 15 reduces to Eq. 14. However, Eq. 15 is much more useful than Eq. 14, since R can be calculated using very sophisticated bottom models or can be represented as a table of experimental values.

The plane wave reflection coefficient can also be used when the sound waves in the water are not plane. This will be illustrated for the single water layer problem as shown in Fig. 4. The generalization to the multi-layer case is straightforward. In the water layer the potential function can be written as the sum of two linear independent solutions, f and g, of the z-separated wave equation. For example if the density is constant and $1/c(z)^2$ is a linear function of depth then f and g are Airy functions. Over the depth from $z = z^*$, as shown in Fig. 4, to the bottom at $z = H$ we consider the water to have the constant sound speed c_b. The potential function will then have the form

$$\Phi(z) = A\ f(z) + B\ g(z), \qquad 0 \leqq z \leqq z^* \tag{16a}$$

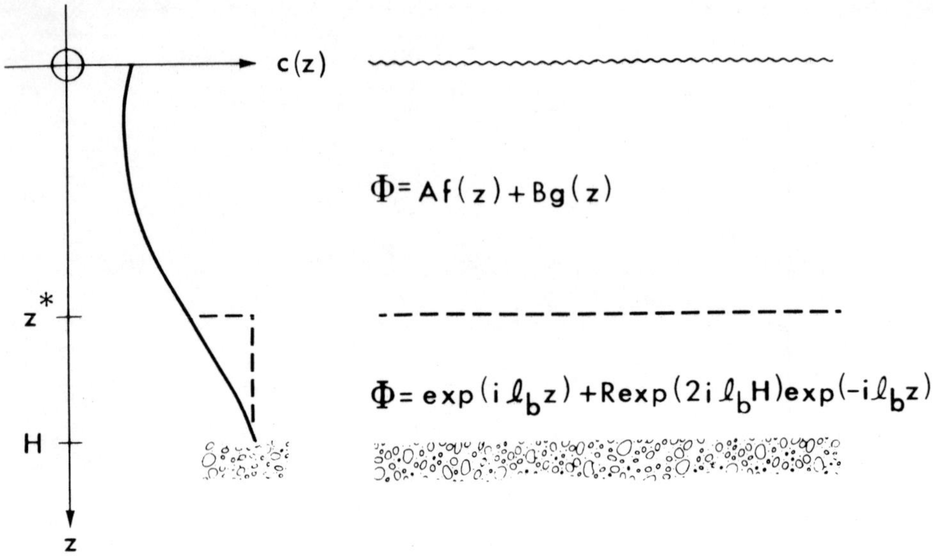

FIGURE 4

WAVE THEORY DEPICTION OF SOUND FIELD;
VARIABLE SOUND SPEED AND
GENERALIZED BOTTOM

$$\Phi(z) = \exp(i\ \ell_b\ z) + R\exp(2i\ \ell_b\ H)\exp(-i\ \ell_b\ z) \quad , \qquad z^* \leqq z \leqq z_b \quad . \tag{16b}$$

As before, using the requirements that $\Phi(z)$ must satisfy the surface condition at $z = 0$ and two interface conditions at $z = z^*$ we can derive a dispersion equation for k. Then by letting $z^* \rightarrow z_b$ the iso-speed layer at the bottom of the channel is removed from the problem. However, the plane wave bottom reflection coefficient R remains in the equations. In effect we have decomposed the general sound field at the bottom into up- and down-going plane waves which are in proportion to the plane wave bottom reflection coefficient.

Bottom reflection functions can be defined in various forms. For example we might define the function as shown in Fig. 5. Here we have a ray tube of cross section δA incident on the bottom at grazing angle γ_o and define a bottom reflection-scattering function R_s by

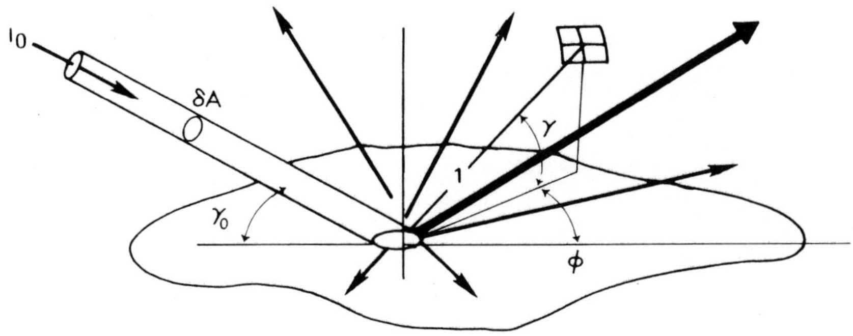

FIGURE 5

BOTTOM REFLECTION COEFFICIENT FOR RAY TUBE

$$I_s = I_o R_s (\gamma_o, \gamma, \Phi) \delta A \quad . \tag{17}$$

where I_s and I_o are the incident and reflected-scattered sound intensities, and γ and Φ are polar angles. In most cases we would probably separate the coherent specular component (i.e., $\gamma \approx \gamma_o$ and $\Phi \approx 0$) from the other parts of the function.

A review of the earlier paragraphs of this paper shows that the R_s defined by Eq. 17 is not the proper coefficient for use in either the ray theory or wave theory models. We can develop a more useful function by grouping the ray tubes so that the incident rays lie along a plane as shown in Fig. 6. We can imagine the incident rays as being generated by using a slit of width $\delta\ell$. This will reduce the scatter because the y components will cancel out if the bottom is uniformly rough. In other words, if in Fig. 5 there is energy scattered in the (γ^+, Φ^+) direction only the component in the (x, z) plane will contribute to the reflection-scattering function of Fig. 6. For Fig. 6 we might define a bottom function by

$$I_s = I_o R(\gamma_o, \gamma) \delta\ell \quad . \tag{18}$$

This is very close to the type of bottom function used in the ray theory. There are three differences which are usually negligible: (i) in the ray theory it may be necessary to use spherical waves near the source rather than plane waves; (ii) the non-specular part is normally omitted in ray theory calculations; (iii) reflections from sub-bottom interfaces will cause lateral displacements

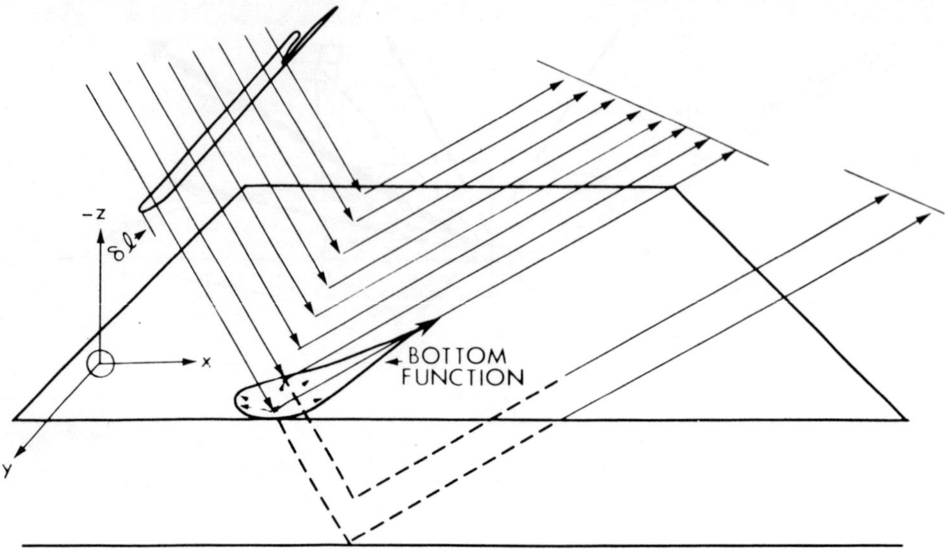

FIGURE 6

BOTTOM REFLECTION COEFFICIENT FOR CYLINDRICAL RAY SLIT

of the rays as shown in Fig. 6. In spite of these problems it is accepted practice to take the specular part of the R to use as the bottom reflection function. Most of the time, the calculated values of the sound field are in reasonable agreement with experimental data.

The plane wave coefficient, R, used in wave solutions includes reflections from sub-bottom interfaces which add with the water-sediment reflections. Again, there is some energy scattered at non-specular angles but this can usually be neglected when calculating the propagating sound field. Besides the scattering problem, the plane wave coefficient does not cover energy directed along the interfaces in waves which are generated when a spherical wave is reflected by the interface. In Fig. 7 a spherical wave reflects from a liquid-liquid interface where the sound speed is greater in the lower layer. Tracing out the rays by Huygen's Principle shows the direct wave, D, the reflected wave, R, and an interface or lateral wave L. There are other types of interface waves. For example at a water-sediment interface a spherical wave reflection generates a Stoneley-wave [4] that has both transverse and longitudinal particle motion. In the lower half of Fig. 7 an apparatus is illustrated that has been used to generate and detect Stoneley waves [5]. The speed of travel of the Stoneley wave is strongly related to the shear modulus of the sediment. Thus, while

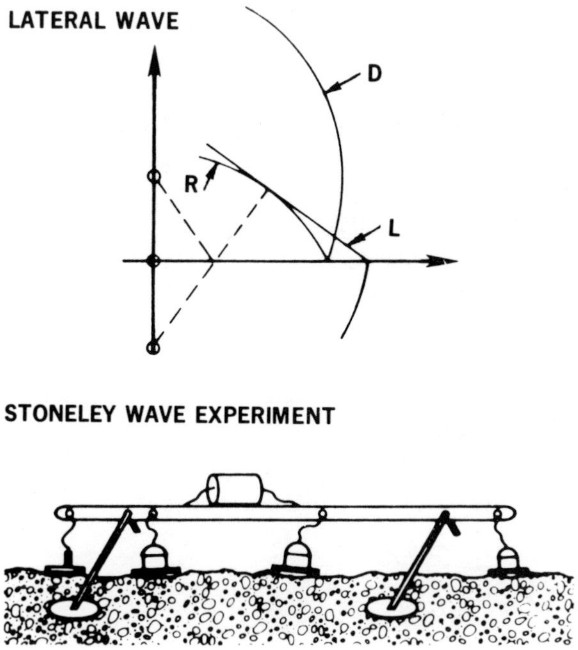

FIGURE 7

BOTTOM REFLECTION FOR A SPHERICAL WAVE FRONT

boundary waves do not contribute significantly to the propagating acoustic field we can use their existence to determine sediment properties. It might be noted that boundary waves can be included in the wave theory solutions by including the appropriate branch line contours. This has not been necessary at sound frequencies currently used by the Navy.

The problems with most wave models are that solutions for transient pulses are very lengthy (this is usually done by use of the Fourier integral representation) and that significant horizontal changes in the water column or sediment parameters are not accounted for in present models. For most calculations of the plane wave bottom reflection coefficient we have used one or both of two standard models. The first model [6] is similar in concept to a transmission coefficient model developed by Thomson [7]. First, there is the water half-space with the potential function Φ_o representing an incident and a reflected wave. Consistent with the previous discussion we would set $\Phi_o = \exp(i\ \ell_o\ z_1) + R\ \exp(-i\ \ell_o\ z_1)$, where R is the reflection coefficient. The ocean sediments are represented as an arbitrary number of isotropic elastic layers with complex Lamé parameters. In sediment layer i there is a function

Φ_1 for the up-and-down going compressional waves and a function Ψ_i for the up- and down-going shear waves. The potential functions can be written as

$$\Phi_i = A_i \exp(i\, a_i\, z_i) + B_i \exp(-i\, a_i\, z_i) \tag{19a}$$

$$\Psi_i = C_i \exp(i\, b_i\, z_i) + D_i \exp(-i\, b_i\, z_i) \tag{19b}$$

If there are N layers then there will be only down-going waves in the bottom layer or half-space. The relevant parameters are:

for the water half-space,

$k = (\omega/c_o) \cos \gamma$, (γ is the grazing angle of the incident wave)

$\ell_o^2 = (\omega/c_o)^2 - k^2$,

for a sediment layer i,

$a_i^2 = \omega^2 \rho_i^2/(\lambda_i + 2\mu_i) - k^2$,

$b_i^2 = \omega^2 \rho_i^2/\mu_i - k^2$.

In the above k is the horizontal wave number, c_o is the sound speed in the water, γ is the grazing angle of the incident sound wave, ρ_i is the density of sediment layer i and λ_i and μ_i are Lamé constants for this layer. The parameters λ and μ are complex numbers. Their imaginary parts, which determine the absorption of the sound waves in the sediment layers, have been found to be essentially independent of frequency. [8] Thus the sediment sound speeds are nearly independent of frequency while the attenuation varies as the first power. In the definitions of the various potential functions there are 4N - 1 unknown constants (R, A_1, B_1, C_1, D_1, A_2, B_2, ..., A_N, C_N) and 4N - 1 interface conditions (continuity of displacements and stress). Solution for R is obtained by standard computational methods.

The next three figures show comparisons of calculations made using this model with experimental data from one of the FASOR Stations. FASOR was a five year program undertaken by the Naval Undersea Center to measure acoustic and environmental data related to sonar operations in forward areas of the Pacific Ocean and contiguous water masses. Figure 8 shows experimental data obtained using 1.8 lb explosive sources and analyzed in a 1/3 octave band at 3.15 kHz. The agreement is reasonably good, which was the case for

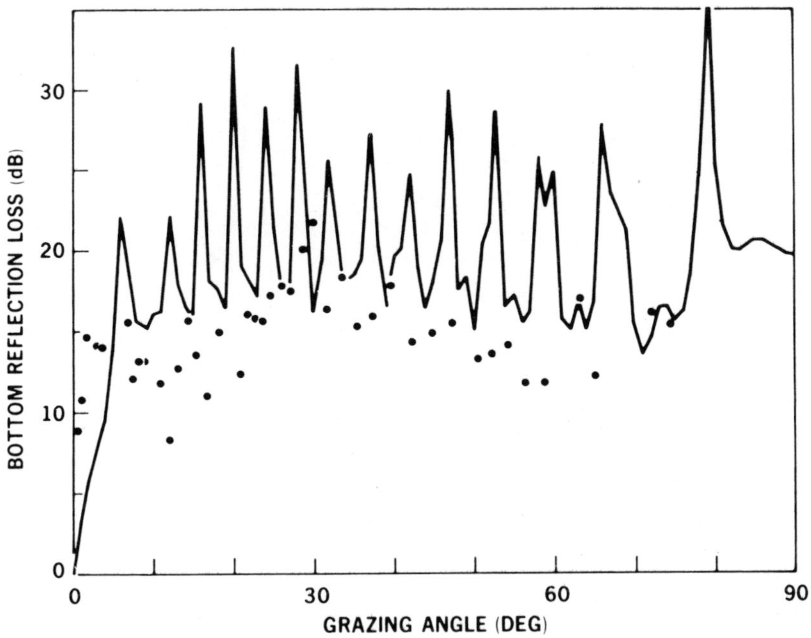

FIGURE 8

BOTTOM LOSS vs GRAZING ANGLE; ISOTROPIC ELASTIC LAYERS
3.15 kHz

over 90% of the FASOR stations. Figure 9 shows data from the same propagation run analyzed at 1.0 kHz, and Fig. 10, at 50 Hz. It is clear that as the frequency is reduced the measured bottom loss becomes less while the calculated loss remains at about the same level. This problem was solved when we included the effects of the positive sound speed gradients in the layers. The gradients are partly due to the compaction of sediment materials as the overburden pressure increases and mostly due to the temperature gradients in the sediment layers. The latter can be accurately determined by heat flow measurements.

The bottom loss model that accounts for the effects of sound speed gradients was developed by H. E. Morris [9]. In this model the square of the index of refraction is taken to be a linear function of depth. In this case the potential function in a sediment layer can be written in terms of Airy functions as

$$\Phi(z) = A \cdot Ai[\xi] + B \cdot Ai\,[\xi \exp(i\, 2\pi/3)]$$

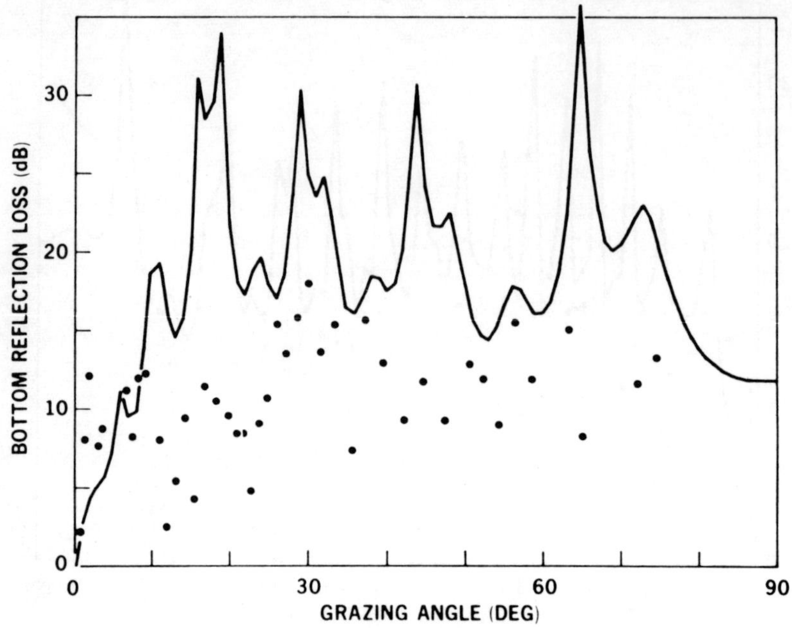

FIGURE 9

BOTTOM LOSS vs GRAZING ANGLE; ISOTROPIC ELASTIC LAYERS
1.0 kHz

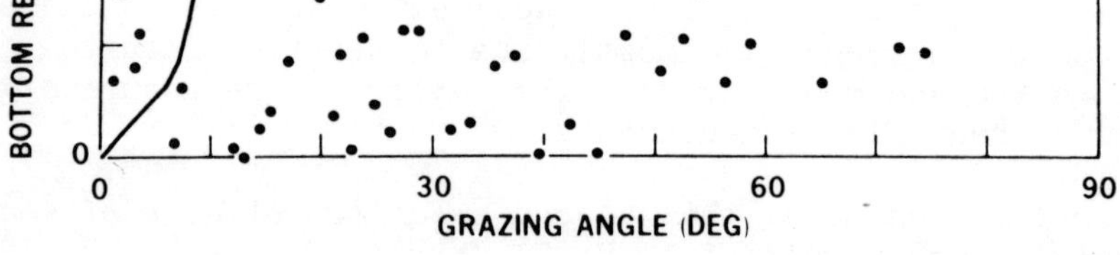

FIGURE 10

BOTTOM LOSS vs GRAZING ANGLE; ISOTROPIC ELASTIC LAYERS
50 Hz

where $$\xi = (k^2 - \omega^2/c^2)/\eta^2 - \eta^2 - \eta z \quad ,$$

and $$\eta = (\omega^2\beta/c^2)^{1/3} \quad , \quad c_o{}^2/c^2 = 1 + \beta z \quad .$$

Solution for the bottom reflection coefficient proceeds in the usual manner as described before.

The last 3 figures show the comparison of the FASOR station data with calculations using the Morris model with sound speed gradients. Figure 11 is for 3.15 kHz, Fig. 12 is for 1.0 kHz, and Fig. 13 shows the comparison at 50 Hz. In contrast to the previous case, there is reasonable agreement between the experimental and calculated values at the lower frequencies.

In summary, the propagation models can be related to sediment properties by means of a bottom reflection coefficient. To date, reasonable fits of calculated and experimental data have required only the use of simple layered models. For detailed agreement on a fine scale we may have to use models that account for scattering or that provide a more complicated representation of the sediment structure.

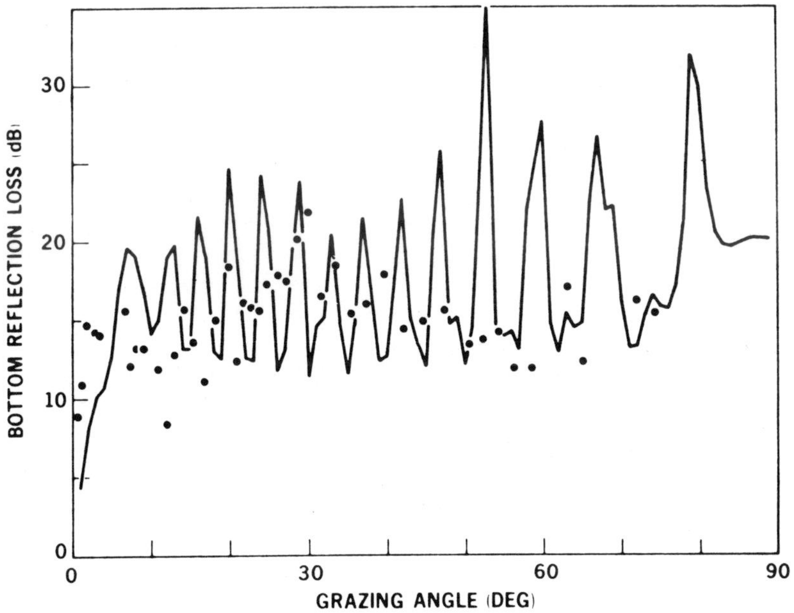

FIGURE 11

BOTTOM LOSS vs GRAZING ANGLE; LINEAR GRADIENT LAYERS
3.15 kHz

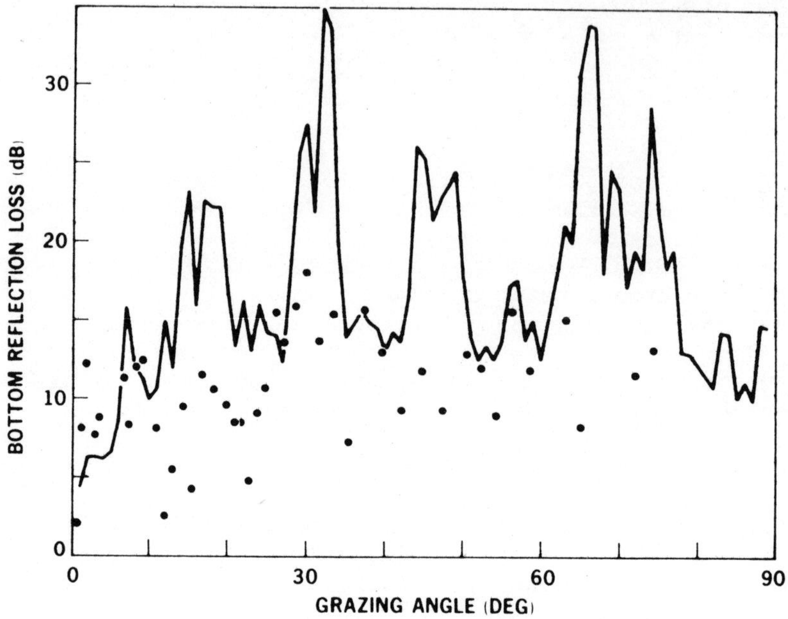

FIGURE 12

BOTTOM LOSS vs GRAZING ANGLE; LINEAR GRADIENT LAYERS
1.0 kHz

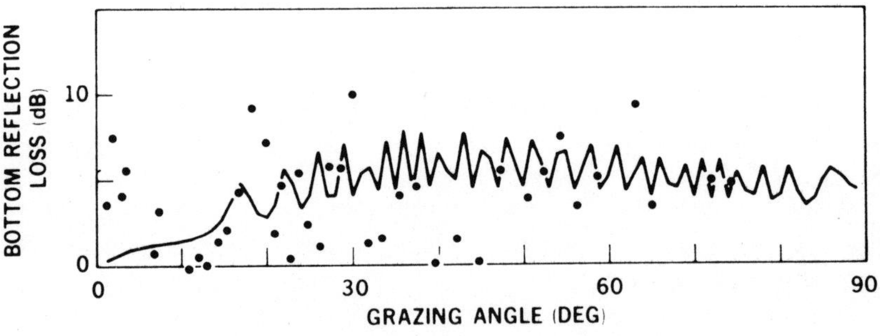

FIGURE 13

BOTTOM LOSS vs GRAZING ANGLE; LINEAR GRADIENT LAYERS
50 Hz

REFERENCES

1. C. B. Officer, Introduction to the Theory of Sound Transmission (with Application to the Ocean), McGraw-Hill, New York (1958), pp. 36-59.

2. H. P. Bucker, "Sound Propagation in a Channel with Lossy Boundaries," J. Acoust. Soc. Am. 48, 1187-1194 (1970).

3. C. L. Pekeris, "Theory of Explosive Sound in Shallow Water," Sec. II in Propagation of Sound in the Ocean, Geological Soc. Amer. Memoir 27 (1948).

4. E. Strick, W. L. Roever, and T. F. Vining, "Propagation of Elastic Wave Motion from an Impulsive Source Along a Fluid/Solid Interface," Phil. Trans. Royal Soc. Lon., Ser. A 251, 455-523 (1959).

5. E. L. Hamilton, H. P. Bucker, D. L. Keir, and J. A. Whitney, "Velocities of Compressional and Shear Waves in Marine Sediments Determined in Situ from a Research Submersible," J. Geophys. Res., 75, 4039-4049, 1970.

6. H. P. Bucker, J. A. Whitney, G. S. Yee, and R. R. Gardner, "Reflection of Low Frequency Sonar Signals from a Smooth Ocean Bottom," J. Acoust. Soc. Am. 37, 1037-1051 (1965).

7. W. T. Thomson, "Transmission of Elastic Waves through a Stratified Solid Material," J. Appl. Phys. 21, 89-93 (1950).

8. E. L. Hamilton, 1972, "Compressional-Wave Attenuation in Marine Sediments": Geophysics 37, 620-646.

9. H. E. Morris, "Bottom-Reflection-Loss Model with a Velocity Gradient," J. Acoust. Soc. Am. 48, 1198-1202 (1970).

NONLINEAR ACOUSTICS AND ITS ROLE IN THE SEDIMENTARY GEOPHYSICS OF THE SEA

THOMAS G. MUIR

Applied Research Laboratories

The University of Texas at Austin, Austin, Texas 78712

ABSTRACT

A theoretical and experimental exposition of the principal nonlinear acoustic phenomena is presented and interpreted in the light of applications to the sediments.

INTRODUCTION

Over the past few years, nonlinear acoustics has become more and more important in the fields of physical and engineering acoustics. Much attention is now being given to practical applications of nonlinear acoustics. The terms "nonlinear and parametric sonar" have found their way into the technical vocabulary. Certain nonlinear phenomena have been identified in acoustical systems previously considered to be linear. As our understanding of nonlinear wave propagation develops with continued research, we may see further implementation of this discipline. The sonic boom is an example of a nonlinear acoustic phenomenon already well documented. The radiation of high intensity noise from jet engines also appears to involve nonlinear effects. An emerging application is the utilization of nonlinear acoustics in high resolution probing of undersea sediments. This development should be expected to provide new measurement techniques in both geophysics and in marine archeology.

Although nonlinear effects are relatively new to most underwater acousticians, nonlinear acoustics is actually quite an old subject. The early history of this discipline has been documented[1] and it is evident that nonlinear acoustics was an important topic in the evolution of classical physics. Truly significant work in nonlinear acoustics dates back to Euler (1755) and features such master practicioners as Lagrange, Poisson, Stokes, Airy, Earnshaw, Riemann, and Rayleigh. These early researchers concentrated on the propagation of plane waves in lossless gases. Recent work has been facilitated by modern technology, especially by the development of high intensity sound sources. At present, the main thrust of new developments in nonlinear acoustics appears to come from the underwater and aeronautical sound communities.

In the treatment to follow, we will be limited to a modern interpretation of nonlinear effects in underwater sound and sedimentary geophysics. The problems of finite amplitude distortion, shock formation, harmonic generation, and saturation will be described. The topic of parametric arrays will be treated in greater detail as it is perhaps the most significant nonlinear effect in practical applications. These important nonlinear phenomena will first be surveyed and described in order to develop a fairly complete understanding of the subject matter. Specific applications to the sediments are treated in the last two sections of the paper.

Most of the experiments presented to illustrate the theory were conducted by the author and his colleagues at Applied Research Laboratories' Lake Travis Test Station. It should be stated that many excellent experiments have been conducted by scientists and engineers throughout the world. The results of many of those experiments are not included here for reasons of convenience only. Finally, theory, experiment, and applications are separated for the purpose of allowing specialists in each of these areas to more easily deal with the subject matter of greatest interest.

THEORY

Since it will be necessary to refer to several theoretical developments in the evolution of nonlinear acoustics that may not be widely appreciated, it is convenient to review some of the more important concepts that are fundamental to the subject matter.

The Equations of Fluid Mechanics

For lossless media, the three concepts of fluid mechanics that form the basis of acoustics are those of continuity, conservation of momentum, and the physical equilibrium or state of the medium. These concepts may be stated mathematically as follows.

Continuity equation

$$\frac{\partial \rho_T}{\partial t} + \nabla \cdot (\rho_T \vec{u}) = 0 \quad , \tag{1}$$

Momentum equation

$$\frac{\partial}{\partial t}(\rho_T \vec{u}) + \rho_T \vec{u}(\nabla \cdot \vec{u}) + (\vec{u} \cdot \nabla)\rho_T \vec{u} + \nabla p = 0 \quad , \tag{2}$$

Isentropic equation of state

$$p = \sum_{n=1}^{\infty} \frac{1}{n!} \left(\frac{\partial^n p}{\partial \rho^n} \right)_{\rho=\rho_o, S} \rho^n \quad , \tag{3}$$

where

$\rho_T = \rho_o + \rho$, total density,

ρ_o = static or equilibrium density,

ρ = excess density,

∇ = the vector differential operator,

$\vec{u}$ = particle velocity,

t = time,

p = acoustic or excess pressure,

n = a summation index, and

S = entropy.

The equation of state shown here is a Maclaurin series expansion of the pressure-density relation. It is usually approximated by retaining only the first two terms in the series, i.e.,

$$p = c_o^2\rho + \frac{1}{2}\frac{c_o^2}{\rho_o}\left(\frac{B}{A}\right)\rho^2 \quad , \tag{4}$$

where

$$c_o^2 = \left(\frac{\partial p}{\partial \rho}\right)_{\rho=\rho_o, S} \quad , \tag{5}$$

and

$$\frac{B}{A} = \frac{\rho_o}{c_o^2}\left(\frac{\partial^2 p}{\partial \rho^2}\right)_{\rho=\rho_o, S} \quad . \tag{6}$$

Equation (5) is the definition of the small-signal sound speed, c_o, whereas Eq. (6) may be taken as the definition of the "parameter of nonlinearity" of the medium, following the nomenclature of Beyer.[2] The results of several measurements indicate that B/A has the value 5.2 for water at 30°C.

Distortion and Shock Formation

A classic problem in nonlinear acoustics is the distortion and formation of shocks in large amplitude waves. This process has also been shown to have a great influence on high amplitude parametric arrays.

The underlying cause of finite amplitude distortion is the dependence of the signal speed on the particle velocity. (The signal speed is defined as the velocity of any specified point on an acoustic waveform.) This dependence may be obtained from the basic equations of fluid mechanics, but it is sufficient to state it here as*

$$\frac{dx}{dt} = c_o + \left(1 + \frac{1}{2}\frac{B}{A}\right)u \quad , \tag{7}$$

where dx/dt = the signal speed, and u = the particle velocity.

*The evolution of this result has been documented in Ref. 1. References 3 and 4 are also recommended for further clarification of the material presented in this section.

Because the particle velocity varies with location on the waveform, points on the waveform in a condensation phase travel faster than those in a rarefaction phase. If the original wave has a sinusoidal shape, this variation in signal speed will cause a cumulative distortion in the wave shape as the wave travels, until a shock forms. The various stages in this process are demonstrated by the waveforms shown in Fig. 1. For spherical waves, the range at which a shock forms is given by

$$\bar{r} = r_o \exp\left(\frac{1}{\beta\epsilon k r_o}\right) \quad , \tag{8}$$

where

$$\beta = \left[1 + \frac{1}{2}\frac{B}{A}\right]$$

$$\epsilon = \frac{u_{max}}{c_o} \text{ , the acoustic Mach number,}$$

$$k = \frac{2\pi}{\lambda} \text{ , the wave number, and}$$

r_o = the radius of the spherical source, or the range at which spherical divergence begins.

The process of shock formation is often characterized in terms of a dimensionless parameter σ, defined (for spherical waves) as

$$\sigma = \beta\epsilon k r_o \ln \frac{r}{r_o} \quad . \tag{9}$$

The initial formation of a vertical discontinuity takes place when σ=1, as is shown in Fig. 1. At this point, the fundamental frequency component of the waveform suffers about 1 dB of attenuation due to the process of distortion. If the amplitude of the initial waveform is high enough to cause further distortion, a "sawtooth" waveform may be obtained at a range $\tilde{r}$ corresponding to σ=3, i.e.,

$$\tilde{r} = r_o \exp\left(\frac{3}{\beta\epsilon k r_o}\right) \quad . \tag{10}$$

At this point, the distortion and the dissipation at the shock front can cause several decibels of attenuation of the fundamental, depending on the role that ordinary dissipation plays in the problem at hand.

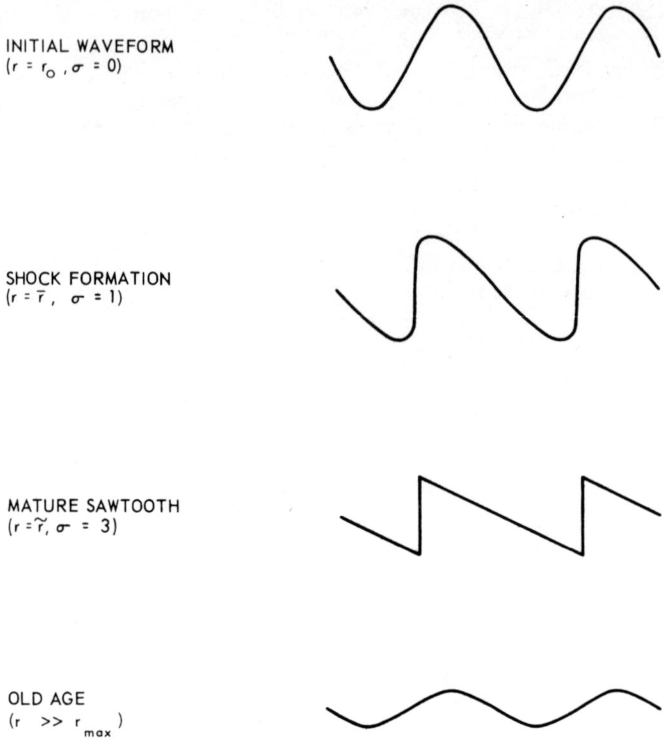

FIGURE 1

FINITE AMPLITUDE WAVEFORMS

Since multivalued waveforms are physically impossible, the tendency for further distortion must be matched by dissipation at the shock front. In this sense, the sawtooth waveform is a stable waveform because it maintains its shape until the dissipation finally reduces its amplitude to a level commensurate with small-signal acoustics.

The transition range between finite-amplitude and small-signal regions has been obtained[3-5] by equating the rates of the two types of dissipation. For spherical waves, this range is given by

$$r_{max}^{(n)} = \frac{\beta \epsilon k r_o}{\alpha_n \left[1 + \beta \epsilon k r_o \ln \left(\frac{r_{max}}{r_o} \right) \right]} , \qquad (11)$$

where n denotes the harmonic number of each Fourier component of the finite-amplitude wave and α_n = the small-signal attenuation coefficient of each component. Note that the factors ϵ and k appear together in product form in each of these equations. This means that nonlinear effects increase with frequency just as they do with source level.

Harmonic Generation

As one might suspect, the nonlinear distortion experienced by a propagating finite-amplitude wave results in the generation of harmonic components. As the waveform becomes steeper and steeper, more and more harmonics are generated.

For relatively strong waves at ranges shorter than r_{max}, an approach known as "weak shock" theory (which is based on perturbation solutions of Eqs. (1) through (3)) can be taken as a reasonably valid method for the analysis of finite-amplitude waveforms. Of prime importance is the problem of describing the behavior of the frequency components in a train of originally sinusoidal waves. Blackstock[3] obtained a general solution of this problem through a Fourier series expansion of certain solutions resulting from weak shock theory that describe the time domain behavior of the waveforms. For spherical waves, this expansion may be stated as

$$p = p_o \left(\frac{r_o}{r}\right) \sum_{n=1}^{\infty} B_n \sin n[\omega t - k(r-r_o)] \quad , \tag{12}$$

where

$$B_n = \left(\frac{2}{n\pi}\right) W_b + \frac{2}{n\pi\sigma} \int_{\Phi_{min}}^{\pi} \cos n(\Phi - \sigma \sin \Phi) d\Phi \quad ,$$

$$W_b = \left(\frac{1}{\sigma}\right) \left[j_o^{-1} \left(\frac{1}{\sigma}\right) \right] U(\sigma - 1) \quad ,$$

$j_o^{-1}\left(\frac{1}{\sigma}\right)$ = the quantity whose zero-order spherical Bessel function is $\frac{1}{\sigma}$, and

$$\Phi_{min} = \sigma W_b \quad .$$

The B_n's in this solution are the Fourier coefficients of the harmonic components of the distorted waveform.

Equation (12) can be used to predict the frequency spectra of the harmonic components as a function of range. Two particularly simple and perhaps useful features of the harmonic radiations can also be extracted from Eq. (12). The first has to do with the relative amplitudes of harmonic components. In the sawtooth region, these amplitudes are related to the amplitude of the fundamental through the factor $1/n$. This means that the maximum relative amplitude of the second harmonic is 6 dB below the fundamental, while the third harmonic is 10 dB down, the fourth harmonic 12 dB, etc. Although absorption reduces the harmonic amplitudes at ranges associated with $r_{max}^{(n)}$, especially in the old age region, they nonetheless have fairly respectable amplitudes in the sawtooth region.

The second feature provided by the theory summarized in Eq. (12) is that each successive harmonic has a progressively smaller beam, with progressively greater minor lobe suppression.[5] Prior to shock formation, for example, the directivity function of the nth harmonic goes as the directivity of the fundamental to the nth power; i.e., for $\sigma_1<1$,

$$D_n(\theta) = [D_1(\theta)]^n \quad . \tag{13}$$

For strong shocks in the sawtooth region, where $\sigma_1>3$ and $r<r_{max}^{(n)}$, the directivities go as

$$D_n(\theta) = D_1(\theta)\,\frac{1+\sigma_1}{1+\sigma_1 D_1(\theta)} \quad . \tag{14}$$

This solution implies that the beams become isotonic in the angular sector that satisfies the $\sigma \geqq 3$ criterion. In actual practice, only a portion of the beam usually satisfies this criterion, so that for strong shocks one must use Eq. (12) and its corollaries to determine the directivity function. This can be effected by normalizing $B_n(\theta)$ with respect to $B_n(0)$. When this is done, the harmonic directivities do in fact increase with harmonic number even when a portion of the beam satisfies the $\sigma \geqq 3$ criterion, as will be shown in the experimental section to follow.

The fact that the harmonic radiations exist at respectable levels in high intensity transmissions (at times regardless of the intent of the designer), coupled with the fact that they can be highly directive, makes their potential use in high-resolution acoustic systems quite attractive.

As was stated above, Eq. (12) is strictly valid for ranges $r<r_{max}$. Its validity at greater ranges may be approximated[4,5]

by allowing B_n to take on the form

$$B'_n = \left[1 - U\left(r-r^{(n)}_{max}\right)\right]B_n$$
$$+ \left[U\left(r-r^{(n)}_{max}\right)\right]B_n\left(r=r^{(n)}_{max}\right)e^{-\alpha\left(r-r^{(n)}_{max}\right)} \quad . \tag{15}$$

In this approximation, B_n is assumed to represent the only absorption agency out to the $r^{(n)}_{max}$ range. At this range, the B_n damping factor is switched out by the step function, $1 - U(r-r^{(n)}_{max})$, and small signal absorption takes over, beginning with the amplitudes specified by the value of B_n at $r=r^{(n)}_{max}$.

Finite-Amplitude Propagation and Saturation

An important consequence of Eq. (12) and its corollaries is the concept of acoustic saturation. By allowing the amplitude at the source to increase indefinitely, it can be shown[4] that the amplitude at the fundamental frequency approaches a finite, limiting value. The result is

$$p^1_{(sat)} = \frac{2\rho_o c_o^2}{\beta k r \ell n\left(r^{(1)}_{max}\infty/r_o\right)} \exp\left[-\alpha_1\left(r-r^{(1)}_{max}\infty\right)\right] \quad , \tag{16}$$

where

$$r^{(1)}_{max}\infty = \int_{p_o\to\infty} r^{(1)}_{max} = \left[\alpha_1 \ell n\left(r^{(1)}_{max}\infty/r_o\right)\right]^{-1} \quad .$$

An interesting feature of Eq. (16) is that it does not depend on the amplitude at the source. Equation (16) can also be interpreted as saying that the medium refuses to sustain and propagate arbitrarily intense sounds and that a limiting value will always be reached, above which more intense sounds are impossible. Note also that the presence of α_1 in the negative exponential and k in the denominator of Eq. (16) serves to increase the saturation limit with decrease in frequency. At very low frequencies, saturation is unlikely, and cavitation becomes the primary limitation.

Parametric Arrays

In 1960, P. J. Westervelt[7] at Brown University developed the theory for a sound source that he called a parametric acoustic array. This type of source allows the creation of highly directive sound at relatively low operating frequencies through the nonlinear interaction of acoustic waves. The original theory treated the problem of the interaction of two plane, collimated, monochromatic primary waves emitted simultaneously from a high frequency projector. A sketch of the geometry is shown in Fig. 2(a). The primary waves were postulated to interact nonlinearly in the medium within a zone of modulation that is bounded by the collimated beam and extends a distance along the acoustic axis determined by the small-signal absorption of the primaries. In this sense, the parametric acoustic array may be thought of as a volumetric end-fired array that occupies a portion of the medium directly in front of the projector.

Experimental evidence of the existence of parametric arrays was first reported in 1963 by Bellin and Beyer[8] and has continued to appear[9-18] since that time.

The engineering significance of the parametric array is due to its great directivity, specifically its narrow beamwidth and the absence of minor lobes in its radiation patterns. The narrow beamwidth is a result of the shape of the zone of modulation. By analogy to the ordinary end-fired array, it can easily be seen that the longer the zone of modulation, the smaller is the beamwidth of the parametric array. Since the zone of modulation is usually very long in comparison to its width, the parametric array produces a radiation much narrower than that which would be produced by the source transducer were it operating linearly at the difference frequency. The parametric array is shaded by virtue of the naturally smooth decay in the conversion of carrier frequency sound to difference frequency sound with increasing distance from the electroacoustic source. This inherent form of array shading, which is due to the absorption of the primary waves, and to diffraction within the zone of interaction, gives rise to an angular response that decays monotonically with increasing angle from the acoustic axis. The beam pattern of the parametric array therefore contains none of the undesirable minor lobe structures common to conventional piston type transducers.

Although the generation of difference frequency sound is of the greatest current interest, it should be pointed out that parametric arrays also generate sound at frequencies equal to the sum of the original primary frequencies.

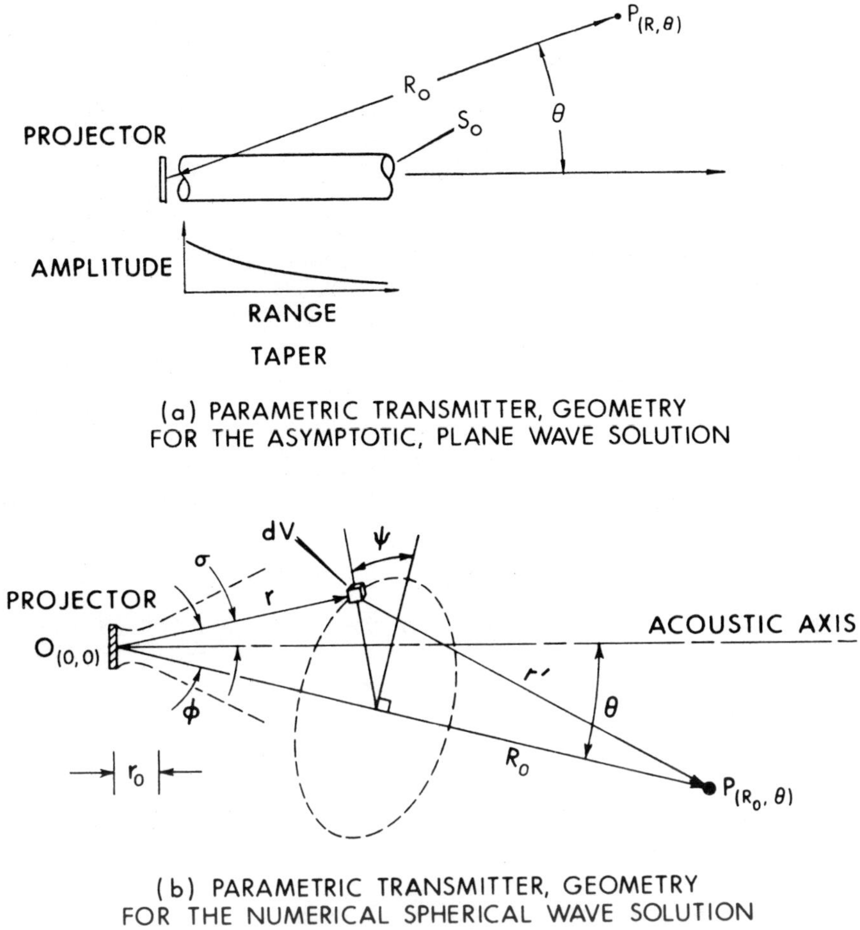

FIGURE 2

PARAMETRIC ACOUSTIC ARRAYS

The theory underlying the parametric array problem has its roots in the basic Eqs. (1) - (3), as do the other problems previously discussed. In working with these equations, Westervelt found an inhomogeneous wave equation appropriate for the description of parametric interaction. This is

$$\Box p_s = - \frac{\beta}{\rho_o c_o^4} \frac{\partial}{\partial t^2} p^2 \quad , \qquad (17)$$

where $\Box$ is the d'Alembertian operator, $=\nabla^2 - 1/c_o^2 \, \partial^2/\partial t^2$, p is the sum of the two primary pressures, and p_s is the amplitude of difference frequency sound. Westervelt[7] solved Eq. (17) for the case of two highly collimated plane wave primaries in the steady state. The result, which is valid at long ranges from the interaction volume, may be written in the form

$$p_s(R_o,\theta) = \frac{\beta\omega_s^2 p_1 p_2 S_o}{8\pi R_o \rho_o c_o^4} \frac{1}{\left[\alpha^2 + k_s^2 \sin^4(\theta/2)\right]^{1/2}} \quad , \qquad (18)$$

where α is the mean carrier or primary wave attenuation coefficient, S_o is the cross-sectional area of the primary beam, p_1 and p_2 are the primary wave amplitudes, and the subscript s indicates difference frequency parameters. The half-power beamwidth of the difference frequency radiation can be obtained from Eq. (18) as

$$\theta_{HP} = 4\sqrt{\frac{\alpha}{k_s}} \quad . \qquad (19)$$

Equation (19) can be used to show that the beamwidth is reduced by reducing the primary frequencies and increasing the difference frequency. This dependence is illustrated in Fig. 3, which plots the half-power beamwidths as a function of the primary and difference frequencies for sea water, using the Schulkin-Marsh absorption formula.[19] Since the assumptions underlying Eq. (19) require highly collimated, plane wave primaries, Fig. 3 should not be used for primaries of large sphericity and beamwidth. In particular, the primary beamwidth should be smaller than the predicted difference frequency beamwidth in order to satisfy these assumptions.

Refinements to the solution represented by Eq. (18) have also been made.[18,20,21] One of these[18] is of practical significance in underwater sound as it allows the extension of theory to spherical wave propagation in the farfield of the primary electroacoustic source(s). This modification utilizes a numerical integration to provide for an exact description of the directivities of these sources. The numerical solution also allows the difference frequency sound field to be predicted within the zone of interaction.

For a circular piston projector, as shown in Fig. 2(b), the numerical solution may be stated as

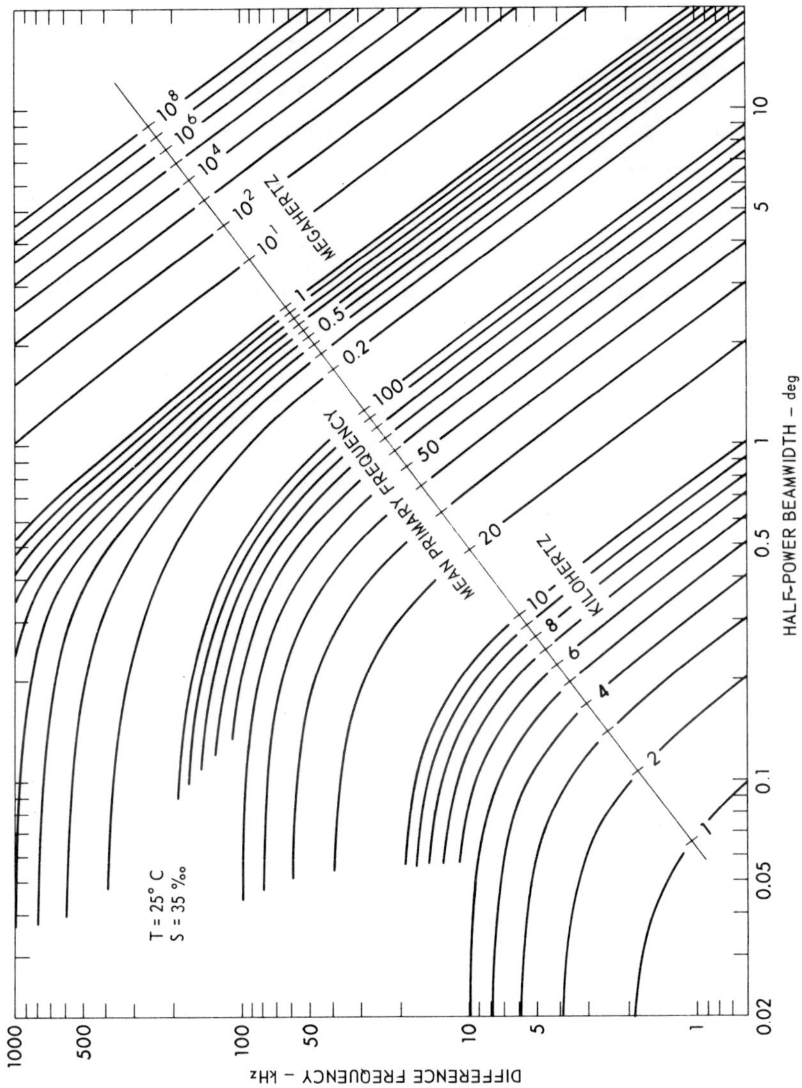

FIGURE 3

NOMOGRAPH OF PARAMETRIC TRANSMITTING ARRAY BEAMWIDTHS
AS CALCULATED FROM THE WESTERVELT FORMULA

$$p_s(R_o,\theta) = \frac{2p_1p_2r_o^2\omega_s^2\beta}{\Pi\rho_o c_o^4 k_1 k_2 a^2} \int_{r_o}^{R_o} \int_0^{\varphi_e+\theta} \int_0^{\Pi}$$

$$\cdot \frac{J_1(k_1 a \sin\sigma(\psi,\varphi,\theta)) J_1(k_2 a \sin\sigma(\psi,\varphi,\theta))}{\sin^2\sigma(\psi,\varphi,\theta)} \qquad (20)$$

$$\cdot \frac{\exp -\left\{[(\alpha_1+\alpha_2) - ik_s]r - (ik_s-\alpha_s)r'(r,\varphi)\right\}}{r'(r,\varphi)} \sin\varphi d\psi d\varphi dr \quad ,$$

where, as a result of some elementary trigonometry,

$$\sin\sigma(\psi,\varphi,\theta) = \left\{[\sin(\theta-\varphi) + \sin\varphi\cos\theta(1-\cos\psi)]^2 + \sin^2\varphi\sin^2\psi\right\}^{1/2},$$

and

$$r'(r,\varphi) = \left(R_o^2 + r^2 - 2rR_o\cos\varphi\right)^{1/2}$$

In this solution, $J_1(\cdot)$ is the first order Bessel function of argument $(\cdot)$ and φ_e is taken to be some effective angle which covers the main concentration of energy in the primary beam. The subscript "s" also describes the generation of sum frequency sound in this solution, depending on the assignment of quantities. The field variables are defined by the geometry of Fig. 2(b).

The closed-form solution of Eq. (20) has not yet been found so that at present one must use numerical integration techniques to treat practical problems. These techniques are nonetheless very straightforward and many computations have been made with Eq. (20) showing good agreement with experiment.

Other numerical approaches[22] have also been used with success, especially in calculations of difference frequency sound pressure levels. Approximate solutions, based on empirical evidence, have also been developed[23] in graphic form for use in design studies.

Parametric Self-demodulation

The parametric array solutions discussed in the previous section are for continuous wave (cw) or steady-state excitation. Nonlinear transients are also created in the parametric interaction of pulsed primaries. These parametric transients were first identified theoretically by Berktay,[9] who called the process "self-demodulation" because the primary pulse demodulates itself into a secondary pulse. Moffett, Westervelt, and Beyer[24] confirmed Berktay's predictions experimentally, and other papers on the subject have also appeared.[25-27]

An example of this process is illustrated in Fig. 4, which shows a primary pulse modulated in a Gaussian envelope with its associated Fourier spectrum. The primary spectrum is arbitrarily centered at 1 MHz with an envelope time constant of $5.8 \times 10^{-10} sec^{-2}$. Energy at the different frequencies contained in the primary spectrum interacts in the parametric process to generate a low frequency transient centered at 60 kHz which has roughly the same pulselength and bandwidth as the primary radiation.

Using essentially the same assumptions as for Eq. (18), a time domain solution of Eq. (17) has been obtained[9] for the self-demodulation process that is valid on the acoustic axis of the system. This is

$$p_s = \frac{\beta P^2 S_o}{16\pi\rho_o c_o^4 \alpha R_o} \frac{\partial^2}{\partial t^2} f^2\left(t - \frac{x}{c_o}\right) , \tag{21}$$

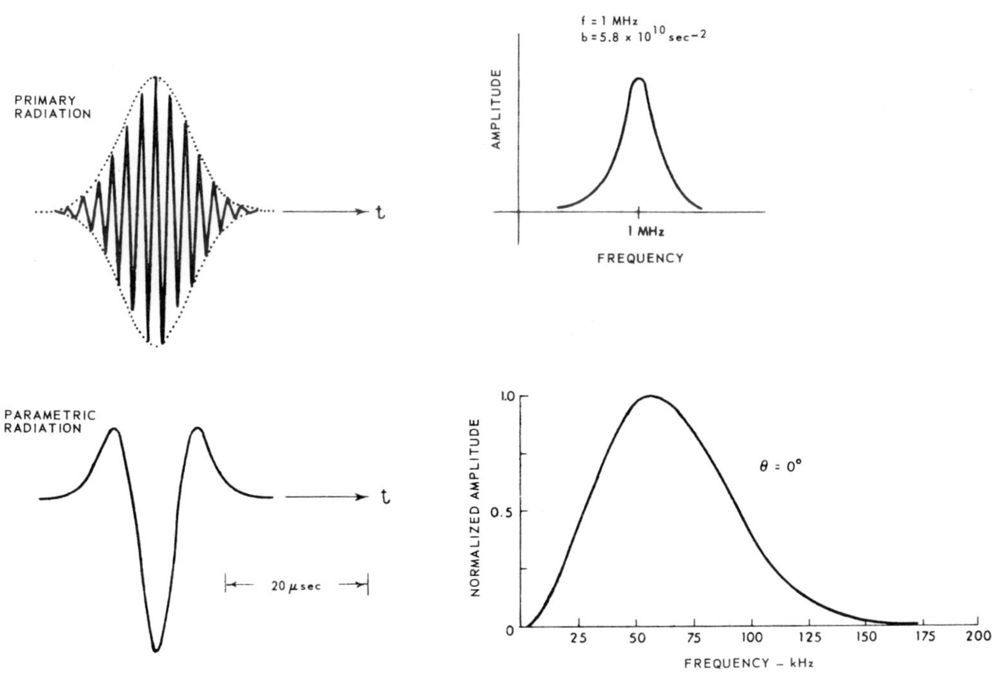

FIGURE 4

PARAMETRIC SELF-DEMODULATION

where P is the carrier pulse amplitude, $f(t-x/c_o)$ is the envelope function of the primary pulse, and the remaining symbols are the same as defined earlier for Eq. (18).

This solution shows the demodulated waveform to be proportional to the second time derivative of the square of the primary envelope function. Although Eq. (21) is valid only on the acoustic axis, the demodulated waveform can be shown to have directivity properties similar to those of steady-state parametric arrays.

The author of the present paper has recently applied Fourier techniques to obtain solutions for the complete field of demodulated waveforms, for the same assumptions as stated for Eq. (18).[28] The frequency domain solution is

$$p_s(R_o,\theta,\omega_s) = W(R_o,\theta,\omega_s)\ F\left\{|\hat{f}(t)|^2\right\} \quad , \qquad (22)$$

where

$$W(R_o,\theta,\omega_s) = -i\left(1+\frac{1}{2}\frac{B}{A}\right)\frac{\omega_s^2 P^2 S_o}{8\pi\rho_o c_o^4 R_o}\left[i\alpha_c + \frac{\omega_s}{c_o}\sin^2\frac{\theta}{2}\right]^{-1} \quad ,$$

and $F\{\cdot\}$ represents the Fourier transformation of $\{\cdot\}$. The time domain solution is merely the Fourier transform of Eq. (22), i.e.,

$$p_s(R_o,\theta,t) = \hat{p}_s(R_o,\theta,\omega_s) \quad . \qquad (23)$$

Computation with Eqs. (22) and (23) show the shape of the demodulated waveform to change with increasing angle from the acoustic axis, approaching a constant value at twice the half-power angle.[28] For the example treated in Fig. 4, the shape changes from that shown to a single cycle of a sine wave at $\theta=2°$. There is also a corresponding downshift in the peak frequency of the Fourier spectrum of the demodulated waveform between $\theta=0$ and $\theta=2\theta_{HP}$.

Parametric Reception

The receiver analogy of the parametric transmitter was recognized in the first paper on parametric arrays,[7] and was shown to exist shortly thereafter.[10] The basic concept is illustrated in Fig. 5, part (a). An incoming signal at some low frequency ω_2 is made to interact with a high intensity "pump" wave of frequency ω_1, where $\omega_1 > \omega_2$. The receiver is tuned to the side band(s) containing the interaction frequency(ies).

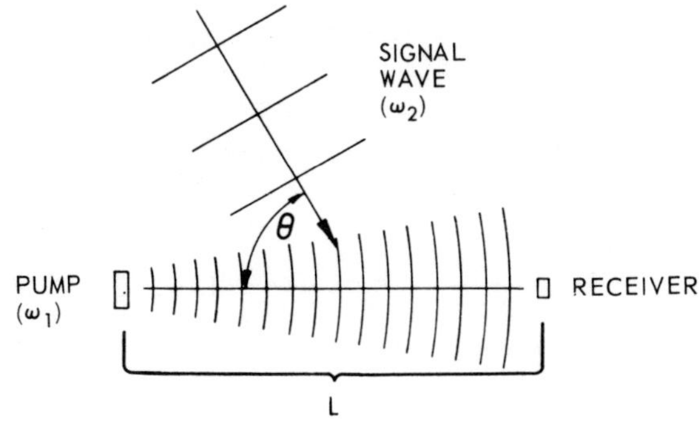

SINGLE-ELEMENT PARAMETRIC RECEIVING ARRAY

(a)

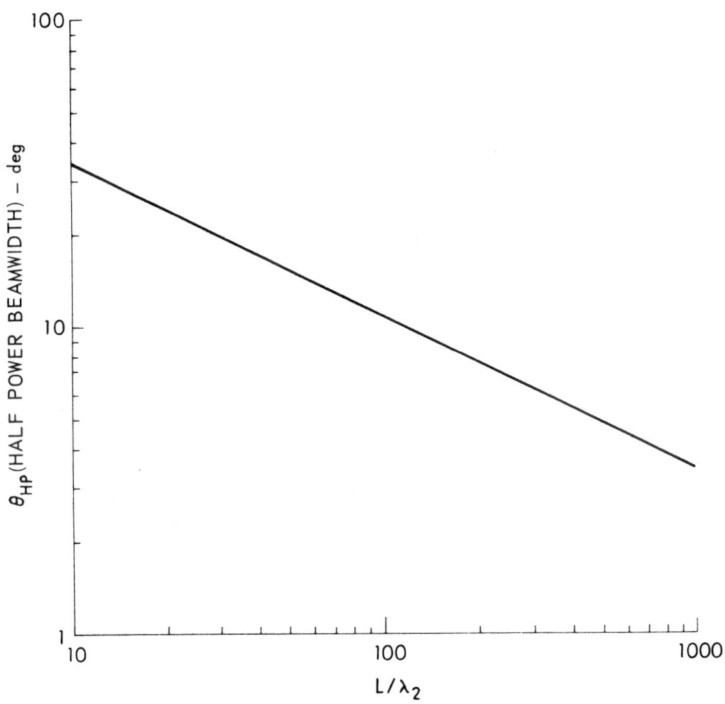

BEAMWIDTH OF A PARAMETRIC RECEIVER
AS A FUNCTION OF LENGTH

(b)

FIGURE 5

PARAMETRIC RECEIVING ARRAY

The theory of parametric receivers has been developed by Berktay,[29] who solved Eq. (17) for several boundary conditions. The result of one such derivation, for the sum and difference frequency pressure amplitudes at a point a distance L in the farfield of the pump transducer (on its acoustic axis), is given in the form

$$p_s(L,\theta) = -\frac{(\omega_1 \pm \omega_2)\beta P_1 P_2}{2\rho_o c_o^3} \exp\{-(\alpha_s + jk_s)L \mp jM\} D(\theta) \quad , (24)$$

where $D(\theta) = \sin M/M$ and $M = k_2 L(1-\cos\theta)/2$. This result is valid for L much greater than the nearfield distance of the pump transducer, and for those situations in which the pump beamwidth is less than that of an end-fired array of length L and frequency ω_2. Some significant features of this result should be pointed out.

1) The solution is independent of the directivity function of the pump transducer.
2) The directivity function of the parametric receiver is independent of the pump frequency, and is the same whether the sum or the difference frequency component is used.
3) The axial response of the parametric receiver is dependent on its length L only through the absorption term.

In general, parametric reception offers a means to develop directive receivers at low frequencies with small transducers. The beamwidth of such a system is equivalent to that of a linear end-fired array of length L. The half-power beamwidth for parametric reception is plotted in part (b) of Fig. 5 as a function of the array length L divided by the signal wavelength, λ_2. The parametric receiver differs from linear end-fire receivers in many respects, the major difference being that only two small transducer elements are required, whereas the linear end-fired array usually requires a large number of elements distributed over its length.

EXPERIMENTS

Since the theoretical results summarized in the previous section predict some interesting effects that are of increasing importance in practical applications of underwater sound, it is appropriate to illustrate these results by comparison to experiment. A number of measurements have been made in fresh water with a 3 in. diam piston projector, operating at fundamental frequencies near 450 kHz. The transmitting apparatus consisted of an oscillator, gate, and power amplifier, and the receiver was made up of the usual components common to transducer calibration work. The measurements to be discussed in this section were conducted with this apparatus.

Saturation Measurements

We begin with some simple measurements on the aforementioned system operating at high power. Some typical data for the amplitude response on the acoustic axis, with range as a parameter, are presented in Fig. 6. The input-output curves shown here are for the fundamental frequency component. They were each measured as a function of input power at six different range stations. The ordinate labeled extrapolated peak source level was obtained from linear theory and was verified by the linearity of data taken at the shortest range, 0.76 yd.

Since only the range was changed for each successive curve, it is clear that medium effects are responsible for the progressive roll-off from linearity in each successive curve. These medium effects are associated with finite-amplitude attenuation caused by shock formation in the high intensity beam. The distortion and shock formation causes energy originally contained in the fundamental component to be translated, in the medium, to harmonics of the fundamental. The increased attenuation at the harmonic frequencies leads to the finite-amplitude attenuation, which is much greater than the ordinary thermoviscous absorption associated with small-signal propagation. The roll-offs in the amplitude response curves of Fig. 6 are clearly due to finite-amplitude attenuation.

As the source amplitude is further increased, these curves approach limiting values, as shown. The phenomenon of acoustic saturation then occurs, a state in which the medium's capacity to sustain arbitrarily intense sounds is exceeded. The theoretical saturation levels are shown as solid dotted segments on the right of Fig. 6. Since the saturation formula was obtained by letting p_o approach infinity, and since the experimental values for this parameter were finite, one should not expect exact agreement between predicted and measured levels. The discrepancy at the three

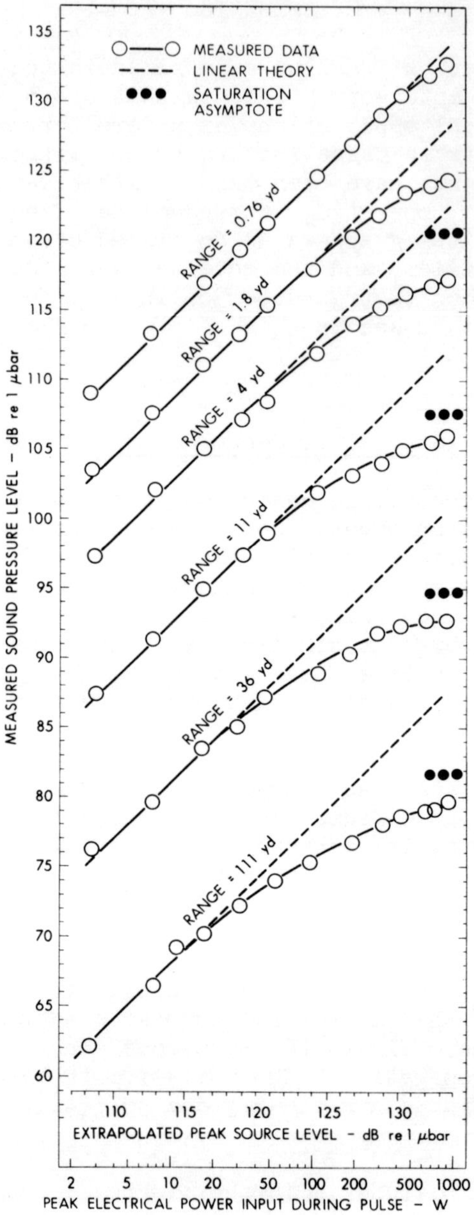

FIGURE 6

AMPLITUDE RESPONSE CURVES SHOWING THE EFFECT
OF EXTRA LOSSES CAUSED BY NONLINEAR EFFECTS
(3 in. diam PISTON PROJECTOR, f = 454 kHz, DI = 36.5 dB, η = 40%

longest ranges, where plateaus in the response curves are best approached, is nonetheless only a few decibels. The approximate nature of Eq. (16), coupled with the fact that no account of nearfield distortion was made in the present calculations, renders this comparison of theory and experiment quite satisfying.

The progressive roll-off from linearity in the input-output curves of Fig. 6 demonstrates an important practical consequence of nonlinear effects in high intensity transmissions. As saturation is approached, one quickly arrives at a region of diminishing return, where, for example, a 1 dB increase in source level may effect only a fraction of a decibel increase in the sound pressure level of the fundamental component at the field point. In this region, further increase in the size of the power amplifier is practically useless, unless the intent is to generate harmonic radiations.

The importance of saturation in measurements involving either the water column or the sediments should not be overlooked. Experiments to be interpreted on the basis of amplitude information are often subject to the effects of finite-amplitude attenuation that are not usually accounted for in data reduction.

Acoustic Field of the Fundamental Component in a Finite-Amplitude Wave

The propagation of spherical waves under the influence of nonlinear effects leading to and including saturation is also of importance in understanding the consequences of finite-amplitude attenuation. Propagation data, abstracted from the amplitude response curves of Fig. 6 for six different levels of transmitted power, are shown in Fig. 7. Here, the solid lines indicate the results of computations with Eq. (12) and its corollaries. The dashed lines represent the approximate extension of farfield theory to the nearfield. The blank space in the propagation curves signifies that long range data were taken at a different time of the year than the other data, when the water was warmer and was therefore slightly less absorptive. Again, satisfactory agreement between theory and experiment was obtained.

The influence of nonlinear effects can be seen in the separation of the curves. At low amplitudes, there is as much separation at the shorter ranges as there is at long ranges. At high amplitudes, however, the separation diminishes with increase in range, demonstrating an amplitude-dependent attenuation. The decrease in the separation of these curves again demonstrates the diminishing return effect; that is, one obtains just about as much SPL at the long ranges by transmitting at a source level of

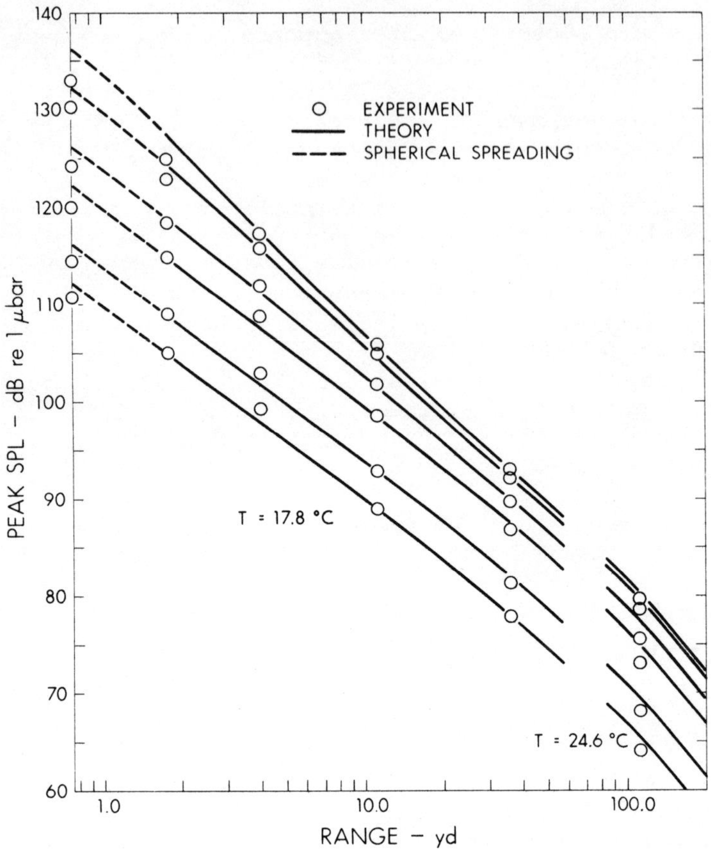

FIGURE 7

PROPAGATION CURVES

130 dB as is obtained by transmitting at a source level of 134 dB. This occurs despite the fact that the transmitted power is increased by a factor of 2.5.

Mutations in the beam patterns of nonlinearly influenced radiations were observed in the present experiments, as they were in previous experiments in air.[6] This phenomenon is illustrated by the measured data presented in Fig. 8, which shows a series of high and low amplitude beam patterns of the 454 kHz piston, measured at the six range stations used to acquire the aforementioned amplitude response data. The high amplitude patterns, taken at a peak source level of 133 dB re 1 μbar at 1 yd, are shown in Fig. 8(a), while the low amplitude patterns, taken at a 1 yd source level of 115 dB, are shown in Fig. 8(b). The first patterns, taken at a range of 0.76 yd, show the effect of a

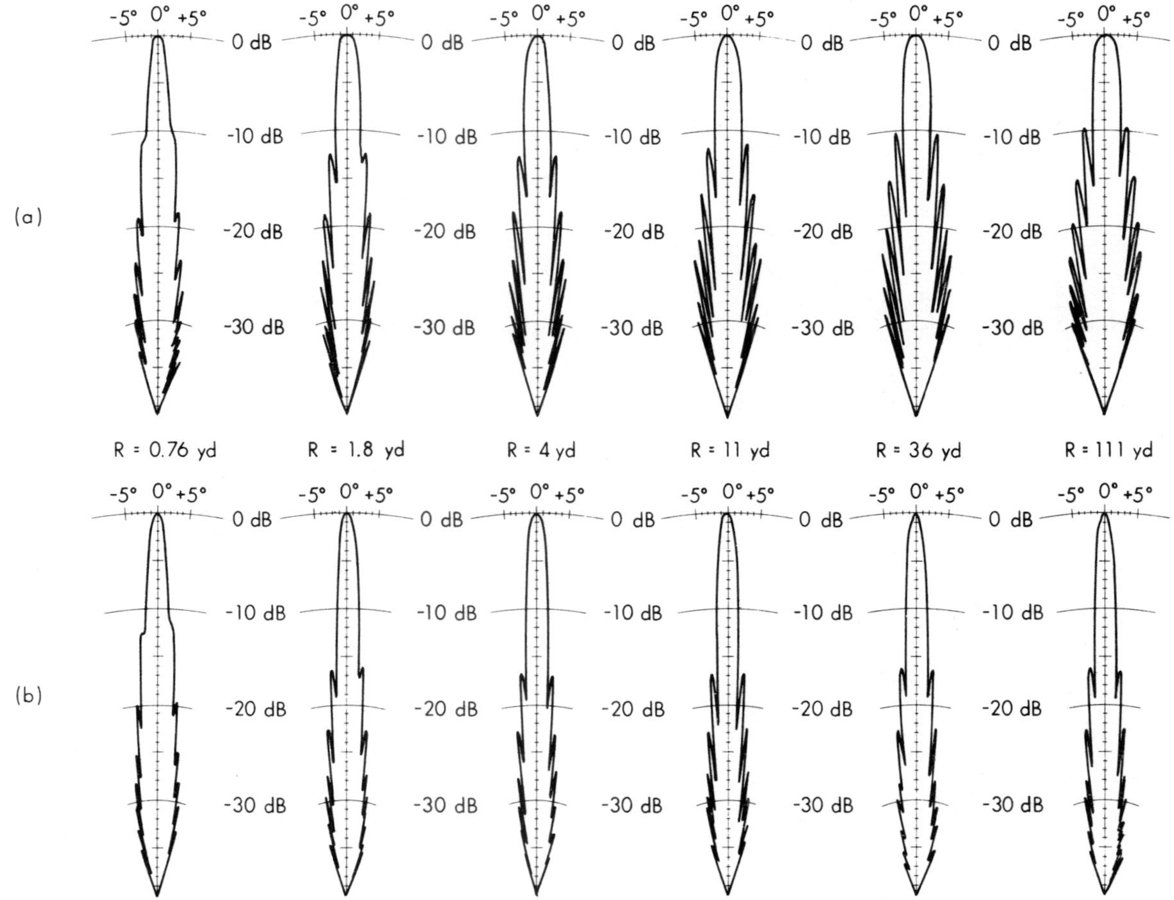

FIGURE 8

BEAM PATTERNS AT VARIOUS RANGES FOR A STRONG AND A WEAK WAVE
(3 in. diam PISTON PROJECTOR, f = 454 kHz, DI = 36.5 dB, η = 40%

measurement in the nearfield of the projector in that the minor lobe structures are not yet fully developed. The onset of the minor lobe is nonetheless about 1 dB higher in the high amplitude pattern than in its low amplitude counterpart. The width of the major lobe is also slightly larger in the high amplitude pattern. This trend continues with increase in range as is evident in the remaining five beam pattern pairs. At the maximum range, an 80% increase in the half-power beamwidth results and the minor lobe suppression decreases by 7 dB. These effects may be explained by noting that the major lobe is in the domain of nonlinear acoustics and hence suffers finite-amplitude attenuation, while the minor lobes are of sufficiently low amplitude to be well approximated by linear theory and to suffer little or no extra attenuation.

Since in many sonar applications, including those designed for sediment work, it is desirable to operate with as narrow a beam and as good a minor lobe suppression as possible, beam broadening and loss of suppression due to nonlinear effects can degrade system operation as much or even more than the loss in amplitude previously described.

Acoustic Field of Finite-Amplitude Harmonic Radiations

The propagation of harmonic components in a finite-amplitude wave is illustrated in Fig. 9. The 3 in. diam piston was driven at a source level of 127 dB re 1 μbar at 1 yd at a fundamental frequency of 450 kHz. The amplitudes of the fundamental and the second and third harmonics were then measured as a function of range. Propagation curves for these harmonics are identified by n=1,2,3, respectively, in Fig. 9.

As can be seen, the harmonics exist and their amplitudes are relatively high, even though this source level-frequency combination was not sufficient to generate a mature sawtooth wave. In a sawtooth wave, the amplitudes of the harmonic components are related to that of the fundamental by the factor $1/n$, where n is again the harmonic number. Beyond the sawtooth region, however, absorption reduces the harmonic amplitudes, as can be seen from the data of Fig. 9. The theory summarized by Eqs. (12) and (15) is quite accurate at the shorter ranges and lower harmonic numbers, and fairly accurate elsewhere.

Directivity measurements made in conjunction with the above described propagation data are shown compared to theory in Fig. 10.*

* For details on the theoretical computation, see the discussion following Eq. (14).

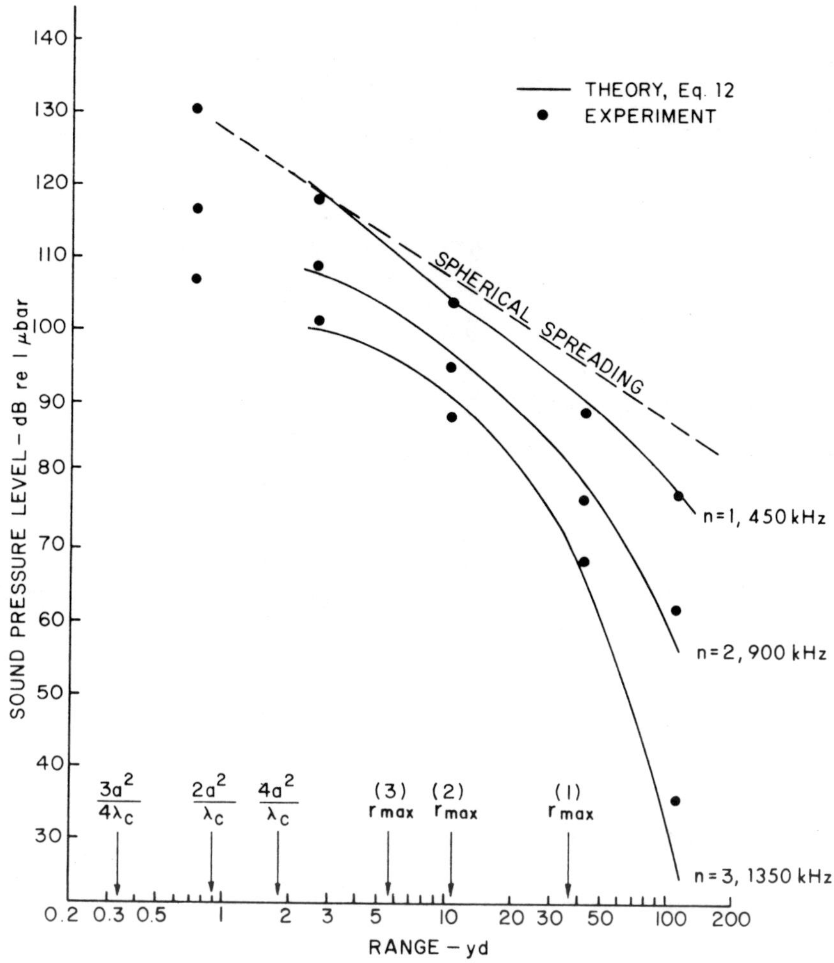

FIGURE 9

PROPAGATION CURVE
HARMONICS IN A SPHERICAL WAVE OF FINITE AMPLITUDE

As can be seen, there is a progressive reduction in beamwidth with increase in harmonic number. This occurs despite the fact that there was some beam broadening due to saturation effects. The minor lobe levels are also reduced with increase in harmonic number.

Each of these trends is most desirable in high resolution sonar. The utilization of harmonic radiations in underwater acoustics should therefore be recognized as a potentially useful technique. It should be pointed out that operation at preshock

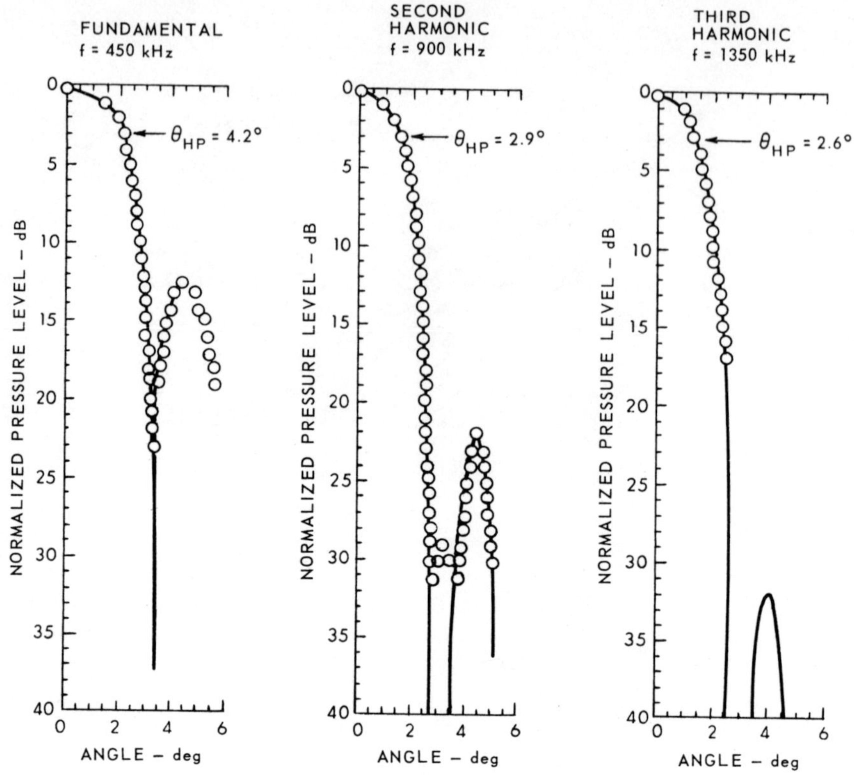

FIGURE 10

BEAM PATTERN DATA FOR STRONG WAVES
ooo EXPERIMENT —— THEORY
FUNDAMENTAL SOURCE LEVEL = 127 dB re 1 μbar at 1 yd
PATTERNS MEASURED AT R = 112 yd

levels (i.e., those for which Eq. (13) is valid) provides for much higher directivity than demonstrated by the example presented here. However, the amplitudes of harmonic components in that regime are correspondingly reduced.

The significance of nonlinear harmonic radiations in sediment technology will be discussed in the final section of this paper.

Parametric Transmitting Array Measurements

Experiments with parametric arrays may be conducted in several ways, as is illustrated in Fig. 11. One of these is to sum the outputs of two oscillators in a linear adder and then apply the

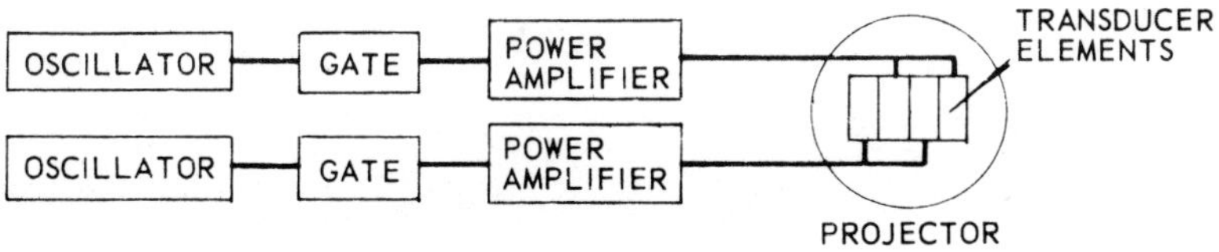

(A) DISCRETE PARAMETRIC TRANSMITTER

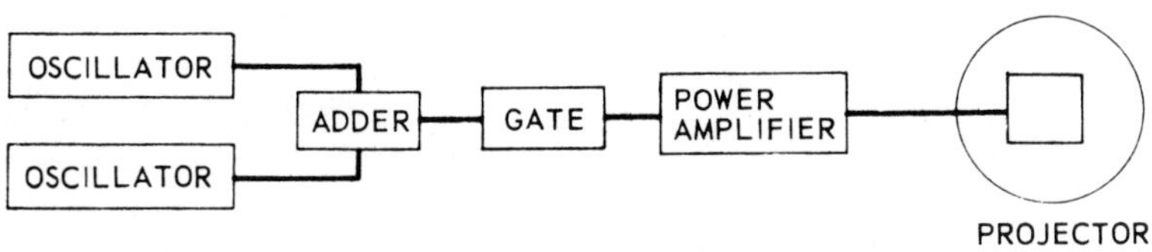

(B) COMPOSITE PARAMETRIC TRANSMITTER

FIGURE 11

METHODS FOR PARAMETRIC EXCITATION

amplified combination signal to a single, linear projector. Another method involves the use of two sets of interleaved transmitting arrays in which each array is allocated a single carrier frequency. The former method was used[18] to obtain the data shown in Fig. 12. For these measurements, primary radiations at 418 and 482 kHz were each transmitted at a source level of 108 dB re 1 μbar at 1 yd. In Fig. 12, axial propagation data for a carrier radiation at a mean frequency of 450 kHz and for the sum and difference frequency radiations are shown compared to theory. The curve for the mean carrier was constructed from nearfield theory for the axial response, and includes spreading and absorption in the farfield. Although both the sum and difference frequency radiations undergo spreading, they experience a noticeable influx of energy from parametric generation out to a range of 50 to 100 yd, where the difference frequency curve is seen to asymptotically approach the spherical spreading curve. The sum frequency curve appears to approach the spreading curve sooner, but this is primarily due to the high absorption at this frequency. The sum frequency data show a considerably higher amplitude than the difference frequency data. This can be explained on the basis of the dependence of secondary amplitude on the square of the secondary frequency as given in Eq. (20). As mentioned earlier, Eq. (20) is valid only for interaction in the farfield of the carrier transducer.

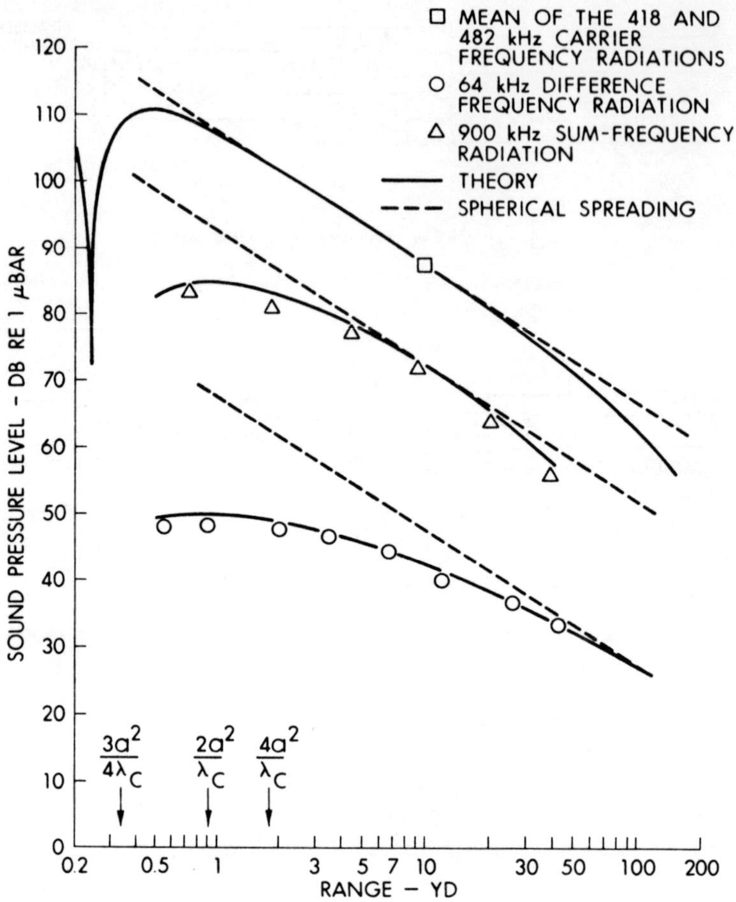

FIGURE 12

PARAMETRIC ARRAY PROPAGATION CURVES

The theoretical curves shown in Fig. 12 were computed with the nearfield parameter $r_o=3a^2/4\lambda_c$. This choice for the nearfield-farfield "boundary" was based on the realization of the farfield half-power beamwidth at this range. The fact that most of the nonlinear interaction took place in the farfield of the present experiment did not make the choice of r_o particularly critical, at least for ranges $R_o \gg r_o$. The agreement between theory and experiment shown in Fig. 12 is within experimental error, which is estimated to be about ±2 dB.

The relative amplitudes of the data shown in Fig. 12 indicate that the parametric conversion process is an inefficient one. This is indeed the case, as efficiencies on the order of 1% are common. It can be shown, however, that the parametric efficiency is directly proportional to the primary power, provided shocks are not formed in the carrier waves. In considering practical applications of parametric arrays, the reduction of primary transducer size must be weighed against the inefficiency of the process. As will be shown, the wide band nature of the parametric process enables one to recapture much of the conversion loss by use of wide band signal processing techniques.

Directivity measurements of the two carriers and of the difference frequency radiation are shown in Fig. 13. The carrier patterns demonstrate the usual minor lobe structures common to linear transducers, while the difference frequency pattern is devoid of minor lobes, in accordance with theory. The narrow width of the difference frequency beam is also evident. For the parameters of the experiment at hand, the difference frequency beamwidth is practically the same as either carrier. Had a linear radiation at a frequency equal to the difference frequency been radiated from this projector in the conventional manner, the resulting beamwidth would have been about eight times larger. This demonstrates the practical utility of parametric arrays in providing high resolution, low frequency sound beams with small apertures. The parametric array is perhaps the most important contribution to sediment technology that has originated in nonlinear acoustics. This point will be expanded in a subsequent section.

The beam patterns of sum frequency parametric arrays are practically identical to those for the second harmonic radiations, due primarily to their similarity in frequency.[18]

As was mentioned earlier, the parametric interaction process is inefficient. One way to overcome this inefficiency is to utilize the wide bandwidth of parametric conversion.[25] This feature is illustrated in Fig. 14, which shows photographs and frequency spectra for the principal waveforms in a parametric array experiment involving cw pulses. The double peaks in the carrier spectra indicate a concentration of energy at the two main carrier frequencies, with some energy in the side lobes. The width between nulls in the main carrier bands is 4 kHz, for the experiment at hand. The width between nulls of the main difference frequency band is also 4 kHz.

These data indicate a bandwidth translation from carrier to difference frequency sound that enables the generation of extremely wide band radiations at low frequencies. It is a simple matter, for example, to design an electroacoustic source having a resonance bandwidth whose half-power points span 15% of the primary frequency, ω_p. The use of two primary bands separated by the mean

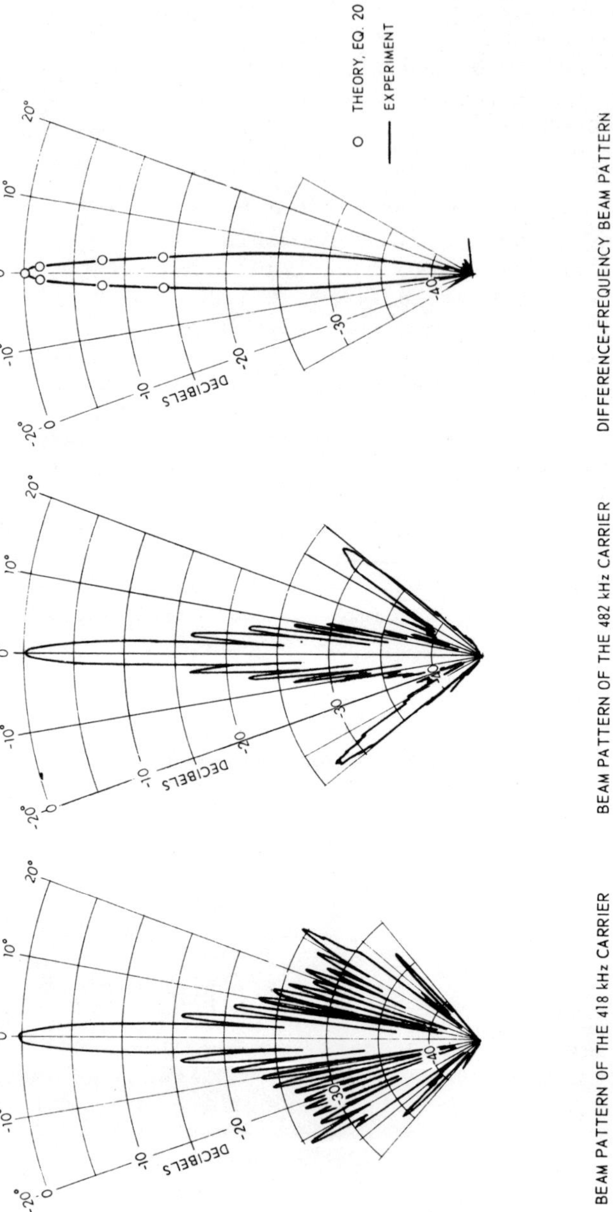

FIGURE 13

PARAMETRIC TRANSMITTING ARRAY DIRECTIVITY PATTERNS

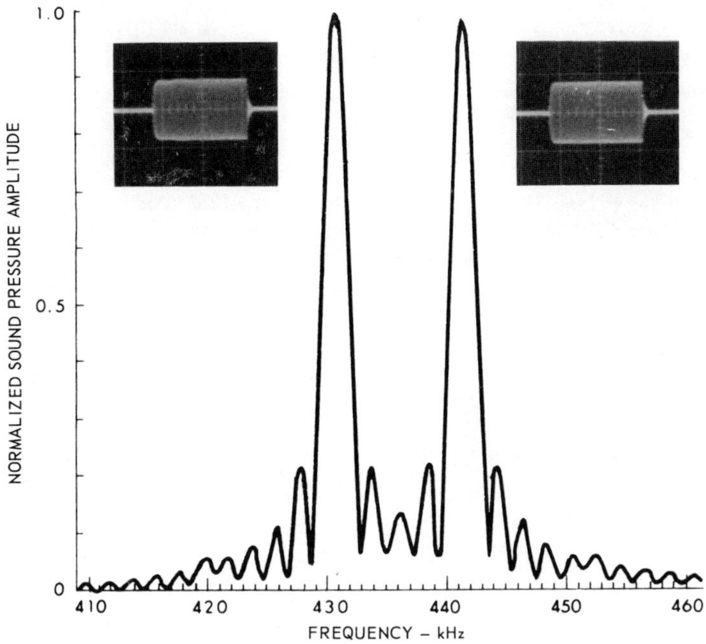

PULSE SHAPE AND FOURIER SPECTRUM OF THE CARRIERS

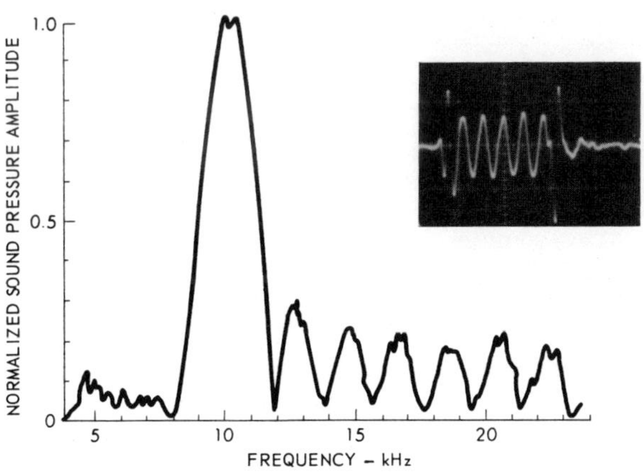

PULSE SHAPE AND FOURIER SPECTRUM OF DIFFERENCE FREQUENCY SOUND

FIGURE 14

TIME BANDWIDTH CHARACTERISTICS OF A PARAMETRIC ARRAY

difference frequency, ω_s, would then lead to a difference frequency bandwidth that can be as wide as 15% of ω_p. This relationship is not rigorous, as the effects of ω_s^2 weighting (see Eqs. (18) and (20)) should really be taken into account; however, the concept of the bandwidth translation is a valid one and it can provide the designer with an option to develop wide band difference frequency signals. Many of the well known wide band signal processing techniques (such as replica correlation, etc.) can then be used to extract the parametric signals from background noise.[25]

The "spikes" on the leading and trailing edge of the difference frequency pulse of Fig. 14 are due to a nonlinear transient effect associated with the interaction of side band energy in the Fourier spectra of the primary pulses. This effect, known as parametric demodulation, was discussed in the theoretical section of this paper for the case of a single primary pulse. Transients like those of Fig. 14 also result from cross-demodulation when two primary pulses are present. Their effect is to accentuate the upper side lobes in the Fourier spectrum of the difference frequency pulse.

Although the frequency domain analysis of these transients follows the reasoning of crosscorrelation (rather than autocorrelation as is the case for self-demodulation of a single primary pulse), the time domain descriptions of these two phenomena are identical. This description simply states that difference frequency transients are generated in proportion to the second time derivative of the square of the primary pulse envelope (see Eq. (22)).

The significance of this argument and of the data of Fig. 14 is that the transients created by nonlinear self-demodulation exist in practically all high intensity pulse transmissions. Like harmonic generation, the demodulated transients are generated in almost all sonar pings, usually irrespective of the designer's intent or knowledge. These transients are most intriguing, especially since they have excellent directivity properties and since their amplitudes can actually be higher than the cw component of conventional difference frequency tones, as is shown in Fig. 14. Comments on the practical utilization of these transients in sediment measurements will be presented in a subsequent section.

Parametric Receiving Array Measurements

Several recent measurements[30] with parametric receiving arrays operating in the water column have been conducted with the apparatus summarized in the sketch of Fig. 5. In these measurements a 5 kHz signal arriving from a remote source was made to interact with a 460 kHz pump signal radiated by a 1 in. diam projector. The resulting 455 kHz difference frequency sound was received by a 1/2 in. diam hydrophone at a range of 19 ft on the axis of the pump projector.

Rotation of the composite pump and receiver apparatus enabled the acquisition of the beam pattern data shown in Fig. 15, part (a). Although the pattern obtained was slightly narrower than predicted by Eq. (21), the axial levels predicted by that solution agreed quite well with the measured values.

It is possible to combine pump-receiver combinations, i.e., parametric receiving elements, in arrays of parametric arrays to perform any number of array functions.[30] Data resulting from the combination of four such elements, identical to the one described above, is shown in part (b) of Fig. 15. Here, the use of four parametric receivers to form a broadside array has effected a sizeable reduction in beamwidth, in accordance with ordinary array theory. This type of array can also be steered electrically by inserting a progressive phase shift in the output of each element. The data shown in part (c) of Fig. 15 demonstrate this technique of beam deflection, showing two beams separated by a steered angle of 6°. Since the steered sector is limited by the width of the single element pattern, the length of each parametric end-fired receiver need not be inconveniently large.

To shed some light on the potential advantages of arrays of parametric receivers, some comparisons to equivalent linear systems should be made. First, consider a 4-element linear system having roughly the same transducer area as the 4-element parametric receiver just described. Unlike the parametric system, the linear system would possess grating lobes of equal sensitivity at $\theta=30°$ and 90°, and there would be no discrimination against signals or noise arriving from both the back of the array and from angles contained in the vertical plane. Next, consider a 4-element linear system with the same acoustic properties as the aforementioned parametric system. This would require four piston transducers whose largest dimension would be 2.5 ft, which amounts to a surface area of 2820 sq in., or roughly 720 times that of the parametric device.

It should not be implied that transducer size and cost are the only factors to be considered in the comparison of linear and parametric systems. Clearly, the electronics associated with active arrays of the parametric type are more complex than those associated with linear arrays. However, the increased electronic complexity need not be prohibitive if the proper choice of parameters is made.

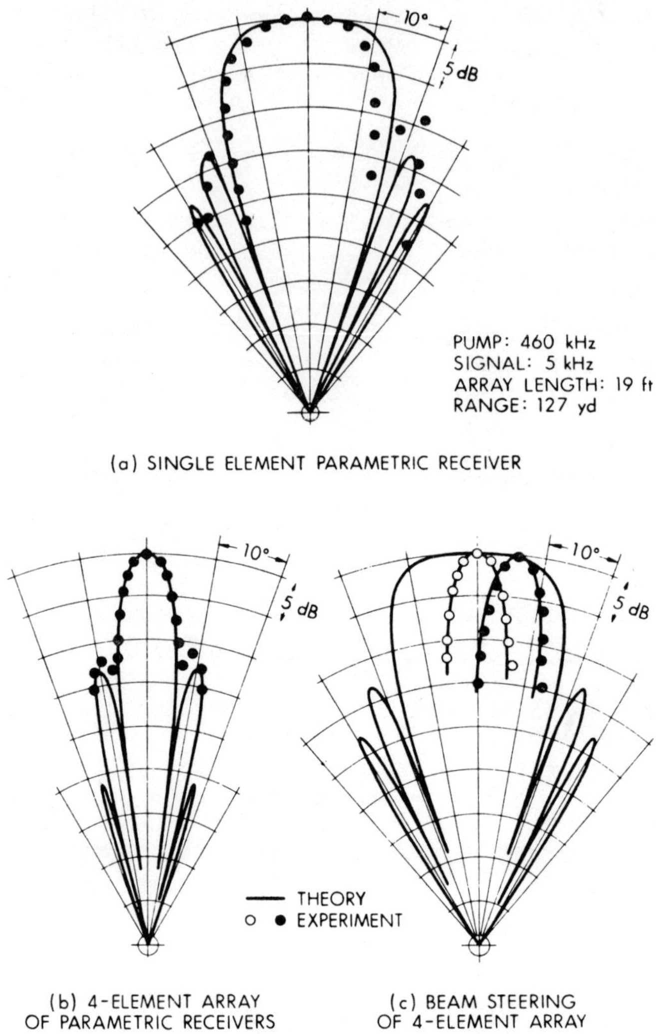

FIGURE 15

PARAMETRIC RECEIVING ARRAY MEASUREMENTS

Although parametric receivers are interesting devices, their application to sediment technology is not as obvious as is the case for parametric transmitters. However, use of the sediment-water interface should not be overlooked as a possible medium for parametric receiver interaction. The practicality of this approach hinges on the nonlinearity of the sediment-water interface.

APPLICATIONS

In this section, we will review several important applications of nonlinear acoustics to measurements in sedimentary geophysics. Most of these applications are a consequence of the phenomena summarized in previous sections of this paper. Some of the topics to be discussed are already proven, so that their inclusion here is entirely justified. Other, less well understood topics have been included primarily because of their potential development as tools for sediment studies and surveys. One assumes a certain risk in speculating on this latter category of applications, and it should be understood that all of them may not develop into feasible techniques. They may, on the other hand, provide useful examples for the evolution of innovative thought on both the physics of the sediments and on instrumentation for sediment measurements.

Parametric Echo Sounding and Subbottom Profiling

The applicability of the parametric acoustic transmitting array to bottom and subbottom profiling was first recognized[14] in 1968 to be a feasible approach to high resolution bathymetry. The parametric transmitter offered all of the desirable features for bottom and subbottom profiling, i.e., a narrow beam at a low operating frequency, with a small projector that could be used from small oceanographic vessels. The only significant disadvantage was the inefficiency of the parametric process; however, this difficulty did not turn out to be prohibitive for echo sounding, as the ocean bottom affords an excellent target with good echo return. It was clear from the onset that deficiencies in the parametric signal-to-noise ratio could be remedied by use of any of a number of wide band signal processors. This is especially true in subbottom profiling, where signal processing is essential to good sediment penetration.

A block diagram of one of the earlier designs is shown in Fig. 16. This system is sufficiently similar to those fabricated and tested in this country[31] and abroad[32] to serve as an example for discussion. The outputs of two FM oscillators are summed, gated into a dual FM pulse, amplified, and applied to a single high frequency projector. This approach is the most economical as it requires only one transmitter; however, care must be taken to insure linearity in the transmitter components so that sound at the difference frequency will not be directly radiated from the projector. This undesirable effect can be minimized by designing the high frequency amplifier and projector components to be very inefficient in the difference frequency band. Difference frequency sound generated in the water reflects from the ocean bottom and returns to a low frequency hydrophone located

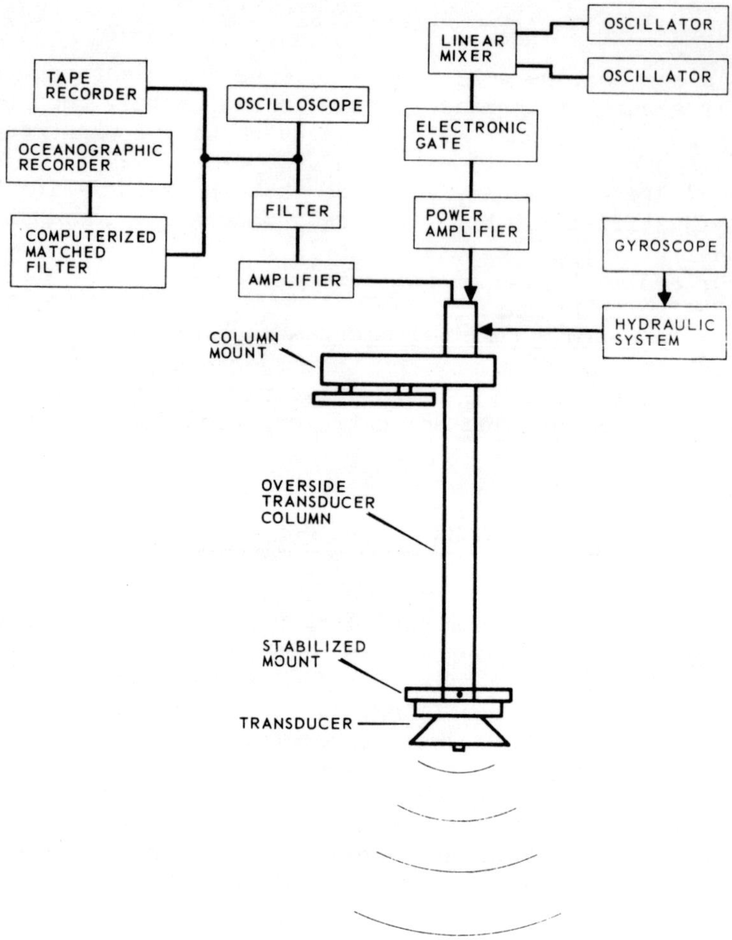

FIGURE 16

BLOCK DIAGRAM OF PARAMETRIC DEPTH SOUNDER

in the same housing as the high frequency source. The received signal is amplified, filtered, and processed by means of a replica correlator. When the difference frequency pulse in the shape of an FM slide is used, the processing gain in the presence of white noise is roughly given by[25]

$$P.G. = 10 \log TW \qquad (25)$$

where T = pulse length, W = bandwidth. The usual components for bathymetric measurements and research are also shown in the sample block diagram, namely an oceanographic recorder and a tape recorder. One method of mechanical mounting of the system transducer is also indicated. This method is applicable to extremely narrow beam systems that may require a stabilized soundhead.

The first at-sea tests of a parametric echo sounder were reported by G. Walsh[31] of the Raytheon Company. The system tested differed from that of Fig. 16, primarily in that the soundheads were towed in a VDS fish to avoid the mechanical stabilization problem. This system, dubbed "FADS" for Finite-Amplitude Depth Sounder, used a mean primary frequency of 200 kHz and a mean difference frequency of 12 kHz, in a 60 msec FM pulse having a bandwidth of 2 kHz. The primaries were transmitted from a 9 in. aperture at a source level of 125 dB re 1 μbar at 1 yd. A difference frequency source level of 88 dB re 1 μbar at 1 yd was produced in a 2.5° beam.

A side by side comparison of this system with a conventional 12 kHz echo sounder, both operating in Narragansett Bay, is shown in Fig. 17.[31] Oceanographic recorder data for the parametric system is shown at left, with the conventional system data presented at right. The superiority of the parametric system is clearly evident in three major features. First, the fish shoals appear as distinct targets in the parametric case, in contrast to their appearance as hyperbolic traces in the data acquired by the conventional system. This is due to the fact that the conventional system beamwidth is so broad as to receive echoes from these targets long before, during, as well as after, the survey vessel passes over the target area. The location and extent of the fish shoals is much more clearly defined with the parametric system. Secondly, the bottom appears as a distinct boundary in the parametric case, with some lateral discontinuities that are not evident in the data for the conventional system. This is due to a substantial amount of averaging within the wide beam of the conventional system that "washes out" many of the important details that are recorded by the narrow beam parametric system. Finally, the remarkable fine structure in the subbottom layering for the parametric data is far superior to that of the conventional system, also due primarily to the latter's large-scale beam averaging.

Considering the fact that both of these systems have radiating apertures of identical size on the survey vessel, and that the increased complexity of the parametric system is purely electronic (and therefore relatively easy to accommodate), one is drawn to the obvious conclusion that a significant advance in echo sounding has been achieved.

Detection of Buried Artifacts in Marine Archeology

The utilization of parametric sonar for search and discovery in marine archeology has also been recognized[33,34] as a unique

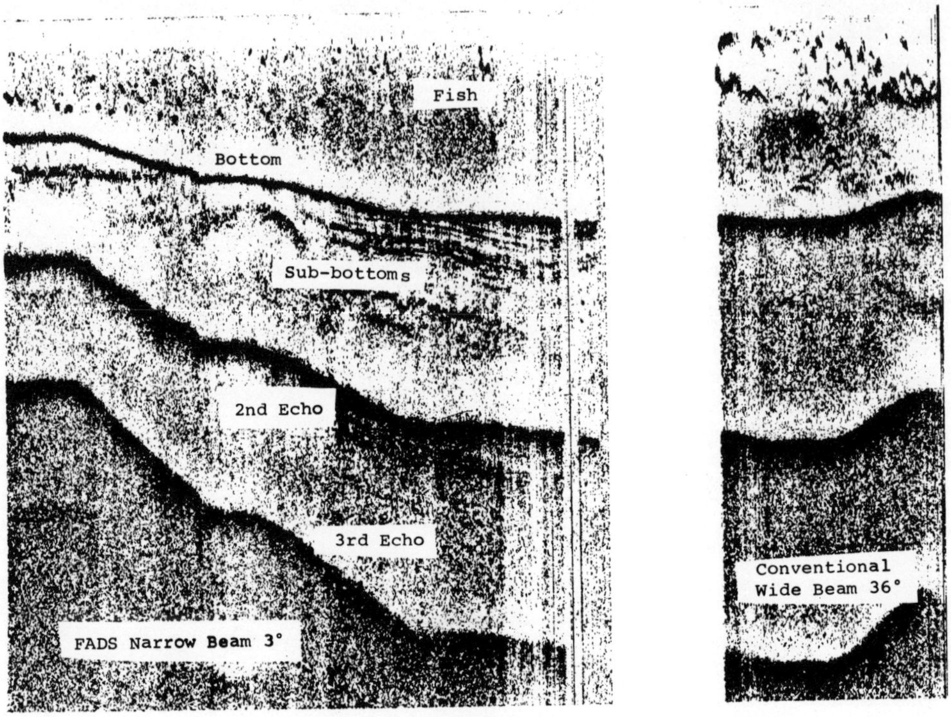

TAKEN FROM: G WALSH, PROC. SPECIALISTS MEETING ON NONLINEAR ACOUSTICS, UNIV. BIRMINGHAM, APRIL 1971, BRITISH ACOUSTICAL SOCIETY.

FIGURE 17

FADS/CONVENTIONAL COMPARISON AT 12 kHz

application of nonlinear acoustics. Archeologists usually locate a potential site by some combination of (1) extensive historical research (2) "tips" from amateur treasure hunters, and (3) by accident. These methods of location are inexact by modern technological standards because of the rather large margins of error associated with ancient navigational records and those of latter day navigators and amateur archeologists as well. In recent years, magnetometers and conventional subbottom sonars have been used in conjunction with improved navigational techniques. Although modern navigational aids are adequate, the sensors now used for artifact detection are limited by poor resolution. Once on site, the recovery of artifacts from the sediments is an expensive, time consuming process, due simply to the fact that the excavators have considerable difficulty in determining exactly where to dig. Acoustic sensors having high resolution beams with frequencies low enough for good sediment penetration are therefore of great practical interest in this field.

This interest, coupled with the success of the echo sounder in penetrating subbottom layers, prompted some field experiments[34] on the detection of buried artifacts. It was decided to study the parametric echo return from a small metal sphere at oblique angles of incidence, to determine the feasibility of a subbottom sonar having a relatively wide searched sector. An aluminum sphere, 5 in. in diameter and not unlike an ancient cannon ball, was buried in a dense sand sediment in the Corpus Christi ship channel. This sediment was classified as a fine, well-sorted sand with a specific gravity of 1.89, an acoustic attenuation of 0.18 dB/ft/kHz, and a sound speed 12% higher than that of the water column. After burial, approximately 6 in. of sand covered the target.

A parametric array generated by the nonlinear interaction of 190 and 210 kHz primaries emitted by a 17 in. diam line-in-cone projector was used to develop a 20 kHz difference frequency radiation having a half-power beamwidth of 2.4° and a pulse length of 230 μsec. This device was used to obtain echoes from the sediment and the target, before and after burial. Some results are shown in Fig. 18, which presents A-scan echo data in both the primary and difference frequency bands. Here, the target was located at a slant range of 11 ft, where it subtended projector grazing angles in the neighborhood of 35°. As can be seen, both the primary and difference frequency radiations produced distinct bottom and target echoes before burial (Fig. 18(a) and (b) and (d) and (e)). When the target is buried (Fig. 18(c) and (f)), only the difference frequency radiation produces a target echo. This is due to the prohibitively high attenuation of the 200 kHz primaries in the sediment.

Although this target was buried to a depth of only 6 in., its detection was made under relatively poor circumstances in that (1) the detection angle was near the so-called "critical" angle of 30°, as predicted from lossless theory, and (2) the proximity of the water-sediment interface provided background reverberation capable of masking target echoes. The detection of this small target in these measurements was therefore encouraging as regards the eventual feasibility of parametric subbottom sonars for use in marine archeology. More recent measurements at even smaller grazing angles (including subcritical angles) also support this contention.[35]

It should therefore be feasible to consider subbottom sonar designs for archeological surveys of the type sketched in Fig. 19. Although several sonar configurations are undoubtedly good candidates for practical implementation, the one shown here is especially adaptable to the small "vessels of opportunity" usually employed by marine archeologists. The soundhead is towed in a spherical fish to minimize the effects of the vessel's pitch and

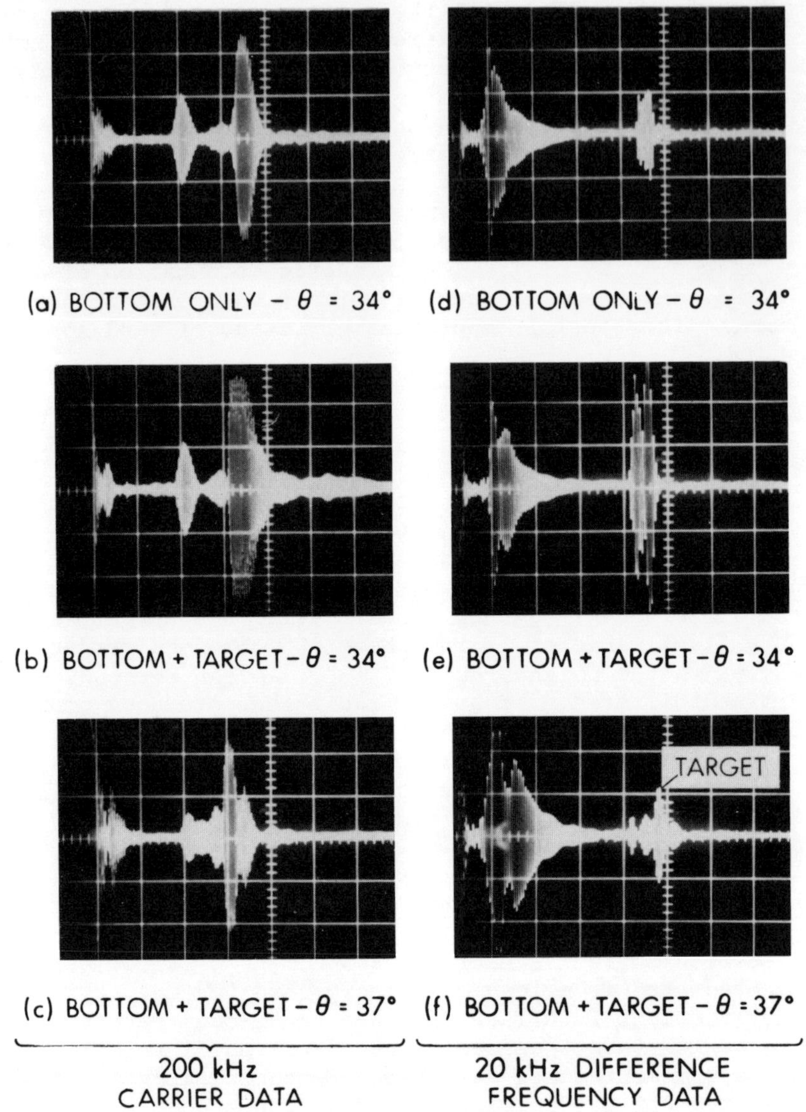

FIGURE 18

PARAMETRIC SONAR DETECTION OF BURIED TARGET

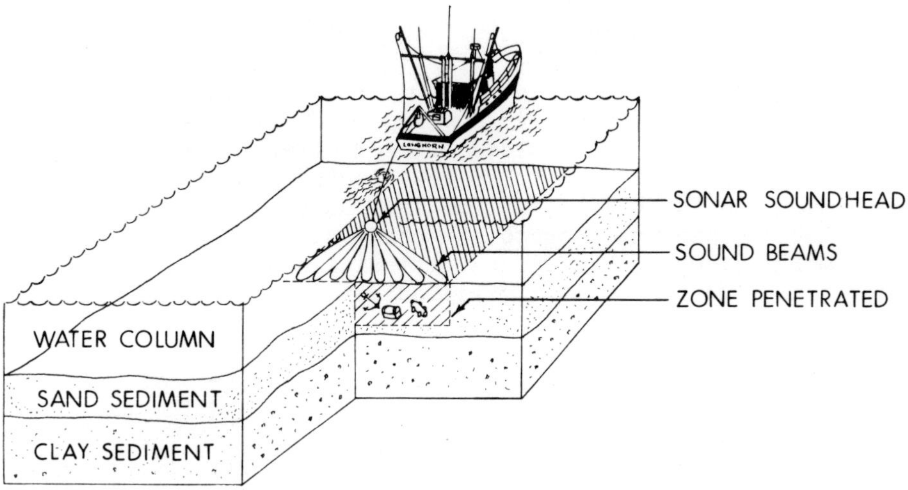

FIGURE 19

CONCEPT SKETCH OF AN ARCHEOLOGY SONAR

roll as well as the noise from its engines and propeller. Sequential formation of a large number of parametric beams can be achieved by several techniques, including mechanical rotation of the transducers,[33] the use of acoustic lenses[36] and/or rotating acoustic prisms.[37] Although sequential beam formation and processing lead to low sonar search rates, this approach greatly simplifies system design as it requires only a single channel sonar. A small VDS hoist, an oceanographic recorder, and a modest amount of electronic hardware for this simple sonar completes the configurational description. It is anticipated that such a sonar would be a significant asset, both in locating artifacts of historical importance and in reducing archeological survey time and expense.

Wide Band Echo Sounders and Subbottom Profilers

In the preceding pages, practical applications of only one nonlinear acoustic phenomenon were discussed. All of these are based on Westervelt's parametric acoustic array and each is in some intermediate state of development. The initial success of these developments indicates that other nonlinear acoustic phenomena may also have important practical significance. It is therefore the author's intent to suggest here some possible future applications that employ both the directivity and the wide band nature of two unexploited nonlinear acoustic phenomena.

1. Parametric Transient Devices

The first potential application employs the transient analog of the parametric array, first discovered by H. O. Berktay.[9] Parametric demodulation of a pulsed carrier offers an excellent means of producing highly directive transients in the medium, as has been discussed in the theoretical section of this paper. These transients can be made to possess the bandwidth of the high frequency primary pulse, with a directivity ultimately limited by the absorption length of the primary radiation in the sense of an end-fired array.

The question arises as to what significance directive, wide-band transients may have in sediment measurements. While the answer to this question may not be immediately obvious, we are reminded that the small toothed whales have, over million of years of evolution, developed highly sophisticated sonars, used in foraging, that utilize directive wide band transients. (A speculative connection between the porpoise "click" and parametric transients has recently been advanced.[38]) Perhaps more germane is the use to which transients have been employed in linear systems analysis, primarily in the determination of the impulse response of unknown networks. It is therefore not unreasonable that a wide band, highly directive transient may be capable of exciting a stratified sedimentary body such as to enable instantaneous measurement of frequency-dependent absorption and perhaps even the trapped mode resonances of sedimentary strata. Such measurements are difficult to make with existing explosive source technology due to the absence of directivity in the radiated impulses and to the troublesome bubble pulse that follows the initial shock.

One possible configuration of the proposed scheme is shown sketched in Fig. 20. A high frequency primary radiation, pulsed in a Gaussian envelope, propagates toward the bottom as a high intensity transmission. The self-demodulation process causes some of the primary energy to distort into a low frequency transient, according to Eq. (21), as illustrated in Fig. 4. At the bottom interface, some of the demodulated transient energy and most of the primary energy returns to the system's hydrophones. The ratio of primary to secondary energy return from this interface provides a measurement of surface roughness, as some of the short wavelength energy will be scattered out of the beam by the stochastic topography. Echo returns from the subbottom structure in the primary and secondary frequency bands should be expected to be quite different, due to the frequency dependent absorption of the individual layers. This absorption could also be measured by Fourier analysis of the time-dependent subbottom returns, which may be normalized to the return from any particular stratigraphic layer, including the sediment-water interface. The low frequency nature of the

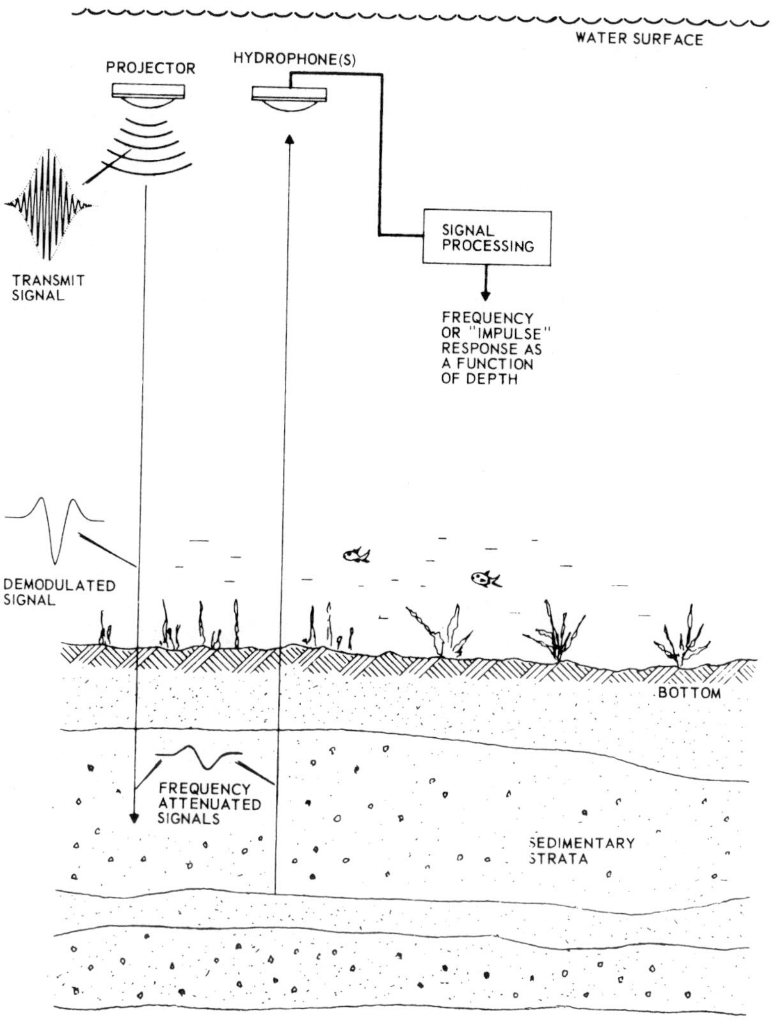

FIGURE 20

SELF-DEMODULATION SEDIMENT PROFILER

demodulated waveform could also be designed to provide a means of measuring the resonances and antiresonances of layers satisfying the conditions of normal mode excitation. That is, if the Fourier spectrum of the demodulated waveform is so designed as to contain radiant energy whose wavelengths are on the order of that of the layer separations, then modal excitation of individual strata might be effected.

The feasibility of using parametric self-demodulation in the above described processes has not yet been developed. Quite obviously, a substantial amount of Fourier signal processing would be required. A computer analysis of magnetic tapes of self-demodulated echo returns would be quite helpful in determining the feasibility of these techniques and in evolving new concepts.

2. Harmonic Radiation Devices

The use of harmonic radiations generated by distortion of a finite-amplitude primary also appears to offer new dimensions in remote acoustic sensing of sediment properties. It appears this method has many of the possibilities of the previous method, with perhaps a simpler receiver.

The process is illustrated in Fig. 21. A relatively low frequency projector transmits a sinusoidal wave of high amplitude in the form of either a cw or FM pulse. The Fourier spectrum of this pulse is localized in the frequency domain according to the sketch shown. As this transmission propagates toward the bottom, its time domain distortion follows that predicted by Eq. (7), with harmonic components generated in the frequency domain according to Eq. (12). The fully shocked state of this transmission appears as a train of sawtooth waves whose Fourier components decay as $1/n$, where n is the harmonic number. At the bottom, some energy in all of the frequency components is specularly reflected back to the hydrophones, with scattering out of the beam becoming more and more important with increase in harmonic number. The ratio of fundamental to harmonic return is again a measure of surface roughness. The increase in directivity of each successive harmonic component [see Eqs. (13) and (14)] would enable the bottom topography to be measured with several simultaneous orders of resolution. This would allow a determination of the degree of resolution required for accurate measurements of topographic relief for different types of bottoms. Within the bottom, the attenuation of each layer could be measured by iterative comparison of harmonic returns from the strata interfaces. Here, the good directivity of the harmonic components is essential as it minimizes echo averaging within the beam. Excitation of modal properties of the strata could possibly be achieved (as discussed for the preceding case of transient excitation), but only if the harmonic frequencies were related to the eigenfrequencies of the stratification.

Like the preceding example for the case of profiling with demodulated waveforms, the concept of profiling with nonlinearly generated harmonics has not yet been investigated. The signal processing requirements for the present example are theoretically quite simple. As is indicated in Fig. 21, a matrix of real-time filters could be employed to perform the basic measurement on the

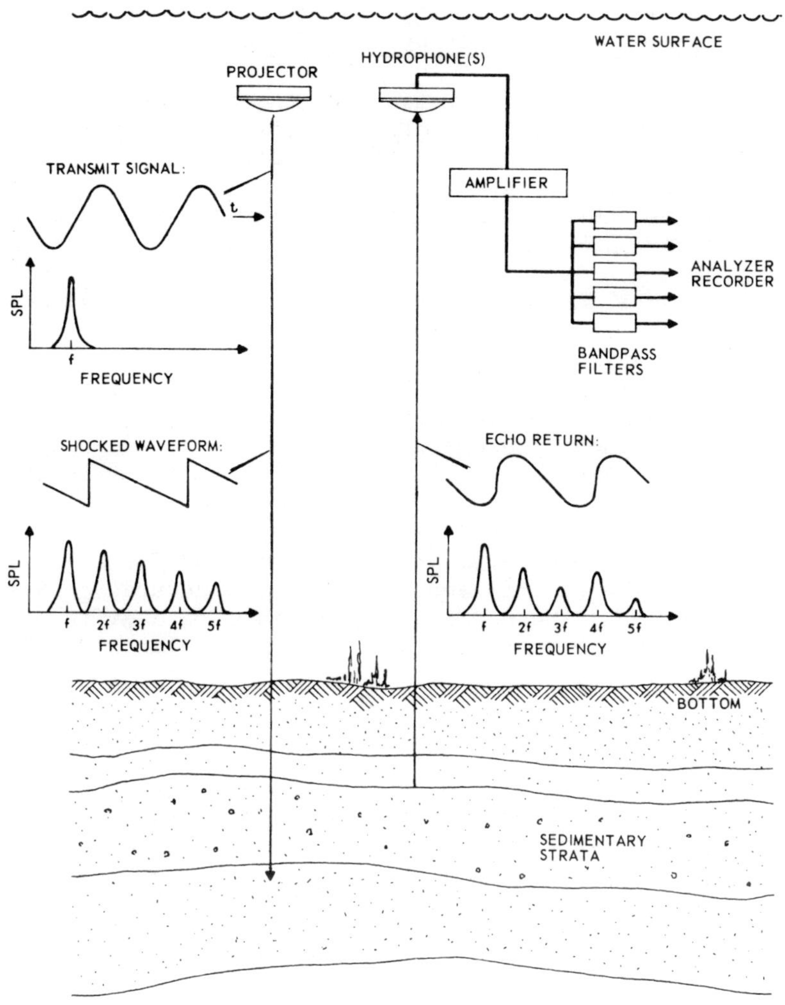

FIGURE 21

NONLINEAR HARMONIC ECHO SOUNDER AND PROFILER

time-dependent frequency spectra. The analysis could also be made on magnetic tapes of harmonic echoes by use of a laboratory based computer. This would be quite helpful in determining the feasibility of the technique and in evolving new profiling concepts.

Acoustic Nonlinearity of the Sediments

Although many static measurements of the stress-strain relationship for different types of sediments have been made well into the nonlinear regime, no measurements of the dynamic nonlinearity have been reported. A major scientific aspect of sedimentary geophysics therefore remains to be investigated. As was the case for water, it is not known whether or not the acoustic nonlinearity of the sediments will be of any engineering significance. This is hardly relevant, as the knowledge of the acoustic nonlinearity is entirely justified on scientific grounds.

There are several approaches to the measurement of acoustical nonlinearity. The most obvious methods employ measurements of one or more of the nonlinear phenomena. This is due to the fact that the parameter of nonlinearity, B/A, appears as a factor in all of the mathematical solutions for these phenomena, as discussed in the theoretical section of this paper. One merely measures all of the unknowns in a given experiment and then solves for B/A. An important shortcoming of this approach is its inaccuracy. Being a second order effect, the nonlinear generation of a difference frequency or a harmonic, etc., is quadratically related to the primary variables. This means, for example, that a 1 dB error in the measurement of a primary wave amplitude could cause a 2 dB error in the determination of B/A.

A more accurate, and perhaps even simpler, method was suggested by Beyer[2] (based on an interpretation made by Rudnick) and has been applied to several fluids.[39] This method is founded on a thermodynamic expansion of the equation of state, and it is sufficient to state it here as

$$\frac{B}{A} = 2\rho_o c_o \left(\frac{\partial c}{\partial p}\right)_{\rho=\rho_o, T} + (2c_o T\beta_p/c_p)\left(\frac{\partial c}{\partial T}\right)_{\rho=\rho_o, p} \qquad (26)$$

where T = temperature, $\beta_p = 1/V\ (\delta V/\delta T)_p$, the isobaric volume coefficient of thermal expansion, V = volume, c_p = specific heat at constant pressure, and all other terms are standard and have been previously defined. Examination of Eq. (26) shows that the acoustic nonlinearity of the sediments could be determined primarily by independent measurements of the variance of sound speed with temperature and pressure. Knowledge of the ambient density and sound speed as well as the isobaric coefficient of volume expansion and the specific heat is also required and these constants may also require measurement in some instances. The crucial acoustic measurements, however, are sound speed measurements and these can be done quite accurately in the laboratory and in the field.

A test cell containing the sample and apparatus for velocity measurements could be successively elevated in temperature and pressure so as to determine the appropriate velocity gradients. The parameter of nonlinearity could then be calculated. The end result would provide valuable information pertinent to the physics of sound in the sediments.

ACKNOWLEDGMENTS

This work was supported by the Office of Naval Research.

REFERENCES

1. D. T. Blackstock, "History of Nonlinear Acoustics and a Survey of Burgers' and Related Equations," in Nonlinear Acoustics, Proceedings of the 1969 ARL Symposium, T. G. Muir, (ed.) (1970).

2. R. T. Beyer, "Parameter of Nonlinearity in Fluids," J. Acoust. Soc. Am. 32, 719 (1960).

3. D. T. Blackstock, "Connection Between the Fay and Fubini Solutions for Plane Sound Waves of Finite Amplitude," J. Acoust. Soc. Am. 39, 1019 (1966).

4. J. A. Shooter, T. G. Muir, and D. T. Blackstock, J. Acoust. Soc. Am. 49, 119(A) (1971); to be published as "Acoustic Saturation of Spherical Waves in Water" (1973).

5. J. C. Lockwood, T. G. Muir, and D. T. Blackstock, "Directive Harmonic Generation in the Radiation Field of a Circular Piston," J. Acoust. Soc. Am. 53, 1148 (1973).

6. C. H. Allen, "Finite Amplitude Distortion in a Spherically Diverging Sound Wave in Air," Ph.D. Thesis, Pennsylvania State University (1950).

7. P. J. Westervelt, "Parametric Acoustic Array," J. Acoust. Soc. Am. 35, 535-537 (1963).

8. J. L. S. Bellin and R. T. Beyer, "Experimental Investigation of an End-Fire Array," J. Acoust. Soc. Am. 34, 1051-1054 (1962).

9. H. O. Berktay, "Possible Exploitation of Nonlinear Acoustics in Underwater Transmitting Applications," J. Sound. Vib. 2, 435-461 (1965).

10. H. O. Berktay, "Parametric Amplification by the Use of Acoustic Nonlinearities and Some Possible Applications," J. Sound. Vib. 2, 462-470 (1965).

11. H. O. Berktay, "Some Proposals for Underwater Transmitting Applications of Nonlinear Acoustics," J. Sound. Vib. 6, 244-254 (1967).

12. V. A. Zverev and A. I. Kalachev, "Measurement of the Scattering of Sound by Sound in the Superposition of Parallel Beams," Sov. Phys. Acoust. 14, 173-178 (1968).

13. H. O. Berktay, "Nonlinear Interactions between Acoustic Waves in Liquids--Possible Applications," Application of Finite Amplitude Acoustics to Underwater Sound, Proceedings of a Seminar held at the U. S. Navy Underwater Sound Laboratory on 27 May 1968, NUSL Rpt. No. 1084, 15-38 (1970).

14. T. G. Muir and J. E. Blue, "Acoustic Modulation of Large Amplitude Waves," J. Acoust. Soc. Am. 46, 227-232 (1969).

15. D. G. Tucker, "A Review of Progress in Underwater Acoustics," Radio and Elec. Engrs., U. D. C. 621-39, 69-84 (1970).

16. R. H. Mellen, D. G. Browning, and W. L. Konrad, "Parametric Sonar Transmitting Array Measurements," J. Acoust. Soc. Am. 49, 932(L) (1971).

17. See, for example, Nonlinear Acoustics, Proceedings of the 1969 ARL Symposium, T. G. Muir, (ed.), Applied Research Laboratories, The University of Texas at Austin (1970).

18. T. G. Muir and J. G. Willette, "Parametric Acoustics Transmitting Arrays," J. Acoust. Soc. Am. 52, 1481-1486 (1972).

19. M. Schulkin and H. W. Marsh, "Sound Absorption in Sea Water," J. Acoust. Soc. Am. 30, 864 (1962).

20. H. O. Berktay and D. J. Leahy, "Farfield Performance of Parametric Transmitters," in publication, J. Acoust. Soc. Am.

21. V. Lannstad, J. Naze, and S. Tjötta, "Nonlinear Interaction of Two Soundwaves," Acta Universitatis Bergensis Series Matematica, (Norwegian Universities Press, Oslo) 12, 1-24 (1964).

22. F. H. Fenlon, "A Recursive Procedure for Computing the Nonlinear Spectral Interactions of Progressive Finite-Amplitude Waves in Nondispersive Fluids," J. Acoust. Soc. Am. 50, 1299-1313 (1971).

23. R. H. Mellen, W. L. Konrad, and D. G. Browning, "Approximate Scaling Laws for Parametric Sonar Transmitter Design," Proc. Brit. Acoust. Soc. Specialists Meeting on Nonlinear Acoustics (1971).

24. M. B. Moffett, P. J. Westervelt, and R. T. Beyer, "Large-Amplitude Pulse Propagation--A Transient Effect," J. Acoust. Soc. Am. 49, 339-343 (1971).

25. T. G. Muir and J. E. Blue, "Transient Response of the Parametric Acoustic Array," in Nonlinear Acoustics, Proceedings of the 1969 ARL Symposium, T. G. Muir, (ed.), Applied Research Laboratories, The University of Texas at Austin.

26. H. M. Merklinger, "High Intensity Effects in the Non-Linear Acoustic Parametric End-Fire Array," Ph.D. Thesis, University of Birmingham (England), June 1971.

27. F. H. Fenlon, "On the Derivation of Zone I Spectra for a Pulsed Finite-Amplitude Source Operating in a Nonviscous Nondispersive Medium," J. Acoust. Soc. Am. 54, 465-473 (1973).

28. T. G. Muir, "A Parametric Model for Porpoise Sonar," Unpublished manuscript, Applied Research Laboratories, The University of Texas at Austin, January 1973.

29. See, for example, H. O. Berktay and J. A. Shooter, "Parametric Receivers with Spherically Spreading Pump Waves," J. Acoust. Soc. Am. 54, 1056-1061 (1973).

30. H. O. Berktay and T. G. Muir, "Arrays of Parametric Receiving Arrays," J. Acoust. Soc. Am. 5, 1377-1383 (1973).

31. G. M. Walsh, "Finite Amplitude Depth Sounder," Proc. Brit. Acoust. Soc. Specialist Meeting on Nonlinear Acoustics (1971).

32. H. O. Berktay, private communication.

33. T. G. Muir, "The Potential of Sonar Surveys in Marine Archeology," Applied Research Laboratories Prospectus, (June 1971).

34. T. G. Muir and R. S. Adair, "Potential Use of Parametric Sonar in Marine Archeology," J. Acoust. Soc. Am. 52, 122(A) (1972).

35. L. A. Thompson and T. G. Muir, "Narrow Beam Sound Fields in a Sand Sediment," Paper V6, 86th Meeting of Acoustical Society of America.

36. T. G. Muir and D. L. Folds, "Parametric Acoustic Lens Sonar," Paper V2, 86th Meeting of Acoustical Society of America.

37. This technique has been successfully used in sonic imaging for biomedical diagnostics. Private communication, D. Ramsey, Stanford Research Institute.

38. T. G. Muir, "A Survey of Several Nonlinear Acoustic Experiments on Traveling Wave Fields," Proc. 1973 Symposium on Finite-Amplitude Effects in Fluids, Tech. Univ. Denmark, IPC Science and Technology Press, Surrey, England (1974).

39. A. B. Coppens et al., "Parameter of Nonlinearity in Fluids," J. Acoust. Soc. Am. 38, 797-805 (1965).

NEAR-BOTTOM ACOUSTIC OBSERVATIONS OF ABYSSAL TOPOGRAPHY AND REFLECTIVITY

PETER F. LONSDALE, ROBERT C. TYCE, AND F. N. SPIESS

University of California, San Diego
Marine Physical Laboratory of the
Scripps Institution of Oceanography
La Jolla, California 92037

ABSTRACT

A deeply towed instrument system has made near-bottom geophysical surveys of patches of the abyssal sea floor. The acoustic instrumentation includes a narrow beam 40 kHz echo sounder, a 110 kHz side-looking sonar system, a 4 kHz seismic profiling system, and an acoustic transponder system for precision navigation. These sensors have provided us with an abundance of qualitative data on the considerable small-scale variability of sea floor topography, acoustic backscattering, and acoustic reflectivity, from a wide range of abyssal environments. A capability for digital processing of 4 kHz seismic profiler data has recently been added: preliminary quantitative results from the northern Samoan Passage region suggest spatial variations (in 4 kHz bottom reflectivity) of over 10 dB for distances as small as 100 m.

Since 1965 the Marine Physical Laboratory of the Scripps Institution of Oceanography has been making detailed geophysical and geological surveys of the abyssal sea floor using a compact, deeply towed instrument package (Fish) which is towed at slight altitude above the bottom by a cable from a surface ship. The package (Fig. 1) has grown in size, complexity, and capability in the past twelve years, and presently comprises up, down, forward, and side-looking sonars, acoustic transponder navigation hydrophones, and a 4 kHz seismic profiler as its acoustic systems, with a battery of stereo, wide-angle, and snapshot television cameras, an oscillating quartz crystal thermometer, and a proton precession magnetometer as accessory sensors. Control and data signals are telemetered bilaterally in the coax core of the armored towing cable. Major developments in shipboard computer

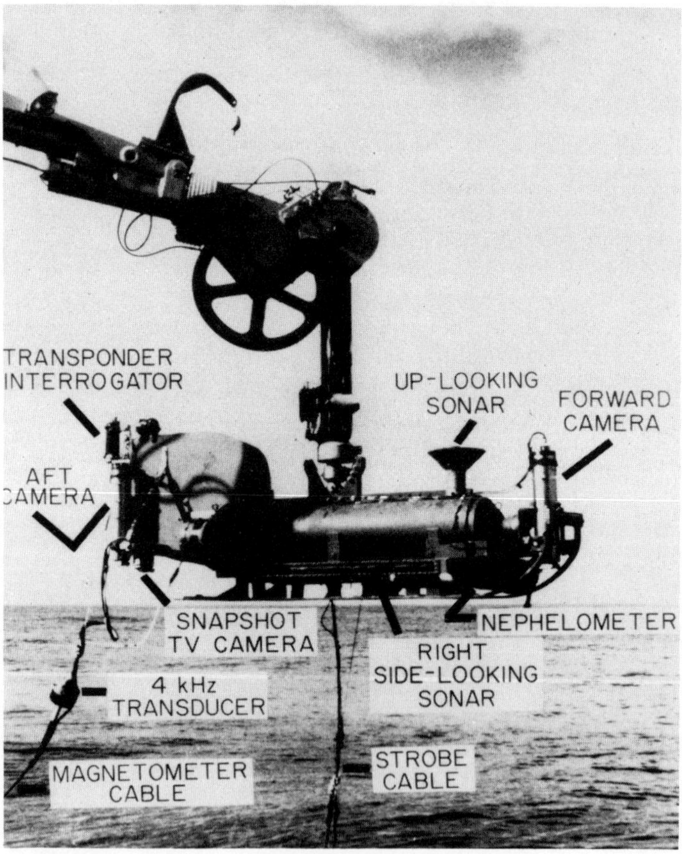

FIGURE 1

THE DEEP-TOW FISH SUSPENDED FROM ITS CRANE BEFORE A LOWERING INTO THE ABYSS.

processing have enabled us to stay abreast of the ever-increasing flow of data which travels up the cable to the shipboard laboratory. A description of the system and summary of the results obtained up until 1969 is included in Spiess and Mudie (1970). Further technical details may be found in Spiess and Tyce (1973).

Limitations of towing speed (approximately 3 km/h) and a 15 km effective range of transponder navigation restrict our activities either to detailed well navigated surveys of small areas or to long satellite navigated profiles. We have operated at depths as great as 7.5 km, and at far-flung, remote sites throughout the Pacific and North Atlantic Oceans (Fig. 2). With careful selection of regions for detailed, deep-tow survey, often after intensive surface-ship reconnaissance, we have studied 100 km^2 patches representative of most of the provinces of the deep ocean, and sites where the effects of important geological processes may be observed with particular clarity. In recent months there has been an emphasis

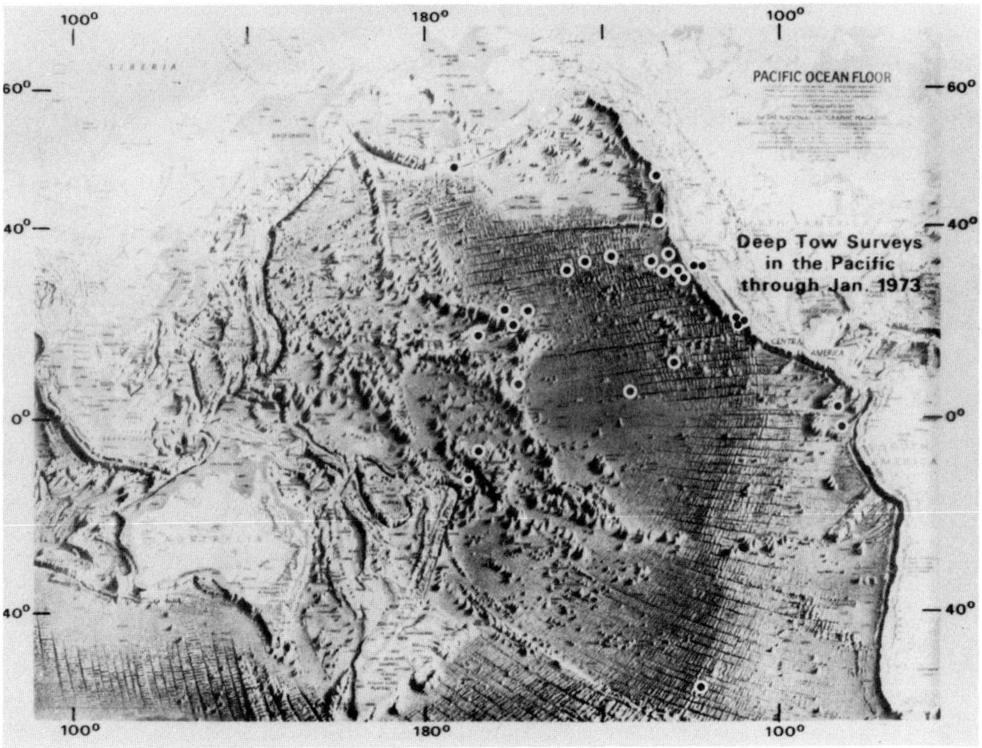

FIGURE 2

DEEP-TOW SURVEY AREAS IN THE PACIFIC OCEAN 1965-1973. DURING THE SUMMER OF 1973 SURVEYS HAVE BEEN CONDUCTED AT SEVERAL LOCATIONS IN THE NORTH ATLANTIC OCEAN.

on studies of magnetic reversals and tectonic processes at spreading centers where new oceanic lithosphere is being created (e.g., Atwater and Mudie, 1973; Mudie et al., 1972a; Mudie et al., 1973) and on problems of abyssal sedimentation, especially erosion by bottom currents (e.g., Johnson, 1972; Lonsdale et al., 1972a; Spiess et al., 1973a, 1973b).

Since review articles and descriptions of instrumentation already exist (Spiess and Mudie, 1970; Spiess and Tyce, 1973) and the results of our most recent shipboard work are being published in accessible journals as soon as they are available, this paper will give merely a brief description of the acoustic components of the deep-tow system, and present some samples of the data obtained. A recent development is the extraction of quantitative information on properties of the deep sea floor from the acoustic signals, and progress in this field will also be discussed.

We use acoustic instrumentation both to gather information on the topography and composition of the deep sea floor and to determine the position of the instrument package. A requirement for navigation is continuous monitoring of Fish depth and the slant ranges to several (2-6) acoustic transponders (McGehee and Boegeman, 1966) tethered to the sea floor in a fixed array. Fish depth is controlled by varying the speed of the towing ship and the length of the towing wire, and is determined by an inverted echo sounder, operating at 23.5 kHz with a 15° beamwidth, which receives echoes from the air-sea interface. The depth of each transponder is taken as the depth of the sea floor minus the length (20 to 100 m) of the taut line tether between anchor and hydrophone. Each unit of the transponder array is interrogated in a sequentially coded pattern at its individually assigned frequency of 10.0, 10.5, or 11.0 kHz by a transducer at the tail of the deep-tow vehicle. All replies are at 12.0 kHz (Spiess et al., 1966). The geometry of the transponder array is initially estimated from the dead-reckoned positions of the ship at the times of deployment. This estimate is soon refined during the deep-tow survey by collecting range information from all available transponders rather than only from the two which are necessary to fix the vehicle position, so that redundant data are available for recalculation of the relative positions of transponders (Lowenstein, 1966). Once the geometry of the array has been acoustically determined, errors in relative position of the instrument package are usually less than 10 m. Spiess and Sanders (1971) report on a survey where the sound velocity in the bottom water was carefully calculated, and the root mean square residual error was 2 m. The ship also interrogates all bottom transponders, and these reply via the deep-tow vehicle's transducer and the tow cable (Fig. 3). Ship navigation is less accurate: Spiess and Sanders (1971) report a root mean square residual range uncertainty of 13 m. Transponder navigation of the ship is, however, essential in permitting both proper maneuvering

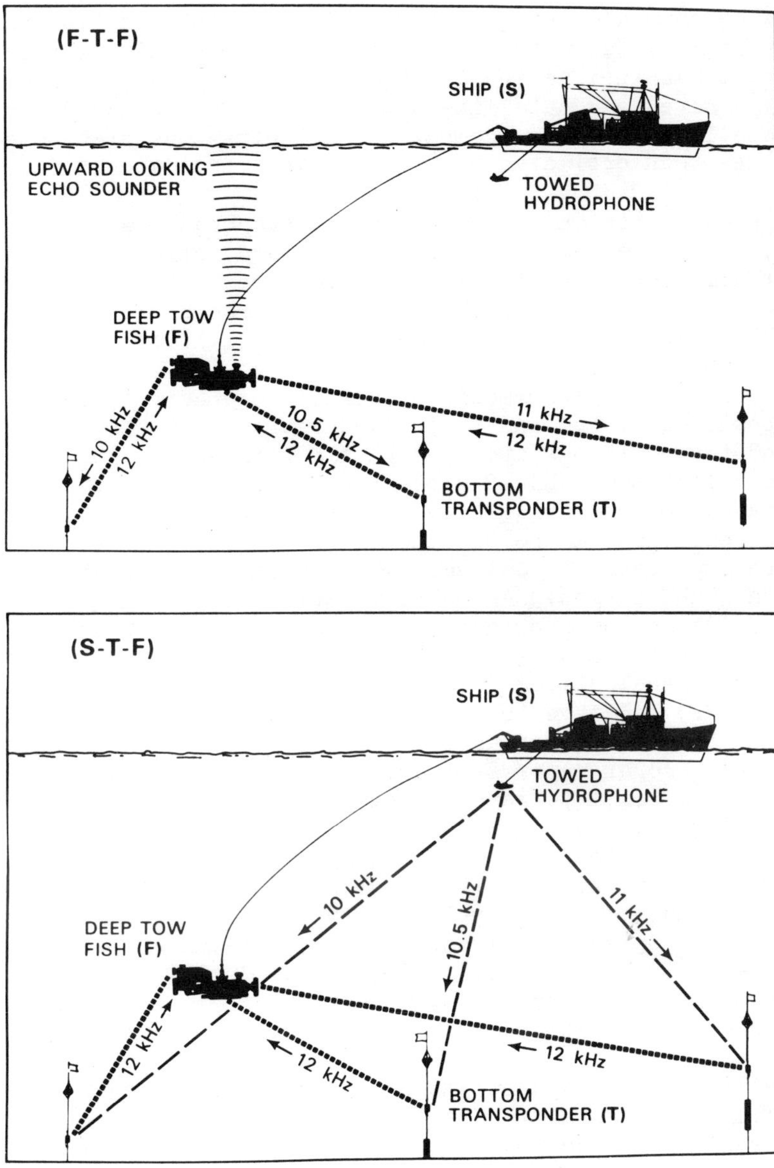

FIGURE 3

TOWING CONFIGURATION. NOTE THAT DIFFERENT PARTS OF THE FIGURE ARE NOT DRAWN TO SCALE. ILLUSTRATES METHOD OF ACOUSTICALLY POSITIONING THE DEEP-TOW FISH (UPPER) AND THE TOWING SHIP (LOWER), RELATIVE TO A FIXED ARRAY OF BOTTOM TRANSPONDERS. FROM BOEGEMAN ET AL. (1972).

of the towing ship and integration of the relative positioning of the transponder system with the position in geographic coordinates provided by the ship's navigation systems. Since 1969, all our towing ships have used satellite navigation. The standard deviation of satellite positions compared to simultaneous transponder positions has been estimated as 450 m (Spiess and Sanders, 1971).

Instrument depth and ranges to all transponders which are not topographically shielded or shadowed by upward refraction of sound (Spiess et al., 1966) is displayed in the laboratory of the towing ship on a standard Gifft wet-paper facsimile recorder. A specially designed attachment permits the ranges to be digitized and sent directly to the shipboard computer (usually an IBM 1800). Calculation of sound velocity corrections, slant range corrections, and the plan position of the instrument and towing ship takes about 10 sec, and the positions are then plotted, in an X-Y coordinate system, on a flat bed CALCOMP plotter. Usually, position fixes are calculated no more frequently than every five minutes, but at times of stress they can be obtained at a 20-sec repetition rate. This can be important for delicate maneuvering while trying to pass over a small target. As multiple-transponder fixes are accumulated during the course of a several-day survey, the estimate of transponder positions is periodically improved, and all pre-existing fixes recalculated (using the transponder ranges stored by the computer) relative to these improved positions. This further facilitates such operations as photographing small objects noticed on side-looking sonar records of previous passes. In order to sample similarly small targets, a relay transponder can be attached to a coring device, and the plan position of the corer may then be monitored, and adjusted, as it descends to the sea floor (Boegeman et al., 1972). We often also take advantage of the precise location of transponders and the reliability of their acoustic recall recovery system (Spiess and Tyce, 1973) by attaching free-fall current meters (Schick et al., 1968) to them during surveys which investigate local patterns of sedimentation (e.g., Lonsdale et al., 1972a; Johnson, 1972; Spiess et al., 1973b).

The altitude of the deeply towed instrument above the sea floor is frequently as little as 10 m for prolonged periods, in typical oceanic water depths of 4000 m. It is measured with a narrow beam 40 kHz echo sounder, whose transmitting unit is a 30-cm-long PZT cylinder oriented fore and aft, and whose receiver is a line of PZT segments mounted athwartships. The overlapping beam patterns result in an effective beamwidth of about 4°, or an effectively insonified area which may be as little as 1 m^2. The transmit pulse of this down-looking system is usually keyed by the returning echo of the up-looking system, so that the trace of its received pulse is a high-resolution profile of the sea floor. This bathymetric profile is one of the most fundamental sets of data collected by the deep-tow instrument. As well as allowing the compilation of contour maps and the recognition of bathymetric lineations (e.g., Luyendyk, 1970; Mudie et al., 1972b) it presents an accurate rendition of steep slopes, which are a confusion of overlapping echoes on surface-ship records. This record has allowed differentiation of faulted and volcanic landforms (Larson, 1972), resolution of the morphology of roughness elements which had previously been known only as trains of regular hyperbolic echoes (Fig. 4), and calculation of the statistics of bottom slope distributions (Spiess et al., 1969; Larson and Spiess, 1970).

Damaging contact of the instrument package or of the dangling strobe light with the sea floor is avoided by maintaining a close watch on the altitude and directly controlling the winch. Early warning of major scarps and mountains is provided by the towing ship's echo sounders. In regions of particularly rough and rocky

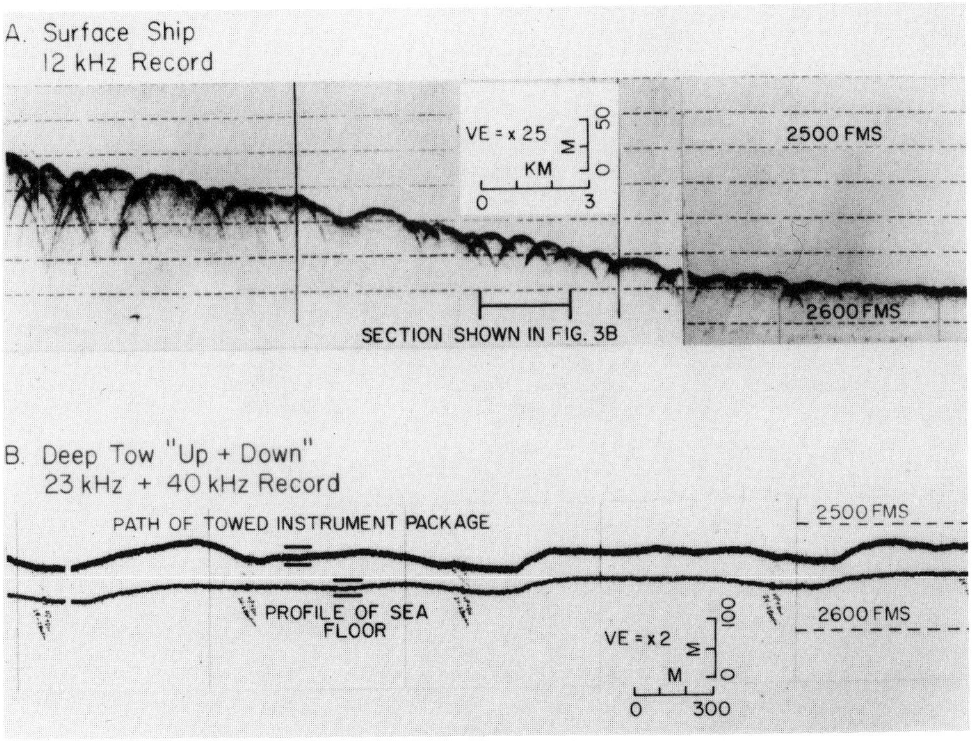

FIGURE 4

A COMPARISON OF SURFACE SHIP WIDE-BEAM AND DEEP-TOW, NARROW-BEAM BATHYMETRIC PROFILES ACROSS SEDIMENTARY RIDGES ON THE APRON OF SEDIMENTS NORTH OF THE SAMOAN ISLANDS (SPIESS ET AL., 1973b). THE SURFACE-SHIP RECORDS CLOSELY RESEMBLE THOSE DESCRIBED FROM THE LOWER CONTINENTAL RISE OFF THE EASTERN UNITED STATES (e.g., BRYAN AND MARKL, 1966; FOX et al., 1968; RONA, 1969), AND OTHER REGIONS WHERE STRONG BOTTOM CURRENTS MOLD RAPIDLY ACCUMULATING SEDIMENTS.

sea floor, a forward-looking sonar is substituted for the narrow beam down-looking transducer. This is a transversely mounted line transducer, also operating at 40 kHz, which provides a relatively narrow beam pattern, oriented along the track of the vehicle, and hence gives warning of approaching obstacles and measures altitude as well. The forward-looking sonar may also search for targets for bottom photography.

A pair of side-looking sonars mounted on either side of the deep-tow vehicle extends our coverage of the sea floor from a profile to a kilometer-wide swath. Each transducer is 121 cm long and operates at 110 kHz. Time-varied gains both in the Fish and aboard ship correct for spreading and grazing angle losses. As with the other acoustic systems, the side-looking sonars usually transmit once per second and telemeter the returned signals up the tow cable for display in the shipboard laboratory on Gifft facsimile recorders. The transmitted signal is an adjustable 400 W (electrical) pulse of 0.1 to 0.4 msec duration with farfield (> 100 m) beamwidths of 3/4° in the horizontal and 60° in the vertical. With a typical Fish speed of 70 cm/sec, a one-second repetition rate results in about two pings for an advance of one beamwidth at 100 m range. Area coverage rate is about one square nautical mile per hour with a resolution of 15 cm in range and 120 cm across the sonar beam.

Variations in acoustic backscattering observed by the side-looking sonars are caused by either of the following types of variation of the sea floor:

1) Changes in slope, either on a small scale, as with regular bedforms (Fig. 5), or on a regional scale, where subtle changes in slope may be apparent as migration of the traces of side lobes produced by destructive interference in the beam pattern (Normark, 1969).

2) Changes in texture of the sea floor, as between mud and sand, ooze and manganese pavements, or undisturbed sediment and artificial debris (Fig. 6).

When slopes become steep enough, or the texture of the sea floor becomes coarse enough, parts of the sea floor become acoustically shadowed, returning no signals because the insonifying beam is intercepted before reaching them (Fig. 7). The degree of shadowing and the detection of slope variations are very sensitive to the Fish altitude. To fulfill the role of recognizing and surveying phenomena intermediate in size between those visible on bottom photographs and those visible on vertical incidence echo sounder and seismic profiles, an altitude of 20 to 50 m is optimal. We have mapped several types of bottom roughness which cannot be identified on vertical incidence profiles (even on the high

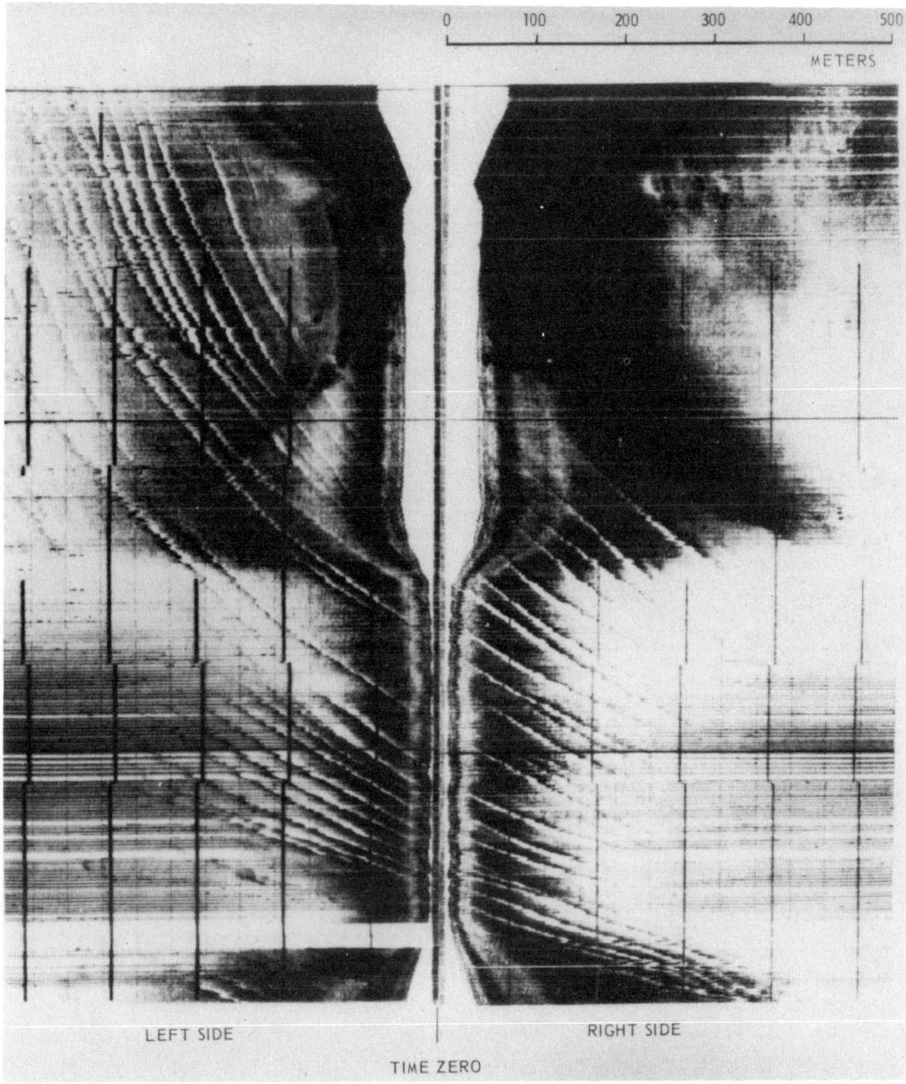

FIGURE 5

PATTERNS ON A PAIR OF SIDE-LOOKING SONAR RECORDS CAUSED BY CHANGES IN SLOPE OF AN OTHERWISE HOMOGENEOUS SEA FLOOR (FROM SPIESS ET AL., 1973b). THE CURVING AND OCCASIONALLY BRANCHING STREAKS REPRESENT EROSIONAL FURROWS CUT BY BOTTOM CURRENTS INTO A LAYER OF CALCAREOUS NANNOFOSSIL OOZE. THE SIDES OF THE FURROWS FACING AWAY FROM THE TRANSDUCERS ARE LIGHT; THOSE FACING TOWARDS THEM SCATTER BACK MORE ACOUSTIC ENERGY AND ARE DARK. SAMOAN PASSAGE, WESTERN EQUATORIAL PACIFIC; WATER DEPTH 5800 m. SIMILAR FEATURES HAVE BEEN MAPPED BY SURFACE-SHIP SIDE-LOOKING SONARS IN THE SHALLOW (3 to 5 m) WATERS OF TIDAL ESTUARIES (DYER, 1970).

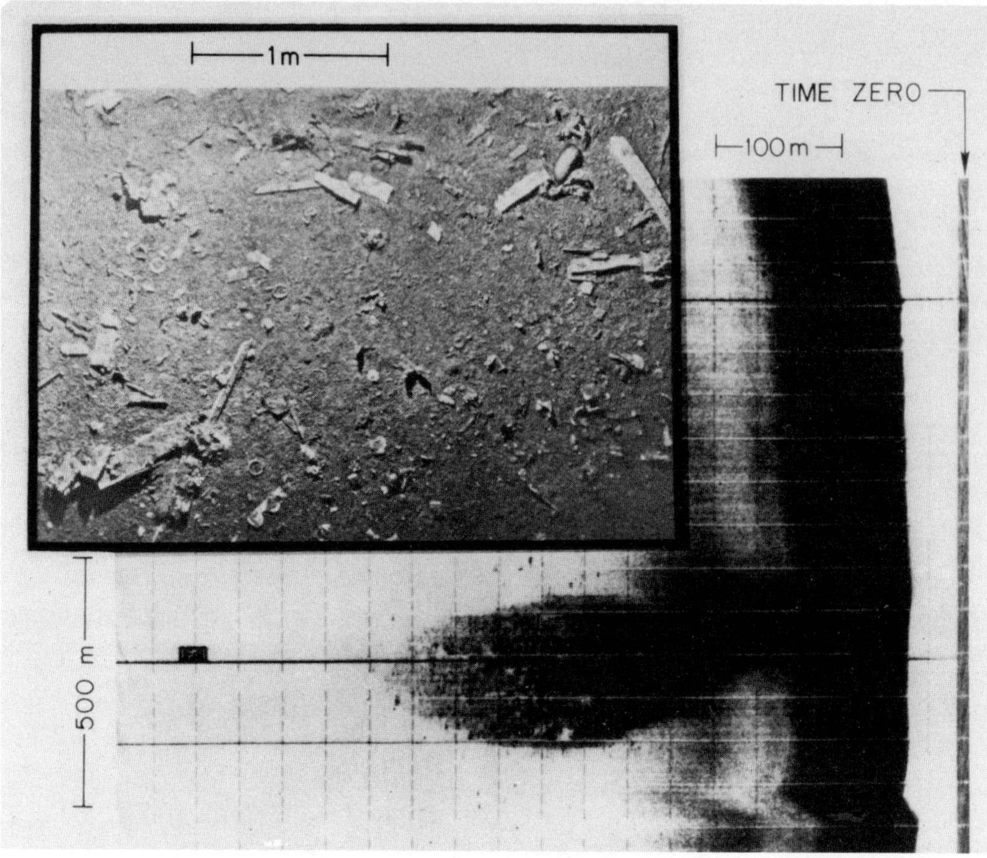

FIGURE 6

VARIATION ON A SIDE-LOOKING SONAR RECORD CAUSED BY CHANGES IN TEXTURE OF THE SEA FLOOR. THE SMOOTH, MUDDY, AND POORLY SCATTERING SURFACE OF THE NITINAT ABYSSAL FAN IS SCARRED WITH A PATCH OF INTENSE BACKSCATTERING CAUSED BY SMALL FRAGMENTS OF METAL STREWN ACROSS THE SEA FLOOR DURING DISPOSAL OF A CHASE MUNITIONS SHIP. PHOTO (INSET) SHOWS THE EXPLODED REMAINS, AND CONFIRMS THAT FEW LARGE SECTIONS REMAINED INTACT. AFTER SPIESS AND SANDERS (1971).

resolution narrow beam profiles collected by the deep-tow instrument) and which had previously gone unrecognized on bottom photographs (Fig. 8). For delineating the trend of features clearly resolved by the vertical profiler (e.g., fault scarps) a greater altitude is better, allowing the feature to be followed out to greater ranges. Conversely, if the intention is to identify the pattern of small objects visible in bottom photographs, then a lesser altitude, which allows the nearby region to be insonified at a smaller grazing angle, may be required.

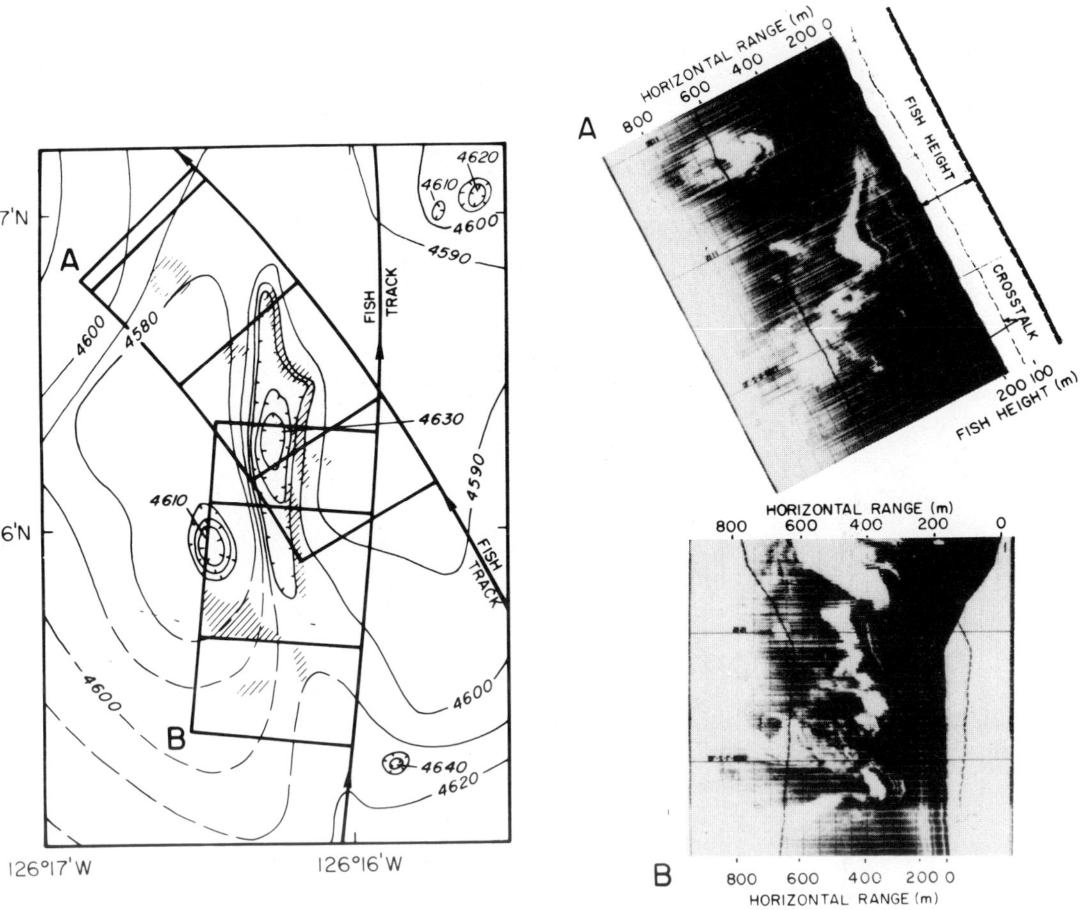

FIGURE 7

ACOUSTIC SHADOWS ON SIDE-LOOKING SONAR RECORDS. SMALL TROUGHS WERE DISCOVERED IN THE SEDIMENTS COVERING ABYSSAL HILLS (4600 m DEPTH) IN THE EASTERN EQUATORIAL PACIFIC. THE CONTOUR MAP ILLUSTRATES THE EXTENT OF COVERAGE OF THE TWO SIDE-LOOKING SONAR RECORDS SHOWN IN INSET. MOST OF THE TROUGH FLOORS ARE IN SHADOW (WHITE) BECAUSE OF THEIR DEPTH (40-60 m) AND THE STEEPNESS OF THEIR WALLS (MORE THAN 30°). FROM MUDIE et al. (1972).

A seismic profiling system was added to the deep-tow instrumentation in late 1966. Until early 1972, this system consisted of a cylindrical projector operating in a nonresonant mode at 3.5 kHz, a smaller cylindrical hydrophone suspended below the

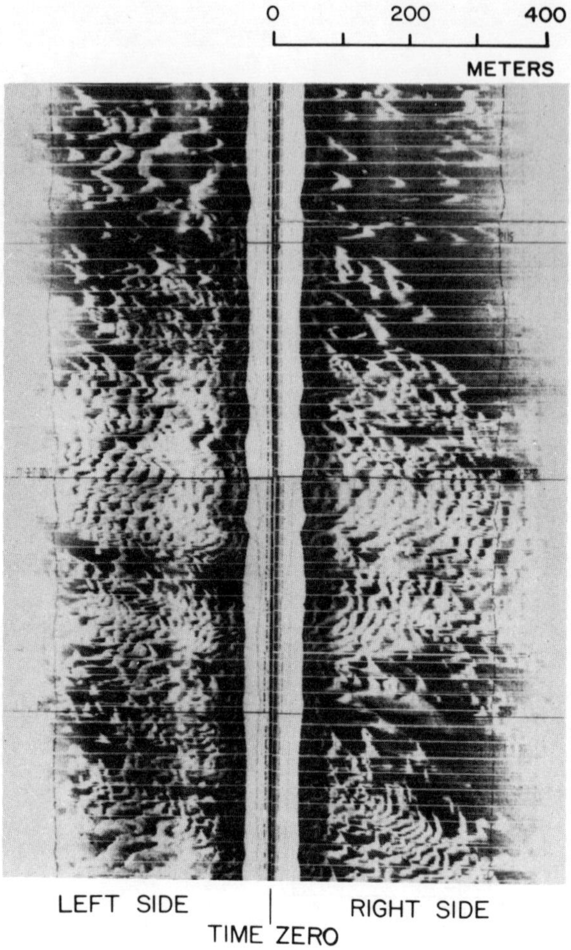

FIGURE 8

PATTERNS ON SIDE-LOOKING SONAR RECORDS CAUSED BY SLOPE VARIATIONS, WITH SOME SHADOWING, AND TEXTURAL HETEROGENEITY. THE WHITE PATCHES ARE CRESCENTIC AND TRANSVERSE DUNES OF POORLY REFLECTIVE FORAMINIFERAL SAND WHICH REST ON AN INTENSELY SCATTERING PAVEMENT OF MANGANESE NODULES. LITHOLOGIES WERE IDENTIFIED BY BOTTOM PHOTOGRAPHY AND CORING. (LONSDALE et al., 1972). THE DUNES AVERAGE 0.5 m IN HEIGHT AND ARE HIGHLY ASYMMETRIC, WITH STEEP AVALANCHE SLOPES ON THEIR LEEWARD (CONCAVE) SIDE, JUST LIKE THE SUBAERIAL BARCHAN DUNES OF ARID LANDS (e.g., FINKEL, 1959). VARIATIONS IN INTENSITY OF BACKSCATTER FROM THE SANDY AREAS ARE RELATED TO CHANGES IN SLOPE ANGLES. THE TRUE SHAPES OF DUNES IN THE FARFIELD IS OBSCURED BY SHADOWING. THIS DUNE FIELD WAS DISCOVERED DURING A SURVEY OF THE CARNEGIE RIDGE, BETWEEN THE GALAPAGOS ISLANDS AND THE ECUADORIAN MAINLAND (SPIESS et al., 1973a). THERE IS NO EVIDENCE OF IT ON SURFACE-SHIP RECORDS.

Fish, and associated Fish and laboratory electronics. This was a somewhat inefficient system, but the close proximity of the instrument to the sea floor suggested that a high-powered system was unnecessary, and small, light weight transducers resonant at about this frequency and capable of withstanding the pressures encountered at abyssal depths were relatively unavailable. In early 1972, a transducer meeting the above requirements was obtained (International Transducer Corporation Model ITC-2011) and substituted for the projector and hydrophone. The transducer is a 17 x 5.1 cm cylinder operated at its resonance of 4 kHz as both transmitter and receiver. The transducer, with axis oriented fore and aft, is trailed on a short bridle (Fig. 1) to decouple it from fish and wire vibration.

Unlike the side-looking sonar system with its time varied gain control, the seismic profiling system was originally designed with the intention of developing a quantitative capability. Therefore the 4 kHz transducer was carefully calibrated by us at the San Vincente Lake calibration facility (Gibson and Lewis, 1967) prior to installation on the Fish. This work generally confirmed achievement of the general specifications supplied by the manufacturer. At 4 kHz the cylindrical transducer is omnidirectional in the plane of symmetry, and in a plane containing the axis has a broadside beamwidth of 70° between -3 dB points with a single, -15 dB, side lobe on the axis. The transmit response, at beam pattern maximum, was determined to be 35 dB above 1 μbar/V at 1 m, with an efficiency of about 40%. The receive response was determined to be 69 dB below 1 V/μbar. This represents a substantial improvement over the previous transducers.

The 4 kHz seismic profiling system produces 500 to 1000 acoustic watts in a pulse 0.5 to 1 msec in duration, utilizing the same transducer for transmission and reception. To permit calibration of the complete system independent of gain settings, a calibrated 4 kHz signal is introduced to the system in place of the transducer output signal for a few milliseconds immediately following each transmit pulse. This signal cycles through 8 calibrated steps spanning 20 dB, with each step being repeated for 16 pulses to provide sufficient visual display of the calibration signal on the facsimile recorder (Fig. 9). Bottom photographs from the transition region shown in Fig. 9 are shown in Fig. 10. After the insertion of the calibration signal, the transducer is again connected to the Fish electronics, which transmit the received signal up the supporting cable with negligible distortion.

In order to display a topographically correct profiler record, the keying for the system has usually been generated by an up-looking sonar tracking gate, or by a differential pressure gauge on the Fish which uses relative pressure fluctuations to determine vertical Fish motion (Spiess and Tyce, 1973) and generates a

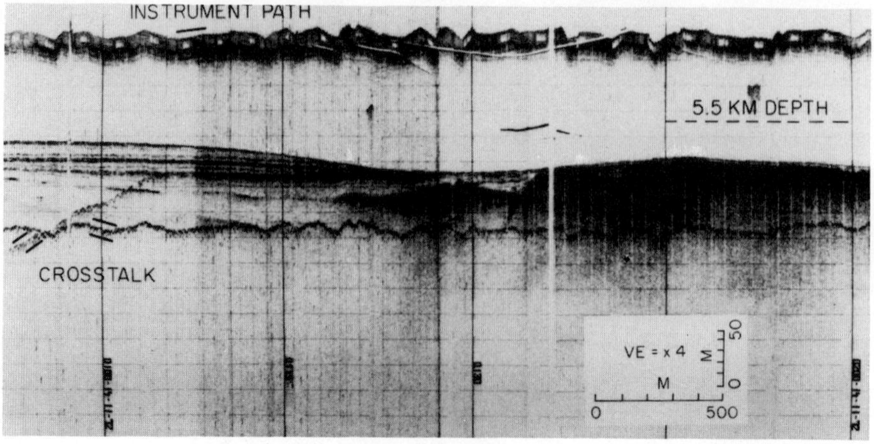

FIGURE 9

PROFILE FROM THE DEEPLY TOWED 4 kHz SEISMIC PROFILER, WITH DIFFERENTIAL PRESSURE GAUGE COMPENSATING FOR DEPTH FLUCTUATIONS OF THE INSTRUMENT PACKAGE. A TRANSITION IS SHOWN FROM AN AREA OF EROSIVE SCOUR (SEE FIGS. 11 AND 12) TO A DOME OF RAPIDLY ACCUMULATING REWORKED SEDIMENTS (FIGS. 13 AND 14). SAMOAN PASSAGE, WESTERN EQUATORIAL PACIFIC; FROM SPIESS et al. (1973b).

transmit keying delay (Figs. 9, 11b and 13b were generated in this latter manner). Aboard ship, the received profiler signal is supplied to a Gifft facsimile recorder for display, but, in addition, the signal is supplied to a computer processing system or alternatively to an analogue recorder for replaying to the computer later. The records reveal the altitudes of reflecting horizons (usually bedding planes) in the top 100 m or so of sediment. They allow recognition of areas of deep-sea erosion, marked by truncation of these reflectors (Johnson, 1972a; Lonsdale et al., 1972a) and small-scale slumping marked by post-depositional deformation (Mudie et al., 1972b). In addition to their obvious usefulness in these abyssal sedimentation studies, the 4 kHz seismic profiles have proved essential in interpreting the magnetic anomalies at ridge crests, by providing the relief of the volcanic layer which underlies thin sediments and which is the source of these anomalies (e.g., Atwater and Mudie, 1973). Larson (1971) used the near bottom profiler to demonstrate that the increase of sediment cover with increasing lithospheric age and distance from the spreading center that had been noted on broad scale surface ship profiles (e.g., Talwani et al., 1971) applied also on a small scale, within the ridge crest province. Similarly, Grow (1973) showed that the acoustically unresolvable stratigraphy of the landward wall of an oceanic trench abutted the stratigraphically undeformed sediment of the trench floor even more abruptly than had been deduced from surface-ship record (Scholl et al., 1968).

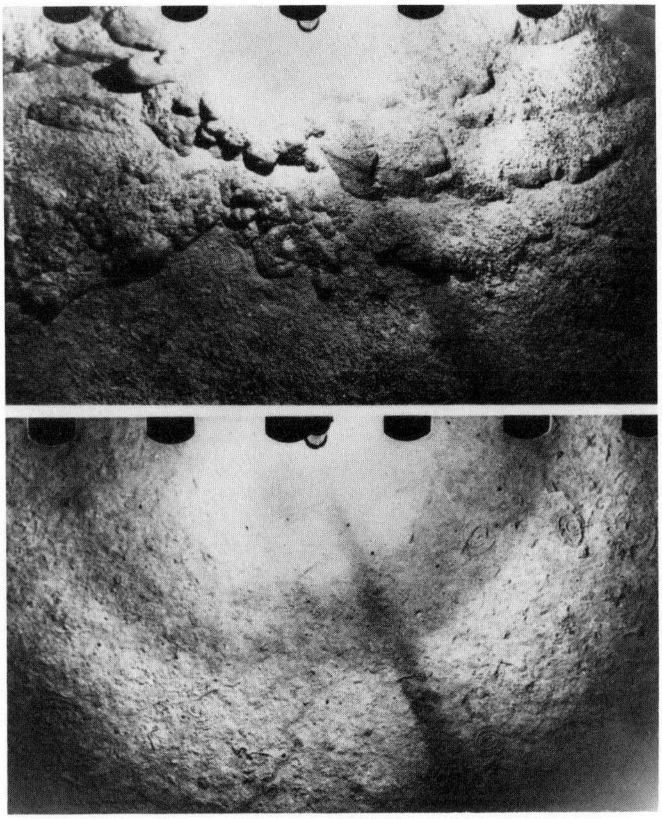

FIGURE 10

BOTTOM PHOTOGRAPHS OF SCOURED ROCK OUTCROPS (a) AND BORROWED SEDIMENT (b) ON EITHER SIDE OF THE TRANSITION BETWEEN NONDEPOSITIONAL AND DEPOSITIONAL TERRAINS SHOWN IN Fig. 9. AREA OF EACH PHOTOGRAPH IS ABOUT 15 m^2.

The computer system now in use for processing seismic profiler data consists of a PDP 11 computer, an analogue to digital converter, two nine-track digital tape drives, and cathode ray tube and electronic printer-plotter displays. The computer digitizes the bandpass filtered 4 kHz signal at 10 kHz, squares and integrates the samples, and stores the resulting energy values for every 0.5 msec interval with a dynamic range of 50 dB. This represents an essential improvement when compared with the 10 to 20 dB of dynamic range typically obtainable on the facsimile recorders, which commonly forces severe clipping of signals from highly reflective subbottom layers in order to provide adequate display of less reflective layers. This effect is readily apparent in the display of the inserted calibration signal (which

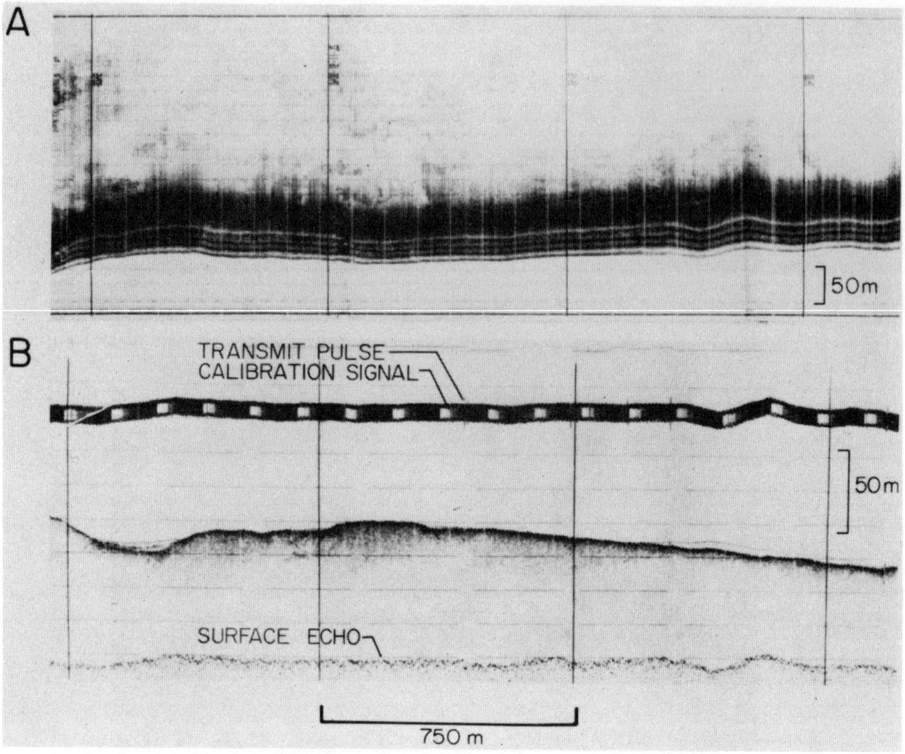

FIGURE 11

(a) LEFT SIDE-LOOKING SONAR RECORD OF A REGION OF SEDIMENT EROSION, PHOTOGRAPHED IN FIG. 10a (b) 4 kHz SEISMIC PROFILE COLLECTED AT THE SAME TIME, WITH THE TRANSMIT AND INTERJECTED CALIBRATION SIGNALS INDICATED. THE DEEP-TOW DIFFERENTIAL PRESSURE GAUGE CONTROLLED THE KEYING TO PROVIDE A DIRECT REPRESENTATION OF TOPOGRAPHY. NOTE ALMOST INVISIBLE PATCHES OF TRANSPARENT SEDIMENT IRREGULARLY OVERLYING A STRONG BASEMENT REFLECTOR.

spans less than the total signal range in 8 steps) where less than half the steps are distinguishable (Fig. 11b). At this time we are in the first stages of processing the digitized data ashore, with later plans for a real-time shipboard display. Much of the accumulated data has yet to be examined.

A number of phenomena control the levels of the observed signal samples: height of transducer off bottom, reflectivity at the sediment/water and internal sediment interfaces, sound velocity, volume reverberation, and absorption--all varying with depth into the bottom. Depending on the use to which the information is to be put, a variety of descriptive numbers can be generated from the raw data. At this time only a simple, first order processing has been implemented. The initial interpretive problems arise because of variations of height of the transducer above the sea floor and ping-to-ping fluctuations of received level which are caused primarily by small changes in orientation of the transducer during pitching of the Fish (as a result of wave induced motion at the towing ship). A correction is thus introduced for spherical spreading, making the corrected relative energy returned from various depths equivalent to that which would have returned from the corresponding region if the source height off bottom had been large compared with the thickness of the sediment column within which appreciable energy is reflected. This corrected received energy number we call the equivalent energy.

At this stage the spherical spreading correction has been made without introduction of sound speed variations in the water and sediment. Considering the magnitudes of these quantities and the fact that most of the contribution is at close to normal incidence, their effect is trivial compared with the observed horizontal variability of reflectivity.

The sequences of equivalent energy values for each return are aligned with respect to the first bottom reflected arrival and averaged from one ping to the next in four-ping blocks. In view of the geometry involved, these individual values should be treated as representation of the average properties of the sediment over horizontal extents of 10 to 20 m. These resulting numbers, then, have been treated in two ways to display the spatial dependence (horizontal and vertical) of the reverberant qualities of the sediment. The first approach (upper part of Figs. 12 and 14) is to generate a dashed line on the plot for every 2 dB step in received averaged "equivalent energy" as a function of depth into the sediment. A thicker dash is used for 10 dB increments. The second approach is to sum the equivalent energy returned from various depth increments within the sea floor. The lower parts of Figs. 12 and 14 plot these curves as a function of distance along the survey track for the sediment thickness intervals 0 to 5 and 5 to 50 m subbottom.

To illustrate the significant (though still preliminary) results of the quantitative 4 kHz profiling system, results from two adjacent but contrasting areas in the northern Samoan Passage region (Spiess et al., 1973b) are presented (Figs. 11 through 14).

The first region is an area of sediment erosion, characterized by a highly reflective bottom of manganese nodules and lithified sediments, irregularly covered by patches of relatively transparent sediment. The second region to be discussed lies about 5 km from the first, and is an area of sediment deposition characterized by

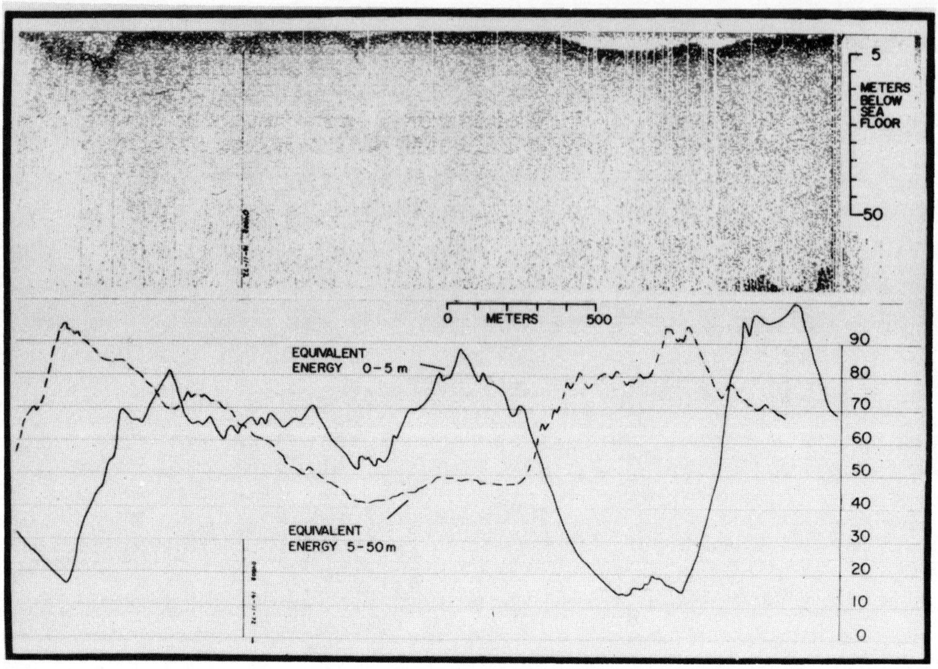

FIGURE 12

(a) --TOP-- A COMPUTER GENERATED CONTOUR PLOT OF NORMALLY REFLECTED EQUIVALENT ENERGY FOR THE 4 kHz SEISMIC PROFILER SYSTEM: PROCESSED 4 kHz DATA FROM THE SAME REGION AS FIG. 11. THE FIRST THIN LINE REPRESENTS THE -25 dB CONTOUR AND THE FIRST THICK LINE, THE -15 dB CONTOUR, WITH FIVE POSSIBLE DIVISIONS BETWEEN.
(b) --BOTTOM-- A COMPUTER GENERATED PLOT FOR THE SAME AREA AS (a), OF EQUIVALENT REFLECTED ENERGY (ON AN ARBITRARY LINEAR SCALE), FOR TWO TRAVEL TIME SECTIONS CORRESPONDING APPROXIMATELY TO 0 to 5 m AND 5 to 50 m BELOW THE SEA FLOOR.

little topographic relief and a bottom of stratified sediment overlying reflective acoustic basement (from the 4 kHz system) at about 50 m below the bottom. Side-looking sonar records collected at the same times have been included in the following discussion. They provide additional information on the backscattering characteristics of the bottom, and allow any backscattered returns on the profiler records, caused by the broad beam pattern of the 4 kHz transducer, to be disentangled from the subsurface signals.

In the first region, where sediment erosion occurs, the side-looking sonar record shows small patches of backscattering contrast in a generally nonscattering background (Fig. 11a) suggesting that the area has a thin sediment cover in most places, even where this

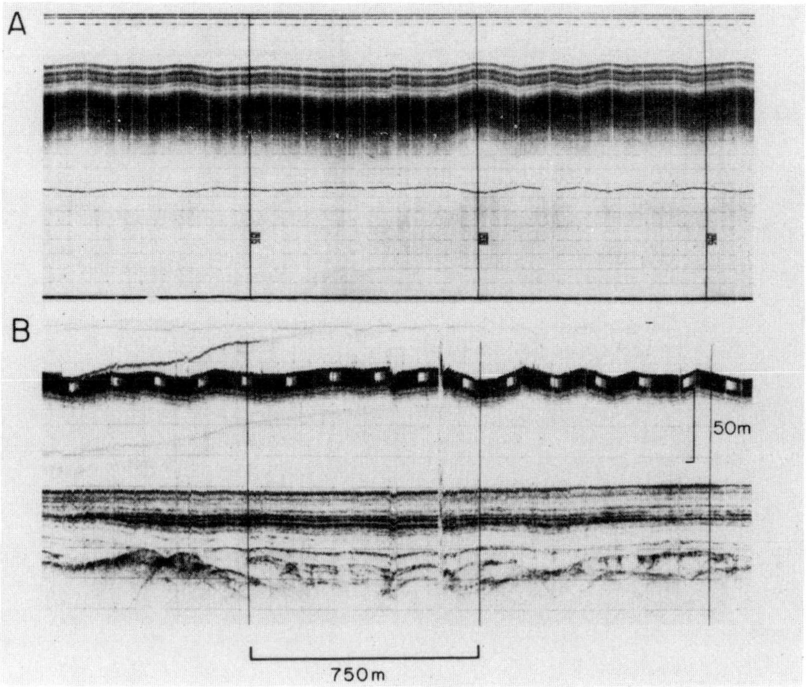

FIGURE 13

(a) RIGHT SIDE-LOOKING SONAR RECORD FOR A REGION OF RAPID SEDIMENT DEPOSITION (PHOTOGRAPHED IN FIG. 10b).
(b) 4 kHz SEISMIC PROFILE COLLECTED AT THE SAME TIME. THE WAVINESS OF THE DEEPER REFLECTORS IS PROBABLY CAUSED BY MOLDING DURING DEPOSITION.

is not readily apparent from the 4 kHz profiler facsimile record (Fig. 11b). From the computer plot of total equivalent energy returned (Fig. 12b) it is apparent that nearly all of it comes from the strong near-surface reflector, with the near-surface value greatly diminishing where surface sediment cover exceeds 5 m. In addition, it is of interest that the sum of the two curves varies by less than 3 dB. Thus, although the energy reflected from the immediate vicinity of the sea floor can vary by 10 dB in a distance of just over 100 m, the total integrated contribution is much less variable. In general the total equivalent energy returned is greater when most of it is from the first 5 m, suggesting that in this case attenuation within the sediment cover is a significant factor, which will be investigated by closer analysis of the data.

The second area, where rapid deposition from a sediment-laden bottom current occurs (Hollister et al., 1973), differs substantially from the first area in the acoustic character of the bottom.

The side-looking sonar record (Fig. 13a) shows virtually no relief or spatial variation in backscattering characteristics over the entire record. The generally striped pattern running along the length of the side-looking sonar records represents the side lobe pattern of the transducers with time varying gains superimposed. The spread of these side lobe bands is often useful for observing slope variations, which tend to compress and expand the bands in the upslope and downslope directions, respectively. The dark vertical lines on all the original records represent 15 minute time lines inscribed and annotated automatically by the laboratory display electronics, and the vertical blank lines indicate transmission interruptions by nonacoustic systems.

The 4 kHz profile record for this depositional area (Fig. 13b) shows a stratified bottom exhibiting substantially more penetration than that observed in the erosional area. In addition, the near-surface reflection is intense compared with most of the subsurface, except for an equally intense reflection from underlying acoustic basement. This is in sharp contrast with the thin, relatively transparent sediment cover overlying very reflective acoustic basement in the first area. Spatial variations in the intensity of the near-surface reflection can be noted along the profiler record, but resolution is poor due to clipping of the more intense signal to fit the Gifft facsimile recorder's dynamic range. However, the computer contour plot for this region (Fig. 14a) shows substantial spatial variability in reflectivity from near-surface and subsurface layers. This plot is on the same vertical and horizontal scales as the corresponding plot for the first area, but with corresponding contours down 10 dB to provide an optimal display. Notable in the contour plot for the second area is the fact that the thick line contours exhibit considerable lateral variation in occurrence just below the surface, though they never intersect the surface as they did in the first region (though admittedly representing a different threshold). This suggests the possibility of a uniformly thin veneer of near-transparent sediment in this region, consistent with the minimal backscattering variability observed on the side-looking sonar record for this area (Fig. 13a).

The computer generated plot of total energy for the second area (Fig. 14b) illustrates the small scale spatial variability of the reflectivity in this region. Since this plot is on the same scale as the corresponding plot for the first area (Fig. 12b), it is apparent that the total equivalent energy returned from both sections is of the same order of magnitude. In the second area, however, the majority of the returned energy is consistently from the 5 to 50 m range, with significant amounts occasionally returned from even deeper. The variations in energy observed in the 5 to 50 m trace are in part due to variations in acoustic basement topography, but topographic fluctuations have little to do with the substantial near-surface variations of several decibels, which

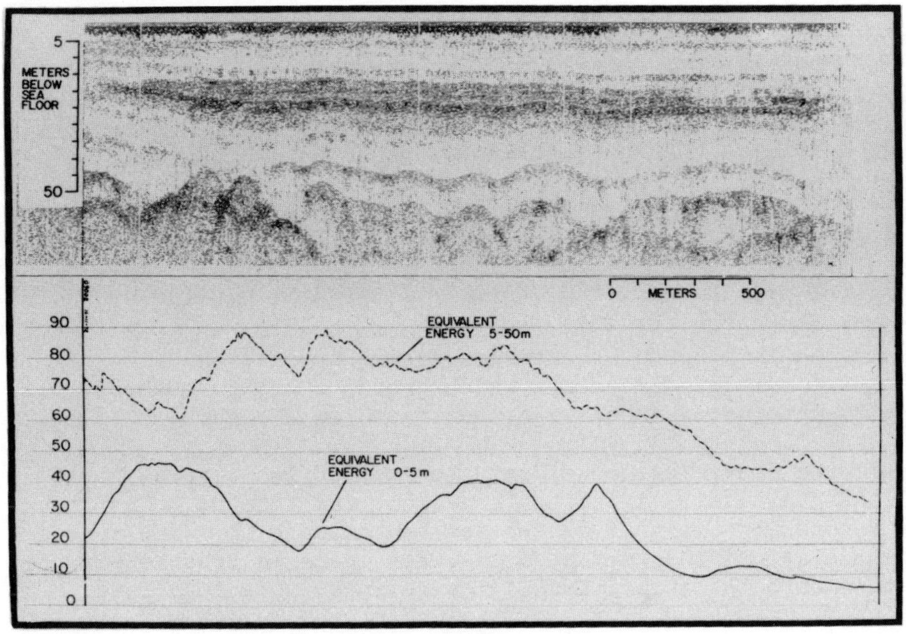

FIGURE 14

(a) --TOP-- A COMPUTER GENERATED CONTOUR PLOT SIMILAR TO FIG. 12a, BUT WITH CORRESPONDING CONTOURS DOWN 10 dB, FOR THE REGION OF FIG. 13.
(b) --BOTTOM-- A COMPUTER PLOT OF EQUIVALENT ENERGY SIMILAR TO AND ON THE SAME SCALE AS FIG. 12, FOR THE REGION OF FIG. 13.

are not easily explained. In addition very little inverse correlation between the two energy traces is suggested for the second area, whereas for the first area just the opposite is the case. It is apparent from the plots that in the second area little of the transmitted energy is being intercepted by the near-surface reflectors, and that the first area is dominated by a single strong reflector.

After more than two dozen surveys the deep-tow group is acquiring an appreciation of the small-scale patchiness of much of the deep sea floor. Continuing work with the seismic profiling system will allow us to appreciate it quantitatively. The Samoan Passage region, from which the records illustrated in Figs. 9 through 14 were collected, is extraordinarily diverse, because of the spatially varying effectiveness of the strong (25 cm/sec) bottom water current which passes through the area and molds the sea floor by erosion and current-controlled deposition (Reid and Lonsdale, 1973). Here, patchiness on the scale of a few kilometers, with thick sediment ponds adjacent to current-blasted manganese pavements, is evident even on surface ship 3.5 kHz records (Hollister et al., 1973), though the small-scale bedforms (e.g., Fig. 5) and reflectivity changes (e.g., Fig. 14a) are not. The substantial small-scale variations in bottom reflectivity discussed above are characteristic of much of this region, but enough quantitative data to allow us to state how widespread such phenomena are throughout the rest of the Pacific have not yet been processed. Speculations on distribution are hampered by our current ignorance of the causes of many of the observed variations in reflectivity. However, several of our surveys in regions generally considered geologically uniform (e.g., abyssal hills) have discovered unpredicted diversity (Fig. 7). Even the surface of abyssal fans, smooth and gently sloping as observed from a surface ship, were found to have subtle depressions which proved a key to understanding the growth patterns of these important features (Normark, 1970).

Until the modern period of intense oceanographic research, the salient characteristic of the sea floor was thought to be its extreme monotony: "In general the sea-bed extends for wide spaces in long undulations and gentle slopes, for agencies ceaselessly labour to smooth the asperities of its surface." (Reclus, 1886). Our dominant impression of the abyssal sea floor is its variability.

Acknowledgments Conduct of the work reported here has involved the cooperative efforts of many people other than the authors, most particularly our deep-tow engineering group under Mr. D. E. Boegeman and the captains and crews of the research vessel THOMAS WASHINGTON. Drs. J. D. Mudie and C. L. Lowenstein have provided a wide range of types of professional support, including advice and criticism on both data collection and analyses for side-looking sonar and 4 kHz reflectivity work. This activity has been sponsored by the Office of Naval Research and the Navy Deep Submergence Program through ONR contract N00014-69-A-0200 and by the National Science Foundation through grants NSF GA-31377X and NSF GD-32021.

REFERENCES

Atwater, T., and J. D. Mudie, A detailed near-bottom geophysical study of the Gorda rise, J. Geophys. Res., 78, 8665-8686, 1973.

Boegeman, D. E., G. J. Miller, and W. R. Normark, Precise positioning for near-bottom equipment using a relay transponder, J. Marine Geophys. Researches, 1, 381-396, 1972.

Bryan, G. M., and R. G. Markl, Microtopography of the Blake-Bahama Region, Lamont-Doherty Geological Observatory Technical Report No. 8, 26 pp., 1966.

Dyer, K. R., Linear erosional furrows in Southampton Water, Nature, 225, 56-58, 1970.

Finkel, H. J., The barchans of southern Peru, J. Geology, 67, 614-647, 1959.

Fox, P. J., B. C. Heezen, and A. M. Harrian, Abyssal anti-dunes, Nature, 220, 470-472, 1968.

Gibson, D. K., and R. P. Lewis, Transducer calibration facility at San Vincente Lake, Scripps Institution of Oceanography Reference No. 67-4, 1967, University of California, San Diego, California.

Grow, J. A., Crustal and upper mantle structure of the Central Aleutian Arc, Geol. Soc. Amer. Bull., 84, 2169-2192, 1973.

Hollister, C. D., D. A. Johnson, and P. F. Lonsdale, Current-controlled abyssal sedimentation, Samoan Passage, equatorial west Pacific, J. Geology (in press), 1974.

Johnson, D. A., Ocean floor erosion in the equatorial Pacific, Geol. Soc. Amer. Bull., 83, 3121-3144, 1972.

Larson, R. L., and F. N. Spiess, Slope distributions of the East Pacific Rise crest, Scripps Institution of Oceanography Reference No. 70-8, 4 pp., University of California, San Diego, California, 1970.

Larson, R. L., Bathymetry, magnetic anomalies, and plate tectonic history of the mouth of the Gulf of California, Geol. Soc. Amer. Bull., 83, 3345-3360,

Lonsdale, P. F., W. R. Normarck, and W. A. Newman, Sedimentation and erosion on Horizon Guyot, Geol. Soc. Amer. Bull., 83, 289-316, 1972a.

Lonsdale, P. F., B. T. Malfait, and F. N. Spiess, Abyssal sand waves on the Carnegie Ridge, abstract presented at the Geol. Soc. Amer. National Meeting, Minneapolis, Minnesota, 13-15 November 1972.

Lowenstein, D. C., Computations for transponder navigation, Proc. National Marine Navigation Meeting, 20-22 January 1966, 305-311.

Luyendyk, B. P., Origin and history of abyssal hills in the north-east Pacific Ocean, Geol. Soc. Amer. Bull., 81, 2237-2260, 1970.

McGehee, M. S., and D. E. Boegeman, Marine Physical Laboratory acoustic transponder, Rev. Sci. Instruments, 37, 1450-1455, 1966.

Mudie, J. D., J. Grow, K. Klitgord, and P. Larson, Shipboard cruise report on Leg two of Expedition South Tow: Studies in the southeast Pacific, Scripps Institution of Oceanography Reference No. 72-66, 40 pp., University of California, San Diego, California, 1972a.

Mudie, J. D., J. A. Grow, and J. S. Bessey, A near-bottom survey of lineated abyssal hills in the equatorial Pacific, J. Marine Geophys. Researches, 1, 397-411, 1972b.

Mudie, J. D., T. Atwater, R. Detrick, K. Klitgord, and K. MacDonald, Shipboard cruise report, Expedition South Tow, Leg six, Scripps Institution of Oceanography Reference, University of California, San Diego, California (in press), 1974.

Normark, W. R., Growth patterns of deep-sea fans, Scripps Institution of Oceanography Reference No. 69-29, 77 pp., University of California, San Diego, California, 1969.

Normark, W. R., Growth patterns of deep-sea fans, Amer. Assoc. Pet. Geologists Bull., 54, 2170-2195, 1970.

Reclus, E., A new physical geography, vol. 2, Air and Ocean, G. S. Virtue and Co., London, 500 pp., 1886.

Reid, J. L., and P. F. Lonsdale, On the flow of water through the Samoan Passage, Physical Oceanography (in press) 1974.

Rona, P. A., Linear "lower continental rise hills" off Cape Hatteras, J. Sed. Petrol., 39, 1132-1141, 1969.

Schick, G. B., J. D. Isaacs, and M. H. Session, Autonomous instruments in oceanographic research, in Marine Sciences Instrumentation, 4, Plenum Press, New York, 203-230, 1968.

Scholl, D. W., R. E. von Huene, and J. B. Ridlon, Spreading of the ocean floor: Undeformed sediments in the Peru-Chile Trench, Science, 15, 869-871, 1968.

Spiess, F. N., M. S. Loughridge, M. S. McGehee, and D. E. Boegeman, An acoustic transponder system, Navigation, 13, 154-161, 1966.

Spiess, F. N., B. P. Luyendyk, and M. S. Loughridge, Bottom slope distributions and implied acoustic bearing errors in abyssal hill regions of the north Pacific, U. S. Navy J. Underwater Acoustics, 19, 1969.

Spiess, F. N., and J. D. Mudie, Small-scale topographic and magnetic features, in The Sea, Vol. 4, Part 1, A. E. Maxwell, ed., 205-250, John Wiley and Sons, Inc., New York, 1970.

Spiess, F. N., and S. M. Sanders, Survey of CHASE disposal area, (NITNATOW), Scripps Institution of Oceanography Reference No. 71-33, 20 pp., University of California, San Diego, California, 1971.

Spiess, F. N., B. T. Malfait, P. F. Lonsdale, and R. C. Tyce, Shipboard cruise report on Leg five of Expedition South Tow, Scripps Institution of Oceanography Reference No. 73-1, 9 pp., University of California, San Diego, California, 1973a.

Spiess, F. N., and R. C. Tyce, Marine Physical Laboratory deep tow instrumentation system, Scripps Institution of Oceanography Reference No. 73-4, University of California, San Diego, California, 1973

Spiess, F. N., P. F. Lonsdale, R. C. Tyce, and J. D. Mudie, Shipboard cruise report on Leg 10 of Expedition South Tow, Scripps Institution of Oceanography Reference No. 73-10, 17 pp., University of California, San Diego, California, 1973b.

Talwani, M., C. C. Windisch, and M. G. Langseth, Reykjanes ridge crest: a detailed geophysical survey, J. Geophys. Res., 76, 473-517, 1971.

REMOTE SEDIMENT CLASSIFICATION POTENTIAL OF REFLECTED ACOUSTIC SIGNALS

D. L. BELL and W. J. PORTER

Submarine Signal Division, Raytheon Company

Portsmouth, Rhode Island

ABSTRACT

It is generally accepted that acoustic echoes received from marine sediments are relatable to a number of sediment geotechnical descriptions. To investigate these relationships, mathematical modeling, preliminary system design concepts, specific instrumentation, and a computer aided analytical approach have been developed. Techniques have been implemented and applied to at-sea data acquisition, employing a towed array, which allow simultaneous remote estimates of compressional wave velocity, attenuation rates, and the reflection coefficient as a function of incident angle. Initial correlations between measured acoustic indices and sediment physical parameters are presented and the potential for remote classification discussed.

INTRODUCTION

Measurement of sediment engineering and geological properties is required to support a spectrum of marine activities, both military and commercial. Historically, these measurements have been made on samples acquired by physical recovery involving coring and other mechanical sampling techniques supported by acoustical data from subbottom profiling and sidescan sonar. Neither the physical nor the present acoustical sampling, however, provides high confidence quantitative data. The sediment sampling devices inevitably cause disturbance of fabric related properties and the commonly employed acoustical techniques yield only qualitative information.

A second area of concern associated with physical sampling is the expense of high density sampling which cannot be afforded by many smaller projects, while limited sampling can result in non-representative data. This latter concern is particularly warranted in near-shore areas where a high degree of small-scale lateral variability is commonly encountered (Bennett et al., 1970; Faas, 1972).

Acoustic energy, which in general can easily penetrate marine sediments, potentially offers a mechanism for rapid and effective measurement of undisturbed marine sediment properties. To date, however, the common line scan records of subbottom profilers and sidescan sonars which depict interfaces between layers and seafloor textural characteristics require skilled interpretation and, although substantial qualitative information is presented, few quantitative statements can be made.

As one approach to increasing the value of acoustics for making remote sediment property evaluations, a joint University of New Hampshire-Raytheon Company Sea Grant project has been undertaken to quantify reflected acoustical signals in terms that are useful to both marine engineers and geologists for continental shelf (<600 ft) applications. Initial research efforts directed towards the parallel development of a field theoretical model and extension of the more simplistic plane-wave model have lead to preliminary design concepts, instrumentation, and a computer aided analytical approach for the remote assessment of marine sediment properties. This approach is based upon measurement of compressional wave velocity, sediment attenuation rates, and the reflection coefficient as a function of incident angle.

The importance of velocity as a marine sediment index has long been appreciated, as it is related to both the physical and engineering properties. Several early laboratory studies examining the velocity-sediment property interrelationships commenced during the 1950's (Laughton, 1954; Hamilton, 1956; Hamilton et al., 1956; and

Sutton et al., 1957). A limited number of field measurements employing classical refraction techniques (Nafe and Drake, 1957) were also made during this period.

Subsequent research activities expanded these efforts in the 1960's with an increase in laboratory investigations (Leslie, 1960; Hamilton, 1963, 1968; Horn et al., 1968; Schreibner, 1968; Hampton, 1966) and field data acquisition utilizing both multiship- and singleship-sonobuoy approaches (Fry and Raitt, 1961; Houtz and Ewing, 1963; Clay and Rona, 1965; Houtz et al., 1968; LePichon et al., 1968).

Complementary research efforts have also pursued velocity measurement by employing in situ probes (Hamilton, 1963), and "up-hole and side-hole shooting" to determine sediment velocities and associated in-place engineering properties (Swain, 1962).

An even larger volume of literature exists treating the reflection of acoustic energy from the sea floor. Representative of those works relating directly to the remote classification of sediments using acoustics are articles by Mackenzie (1960), Cole (1965), Breslau (1965, 1967), Smith and Li (1966), Hamilton (1970), and Pawlowicz (1971).

TECHNICAL APPROACH

Theoretical Development

Initial modeling has assumed a structure composed of a liquid overlying a series of parallel viscoelastic solid layers. For acoustic waves propagating within this structure, displacement potentials will satisfy the wave equations within each layer, the source conditions, and the layer boundary conditions.

Formal solutions to the wave equations are expressed as integrals over an appropriate wave number which is dependent upon the interrelationships between each layer's density and governing elastic constants (Lamé parameters). Explicit solutions are then obtained by evaluating these formal integrals in some approximate limit based on the specific problem.

The field theory treatment involves dispersion theory in which the formal integral is evaluated by contour integration in the complex wave number plane, the solution being expressed as a sum of residues and branch-line integrals. Analytically, the residues at the poles within the complex plane constitute normal modes with the branch-line integrals conceptually associated with the critical angle refraction paths.

The formal field theory treatment is warranted in cases where the duration of the acoustic pulse is long compared with the time intervals between possible innerlayer arrivals. In this case, interference occurs between arrivals and a steady-state solution is applicable.

For those situations where interference among the arrivals within the layered structure is limited, solutions to the formal integral can be obtained in the ray acoustic limit where each arrival is identified as a specific interface reflection. Observation of each arrival's time, amplitude, and phase thus allows investigation of a particular layer's physical properties. In addition, velocities and layer thickness can be determined by simple geometric ray path computations based on either the T^2X^2 approach or on Snell's Law.

As a large percentage of the sediment layers can be considered time resolvable with the short pulse lengths employed in subbottom profiling equipment, a ray acoustic approach has been applied. Accordingly, the expression for the complex reflection coefficient of an obliquely incident plane wave in a liquid (the ocean media) impinging onto the boundary of an elastic layer with absorption (sediment) has been employed directly:

$$\tilde{R}(Z_o) = \frac{(\tilde{Z}_\ell - \tilde{Z}_t)\cos^2 2\tilde{\beta}_t + \tilde{Z}_t - Z_o}{(\tilde{Z}_\ell - \tilde{Z}_t)\cos^2 2\tilde{\beta}_t + \tilde{Z}_t + Z_o}$$

where $$Z_o \equiv \rho_o C_o/\cos\theta_o \quad ,$$

$$\tilde{Z}_\ell \equiv \rho_s \tilde{C}_\ell/\cos\tilde{\beta}_\ell = \rho_s |C_\ell|(1 + i\tan\theta_\ell)/\cos\tilde{\beta}_\ell$$

$$\tan\theta_\ell \equiv k_\ell |C_\ell| / \sqrt{1 + (k_\ell |C_\ell|)^2} \quad ,$$

defined as the complex compressional impedance of the sediment, and

$$\tilde{Z}_t \equiv \rho_s \tilde{C}_t/\cos\tilde{\beta}_t = \rho_s |C_t|(1 + i\tan\theta_t)/\cos\tilde{\beta}_t$$

$$\tan\theta_t \equiv k_t |C_t| / \sqrt{1 + (k_t |C_t|)^2} \quad ,$$

the complex shear impedance of the sediment.

Model Development

The observed values (magnitude and phase) of the complex plane-wave reflection coefficient $\tilde{R}$ will be dependent on the following five unknowns at any given value of Z_o:

- $|C_\ell|$ = Compressional wave velocity
- α_ℓ = Compressional wave attenuation coefficient
- $|C_t|$ = Shear wave velocity
- α_t = Shear wave attenuation coefficient
- ρ_s = Sediment wet density.

In a test situation represented schematically in Fig. 1, the value of $|C_\ell|$ can be determined by velocity profiling, α_ℓ obtained by observing the frequency dependence in the relative normal incident reflection coefficients, and the value of ρ_s from normal incidence reflection coefficient and the measured $|C_\ell|$ values.

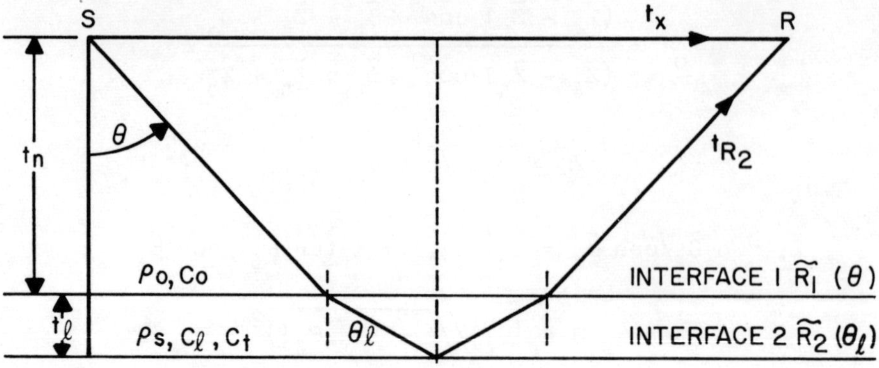

FIGURE 1

IDEALIZED TEST REPRESENTATION

Specifically, $|C_\ell|$ may be estimated through the iterative solutions to

$$2t_N \tan\theta + 2t_\ell \tan\theta_\ell = t_x$$

$$2t_N \sec\theta + 2t_\ell \sec\theta_\ell = t_{R_2}$$

$$\theta_\ell \approx \sin^{-1}(|C_\ell|/C_\omega \sin\theta)$$

and α_ℓ through its effect on the observed amplitude

$$R_{ob}(t_{r_2}, \theta = 0) \equiv R_o = |\tilde{R}_2(0)|\ (1 - |\tilde{R}_1(0)|^2) \cdot C^{-2d_\ell \cdot \alpha_\ell(\omega)}$$

$$d_\ell = |C_\ell| \cdot t_\ell$$

$$\alpha_\ell(\omega) \equiv k_\ell \cdot \omega$$

$$\frac{1}{R_o}\frac{dR_o}{d\omega} = -2d_\ell \cdot k_\ell$$

As $|C_\ell|$ and t_ℓ are known, k_ℓ and hence α_ℓ can be calculated.

The observed magnitude $|\tilde{R}|$ and phase ϕ_R of $\tilde{R}$ will therefore be dependent on only the two remaining unknowns of C_t and α_t as a function of Z_o. In principle, these parameters may now be estimated through the iteration of $|R|$ and ϕ_R in the equations below. Letting

$$\tilde{R}(Z_o) = \frac{(R_\ell^2+I_m^2) - Z_o^2 + i2Z_oI_m}{(R_\ell^2+I_m^2) + Z_o^2 + 2R_\ell Z_o}$$

then

$$|\tilde{R}|(Z_o) = \frac{\sqrt{\left[(R_\ell^2+I_m^2) - Z_o^2\right]^2 + (2Z_oI_m)^2}}{\left[(R_\ell^2+I_m^2) + Z_o^2 + 2Z_oR_\ell\right]}$$

and

$$\phi_R(Z_o) = \tan^{-1}\left[\frac{2Z_oI_m}{(R_\ell^2+I_m^2) - Z_o^2}\right]$$

where

$$R_\ell \equiv \text{Real part of } \left\{(\tilde{Z}_\ell-\tilde{Z}_t)\cos^2 2\tilde{\beta}_t + \tilde{Z}_t\right\}$$

$$I_m \equiv \text{Imaginary part of } \left\{(\tilde{Z}_\ell-\tilde{Z}_t)\cos^2 2\tilde{\beta}_t + \tilde{Z}_t\right\}$$

and

$$\tilde{Z}_\ell = \tilde{Z}_\ell(|C_\ell|,\alpha_\ell)$$

$$\tilde{Z}_t = \tilde{Z}_t(|C_t|,\alpha_t)$$

$$\tilde{\beta}_t = \tilde{\beta}_t(|C_t|,\alpha_t)$$

and hence

$$R_\ell(Z_o;|C_\ell|,\alpha_\ell) = R_\ell(|C_t|,\alpha_t)$$

$$I_m(Z_o;|C_\ell|,\alpha_\ell) = I_m(|C_t|,\alpha_t) \qquad \text{only.}$$

In practice, however, determination of $\phi_R(Z_o)$ is at best difficult and in most cases attempted estimates of C_t and α_t have proven unsatisfactory.

As an alternate approach, C_t and α_t may also be estimated based on the angle of occurrence and magnitude of $\tilde{R}(Z_o)$ minima. Under this condition, the resistive component of the sediment's impedance R_ℓ is canceled by the ocean's impedance Z_o (i.e., $R_\ell = Z_o$), and the two equations to be solved by iteration reduce to:

$$R_\ell(|C_t|,\alpha_t) = Z_o(\theta_m)$$

$$I_m(|C_t|,\alpha_t) = \frac{2Z_o\ |R_{min}|_{\theta_m}}{\sqrt{1 - |R_{min}|^2_{\theta_m}}}$$

Knowledge of $Z_o(\theta_m)$ and $|R_{min}|_{\theta_m}$ thus allows $|C_t|$ and α_t to be estimated.

Data Acquisition System

Based upon the preceding model, it appeared that measurement of normal and oblique incidence reflection coefficients together with the associated layer compressional wave propagation velocity and attenuation rates would yield sufficient data to afford meaningful judgments of sediment characteristics. The previous work also indicated that the shear wave velocity in a seafloor layer might be estimated based upon the characteristics of the oblique incidence reflection coefficient. More specifically, the exact angle and magnitude of the model predicted minima in the angular reflection response is relatable to the interaction between the magnitudes of the compressional and shear wave velocities and their associated attenuation coefficients.

To collect data to exercise this model, a multichannel towed array was designed and constructed. The array approach was adopted for three primary reasons. First, both determination of the angular dependence of the reflection coefficient and of the compressional wave velocity employing either the classical T^2X^2 or Snell's Law approach is well served by simultaneous data acquired at a number of reflection angles (Fig. 2). Second, the towed array essentially fixes the individual source-receiver pair geometries in space rather than necessitating recomputation of a time varying geometry as required by either a two-ship or single-ship sonobuoy approach. The towed array additionally allows almost complete

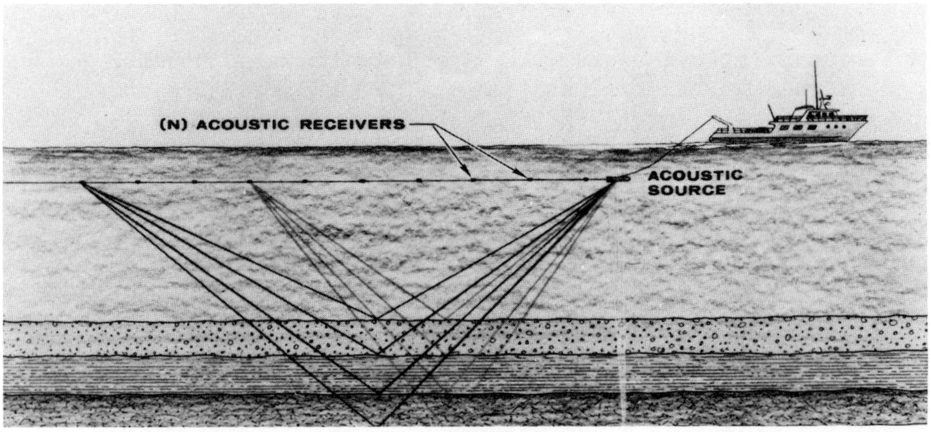

FIGURE 2

REMOTE SEDIMENT CLASSIFICATION DATA ACQUISITION

decoupling of the sensors (source and receivers) from both ocean surface and ship motion. The third reason was one of economics in that neither a second vessel nor the sophisticated communications for data transfer and radio controlled timing link were required. The array approach, moreover, allows more continuous data acquisition operations and simplified logistics, and could be operated by fewer personnel.

The array incorporates a series of discrete receiving elements which simultaneously sense the reflected acoustic response of each sedimentary layer over a range of incidence angles. Under normal conditions this range generally includes angles from zero (normal incidence) up to sixty degrees. To improve data quality, the array also incorporates specialized mechanical damping mechanisms and self-contained signal processing, the latter primarily directed towards S/N improvement. Received signals are further processed by the shipboard portion of the system employing a processor which substantially increases the effective dynamic range of the acoustic signals recorded for subsequent computer analysis.

To date, the prototype systems have only been evaluated in relatively shallow water (≦150 feet), although there appears to be no technical limitation to increasing the operational capability to include areas of significantly greater water depths. Similarly, the subbottom penetration depths in excess of 60 feet now being achieved could also be extended. It should be noted, however, that the total penetration depths achievable are not specifically predictable and, in common with all other acoustic systems, are a function of the sediment characteristics.

DISCUSSION

A series of at-sea tests were conducted in Massachusetts Bay during January 1973, to evaluate the functionality of the array concept for acquisition of remote acoustic sediment assessment data. Initial data assessment has been directed towards determination of array acquisition system capability to obtain reliable compressional wave velocities and reflection coefficients as a function of incidence angle. An example of the velocity and normal incidence reflection coefficient analyses for two sequential sets of 15 pulses is presented in Fig. 3[A] and 3[B], respectively. The associated oblique incidence reflection coefficients for this same site are presented in Fig. 4.

In Fig. 3 it should be noted that the values within the solid blocks are the averages of the measured acoustic indices which appear to the right. For velocity, this average is itself the average of the compressional wave velocities calculated by ensemble averaging 15 pulses at each of ten receivers. The normal and oblique incidence reflection coefficients are similarly based on multiple pulse ensemble averages. The lower portions (dotted) of each of these boxes presents an estimate of the sediment physical properties based upon these acoustic indices using the relationships developed by Hamilton (1969). It should further be noted that the primary intent in making these estimates at this time is to assist in determining the magnitude of potential variance in acoustic estimates of sediment properties rather than to make a sediment classification, per se.

Examination of the data presented in Fig. 3, which was acquired over a silty clay site in Massachusetts Bay, indicates average sediment layer velocities of roughly 4920 ft/sec. These values correlate well with the average compressional wave velocity compiled by Hamilton (1969) for this class of sediment (Table 1). The velocity estimates calculated at each receiver generally exhibit one-sigma precision of approximately two percent with sequential average velocity for the first several layers varying by less than one percent.

Good correlation also exists with Table 1 data for the normal incidence reflection coefficients although the standard deviation is slightly larger (i.e., approximately five percent). The greater variability may be attributable to interface roughness which is unaccounted for in the present model. It is further anticipated that this effect would become more pronounced with decreasing incidence angles such as may be observed in Fig. 4. A second aspect of the data presented in Fig. 4 is the apparent occurrence of a minimum at approximately 30°-35° as suggested by the previously discussed model. Specific conclusions, however, should not be drawn from the limited data presented in this figure, as it is necessary

ACOUSTIC ASSESSMENT PROGRAMS, RAYTHEON COMPANY

MARINE SEDIMENTS

SEA TEST NO.513024 LINE SEGMENT 420-428 DATE 03.15.73

TIME 17.23.53

[A]

AVERAGE VELOCITIES 15 PULSES,,(FT/MS)

	AVE.	SIGMA	MAX.	MIN.	RECEIVER NUMBER 1	2	3	4	5	6	7	8	9	10
WATER COLM	4.837	.025	4.875	4.789	4.801	4.839	4.789	4.837	4.840	4.850	4.875	4.861	4.833	4.841
LAYER NO.1	4.918	.049	5.027	4.848	4.923	4.965	4.909	4.909	4.848	4.899	4.896	4.883	5.027	4.920
LAYER NO.2	5.660	.069	5.798	5.568	5.628	5.673	5.742	5.683	5.616	5.594	5.645	5.568	5.798	5.649
MEAN G DIA	8.19	.53	8.88	7.31	8.88	8.40	8.46	8.46	7.36	8.22	8.16	7.91	7.31	8.77

[B]

	AVE.	SIGMA	MAX.	MIN.	PULSE NUMBER 1	2	3	4	5	6	7	8	9	10	11	12	13	14	15
PING REFL	.127	.008	.137	.105	.135	.129	.125	.124	.137	.105	.124	.124	.137	.124	.129	.117	.137	.127	.123
PING BLOS	18.0		19.6	17.2	17.4	17.8	18.1	18.1	17.2	19.6	18.1	18.1	17.2	18.1	17.8	18.6	17.2	18.0	18.2
POROSITY	81.43	1.28	84.67	79.82	80.17	81.09	81.67	81.79	79.82	84.67	81.79	81.79	79.82	81.79	81.09	82.82	79.82	81.44	81.90
DENSITY	1.325	.022	1.353	1.270	1.347	1.331	1.321	1.319	1.353	1.270	1.319	1.319	1.353	1.319	1.331	1.301	1.353	1.325	1.317
WATER DPTH	114.7	.4	115.3	113.8	115.3	115.1	115.3	115.1	115.1	114.8	114.3	114.2	113.8	114.6	114.5	114.8	114.6	114.8	114.6

ACOUSTIC ASSESSMENT PROGRAMS, RAYTHEON COMPANY

MARINE SEDIMENTS

SEA TEST NO.513024 LINE SEGMENT 420-428 DATE 03.15.73

TIME 17.24.53

[A]

AVERAGE VELOCITIES 15 PULSES,,(FT/MS)

	AVE.	SIGMA	MAX.	MIN.	RECEIVER NUMBER 1	2	3	4	5	6	7	8	9	10
WATER COLM	4.834	.026	4.865	4.780	4.865	4.780	4.802	4.836	4.831	4.842	4.861	4.845	4.833	4.841
LAYER NO.1	4.919	.048	5.025	4.845	4.900	5.025	4.923	4.965	4.909	4.845	4.896	4.887	4.925	4.913
LAYER NO.2	5.686	.058	5.800	5.615	5.800	5.615	5.628	5.673	5.735	5.680	5.690	5.645	5.745	5.650
MEAN G DIA	8.23	.56	8.95	7.32	8.25	7.34	8.88	8.40	8.46	7.32	8.16	7.98	8.95	8.56

[B]

	AVE.	SIGMA	MAX.	MIN.	PULSE NUMBER 1	2	3	4	5	6	7	8	9	10	11	12	13	14	15
PING REFL	.130	.005	.137	.118	.118	.129	.127	.131	.130	.131	.137	.132	.130	.135	.125	.126	.126	.132	.135
PING BLOS	17.8		18.6	17.3	18.6	17.8	17.9	17.6	17.7	17.7	17.3	17.6	17.7	17.4	18.1	18.0	18.0	17.6	17.4
POROSITY	80.99	.72	82.75	79.88	82.75	81.08	81.32	80.72	80.96	80.84	79.88	80.60	80.96	80.24	81.68	81.56	81.56	80.60	80.12
DENSITY	1.333	.012	1.352	1.303	1.303	1.331	1.327	1.338	1.334	1.336	1.352	1.340	1.334	1.346	1.321	1.323	1.323	1.340	1.348
WATER DPTH	114.5	.7	115.7	113.4	113.4	113.6	114.7	115.2	115.7	115.2	114.4	114.1	114.1	114.2	114.2	115.2	115.2	114.7	113.4

FIGURE 3

VELOCITY AND NORMAL INCIDENCE REFLECTION COEFFICIENTS

to process significantly more data prior to establishing the validity and predictive ability of this model indicated phenomenon.

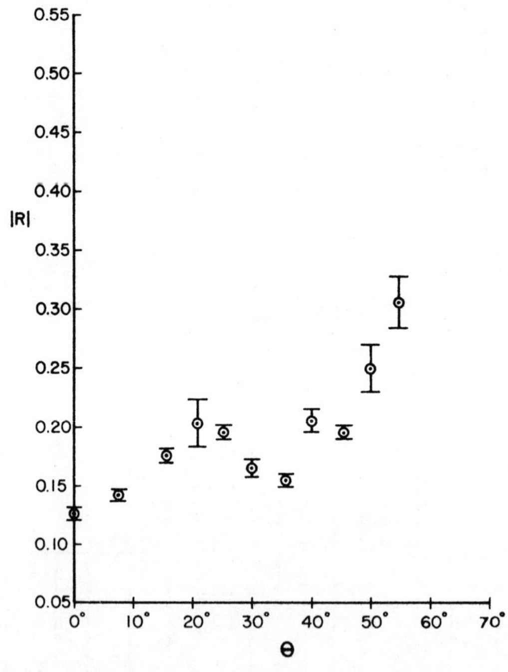

FIGURE 4

REFLECTION COEFFICIENT VERSUS REFLECTION ANGLE

CONCLUSIONS

Consistent measurements may be presently made on a limited number of acoustic index parameters, namely compressional wave velocity and angular reflection coefficients. While the potential exists for sediment classification based upon the velocity and normal incidence reflection coefficients (Table 1), the subsequent availability of C_t and α_t estimates potentially derivable from minima in the oblique incidence response should greatly increase discrimination ability within the range of commonly encountered marine sediments.

The additional significance of the Table 1 presentation is that, while sediment acoustical properties do not vary over a wide range, they do change in a relatively consistent progression allowing assessments to be made if accurate measurements can be obtained.

An additional benefit also accrues from the present automated reflection coefficient data processing in that the high sampling density afforded by acoustic techniques allows a measure of sediment variability (or consistency) to be ascertained by examining the statistics of sequential acoustic bottom reflections. This same data could also be plotted on an area map to assess sediment variability within the geographic section of interest.

SEDIMENT TYPE	GRAIN DIAMETER MEAN (mm)	SAND %	SILT %	CLAY %	DENSITY (G/CC) AVG.	POROSITY AVG.	VELOCITY M/SEC AVG.	REFLECTION COEFF. AVG.
Sand								
Coarse	0.530	100.0	-	-	2.03	38.6	1836	0.4098
Medium	0.376	99.8	0.2		2.01	39.7	1749	0.3835
Fine	0.153	88.1	6.3	7.1	1.98	43.9	1742	0.3749
Very fine	0.090	83.9	13.0	2.9	1.91	47.4	1711	0.3517
Silty sand	0.073	65.0	21.6	13.4	1.83	52.8	1677	0.3228
Sandy silt	0.036	34.5	51.2	14.3	1.56	68.3	1522	0.2136
Sand-silt-clay	0.018	32.6	41.2	26.1	1.58	67.5	1578	0.2504
Clayey silt	0.006	6.1	59.2	34.8	1.43	75.0	1535	0.1767
Silty clay	0.003	5.3	41.5	53.6	1.42	76.0	1519	0.1586

Table adapted from "Sound Velocity Elasticity and Related Properties of Marine Sediments" NUC - San Diego, 1969 Reports NUC TP 143, 145, 163

TABLE 1

SEDIMENT PROPERTIES, CONTINENTAL TERRACE (SHELF AND SLOPE) ENVIRONMENT

FUTURE WORK

Continuing analysis of the data recently acquired with the prototype system is underway. In addition to the computer processing refinements being incorporated in the present analytical procedure, the major goal of these investigations is establishment of the confidence limits which should be placed on remote acoustic sediment property determinations. The continuing need also exists for correlation with a more extensive data bank of physical and engineering properties, which, together with efforts to improve the system performance, form the basis and direction of our near term research activities.

Acknowledgment

This work is part of a continuing study being conducted under a joint University-Industry Sea Grant Program entitled "The Science and Technology of Utilizing the Bottom Resources of the Continental Shelf." The authors wish to express their appreciation to the Office of Sea Grant Programs of NOAA, The University of New Hampshire, and The Submarine Signal Division of The Raytheon Company, all of which contributed financial support to this Sea Grant project.

REFERENCES

Bennett, R. H., G. H. Keller, and R. F. Busby, 1970. Mass Property Variability in Three Closely Spaced Deep-Sea Sediment Cores: Jour. Sed. Pet. 40, 3:1038-1043.

Breslau, L. R., 1965. "Classification of Sea-Floor Sediments with a Shipborne Acoustical System," Le Petrole et la Mer, No. 132, Woods Hole Oceanographic Institution Contribution 1678, p. 9.

Breslau, L. R., 1967. "The Normally-Incident Reflectivity of the Sea Floor at 12 Kc and Its Correlation with Physical and Geological Properties of Naturally-Occurring Sediments," Woods Hole Oceanographic Institution Reference 67-16.

Clay, C. S., and P. A. Rona, 1965. "Studies of Seismic Reflections from Thin Layers on the Ocean Bottom in the Western North Atlantic," J. Geophys. Res., 70, 855-869.

Cole, B. F., 1965. "Marine Sediment Attenuation and Ocean-Bottom Reflected Sound," J. Acoust. Soc. Am. 38, 291-297.

Faas, R. W., 1972. "Mass Physical and Engineering Properties of Some York River Sediments," in Environmental Framework of Coastal Plain Estuaries; a Symposium on Estuaries, B. W. Nelson (Ed.), Geological Society of America Memoir 133.

Fry, J. C., and R. W. Raitt, 1961. "Sound Velocities at the Surface of Deep Sea Sediments," J. Geophys. Res., 66, 589-597.

Hamilton, E. L., 1956. "Low Sound Velocities in High-Porosity Sediments," J. Acoust. Soc. Am. 28, 16-19.

Hamilton, E. L., G. Shumway, H. W. Menard, and C. J. Shipek, 1956. "Acoustic and Other Physical Properties of Shallow-Water Sediments Off San Diego," J. Acoust. Soc. Am. 28, 1-15.

Hamilton, E. L., 1963. "Sediment Sound Velocity Measurements Made In Situ From the Bathyscaph TRIESTE," J. Geophys. Res., 68, 5991-5998.

Hamilton, E. L., 1968. "Sound Velocity, Elasticity, and Other Mass Physical Properties of Marine Sediments, North Pacific," (abstract), Trans. Amer. Geophys. Union, 49, p. 221.

Hamilton, E. L., 1969. "Sound Velocity, Elasticity, and Related Properties of Marine Sediments, North Pacific, Part 1: Sediment Properties, Environmental Control, and Empirical Relationships," Naval Undersea Research and Development Center Tech. Publ. 143.

Hamilton, E. L., 1970. "Reflection Coefficients and Bottom Losses at Normal Incidence Computed from Pacific Sediment Properties," Geophysics, 35, 995-1004.

Hampton, L., 1966. "Acoustic Properties of Sediments," DRL-A-254, Defense Research Laboratory, The University of Texas, p. 81.

Horn, D. R., B. M. Horn, and M. N. Delach, 1968. "Correlation between Acoustical and Other Physical Properties of Deep-Sea Cores," J. Geophys. Res., 73, 1939-1957.

Houtz, R. E., and J. I. Ewing, 1963. "Detailed Sedimentary Velocities from Seismic Refraction Profiles in the Western North Atlantic," J. Geophys. Res., 68, 5233-5258.

Houtz, R. E., J. I. Ewing, and X. LePichon, 1968. "Velocity of Deep-Sea Sediments from Sonobuoy Data," J. Geophys. Res., 73, 2615-2641.

Laughton, A. S., 1954. "Laboratory Measurements of Seismic Velocities in Ocean Sediments," RCL Proceedings, Series A.

LePichon, X., J. Ewing, and R. E. Houtz, 1968. "Deep-Sea Sediment Velocity Determination Made While Reflection Profiling," J. Geophys. Res., 73, 2597-2614.

Leslie, C. B., 1960. "Measurement of Ocean Bottom Characteristics With a Short Acoustic Tube," NAVORD Report No. 6809, U. S. Naval Ordnance Laboratory, Maryland, p. 45.

Mackenzie, K. V., 1960. "Reflection of Sound from Coastal Bottoms," J. Acoust. Soc. Am. 32, 221-231.

Nafe, J. E., and C. L. Drake, 1957. "Variation With Depth in Shallow and Deep Water Marine Sediments of Porosity, Density, and the Velocities of Compressional and Shear Waves," Geophysics, 22, 523-552.

Pawlowicz, E. F., 1971. "Ocean Engineering Significance of Marine Seismic Reflection Profiling Technology." Tech. Note N-1157, Naval Civil Engineering Laboratory.

Schreiber, B. C., 1968. "Sound Velocity in Deep Sea Sediments," J. Geophys. Res., 73, 1259-1268.

Smith, D. T., and W. N. Li, 1966. "Echo-Sounding and Sea-Floor Sediments," Marine Geology, 4, 353-364.

Sutton, G. H., H. Berckheimer, and J. E. Nafe, 1957. "Physical Analysis of Deep Sea Sediments," Geophysics, 22, 779-812.

Swain, R. J., 1962. "Recent Techniques for Determination of 'In-Situ' Elastic Properties and Measurement of Motion Amplification in Layered Media," Geophysics, 27, No. 2, 237-241.

MEASUREMENT OF VISCOELASTIC PROPERTIES OF SEDIMENTS USING A TORSIONALLY VIBRATING PROBE

R. S. ANDREWS, Department of Oceanography, and

O. B. WILSON, JR., Department of Physics and Chemistry

Naval Postgraduate School, Monterey, California 93940

ABSTRACT

The torsional vibrations of a cylindrical probe immersed in a soft sediment generate shear waves which react upon the probe. Measurements of the radiation impedance due to this reaction provide a method for estimating the dynamic rigidity of the sediment. An experimental model of such an instrument, which is operable at frequencies in the range from 900 to 3300 Hz, is described.

The results of some preliminary measurements in soft, saturated sediments, both in situ and in the laboratory, are presented. These results are used to estimate a shear wave speed which is compared with measurements of the speed of surface waves of the Stoneley type. Agreements are better than an order of magnitude. This points out a need for improvements in the precision of the method.

INTRODUCTION

In recent years, attention has been given to the development of improved models for prediction of the sound reflectivity characteristics of the ocean floor. Some of these models have included the effects of shear elasticity or rigidity of the sediment and the effects of the dissipative processes (Bucker et al., 1965; Hamilton, 1972; Akal, 1972). Experimental determinations of the shear moduli which describe both the elastic and the anelastic or dissipative properties of a variety of materials, including relatively rigid sediments, have been carried out by a number of investigators using direct wave propagation methods (White, 1965). For very soft sediments, the propagation speed for shear waves is so small that the direct observation of the wave propagation effects, which enable determination of wave speed and/or the absorption coefficient, becomes very difficult at frequencies of interest in active sonar systems.

We have explored the use of a radiation impedance method for determining the dynamic rigidity of very soft sediments. The method is an adaptation of one developed by Mason (1947) and McSkimin (1952) for observing the viscoelastic properties of polymer solutions. In this method, the dynamic loading effects of material in contact with the side walls of a cylindrical rod which is oscillating in torsion can be interpreted to obtain the complex shear modulus of the material.

A progressive series of developments and tests of this method have been carried out. This paper will present a description only of the currently used instrumentation. Details of earlier work using this method have been reported by Wilson and Andrews (1971). The results of some preliminary in situ measurements using this method are reported and discussed. Also reported are some of the results of a concurrent set of experiments which have been conducted for the purpose of determining the validity of this method. These experiments involve measurement of the speed of propagation of interface waves of the Stoneley type and the speed and absorption of compressive waves.

DESCRIPTION OF THE METHOD

The torsionally oscillating probe method used for measuring the dynamic shear properties of soft sediments reported here has been developed from methods described by Mason (1947) and McSkimin (1952) for the study of viscoelastic polymer solutions. The essential features of the apparatus are: (1) a cylindrical surface which is made to oscillate in torsion with simple harmonic motion; (2) the walls of this cylinder, when placed in contact with the medium being studied, will generate shear waves which propagate into the medium; (3) the mechanical oscillator containing sensors which permit the determination of the reaction to the radiation of these shear waves, the radiation impedance, which may be used in the calculation of the complex shear modulus of the medium.

In the experiments described by Mason, the effect of the radiation reaction upon the resonant frequency and the 'Q' of the torsional mechanical oscillator, a suitably prepared piezoelectric crystal, was observed. McSkimin described the use of traveling torsional waves in a cylindrical rod immersed in a viscoelastic medium. The speed of propagation and the absorption of the torsional waves in the rod are influenced by the shear wave loading at the walls.

We have adapted both of these methods for studying the properties of soft sediments in the laboratory. We found that problems of temperature control, mechanical stability, and of the measurement of resonance frequency when the oscillator was heavily damped made these methods unsuitable for use in situ.

The method now employed is one in which a small section at the end of a torsionally vibrating system is instrumented to measure the torsional mechanical impedance at that point. When inserted in the sediment, changes in the torsional mechanical impedance caused by the waves generated in the sediment at the probe end can be interpreted and the complex shear modulus can be estimated.

DESCRIPTION OF APPARATUS

Figure 1 is a sketch of the viscoelastometer. It consists of (1) a metallic head 3 inches (7.6 cm) long, with vanes, which executes torsional oscillations and generates shear waves in the sediment and which is tapered to minimize disturbance of the sediment upon insertion; (2) an electro-dynamic (moving coil) angular velocity sensor enclosed within the oscillating head; (3) a torque sensor consisting of a 3-inch (7.6 cm) long cylindrical barium titanate torsional transducer designed to produce an electrical signal which is proportional to torsional stress; (4) a stainless steel tube 26 inches (66 cm) long and 0.75 inches (1.9 cm) outside diameter; (5) a hollow cylindrical barium titanate transducer 3 inches (7.6 cm) long fabricated to produce torsional motion when driven electrically; (6) a massive supporting flange which provides

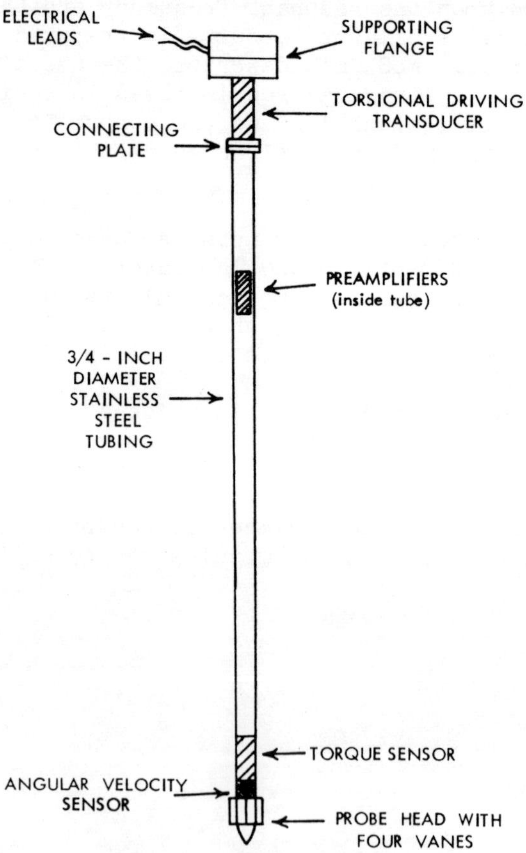

FIGURE 1

THE VISCOELASTOMETER

a reaction for the drive transducer which tends to cause the upper end of the drive transducer to be a nodal plane so that the length from that point to the vaned head is nearly one-quarter of a wavelength at the first resonant mode. The frequency for the fundamental mode of mechanical resonance is about 900 Hz. Overtone operation at frequencies of about 2200 and 3300 Hz also provides useful information. Since the mechanical system is not uniform in its geometry or elastic properties, the overtones are not harmonically related.

An oscillator and power amplifier provide a steady state excitation of the driver transducer at a frequency corresponding to a chosen mode for mechanical resonance. The applied voltage is adjusted to provide a satisfactory signal-to-noise ratio from the torque and angular velocity sensors, and varies with the mechanical damping due to contact with the sediment. The amplitude of the displacement at the surface of the probe due to torsional motion is estimated to be less than 1 micron during measurements in the sediment. This was determined by linear extrapolation from observations made with a calibrated light microscope while the drive transducer was excited by a larger than normal applied voltage.

Care in preserving axial symmetry was exercised during construction in order to control coupling between torsional modes of oscillation and flexural vibrations. A small phonograph pickup was used to assure that the motion was purely torsional.

The Probe Head

The head of the probe consists of an aluminum cap 1.75 inches (4.5 cm) long, 0.75 inches (1.9 cm) outside diameter and tapered at one end for insertion into the sediment with a minimum disturbance. Four tapered vanes about 1.5 inches (3.8 cm) long, 0.2 inches (0.5 cm) wide and 0.03 inches (0.08 cm) thick, located in quadrature, are fared in at the lower part of the head. The head is hollow for the placement of the velocity pickup and is attached with a screw-thread fitting and "o-ring" seal to permit access to the pickup and to permit changing of the vane section to heads of various vane size.

In designing the head, the selection of an optimal length was constrained by the fact that the length of the head and the torque sensor should be small compared to the resonant wavelength of the torsional standing-wave so that the variations in torque and angular velocity along this region would be negligible. Coupled with this constraint was the additional restriction that the head be large enough to enclose a velocity sensor of practical dimensions. The tapered vanes are used to apply a static compression on the transducer and, thereby, provide additional mechanical strength at the joints. All electrical wiring from the torque and velocity sensors is carried internally through the device with egress through a hole at the side of the mass.

Small, integrated circuit preamplifiers for the torque and velocity sensors are located inside the tubing. After assembly of the probe, the external surfaces of both ceramics were coated for electrical insulation and waterproofing. The torque sensor was covered also by an electrostatic shield.

CALIBRATION AND INTERPRETATION OF INSTRUMENTAL DATA

An equivalent electrical circuit for the torsional impedance sensors and the probe is presented in Fig. 2. Figure 3 presents a block diagram of the electronic instruments used in the measurements. The following describes the procedures used in determining the relationships between the observed electrical quantities and the viscoelastic parameters of the sediment in which the probe is imbedded.

Let Z represent the torsional mechanical impedance which is defined by:

$$Z = \frac{T}{V} e^{j\theta} \quad , \tag{1}$$

where T is the torque, V is the angular velocity, and θ is the phase angle between them. Let V_T and V_V represent the voltages from the torque and velocity sensors, respectively, and θ_o represent the observed phase angle between them. It is assumed that the relationship between Z and V_T/V_V is a simple, linear one with a constant of proportionality, K, and an associated phase angle, θ^*, which will include effects of phase shifts in the preamplifiers:

$$Z = |K| e^{-j\theta^*} \left|\frac{V_T}{V_V}\right| e^{j\theta_o} \quad . \tag{2}$$

This requires that $\theta = \theta_o - \theta^*$. (3)

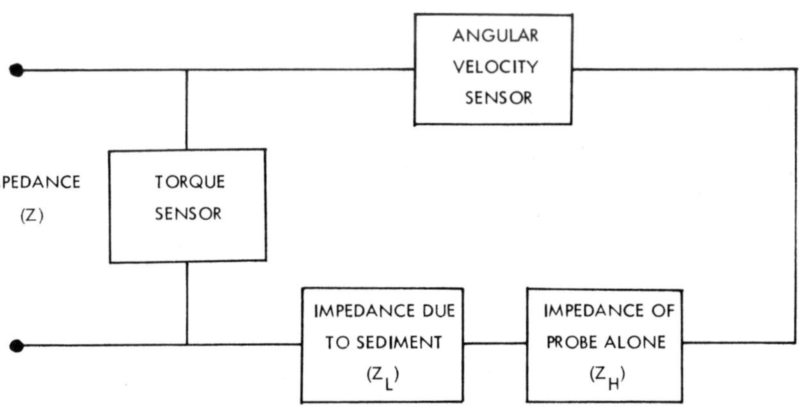

FIGURE 2

EQUIVALENT ELECTRICAL CIRCUIT FOR TORSIONAL IMPEDANCE INSTRUMENT

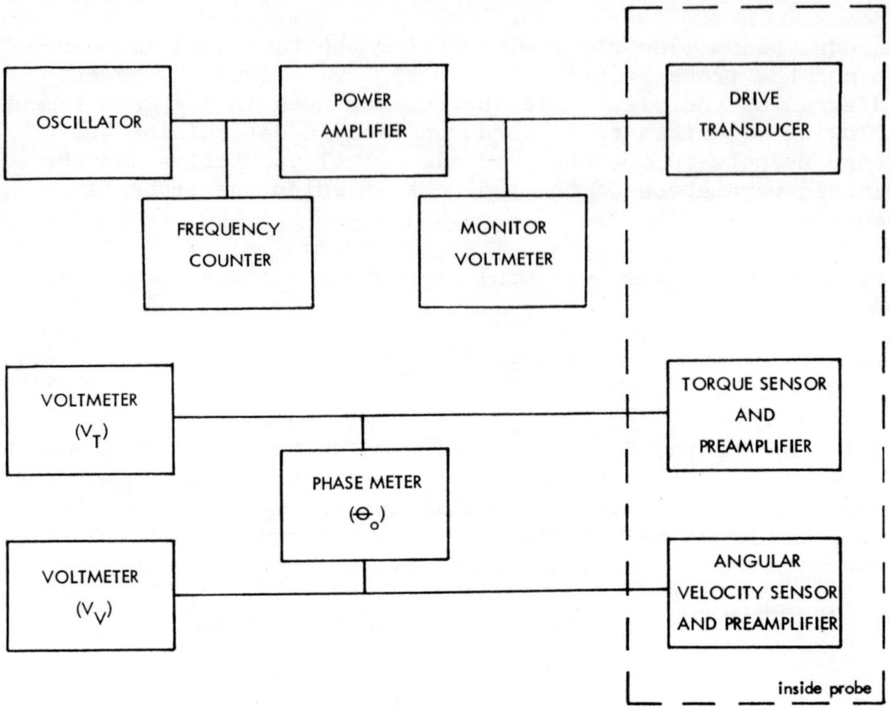

FIGURE 3

BLOCK DIAGRAM OF ELECTRONIC INSTRUMENTATION

The calibration of the torque and velocity sensors is carried out by clamping to the probe several additional masses of known moments of inertia. It is assumed that when the probe is vibrating in air, the torsional mechanical impedance is purely reactive. Since all elements in the sensors and in the probe are relatively small compared to the wavelength for torsional waves in these elements at the frequencies used, the reactance is purely inertial. In this case:

$$Z = j\omega(I_H + I_S) \quad , \tag{4}$$

where I_H and I_S are moments of inertia of the probe head and the added-on standards, respectively. By carrying out a sequence of measurements of V_T, V_V, θ_o at the various frequencies for several known values of I_S, values for K, θ^*, and I_H are determined by solving a system of simultaneous equations.

For the case where the probe is in the sediment, the torsional mechanical impedance due to the sediment, Z_L, is calculated by subtracting the impedance of the probe alone from the measured impedance with the probe in the sediment; i.e.,

$$Z_L = R_L + j\,X_L = Z - j\omega\,I_H, \tag{5}$$

so that:

$$R_L = |Z|\cos\theta \quad ; \quad X_L = |Z|\sin\theta - \omega\,I_H\,. \tag{6}$$

A second part of the calibration procedure is the determination of the relationship between the viscoelastic parameters of the sediment and the measured values of torsional mechanical impedance of the probe due to its contact with the sediment. This relationship is, of course, quite sensitive to the details of the wave motion in the neighborhood of the probe and also to the details of the viscoelastic model chosen.

Mason (1947) and McSkimin (1952), in their use of this approach in studying viscoelastic polymers, assume that the shear waves generated by the torsionally oscillating cylinder are very rapidly damped as they propagate away from the cylinder, so that the waves may be treated essentially as plane shear waves. This assumption seems to be justified for the materials they studied. McSkimin (1952) developed an expression which corrected the acoustic impedance in these waves for the cylindrical divergence effects.

McSkimin (1964) shows that, for the short wavelength limit, the following relationship exists between the components of the complex shear modulus, $G = G' + jG''$ (Voigt model), and the mechanical impedance per unit area of shear wave front, $Z_o = R_o + j\,X_o$, is:

$$G' = \frac{R_o^2 - X_o^2}{\rho} \;;\quad G'' = \frac{2R_oX_o}{\rho}\,, \tag{7}$$

where ρ is the density.

The relationship between the measured torsional mechanical impedance, Z_L, and the impedance per unit area of shear wave, Z_o, is needed. It is assumed that the torsional probe is a cylinder of radius "a" which has a side wall of area A in contact with the sediment. The shear mechanical impedance, Z_M, the ratio of shearing force to velocity, is given by

$$Z_M = A\,Z_o\,. \tag{8}$$

The torsional mechanical impedance, the ratio of torque to angular velocity, Z, is related to Z_o by

$$Z = a^2 Z_m = A a^2 Z_o \ . \tag{9}$$

The probe system could be, in principle, calibrated in an empirical manner by conducting measurements of both the speed of propagation and the absorption coefficient for shear and/or surface waves. In the auxiliary experiments conducted here, the precision of measurement of surface wave amplitudes did not permit calculation of the absorption coefficient.

SUPPORTING EXPERIMENTS

In addition to making measurements of dynamic rigidity in the sediments with the viscoelastometer, the following experiments were conducted: a) compressional wave speed and absorption (attenuation) measurements using a transducer-hydrophone array similar to that described by Hamilton (1963, 1972); b) surface wave speed measurements using a blasting cap-geophone array similar to that used by Hamilton et al. (1970). These experiments were initially conducted in a large, rectangular laboratory tank containing water-saturated kaolinite clay; in situ experiments were made in clayey silt tidal flat deposits (Elkhorn Slough, California) and sandy silt deposits in water depths of 30 ft (9 m) (Monterey Harbor, California).

Compressional Wave Measurements

In the laboratory clay tank, Martinek (1972) generated compressional waves with a piston-type electrostatic transducer at frequencies of 100, 125, and 150 kHz. The signal was received at a barium titanate hydrophone, positioned at various ranges from 50 to 100 cm from the source, and was fed into a boxcar integrator which enhanced the signal-to-noise ratio. The compressional wave speed was determined from the total travel time of the signal over the separation distance between source and receiver, and the absorption coefficient was calculated from a comparison of the amplitudes of the signal peaks measured at various ranges and corrected for spherical divergence.

The in situ measurements were made at frequencies of 10, 20, and 30 kHz with two hydrophones mounted on probes spaced 1 m apart and in line with a transducer source. Compressional wave speed was determined by comparing the time for the sound pulse to travel the distance between the two hydrophone probes when in water just above the bottom and when pushed into the sediment. The absorption coefficient was measured by comparing the relative signal peak amplitudes at the two hydrophones when the two probes were located in the water just above the bottom and when the assembly was pressed into the sediment. The assumption is that the difference in the relative amplitudes is due to absorption in the sediment.

Surface (Interface) Wave Measurements

Surface waves were generated in the clay tank using a small electrodynamic vibrator which caused vertical oscillations of a disk in contact with the sediment (Walsh, 1971). The wave motion was sensed using four vertically polarized velocity geophones placed on the surface of the clay at ranges up to 50 cm from the source. Observations of the waves were made at source frequencies from 75 to 300 Hz. In situ surface waves were generated using blasting caps as a source and detected using similar geophones in various linear configurations (separation distances and location of the source within or at the ends of the geophone array). In general, a spacing of 0.75 m between geophones was found to give the best results. The dominant frequency of the surface waves recorded was about 20 Hz.

Signals from the geophones were amplified and recorded with a visicorder. Surface wave speed was determined by measuring the travel time of similar phases of a wave form between geophones at various ranges from the source. Shear wave speed for the sediments (V_s) is computed from the surface wave speed (V_i), assuming these waves to be Stoneley-type, by the formula:

$$V_s = V_i/A \ , \tag{10}$$

where A is the coefficient of proportionality established by Strict and Ginzbarg (1956) as a function of the ratio of the density of the near-bottom water to the sediment density. This shear wave speed may then be compared with the shear wave speed calculated from the measured dynamic rigidity as an independent check of the validity of the viscoelastometer.

Sediment Mass Physical Properties

Wet density of the sediment was determined by weighing a known volume of the original water-saturated sediment. This sample was then dried in an oven at 105°C for 24 hours for porosity determination. Grain size analysis was conducted using the sieve-pipette method.

PRELIMINARY RESULTS AND DISCUSSION

Some preliminary results of measurements with the torsionally oscillating probe on four sediment types are presented in Table 1. In the first two cases, measurements were conducted in situ while the measurements in the beach sand, confined in a large tub, and the kaolinite, in a large tank, were conducted in the laboratory. The tabulated values include the real part of the measured torsional mechanical impedance presented to the probe by contact with the sediment and estimates of the real component of the shear modulus; also tabulated are estimates of the shear wave speed. Some limitations on the interpretation of these results are described and discussed below.

Some of the other acoustic properties of these materials from the supporting experiments are presented in Table 2. The shear wave speed listed is calculated from the surface wave measurements.

Sediment Type	Density, g/cc	Depth Below Surface, cm	Frequency, Hz	Measured Torsional Impedance (R_L)	Estimated Shear Modulus, Real Part (G'), 10^6 dynes/cm^2	Estimated Shear Wave Speed, m/sec
kaolinite clay (laboratory tank)	1.4	15	883	9.8	0.9	8
			2119	8.3	0.7	7
			3313	20	4	17
		15*	887	14	2	12
			2122	12	1.5	10
			3325	24	5.7	20
clayey silt (Kirby Landing, Elkhorn Slough tidal flat)	1.38	2	831	0.72	0.05	2
		6	845	14	2	12
		12	915	24	6	21
		2	2509	3	0.1	3
		6	2491	76	60	66
		12	2522	150	230	130
sandy silt (Monterey Harbor Pier, about 30 ft)	1.87	12	834	34	8	21
		12	2497	35	8	21
		18	2520	43	14	27
medium beach sand (Del Monte Beach, Monterey, in laboratory container)	1.75	2.5	927	28	6	19
		2.5	2136	16	2	11
		2.5	3368	14	2	11
		10.5	973	34	9	23
		10.5	2133	87	60	59
		10.5	3367	32	8	21

*measurements made after probe was in place in the sediment for a period of 3 hours

TABLE 1

SOME TYPICAL RESULTS OF MEASURED RIGIDITY

Sediment Type	Compressional Wave Speed, m/sec Sediment	Compressional Wave Speed, m/sec Water	Interface Wave Speed, m/sec	Shear Wave Speed, m/sec#	Compressional Wave Absorption Coefficient, dB/m/kHz
kaolinite clay (Laboratory tank)	1422	1479*	-	5	0.033 at 100 kHz 0.034 at 125 kHz 0.033 at 150 kHz
clayey silt (Kirby Landing, Elkhorn Slough tidal flat)	1466	1476	28.5	33	0.10 at 10 kHz 0.18 at 20 kHz 0.16 at 30 kHz
sandy silt (Monterey Harbor Pier, about 30 ft)	1689	1503	58.5	65	1.61 at 10 kHz 0.99 at 20 kHz 0.82 at 30 kHz
medium beach sand (Del Monte Beach, Monterey, in laboratory container)	1428	1522**	-	-	-

*calculated for fresh water at 19.9°C
**calculated for sea water at 20°C, 35‰
#calculated from surface wave measurements

TABLE 2

IN SITU PROPERTIES OF SEDIMENTS

Dynamic Rigidity

Computation of the real and imaginary parts of the shear modulus depends upon a number of procedures of measurement and calculation:

a) Measurement of the magnitude and the phase of the torsional mechanical impedance of the probe under two conditions; first, with no load, and second, with the probe in contact with the sediment.

b) The computation of the difference in the mechanical impedances measured in a).

c) Determining an impedance per unit area for the shear waves generated in the sediment, using the results of b) and a geometric model of the waves generated by the torsional motion.

d) The computation of the complex shear modulus components using the results of c).

The limitations which these procedures impose on the results are discussed below in the same order.

Within very small limits, the only contribution to a resistive component of the torsional impedance is due to the contact of the probe with the sediment. Although the accuracy with which the calibration of the impedance sensors is carried out is estimated to be not better than about 5%, precision in values of the R_L is estimated to be about 4% for a typical case.

The computation of the reactive part of the torsional impedance of the probe, $X = |Z| \sin \theta$, is probably precise to within about 4 to 5%. However, calculation of the reactive part of the impedance arising from contact with the sediment involves subtracting from the measured reactance, the reactance due to inertia of the probe alone (Eq. 6). It is estimated that the uncertainty in X_L is about 6% of the value of $|Z|$. In a number of cases the value of X_L is smaller than this uncertainty. Thus, due to the relatively large inertial reactance of the probe, a small uncertainty in $|Z|$ or in I_H leads to a very large uncertainty in X_L. For this reason, the values of X_L are not tabulated.

Although the problem of developing a precise theoretical relationship between the measured components of Z_L and the components of the complex shear modulus is a formidable one, approximate relations exist (Eqs. 7 and 9) and were used. Since the values of X_L are imprecise, the calculation for the estimates for the shear modulus assumed that X_L is small compared to R_L.

For the sediments studied here, a shear modulus of the order of 10^7 dynes/cm^2 and a density of about 1.5 g/cc may be typical. At the lowest frequency for the operation of the probe, about 900 Hz, the wavelength of the shear waves is of the order of 3 cm. At the highest frequency, 3.3 kHz, this wavelength is about 1 cm. Since the diameter of the probe is of the order of 3 cm, the assumption of plane waves is not very good.

Two additional factors which contribute to variability in the measurements are important. The first is the variability or inhomogeneity in the physical properties of natural materials. A second effect is the sensitivity of physical properties in soft sediments to the recent history of mechanical disturbance. This may influence our results in two ways. The first is a direct one. This is, the properties are themselves changed. The second can arise if the nature of the bonding between the oscillating probe and the bulk of the sediment is influenced by mechanical disturbance.

Compressional Wave Measurements

Table 2 shows the measured values of compressional wave speed for the four sediment types and for the water overlying the sediment. The measurements have a probable error of ±2%.

Calculations of the absorption coefficient are made with the assumptions of similar geometry for propagation of sound in the water and in the sediment and that the signal pressure follows the inverse square law as a function of range between the two receivers. Further, assumptions are made that the absorption of sound in water is negligible over the path lengths considered and that the relative sensitivities of the two hydrophones are the same in water and in the sediment. The absorption (a) in the sediment is found by the equation:

$$a = \frac{1}{x} 20 \log \left[\frac{P_{2s}/P_{1s}}{P_{2w}/P_{1w}} \right], \quad \text{dB/m} \tag{11}$$

where P_{2s}/P_{1s} is the relative sound pressure for a similar phase of the wave form between the hydrophones (subscript "1" indicating the hydrophone 1 m from the source and "2" the hydrophone 2 m from the source) in the sediment and P_{2w}/P_{1w} the relative sound pressure for the hydrophones in water above the bottom. The distance between the hydrophones (x) is equal to 1 m for our experiment. The absorption coefficient (k), listed in Table 2, is related to a by the formula:

$$k = a/f, \quad \text{dB/m/kHz} \tag{12}$$

where f is the sound frequency in kilohertz (Hamilton, 1972). The maximum amplitude peaks for each wave form were generally used to determine the pressure ratios. Further work on measuring the imaginary part of the torsional mechanical impedance (X_L) is necessary to demonstrate its value in predicting the absorption.

Surface (Interface) Wave Measurements

One of the major problems in conducting these experiments is determination of the proper phase of the wave form in the records from each geophone. The interface wave speeds and the corresponding calculated shear wave speeds (from tables of Strict and Ginzbarg, 1956, which assume Stoneley-type interface waves) are shown on Table 2. Lasswell (1970) estimated a precision of about ±15% for these values based on the uncertainty of reading the records and errors in measured distances between geophones. Comparison of the shear wave speeds for the various sediment types as calculated from the real part of the estimated shear modulus (Table 1) and from the interface wave measurements (Table 2) shows some agreement between the two methods. However, it may be concluded that the interface wave measurements are not sufficiently sensitive at present to validate the rigidity data.

CONCLUSIONS

The present model of the torsionally oscillating probe appears to be able to measure the real part of the torsional mechanical impedance due to the contact of the probe in the sediment with an accuracy of the order of 5%. Only the order of magnitude of the reactive part of this impedance is determinable. This causes the calculated values of the real part of the shear modulus to be valid only to within an order of magnitude. This is confirmed by comparing shear wave speeds calculated from the torsional probe data and those calculated from the surface wave experiments.

There are some prospective means for improving the performance. Redesign of the probe can still further reduce the inertia of the probe. The operating frequency can be increased. The precision of measurement of sensor signals can be improved significantly. The details of the shear wave field generated by the torsional oscillations could be understood by both analytical and experimental studies. However, the torsionally oscillating probe, in its present form, is not yet fully developed.

Acknowledgements

We gratefully acknowledge the support of the Office of Naval Research, Ocean Science Division, and the Naval Postgraduate School Foundation Research Program. The contributions made by our students in the conduct of their thesis research are gratefully acknowledged: G. E. Bieda, R. J. Cepek, S. R. Cohen, G. A. Engel, J. R. Hutchins, S. B. Kramer, J. B. Lasswell, C. A. Martinek, G. E. Voelker, and W. E. Walsh, Jr. Mr. Donald E. Spiel provided much technical assistance and critical comment in the design and construction of much of the apparatus and in the conduct of the experiments. We also wish to express appreciation for the discussion and encouragement given us by Dr. E. L. Hamilton and Dr. Ray Smith.

REFERENCES

Akal, T., The relationship between the physical properties of underwater sediments that affect bottom reflection, Marine Geology, 13, 251-266, 1972.

Bucker, H. P., J. A. Whitney, G. S. Yee, and R. R. Gardner, Reflection of low-frequency sonar signals from a smooth ocean bottom, J. Acoust. Soc. Am., 36, 1037-1051, 1965.

Hamilton, E. L., Sediment sound velocity measurements made in situ from bathyscaphe Trieste, J. Geophys. Res., 68, 5991-5998, 1963.

Hamilton, E. L., H. P. Bucker, D. L. Keir, and J. A. Whitney, Velocities of compressional and shear waves in marine sediments determined in situ from a research submersible, J. Geophys. Res., 75, 4039-4049, 1970.

Hamilton, E. L., Compressional-wave attenuation in marine sediments, Geophysics, 37, 620-646, 1972.

Lasswell, J. B., A comparison of two methods for measuring rigidity of saturated marine sediments, M.S. thesis, Naval Postgraduate School, Monterey, Calif., 1970.

Martinek, C. A., Compressional wave speed and absorption measurements in a saturated kaolinite-water artificial sediment, M.S. thesis, Naval Postgraduate School, Monterey, Calif., 1972.

Mason, W. P., Measurements of the viscosity and shear elasticity of liquids by means of a torsionally vibrating crystal, Trans., Amer. Soc. Mech. Eng., 69, 359-370, 1947.

McSkimin, H. J., Measurements of dynamic shear viscosity and stiffness of viscous liquids by means of traveling torsional waves, J. Acoust. Soc. Am., 24, 355-365, 1952.

McSkimin, H. J., Ultrasonic methods for measuring the mechanical properties of liquids and solids, in Physical Acoustics, Vol. 1A, edited by W. P. Mason, pp. 271-334, Academic Press, N.Y., 1964.

Strict, E., and A. S. Ginzbarg, Stoneley wave velocities for a fluid-solid interface, Bull. Seismol. Soc. Amer., 46, 281-292, 1956.

Walsh, W. E., Jr., The use of surface wave techniques for verification of dynamic rigidity measurements in a kaolinite-water artificial sediment, M.S. thesis, Naval Postgraduate School, Monterey, Calif., 1971.

White, J. E., Seismic Waves: Radiation, Transmission, and Attenuation, 302 pp., McGraw-Hill Book Co., Inc., N.Y., 1965.

Wilson, O. B., Jr., and R. S. Andrews, Measurement of the dynamic rigidity of sediments, in Proceedings: The International Symposium on the Engineering Properties of Sea Floor Soils and their Geophysical Identification, Seattle, Wash., 75-94A, 1971.

A METHOD FOR MEASURING IN SITU ACOUSTIC PROPERTIES DURING SEDIMENT CORING

AUBREY L. ANDERSON AND LOYD D. HAMPTON

Applied Research Laboratories

Austin, Texas 78712

ABSTRACT

A system developed for attachment to sediment corers in order to obtain an in situ sound speed profile during a coring operation is described. The system uses two electroacoustic transducers mounted in the cutting head of the corer and associated electronic circuitry to measure the travel time of an acoustic pulse traversing the diameter of the sediment core. Results of laboratory and field tests are presented. There is an ongoing study of the feasibility of expanding the sound speed measurement system capabilities to include a measure of sediment acoustic attenuation and internal volume scattering.

I. INTRODUCTION

Sediment acoustical and mechanical properties, which are measured in the laboratory for either constructed sediments or cores, require correction to determine their in situ values. Limits to the capability of obtaining undisturbed samples for such measurements contribute to uncertainty about actual in situ values. Hamilton (1971) has given a technique for extrapolating laboratory sound speed measurements to in situ values. His method is based on laboratory and in situ measurements of surficial sediments of the North Pacific. Extending this technique to other regions or to sediments deeper than the surface layer remains to be tested. Furthermore, the presence of gas either as bubbles or as a hydrate (clathrate) complicates any attempt to extrapolate to sea floor conditions. In situ measurements of sound speed by probing have previously been limited to surficial sediments (Hamilton, 1963; Bennin and Clay, 1967; Lewis et al., 1970).

Sediment cores are commonly taken to depths of 20 m and more in the bottom. Applied Research Laboratories (ARL), The University of Texas at Austin (UT), has developed a system, for attachment to corers, to obtain an in situ sediment sound speed profile during coring operations. This profilometer system is described below. Laboratory measurements of sound speed using the profilometer are compared with three probe measurements and with predictions of sound speed based on porosity. Field measurements of sound speed are described. The results indicate that the profilometer is not only useful as a logging tool, indicating extent of layering and core complexity, but that it also provides accurate values for in situ sound speed. An example of measurement of the progressive changes in sound speed values (deterioration) of a stored sediment core is given. The last section describes an expansion of the profilometer's capabilities to allow measurement of additional sediment parameters.

II. INSTRUMENT DESCRIPTION

The profilometer system uses two piezoelectric transducers mounted in the cutting head of the corer and connected to the electronics housing by coaxial cables. An acoustical pulse is generated by the transmit transducer. This pulse traverses the sediment contained in the cutting head at that instant and is received by the second transducer which is diametrically across the cutting head from the projector. The time delay resulting from the pulse traversing the sediment in the cutter head is converted to a voltage output which for experimental purposes is presently sent to the surface by a cable and recorded on magnetic tape. This output voltage is calibrated to give a measurement of the speed of propagation of sound in the sediment at the pulse carrier frequency.

A nominal carrier frequency of 190 kHz was chosen as a compromise between attenuation and timing accuracy. Attenuation by scattering and absorption in the sediment decreases with decreasing frequency but the accuracy with which the leading edge of a pulse can be measured also decreases. A repetition rate of 200 pulses per second is used. This repetition rate gave a spacing between measurements of about 2 in. (5 cm) when the technique was used to obtain a 30 ft (9 m) penetration with a piston corer. Total penetration time was slightly less than 1 sec.

Figure 1 shows a block diagram of the electronics system. The system consists of two basic sections, one to generate a pulsed sine wave with controlled phase and the other to amplify the received signal, filter it, and convert the time delay between transmission and reception to an analog voltage.

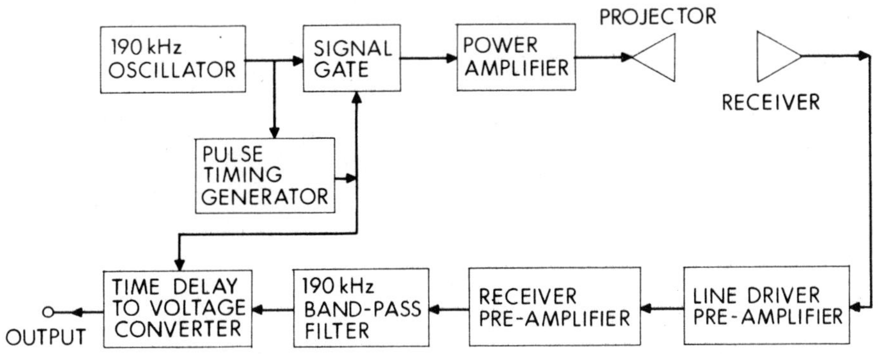

FIGURE 1

SYSTEM BLOCK DIAGRAM

III. LABORATORY MEASUREMENTS OF SOUND SPEED

The feasibility of using the sound speed profilometer was first confirmed by preliminary testing under controlled conditions in the laboratory. Figure 2 compares two sound speed profiles of a 1 ft (0.3 m) layer of sediment in a test tank. The profile labeled "probe measurement" was made with three probes (one projector and two receivers at different distances) lowered incrementally into the sediment. Time delay between the two received pulses was measured with an interval timer, and the sound speed calculated. The profile labeled "dynamic measurement" was made with the profilometer. It is a plot of sound speed calculated from the profilometer voltage output obtained as the coring head was penetrating the sediment after free fall from a height of 4 ft (1.2 m).

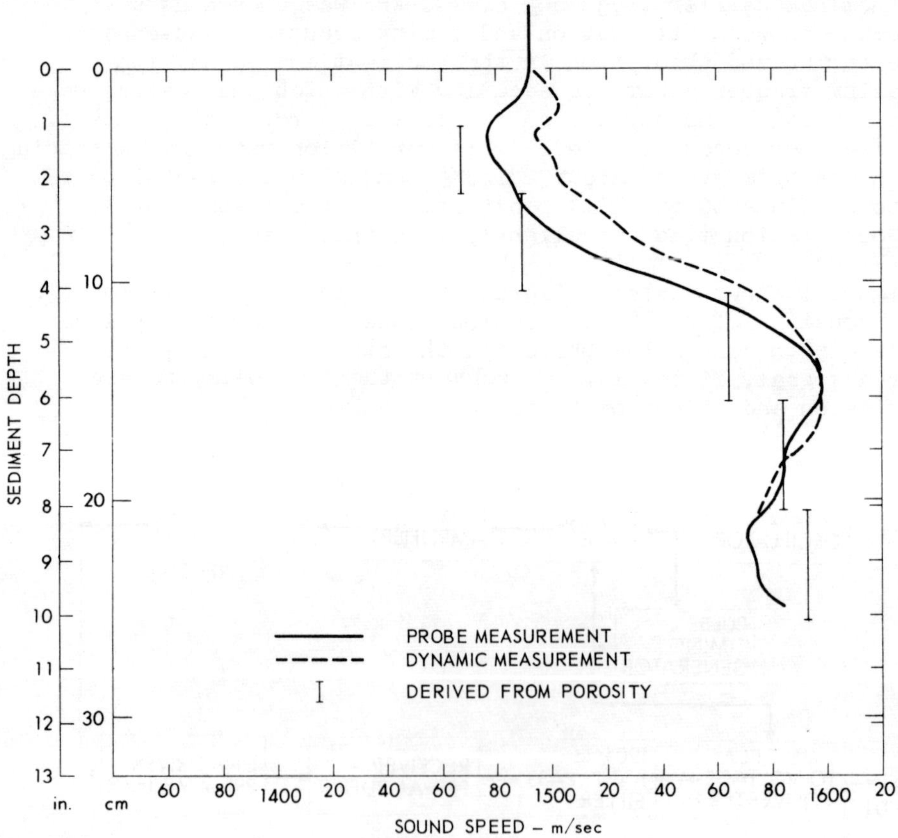

FIGURE 2

SOUND SPEED IN GRADUATED SEDIMENT
T = 23°C

The sediment for which sound speed is shown in Fig. 2 was sampled by coring with a 1 1/4 in. (3.2 cm) diam tube. Subsamples of these core samples were analyzed for porosity. The results are summarized in Table I. Akal (1972) has summarized the relationship between porosity and relative sound speed (ratio of sediment sound speed to water sound speed) for more than 450 marine sediment cores, based on data drawn from numerous authors. He fitted a second order equation to these data and presents a graph of the resulting curve. This curve was used to determine the values of sound speed shown in Table I for the measured porosities. These values are shown in Fig. 2 for comparison with the measured sound speeds. The absolute values of the measured sound speeds are well within the bounds of the observed sound speeds reported by Akal (about ±75 m/sec).

TABLE I

WATER CONTENT, POROSITY, AND SOUND SPEED PROFILE OF LABORATORY SEDIMENT

Sediment Depth Interval (cm)	Water Content (% of Dry Weight)	Porosity (%)	Sound Speed from Akal (1972) (m/sec)
2.5 - 5.7	102.2	73.0	1466
5.7 - 10	62.6	62.4	1490
10 - 15	37.3	49.7	1564
15 - 20	33.6	47.1	1584
20 - 25	32.2	46.0	1593

IV. IN SITU MEASUREMENTS OF SOUND SPEED

Figure 3 shows a sound speed profile obtained in about 30 fathoms (55 m) of water aboard RV ALAMINOS on station off the Mississippi delta in the Gulf of Mexico. The top trace is the unintegrated output of an accelerometer attached to the coring tool and is used as indication of corer entry into the sediment and end of corer motion. The uniform (unlayered) nature of this soft sediment is shown by the sound speed profile. Analysis of the cored material confirmed the unlayered nature of the sediment.

Cores were obtained in cooperation with the UT Marine Sciences Institute using a shallow water coring rig designed by Dr. E. W. Behrens. This device pushed the 3 in. (7.6 cm) diam core into the bottom with block and tackle rigging on a shallow draft platform. Using this equipment, cores were obtained in Baffin Bay and Redfish Bay on the Texas Gulf Coast near Corpus Christi, Texas.

Three sound speed profiles thus obtained in Baffin Bay are compared with core lithology in Fig. 4. The three cores were taken in a line and spaced approximately 2 ft (0.6 m) apart making a maximum separation between cores of only 4 ft (1.2 m). However, both the sound speed profiles and lithology diagrams show a large variation in layering between the cores. Some layers even show a variation across the 3 in. (7.6 cm) diam of the core. Lines on Core 3, Fig. 4, indicate the degree of correlation between the lithology and sound speed record. Cores 1, 2, and 3 are representative of the cores taken from this area in that they consist of alternating layers of mud, shell, and fine sand with intermixing of the three types to some degree in each layer.

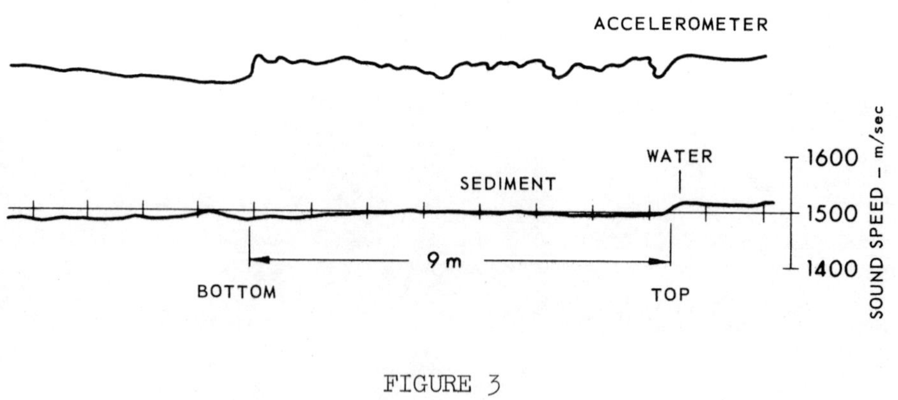

FIGURE 3

SOUND SPEED PROFILE
MISSISSIPPI DELTA, GULF OF MEXICO

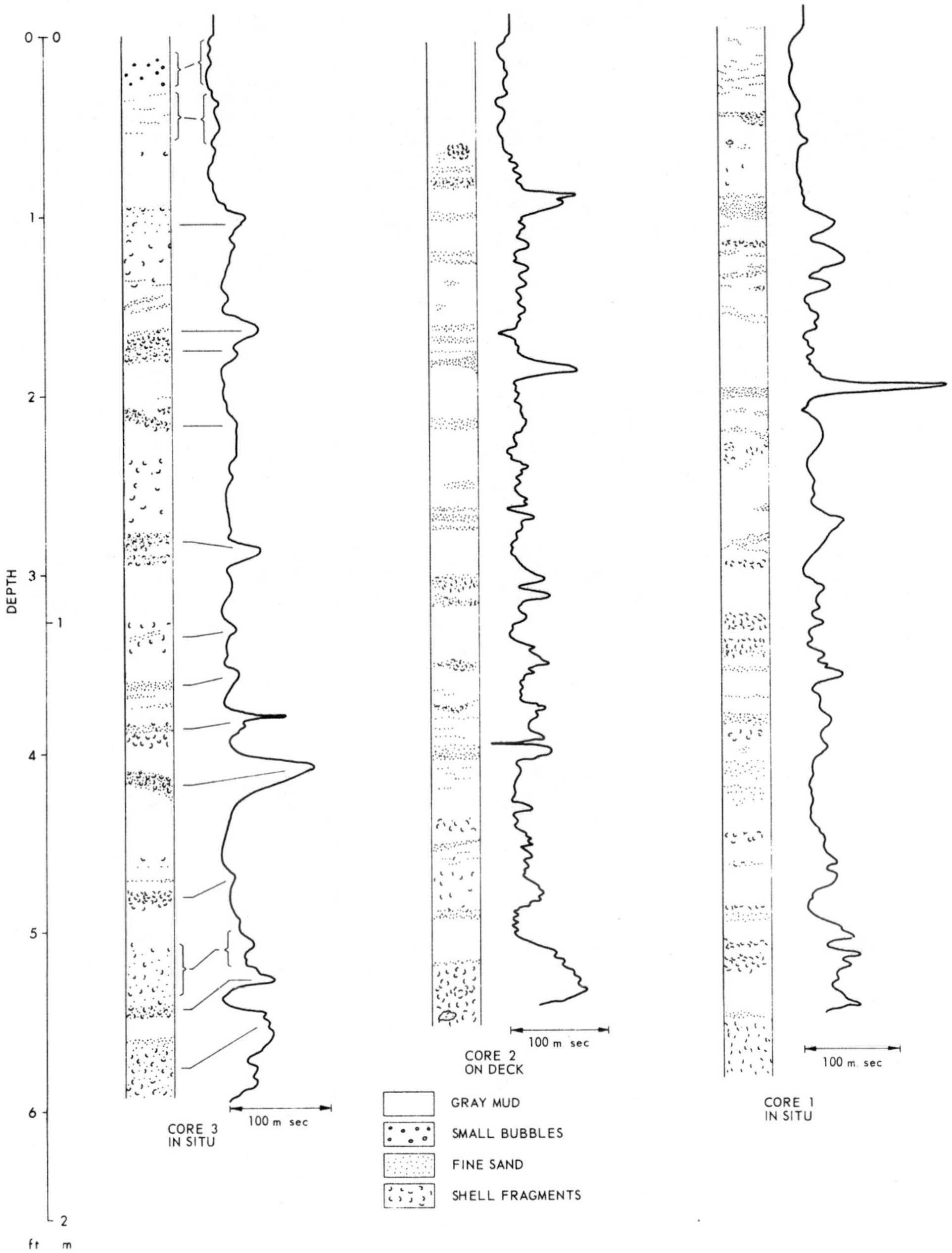

FIGURE 4

CORRELATION OF LITHOLOGIES AND SOUND SPEED
PROFILES OF CORES FROM BAFFIN BAY, TEXAS
August 1972

In Fig. 4 the water-sediment interface is located at the top of the figure and is marked as zero on the depth scale. The top 6 to 8 in. (15 to 20 cm) was a light fluffy mud containing gas and showed a sound speed lower than that of the overlying water. At about 1 ft (0.3 m) depth, the first high sound speed layer was encountered, and it consisted of a mixture of mud, sand, and shell. These top shell and sand layers as well as deeper ones are seen from Fig. 4 to be uneven, discontinuous, and quite often hard to distinguish. However, there is a good correlation between the amount of sand and shell in the cores and the value of sound speed. Sound speed increases with increasing percentage of sand and shell, with sand content the more dominant feature. The next distinctive layer encountered varied in depth from 1 3/4 ft (0.5 m) in Core 3 to 2 ft (0.6 m) in Core 1 and consisted of a very sandy layer about 2 to 3 in. (5 to 7.5 cm) thick. This layer seemingly slopes downward from Core 3 to Core 1. The layer is distinguished mainly by a sound speed that is higher than that of the other layers in the cores. This is especially true for Core 1. For the next 2 to 3 ft (0.6 to 0.9 m), the layering consisted of sandy, shelly layers which correlate with high sound speed, interspersed with muddy layers having a sound speed only slightly larger than that of water. Many of the sandy layers are too thin to have been detected by the sound speed profile and so do not show up on the profile. Finally, at a depth of 5 to 6 ft (1.5 to 1.8 m), a layer consisting of almost pure shell was encountered which could not be penetrated by the corer. This bottom layering was distinctive for all the cores taken in the area and shows up on the sound speed profiles as a broad, medium high speed peak with a dip in the middle. This distinctive layer occurred in all 15 profiles taken in this region as shown in Fig. 5.

The 15 cores were taken in groups with approximately 1 to 2 ft (0.3 to 0.6 m) separation between cores within a group. All cores were taken from an area with a radius of approximately 30 ft (9 m). The groupings were: Cores 1, 3, and 4; 5 and 6; 7, 8, and 9; 10, 11, and 12; 13, 14, 15, and 16.

It is possible from the sound speed profiles to identify the distinct bottom layer as continuous throughout the sampled region. Note also that conclusions about complexity of the area, dimensions of features, and extent of layering in the sediment can be drawn from these profiles without supplemental information.

A core sound speed profile which is of different character than either the homogeneous cores from the Gulf of Mexico (Fig. 3) or the complex Baffin Bay cores (Figs. 4 and 5) is given in Fig. 6. This core was taken in Redfish Bay, Texas, on 9 May 1972. The sound speed profile shown is representative of measurements on 7 cores taken at this site. The major character of the Redfish Bay cores is slow changes of sound speed with depth, in particular a

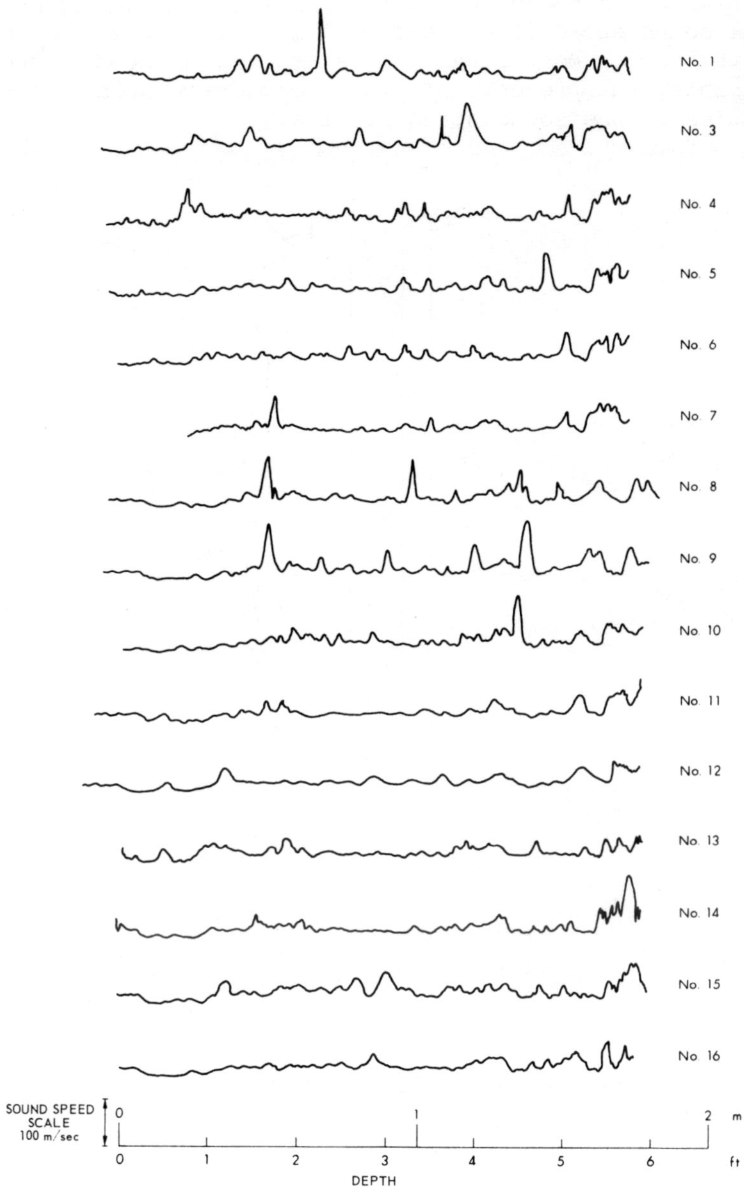

FIGURE 5

IN SITU SOUND SPEED PROFILES FOR FIFTEEN CORES IN SEQUENCE FROM BAFFIN BAY, TEXAS August 1972 uniformly high sound speed in the first 1 1/2 ft (0.45 m) of depth and gradual sound speed decrease from 1 1/2 ft (0.45 m) to the bottom of the core. Many large shells are scattered throughout the core, explaining the presence of fine structure in addition to the broad features in the sound speed profile.

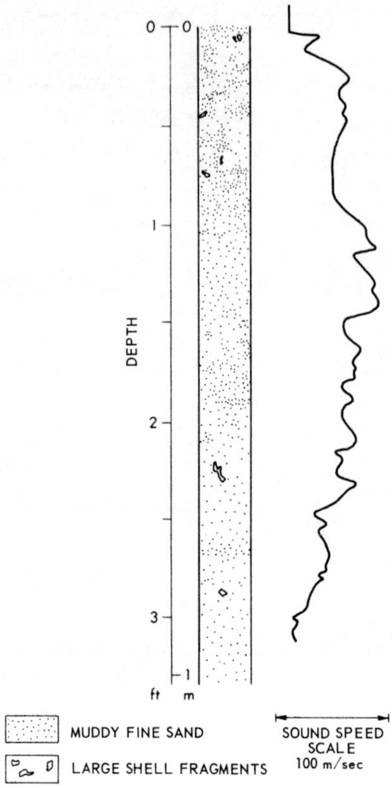

FIGURE 6

CORRELATION OF LITHOLOGY AND SOUND SPEED PROFILE OF CORE 2 FROM REDFISH BAY, TEXAS May 1972

V. COMPARISON OF MEASUREMENTS

In Section III profilometer measurements obtained in the laboratory were compared with measurements obtained with another set of probes and with predictions from porosity measurements. For the RV ALAMINOS core (Fig. 3) the porosity of the sediment, when compared with Akal's (1972) best fit curve relating porosity and sound speed, predicts a sound speed in the sediment about 15 m/sec lower than in the water with variations along the core of about 15 m/sec. These values are in good agreement with the measured profile in Fig. 3.

For the most part the complexity of the lithology of cores from Baffin Bay and Redfish Bay precluded meaningful sampling of portions of the core for sound speed prediction from lithology; however, the high speed layer at about 2 ft (0.6 m) in Core 1 (Fig. 4) was almost pure sand. Grain size distribution was determined for this material by sieve analysis. Geologically, the layer is described as poorly sorted muddy very fine sand with a mean grain diameter of 0.094 mm (3.4 Φ). Hamilton (1970) gives a graph of sound speed versus mean grain diameter which has been measured for sediments of the North Pacific. The measured sound speed of 1660 m/sec for the sand layer of Core 1 is well within the bounds of 1640 to 1700 m/sec reported by Hamilton for material of the same mean grain size. Thus in each measurement, at the laboratory, from the Gulf of Mexico and from Baffin Bay, the dynamic profilometer sound speed is in good agreement with values from previous measurements.

One set of Baffin Bay cores was used to compare in situ sound speed values with measurements on the cored sediment after various periods of storage. Figure 7 shows sound speed profiles made in Core 1 from Baffin Bay (the same one that is shown in Fig. 4). The top profile labeled "in situ" was made with the transducers mounted on the corer. The other three profiles were made using transducers that are designed to slide on the core liner. The profile labeled "on deck" was made approximately 3 hours after the core was taken. The two lowest profiles were made the first and second days after removing the cores to the laboratory.

Notice that the absolute value of sound speed changes somewhat from one determination to the next, and the major highlights show some variation in each profile. Significantly, sound speed in the sand layer (discussed above) is considerably lower in the laboratory measurements through the core liner: 1550 m/sec, which is lower than any of the Hamilton values. Thus, the in situ profilometer measurement provides a better value than the in lab measurement.

The large excursions in the bottom two profiles in Fig. 7 are the result of large attenuation of the signal beyond the capabilities of the AGC system. This attenuation was the result of either gas generation in these layers or physical separation of the material of the core during transportation or possibly both. Profiles taken later than these showed even greater deterioration.

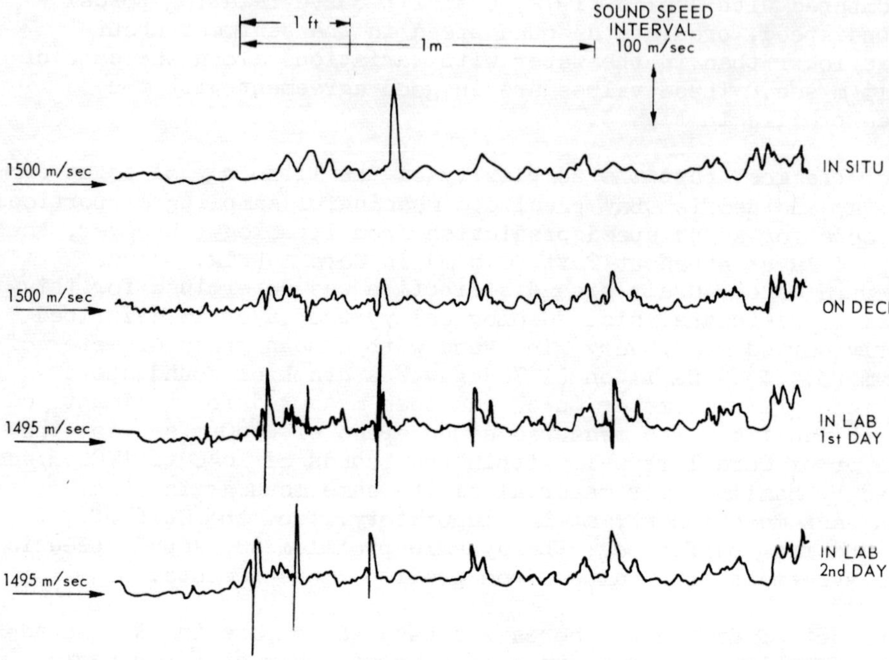

FIGURE 7

SOUND SPEED PROFILES OF CORE 1
FROM BAFFIN BAY, TEXAS, AUGUST 1972

VI. MEASUREMENT OF OTHER PARAMETERS

The automatic gain control feedback voltage is a measure of the insertion loss of the material between the profilometer transducers at any instant. This voltage has been separately recorded in a limited number of coring tests. Figure 8 compares profiles of sound speed and relative received signal amplitude (as measured by the AGC feedback voltage) for a single core from Baffin Bay, Texas. Increased sound speed in these cores is well correlated with increased sand content. As Fig. 8 demonstrates, the increased sound speed layers measured are also well correlated with decreases in the received signal amplitude. Thus, though work has not yet been devoted to reducing this information to a numerical measure of acoustical attenuation, the expected increase of attenuation with sand content is observed.

The internal acoustical volume scattering strength of sediments is not documented in the literature. The acoustical scattering cross section of gas bubbles in water is very large at resonance, more than one thousand times the physical bubble cross section. Bubble scattering cross section in sediments is not known, but should be large at resonance. To test the feasibility of adding measurements of this parameter to the profilometer, a third transducer was added to the cutter head. This transducer was placed so that its acoustical axis intersected at 90° the primary acoustical path joining the sound speed measurement transducer pair. Using the third transducer as a receiver, acoustical energy which is side scattered out of the principal transmission path in the sediment would be measured.

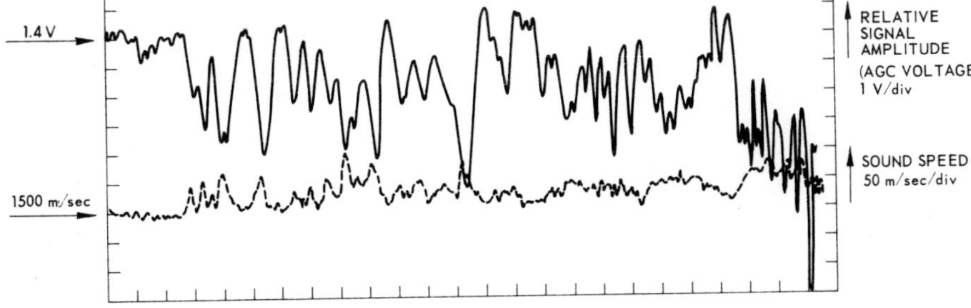

FIGURE 8

RELATIVE SIGNAL AMPLITUDE AND SOUND SPEED PROFILES
FOR A CORE FROM BAFFIN BAY, TEXAS

The acoustical feedover to the volume scatter transducer in clear water was measured to be about 14 dB below the direct path signal. Insertion of the cutter head into saturated sand sediments and muddy sand sediments in the laboratory indicated that the volume scattered levels were below the acoustical feedover levels. This measurement will therefore require greater isolation between the two acoustic paths.

An encouraging indication was obtained from a series of measurements with the 3-transducer arrangement in water. Gas bubbles, generated by hydrolysis, were allowed to rise into the region which is common to both acoustical paths. Side scattered acoustical levels were above the acoustical feedover for some bubble sizes and concentrations. Thus, a sufficient reduction of the feedover level by change of frequency, different transducer arrangement, or acoustical baffling should allow measurement of volume scattering in sediments. Also, the high levels of scattering from bubbles in water suggest the possibility of internal volume scattering measurements as useful indications of gas bubbles in the sediment.

ACKNOWLEDGEMENTS

Much of the development of, and experimentation with, the profilometer was carried out by Mr. D. J. Shirley. The work was sponsored by Office of Naval Research, Code 480.

BIBLIOGRAPHY

Akal, T., "The relationship between the physical properties of underwater sediments that affect bottom reflection," Marine Geology, 13, 251-266, 1972.

Bennin, R. S., and C. S. Clay, "Development of an in situ sediment velocimeter," Technical Report No. 131, Hudson Laboratories, Columbia University, New York, 1967.

Hamilton, E. L., "Sediment sound velocity measured in situ from bathyscaph TRIESTE," J. Geophys. Res., 68, 5991-5994, 1963.

Hamilton, E. L., "Sound velocity and related properties of marine sediments, North Pacific," J. Geophys. Res., 75, 4423-4446, 1970.

Hamilton, E. L., 1971, "Prediction of in situ acoustic and elastic properties of marine sediments": Geophysics, v. 36, p. 266-284.

Lewis, L. F., V. A. Nacci, and J. J. Gallagher, "In situ marine sediment probe and coring assembly," NUSL Report No. 1094, U.S. Naval Underwater Sound Laboratory, New London, Connecticut, 1970.

ACOUSTIC ESTIMATES OF THE TOPOGRAPHY AND ROUGHNESS SPECTRUM OF THE SEA FLOOR SOUTHWEST OF IBERIAN PENINSULA

C. S. CLAY AND WING K. LEONG*
Geophysical and Polar Research Center
Department of Geology and Geophysics
University of Wisconsin-Madison
Middleton, Wisconsin 53562

ABSTRACT

Echo sounding records were used to identify roughness features in the wavelength ranges 6 km to 12 km, 2 km to 6 km, 0.6 km and 3 km, and less than 1 km. The latter features are not resolved at 3.5 km ocean depths. The character of the echo sounding trace was used to find the location of microtopography or small features. Within each of the spectrum regions, the rms roughness has been estimated. We found some long wavelength features having small roughness and others that were smooth. Acoustic implications will be discussed. The area studied in detail is 32°N to 40°N and 9°W to 16°W. Marine Geophysical Survey (1965-66), Naval Oceanographic Office, and Lamont-Doherty Geological Observatory data were used.

* Now at Chevron Oil Company
Western Division
P. O. Box 599
Denver, Colorado 80201

1. INTRODUCTION

Our intention is to make a heuristic combination of scattering theory and conventional bathymetric and marine geophysical records. The results of our efforts are quantitative estimates of the spectra of the microtopography of the sea floor. The latter, densities and sound speeds of subbottom layers, and scattering theory can be used to calculate reflectivity or scattering coefficients of the sea floor. Since these coefficients depend upon both the measurement system and the sea floor, theory is needed to transfer results from one acoustic system to another.

To use sonar systems to measure remotely the properties of the sea floor, it has seemed to us to be of paramount importance to know what we are measuring and how to relate the measurements to what's there. To this end, we examine carefully the concepts of reflection and scattering processes. We will apply these to the estimation of sea floor roughness. The results should be regarded as preliminary in that we are testing a new technique. After presentation of the results, we will use the theory to describe the kind of measurements needed to improve the estimates.

We know that the reflection coefficient over a plane layered bottom depends upon the densities, sound speeds, and losses of the subsurface layers. When the bottom is roughened, local reflections may still be given by the local structure; however, the signal at the receiver is likely to be a confused mess of interferences. If the signal transmission is repeated for exactly the same source and receiver positions (and same sound speed structure in the ocean), the received signal will be identical with the first. Small translations of the measuring system between transmissions cause the signals to change for each transmission. The amount of change depends upon the roughness of the sea floor and the measuring system. Over a section of sea floor which can be described as randomly rough, the ensemble of signals has the appearance of being random.

Over the years, terms such as "reflection coefficients," "scattering coefficients," and the general catch all, "bottom loss," have been much used and, too often, interchangeably. Intuitively, reflections are measured when the interface is plane, and scattering coefficient is measured when the surface is rough.

In physics, the reflected signal is the incident signal multiplied by a reflection coefficient $\mathcal{R}$. Normally, the interface is plane and translation of the measurement system parallel to the interface gives the same result. Over a slightly rough interface, it seems reasonable to extend the definition by defining a coherent reflection coefficient $\langle\mathcal{R}\rangle$ which is obtained by measuring the average of an ensemble of reflected signals (< > indicates average). For transient signals, $\langle\mathcal{R}\rangle$ is obtained by measuring the average transient for many repeated transmissions when the measuring system is translated between each transmission. In the specular directions, the source distance is R_1, receiver distance is R_2, source signal p_o at R_o, and the average signal is

$$\langle p\rangle = \frac{p_o R_o}{(R_1+R_2)} \langle\mathcal{R}\rangle \tag{1.1}$$

The mean square signal $\langle p^2\rangle$ is measured by squaring the received signal and averaging many transmissions. Again, the measuring system is translated between each transmission. The result of the measurement is a scattering coefficient. For source direction θ_1 and receiver direction θ_2, angular frequency ω, an insonified area A, the scattered signal $\langle p^2\rangle$ is

$$\langle p^2\rangle = \frac{p_o^2 R_o^2 A}{R_1^2 R_2^2} S(\theta_1, \theta_2, \omega) \tag{1.2}$$

where S is the scattering coefficient. Over a very rough surface and at very high frequencies, each facet scatters independently and $\langle p^2\rangle$ depends upon S geometrically, i.e., the directions of the incident and scattered sound. The maximum value $\langle p^2\rangle$ can have is given by the reflection from a perfectly smooth surface. To estimate maximum $\langle p^2\rangle$ we square p in Eq. 1.1 and let $\langle\mathcal{R}\rangle = \mathcal{R}$. If only $\langle p^2\rangle$ were measured, then we would have no knowledge of $\langle\mathcal{R}\rangle$ and Eq. 1.2 should be used to analyze data. S has the limits of

$$S(\theta_1, \theta_2, \omega) \leqq S \leqq \frac{R_1^2 R_2^2}{A(R_1+R_2)} \mathcal{R}^2 \tag{1.3}$$

S obviously depends upon the roughness and the correlation function of the elevations of the scattering interface. For the smooth surface, and by implication the partially smooth surface, S also depends upon the measurement system, i.e., the beamwidth and geometry.

The special conditions of marine geophysical experiments require a formulation of the scattering problem which includes spherically diverging wave fronts, a wide range of surface roughness, and arbitrary correlation functions. Since this theory is needed for advanced work and interpretation of data as well as the exposition here, a condensation of Clay's derivations (Clay, 1971 and 1972) is given in Appendix I.

A statistical description of the sea floor can be used in several ways. The spectra of the roughness are quantitative measures of the interface. Sometimes the reflection of sound at the sea floor is used as a link in an acoustic communication system. If the system requires a faithful reproduction of the signal, then the coherent reflection coefficient should be used. If the presence or absence of signal is sufficient, then the scattering function is needed. Even though individual features cannot be resolved in measurements from the surface, the spectra of the group of features on the bottom can be measured.

2. COHERENTLY REFLECTED SOUND AND SEA FLOOR ROUGHNESS ESTIMATES

Scattering and reflection theories usually require assumptions such as the following: the interface is an average plane surface and its roughness has the same statistical description in one area as in any other area. Actually, the bottom of the ocean is quite variable and its slope and roughness change from one place to another. Since the sonar system insonifies a small area on the bottom, roughness is measured relative to the average plane interface in the insonified area. Except when features on the bottom interfere in the path of sound transmission, features outside of the illuminated area are not observed. By making a number of measurements within a physiographic province, an average rms roughness of the bottom elevations, σ, can be measured.

To illustrate the character of the coherent reflection, the relationship of the signals reflected at a rough surface and the roughness of the surface has been demonstrated in a laboratory water tank experiment. As shown on Fig. 2.1, the transducer transmitted a short 200 kHz ping and the signal reflected from the water surface was received at the same position. The echoes were used to intensity modulate the oscilloscope. The sweep was delayed and expanded to display the phases of the reflected signals, i.e., white is plus phase and dark is minus phase. The echoes were photographed by moving the film in the camera very slowly in order to give a record like a standard graphic recorder. The 2 seconds of record show about 200 transmissions.

The surface was roughened by blowing wind over it. The duration of the ping was very short compared to the wave period, 0.1 sec, so that each sound transmission gives a "snapshot" of the surface. The effect of increasing roughness of the surface is very apparent. The phases of the ensemble of signals were jumbled at large roughness and averages of the instantaneous signals at constant arrival times (for example 537 μs, 543 μs, etc.) are nearly zero at $k\sigma = 1.08$. In addition to having incoherent phases from transmission to transmission, the signal lasts longer and has an irregular tail.

The elevations of the wind blown surface had a nearly Gaussian probability density function (PDF) and the coherent component of the reflection fitted A Eq. (19) in the Appendix quite well. (Abbreviations like "A Eq. 19" will be used to designate equations in the Appendix.) The coherent component of the signal is

$$\langle p\rangle/p_o = \exp[-2k^2\sigma^2\cos^2\theta] \qquad (2.1)$$

where $\langle p\rangle$ is the coherently reflected signal, p_o is the signal reflected from a smooth surface, k is the wave number $2\pi/\lambda$, θ is

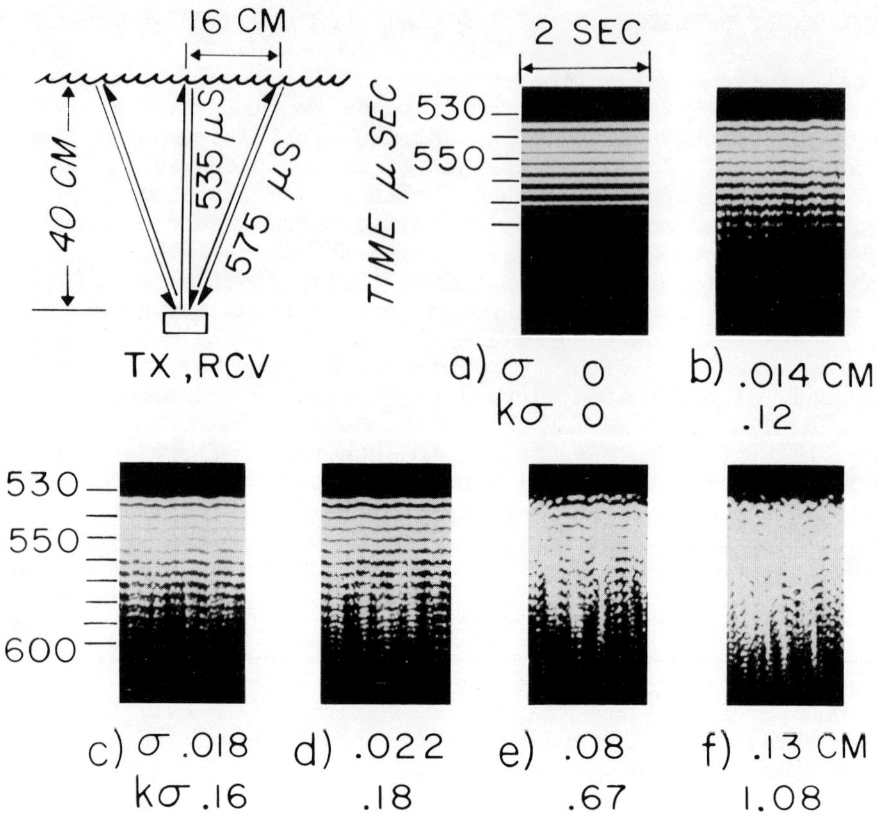

FIGURE 2.1

Photographs showing the time expansion of pings scattered at surface having different roughness. The scope sweep was delayed and expanded to show the individual phases: positive phase is light and negative phase is dark. Echo characteristics for smooth to slightly rough surfaces are represented by (a), (b), (c), and (d); moderately rough surfaces by (e); and rough surfaces by (f) (Leong, et al., 1971).

the incident angle, k cos $\theta = \gamma$ in the Appendix, and σ is the rms roughness of the surface. A graph of Eq. (2.1) is shown on Fig. 2.2. As the coherent component decreases, the incoherent component increases. The shaded area is intended to be a qualitative indication of the presence of the scattered components. Actual calculations of the scattered signals are given in the next section.

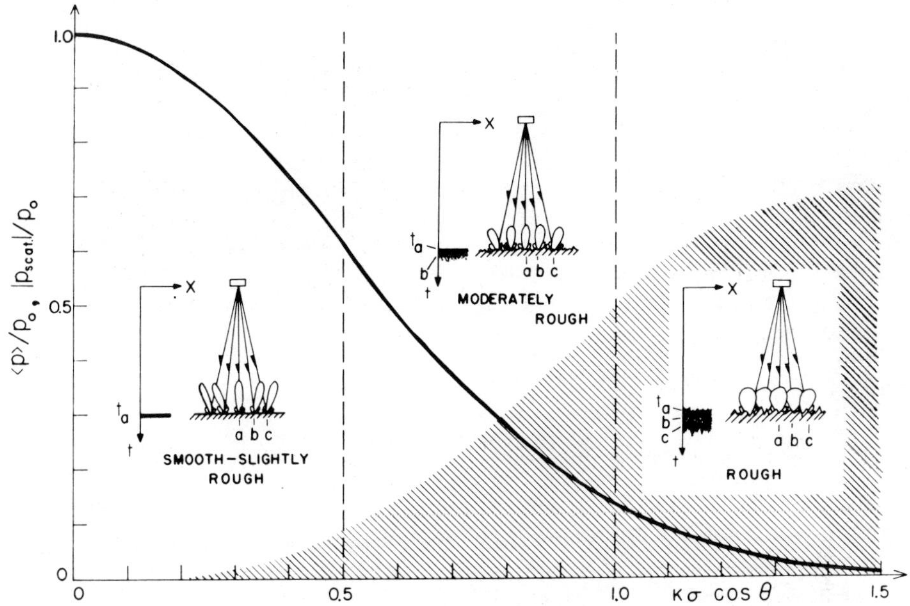

FIGURE 2.2

The expected mean signal as a function of $k\sigma \cos\theta$. The solid curve, $\langle p \rangle / p_0$, is from Eq. 2.1 and the shaded area shows a crude estimate of the incoherently scattered signals, $\langle p^2 \text{ scattered} \rangle^{1/2} / p_0$ from rough surfaces. The three sketches show the echo soundings and ray paths from bottoms having different roughnesses. The beam patterns at a, b, and c indicate the directivity of the scattered signals. For the smooth to slightly rough bottom, sound at a reflects back to the transducer (shown by a small rectangle), while paths at b and c reflect in other directions. For the moderately rough bottom, sound is scattered back from a and b to the transducer. For the rough bottom the scattering pattern is still broader and sound is scattered to the transducer from a, b, and c.

By analogy with the data shown on Fig. 2.1, one would expect echo sounding records over bottoms having different roughness to be like the sketches inset in Fig. 2.2. The increase of the duration of the echo sounding signal is caused by the increased travel time for signals incoherently scattered at b and c. The quality of the reflection trace can be used to estimate upper and lower limits for σ. The coherent component is less than 0.5 for $\sigma > 0.6$, and we use this as an estimation limit. For a smooth thin trace on the 12 kHz and 3.5 kHz echo sounders it means that σ is less than 0.01 m and 0.06 m respectively. At seismic profiling frequencies, 25 to 100 Hz, often the phases of the signal are displayed, and it is easy to determine when they become incoherent as roughness increases (like Fig. 2.1).

For example, we use data taken over the same area with three systems, Fig. 2.3. The smooth trace on the seismic profile indicates σ is less than 2 m (upper signal frequency is about 100 Hz). The scatter in the 3.5 kHz data indicates σ is greater than 0.06 m in the center of the record. At the far left of the 12 kHz data, the trace becomes very narrow and here σ is estimated to be less than 0.01 m. On the seismic data, the subbottom feature which comes up to the right shows that the signal from that interface is incoherently scattered and σ is estimated to be greater than 2 m.

Most marine seismic reflection measurements require that the bottom be smooth enough so that $k\sigma$ is much less than one. A close inspection of wide angle seismic profiles will reveal that this is the normal quality of data required to make interval velocity determinations.

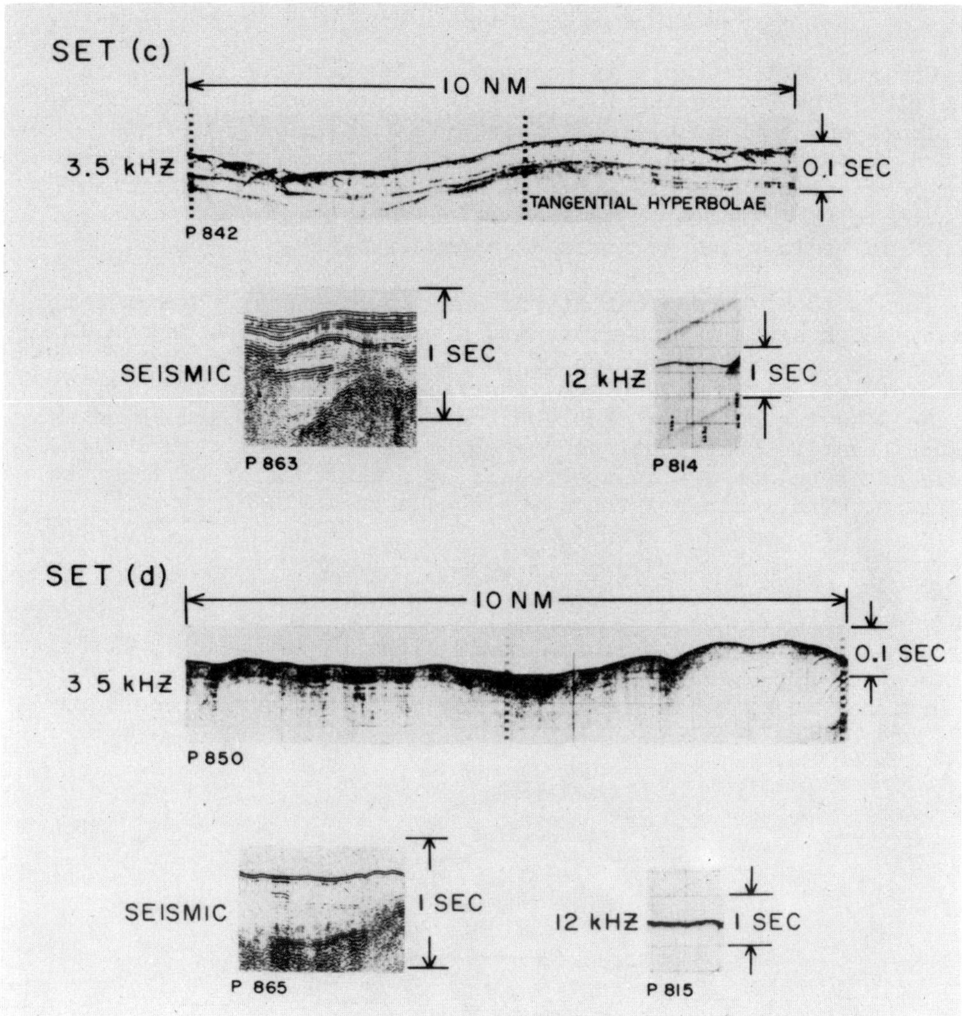

FIGURE 2.3

Bottom and sub-bottom profiles selected from the Kane 9, 1968 cruise. Profiles within each set were taken from the same rough area. The change of bottom roughness is apparent from the 12 kHz profile. All travel times are two-way.

3. SOUND SCATTERED AT THE SEA FLOOR

Our scattering theory is based on an application of Huygens principle: The scattering interface is illuminated by the incident wave front and each subelement of area is a source of wavelets. The travel time, or phase, of the signal scattered at a subelement is proportional to the distance from the source to the subelement and then to the receiver. The analytical calculation of scattered signals is tedious and is given in Appendix I.

The incident wave front at the sea floor is nearly spherical in many geophysical measurements and this places the phenonoma in the Fresnel diffraction region of physical optics. The phase zones for reflection at a plane interface are shown on Fig. (3.1). The zones having positive contributions to the signal are white and the negative contributions are shaded. For simplicity a point source and receiver are on the z-axis at a distance of R_1 from the x-y-plane. The radius of the first zone a is given by

$$a = (\lambda R_1/2)^{1/2} \tag{3.1}$$

where λ is the wavelength of the sound signal. In the theoretical treatment of Appendix I, it was convenient to define a quantity proportional to a, x_f in A Eq. 14. For both source and receiver at R_1 and on the z-axis,

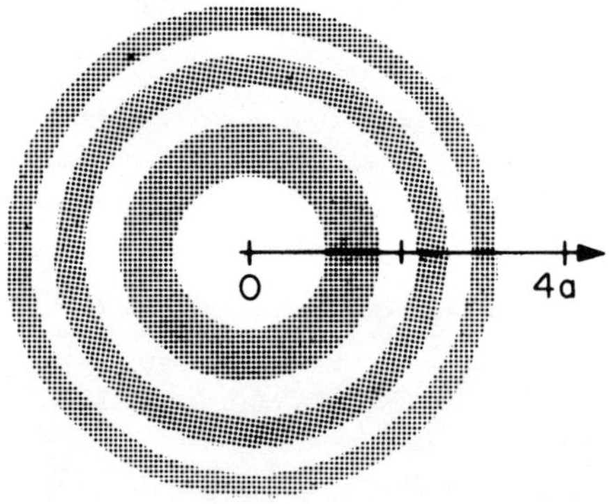

FIGURE 3.1

Phase zones for reflection. The white areas are, relative to the origin, the + phases of the contributions to the signal. Shaded areas are the - phases. The radius of the first zone, a, is given by Eq. (3.1).

$$x_f \equiv (R_1/k)^{1/2}, \qquad k = 2\pi/\lambda \tag{3.2}$$

$$x_f^2 = a^2/\pi \tag{3.3}$$

The Fresnel zone consideration is important because if the bottom is flat over the first zone, then the scattered signal will be large, regardless of how rough the bottom is elsewhere.

The effect of a rough surface is to cause the phases of the contributions within a single zone to differ. If we have n components and they are in phase, then the signal p is proportional to n, or $\langle p^2 \rangle$ is proportional to n^2. For large random phase differences, the components add as signal squared; thus, $\langle p^2 \rangle$ is proportional to n.

For visual examples, a map view of several rough surfaces are sketched as having two elevations, $\lambda/8$ unshaded and $-\lambda/8$ shaded, in Fig. 3.2. The relative path difference for a path from source to receiver is $\lambda/2$, or + phase and - phase of the signal. The Fresnel zones are overlayed on the roughness so that the combination of the phase of the Fresnel zone and roughness zone is apparent. White areas are the combination of + Fresnel zone and + roughness phase and - Fresnel zone and - roughness phases. Shaded areas correspond to - times + and + times -. Examples of large scale roughness and small scale roughness are shown. For the large scale roughness, the dimensions of the plus and minus phases are about those of the first Fresnel zone; hence, many components of the scattered signal would have the same phase. Let us assume that there are n wavelet sources in the first Fresnel zone. For this drawing, Fig. (3.2), most of the sources would have the same phase, and the maximum of $\langle p^2 \rangle$ approaches proportional to n^2. When the dimensions of the roughness are much smaller than the first Fresnel zone, the components add as the sums of squares, and $\langle p^2 \rangle$ is proportional to n. For both cases, the contributions to $\langle p^2 \rangle$ from the outer zones would probably increase as n.

Often, the scattered signal $\langle p^2 \rangle$ is assumed to be proportional to the insonified area. From the figure it is evident that the dimensions of the roughness must be much less than the dimension of the first Fresnel zone for this to be a good approximation. For the other case, one would expect $\langle p^2 \rangle$ to have a different dependence upon the insonified area. Correspondingly, the scattering cross section would depend upon the insonified area. The wind blown waves for the sound scattering data (Fig. 2.1) had correlation distances which were about the same as the first Fresnel zone (Clay and Sandness, 1971). The scattering cross section is shown as a function of the beamwidth in Fig. 3.3. The correlation function of the surface elevations is also shown on Fig. 3.3.

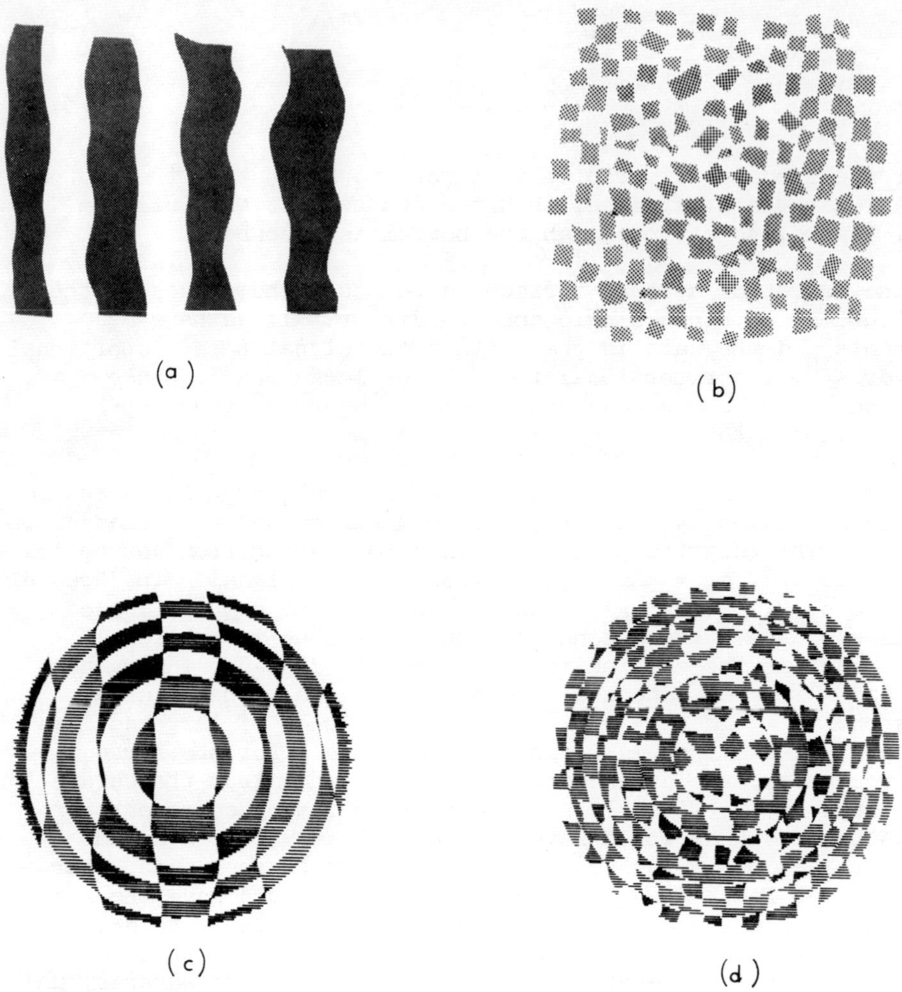

FIGURE 3.2

Combination of simplified rough bottoms and the phases of Fresnel zones.

a) Roughness having spatial dimensions about the diameter of the first Fresnel zone.
b) Roughness having spatial dimensions much less than the diameter of the first Fresnel zone.
c and d) Combinations of roughness and Fresnel zone.

White is plus phase and shaded is the minus phase.

The large scale roughness of Fig. 3.2 was drawn to represent this experiment.

We can gain more insight concerning the interrelationships of roughness, illuminated area, signal frequency, and water depth by using simplified expressions for the scattering function. Often, echo sounding data are taken near vertical incidence and a single transducer is used for transmission and reception. Also, the beamwidths are the same in the x and y directions. In addition, numerical evaluations of A Eq. (54) for many cases have shown that the upper limits of u_{11} and v_{11} are negligible compared to the lower limits and can be ignored. The sub areas beyond the first, S_{12}, S_{22}, etc., are also negligible. For the 1,1 sub area and dropping subscripts

$$S \simeq \frac{k^2 \mathcal{R}^2 F^2}{32} A_x A_y uv \tag{3.4}$$

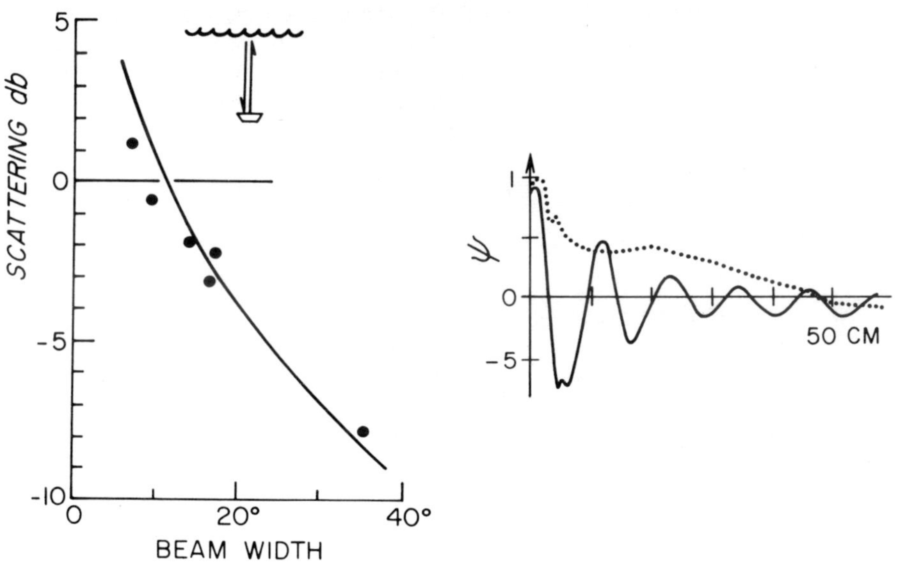

FIGURE 3.3

Scattering function. The data were taken at 200 kHz at depths of 50 cm. σ = 0.14 cm, $k\sigma$ = 1.1. The geometry and signals are shown on Fig. 2.1. The dotted line is the cross wind correlation function and the solid line is the up-down wind correlation function.

and for the vertical incidence and symmetry mentioned above

$$A_x = A_y \qquad u = v \quad ,$$

$$\theta_1 = \theta_2 = 0, \qquad F = \cos\theta = 1, \qquad k = \gamma \tag{3.5}$$

$$\alpha = \beta = 0$$

$$S \simeq \frac{k^2 \mathcal{R}^2}{32} A^2 u^2$$

We have to assume a statistical description of the sea floor to proceed. Without having data to support the assumption, the elevations of the sea floor are assumed to have a Gaussian PDF. Note that the Fourier transformation of the coherent reflection gives the PDF of the surface (A Eqs. (17) and (18)). The surface is assumed to have a rms roughness σ and a correlation function ψ in the insonified area. Direct measurement of ψ basically requires that the elevations of the bottom be measured at extremely high spatial resolution. Luyendyk (1969) reported the spectrum analysis of data taken by means of the Marine Physical Laboratories deep towed fish. Although it is for a different ocean, we have transformed the data to obtain the spatial correlation function shown on Fig. (3.4). ψ is empirical, and the simplest fitting function such as shown on Fig. (3.4) is a linear approximation:

$$\psi = 1 - |\xi|/L_1 - |\eta|/L_1 \qquad 0 \leqq |\xi| \leqq L_1$$

$$0 \leqq |\eta| \leqq L_1 \tag{3.6}$$

$$\psi = 0 \text{ otherwise}$$

where ξ and η are the spatial displacements for the correlation measurements. At large depths x_f^4/X^4 is much less than one and can be ignored. Thus A_x, B_x, and s in A Eqs. (55-58) and (45) are the following:

$$A = 2^{1/2}\, x_f^2/X$$

$$B = 4k^2\sigma^2/L_1 \tag{3.7}$$

$$x_f^2 = R/k$$

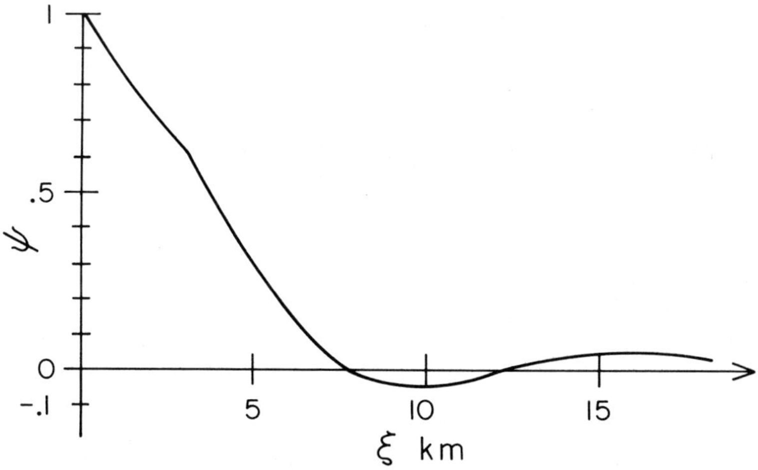

FIGURE 3.4

Spatial correlation function for an area on the sea floor. The correlation function was calculated from a spectrum of an area on the sea floor which was given by Luyendyk (1969). The area is at 125°44'W, 32°23'N and at 4400 m depth.

$$s = AB/2$$

$$s = 8^{1/2}\ k^2\sigma^2\ \frac{x_f}{X}\ \frac{x_f}{L_1} \tag{3.8}$$

$$s = 8^{1/2}\ k\sigma^2 R(XL_1)^{-1} \tag{3.9}$$

The algebraic approximation for u, A Eq. (60) is

$$u \leq \frac{2\pi^{-1/2}}{s+(s^2+4/\pi)^{1/2}} \tag{3.10}$$

The substitution of A into Eq. (3.5) gives

$$S \simeq \frac{\mathcal{R}^2}{16}\ \frac{R^2}{X^2}\ u^2 \tag{3.11}$$

where the beamwidth $\Delta\varphi \equiv X/R$. Numerical evaluation of S as a function of s or the parameters in Eq. (3.9) is basically subjected to the restriction that $4\ k^2\sigma^2 > 1$ because the outer scattering areas were ignored. The graphs shown on Fig. 3.5 were based upon a 3.5 kHz echo sounder. The dependence of S upon σ and the correlation distance L_1 is shown. At 3 km depth, the radius of the first Fresnel zone is about 25 m. The limiting values of S can be obtained by noting that

$$\lim_{s \gg 1} u = \pi^{-1/2}/s \tag{3.12}$$

$$\lim_{s \to 0} u = 1 \tag{3.13}$$

Correspondingly, the limiting values of S are

$$\frac{\mathcal{R}^2 L_1^{\ 2}}{128\pi k^2 \sigma^4} < S < \frac{\mathcal{R}^2 R^2}{16X^2} \tag{3.14}$$

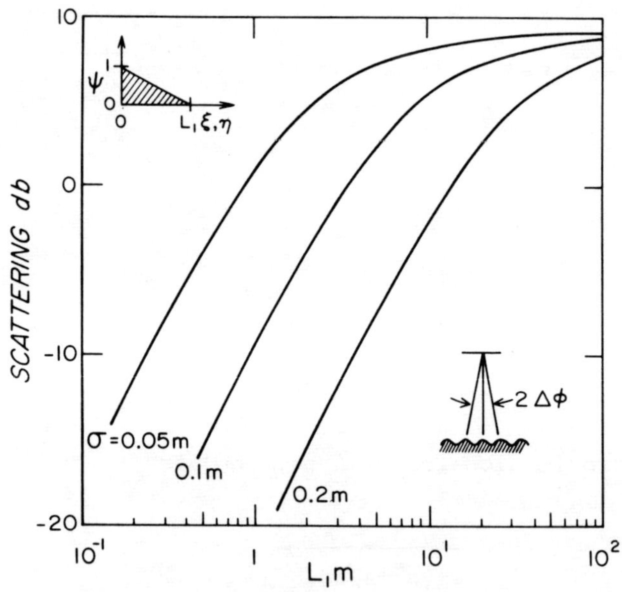

FIGURE 3.5

Scattering function for 3.5 kHz echo sounder. Half beamwidth $\Delta\varphi = 5°$. For convenience, the reflection $\mathcal{R} = 1$. $k = 14.7\ m^{-1}$; $k\sigma = 0.735$, $\sigma = 0.05$ m; $k\sigma = 1.47$, $\sigma = 0.1$ m; $k\sigma = 2.94$, $\sigma = 0.2$ m.

For small L_1 and large $k\sigma$, S tends to be independent of the dimensions of the illuminated area. This is the incident plane wave approximation. Large L_1 and small s gives the upper limit and S depends upon the illuminated area. The average signal $\langle p^2 \rangle$ has the following limits (obtained with the aid of Eq. (1.2)).

$$\langle p^2 \rangle = \frac{P_{so}^2 R_o^2}{R_1^2 R_2^2} AS \tag{3.15}$$

where $A \equiv 4XY$

$$\frac{P_{so}^2 R_o^2 \mathcal{R}^2 A}{32\pi R^4 k^2 \sigma^4} < \langle p^2 \rangle < \frac{P_{so}^2 R_o^2 \mathcal{R}^2}{4R^2} \tag{3.16}$$

The message in Eqs. (3.14) and (3.16) is that, depending upon σ and ψ for the sea floor, both S and $\langle p^2 \rangle$ can have a dependence upon the beamwidth of the measuring system A and the geometry of the measurement.

4. SPECTRAL AND TOPOGRAPHIC CHARACTER OF THE SEA FLOOR

The spectrum estimates to be presented in this section were made by hand analysis. The echo soundings and seismic profiles were taken from the sea surface. The roughness of the bottom was estimated in the following wave length Λ intervals:

ΛI < 0.2 km: roughness estimate σ
ΛII 0.2 km $< \Lambda < 2$ km: relief r
ΛIII 2 km $< \Lambda < 12$ km: relief r
ΛIV 12 km $< \Lambda < 60$ km: relief r

The estimates of roughness, σ or r, are for each wave length interval within a given region. The results are presented as spectra for each region.

Interval ΛI includes features that are not resolved by the standard echo sounders on surface ships. At oceanic depths of about 3-4 km, the width of the illumination area subtended by a 30° beam is about 2 km. Typically, these features include rock outcrops, erratics, small vertical faults, and sediment waves (Heezen and Hollister, 1971). Although none of these features can be profiled by conventional echo sounding techniques, the returned signal from a microtopographically rough bottom is one of prolonged echo. (Examples of smooth and rough bottom echo traces are shown on Fig. 4.1). The technique described in Section 2 was used to estimate σ, the rms roughness. If narrow sonar beam and short ping length signals were used, regions of prolonged echoes may give rise to small hyperbolas (a hyperbolic echo trace consists of signals reflected from a common point reflector: echoes from the sides of the pointed feature are not recorded). Some of these features are too big to be displayed by deep sea photographs, and a deep towed instrument such as that described by Lonsdale and Spiess, 1973, is probably the best way to study them.

Interval ΛII contains bottom features which have wave lengths less than 2 km and relief less than 6 m (the measurable relief of the features is denoted by r, so as to distinguish it from σ which is estimated from the character of the echoes). Although the presence of these features in the Interval ΛII wavelength range is discernible on the echograms, their shapes are not.

Interval ΛIII includes sea floor features which are readily observable on conventional echograms and have topographic wave lengths greater than the width of an area illuminated by an echo sounder. These bottom features have wave lengths between 2 and 12 km and relief less than 200 m.

Interval ΛIV contains large-scale features having wave lengths between 12 and 60 km and relief up to 1 km.

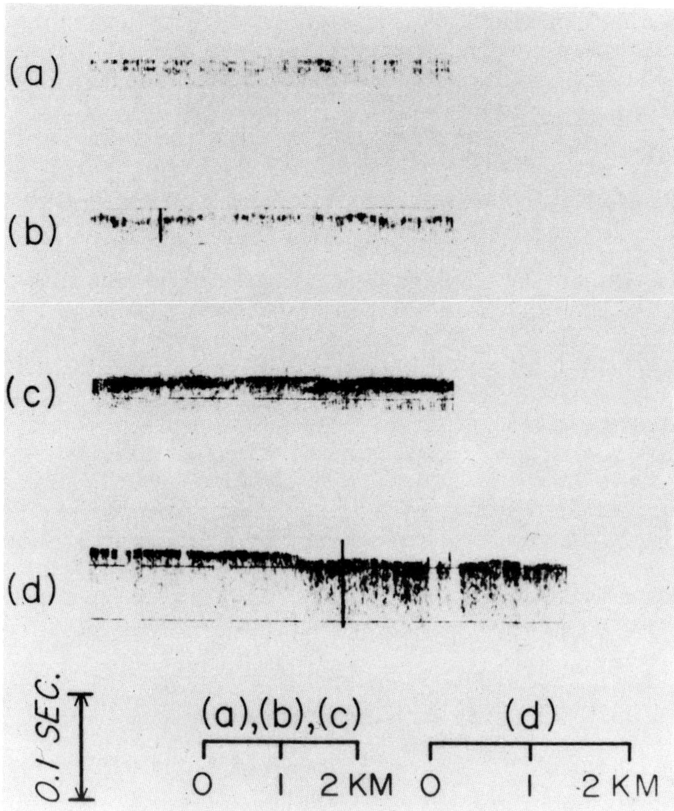

FIGURE 4.1

Echograms showing types of smooth and rough microrelief. Record (a) was taken from a smooth bottom (34.6°N, 11.2°W), as indicated by the presence of coherent sub-bottom reflections; record (b) was taken from a rough bottom (33.5°N, 11.6°W), $r<6$ m; record (c) was taken from a rough bottom and possibly sub-bottom (37.2°N, 12.5°W), $\sigma>0.03$ m; and record (d) was taken from a smooth and rough bottom (32.4°N, 12.9°W), $\sigma>0.03$ m, $r<6$ m (U.S. Naval Oceanographic Office records).

The work is based mainly on an analysis of the echo soundings and seismic profiles (12 kHz, 3.5 kHz, and 50-500 Hz), and bottom sediment data collected from an area southwest of the Iberian Peninsula (Figs. 4.2 and 4.3). The tracks and physiographic provinces are shown on Fig. 4.2. Most of the data were furnished by the U.S. Naval Oceanographic Office (NAVOCEANO) and the Lamont-Doherty Geological Observatory (L-DGO). The quality of the sonic records and the accuracy of navigation range from poor to good. The bathymetric chart, Fig. 4.3, is based upon charts furnished by Dr. A. Laughton of the National Institute of Oceanography of Great Britain.

Since most of the input data are graphic, the results are also graphic. Description of the essentially visual material is kept as brief as possible. Published sources are referenced in Appendix II.

The results of the spectral analysis of the bathymetric and geophysical data are shown on Fig. 4.4. The area has been subdivided into physiographic provinces in which the character of the data were judged to be about the same. The bar graphs indicate the spectrum estimates within each wave length interval. The lower letters indicate a visual impression of the small-scale topography. In the mountainous regions, the g type of topography, the traces were covered with hyperbolas, and small features in intervals ΛI and ΛII could not be interpreted.

Five charts (Figs. 4.5 - 4.9) have been constructed to depict the topographic character of the sea floor, beginning with ΛIV, the largest relief and long wave length features (Fig. 4.5), and ending with the ΛI and ΛII, the superimposed microrelief and short wave length features (Fig. 4.9). Echograms showing the small-scale features, ΛIII, of Fig. 4.5 are given in Fig. 4.10.

For the most part, the construction of the topographic character maps is self-explanatory (see individual figure captions). Where data are lacking, the roughness boundaries were drawn to conform with the regional contours and bottom slopes. Where bottom character changes rapidly, the roughness anomalies were drawn as small patches along the tracks rather than to extrapolate the data far beyond the track limits.

In areas where large hyperbolas predominate, rough microtopography could not be detected on the echo sounding records which were taken from the surface. An effort was made to identify microrelief in areas where the bottom echoes form wave-like echo traces. Some of these echoes could have come from sharp reflectors such as sediment waves and vertical faults.

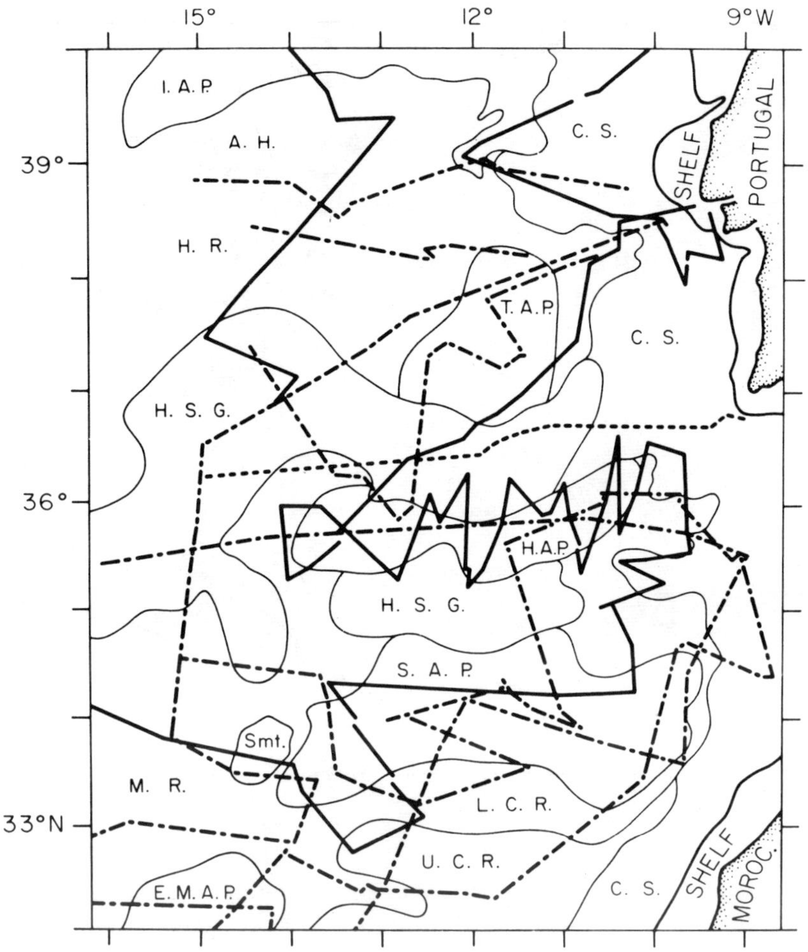

FIGURE 4.2

Location map showing ships tracks and physiographic provinces. Solid lines are tracks from Vema 27 cruise, 1969 (L-DGO); long and short dashed lines are tracks from the Atlantic Seal and Arctic Seal, 1965-1966 (U.S. Naval Oceanographic Office); and short dashed line is a track from the Gibbs, 1966 (OP240, Brakl, 1968). Province boundaries were modified after Plate 4, SP95-5-5 (U.S. Naval Oceanographic Office, 1967a). Symbols used: A.H., abyssal hills; C.S., continental slope; E.M.A.P., East Madeira abyssal plain; H.A.P., Horseshoe abyssal plain; H.R., Horseshoe rise; H.S.G., Horseshoe seamount group; I.A.P., Iberian abyssal plain; L.C.R., lower continental rise; M.R., Madeira rise; S.A.P., Seine abyssal plain; Smt., seamounts; T.A.P., Tagus abyssal plain; and U.C.R., upper continental rise.

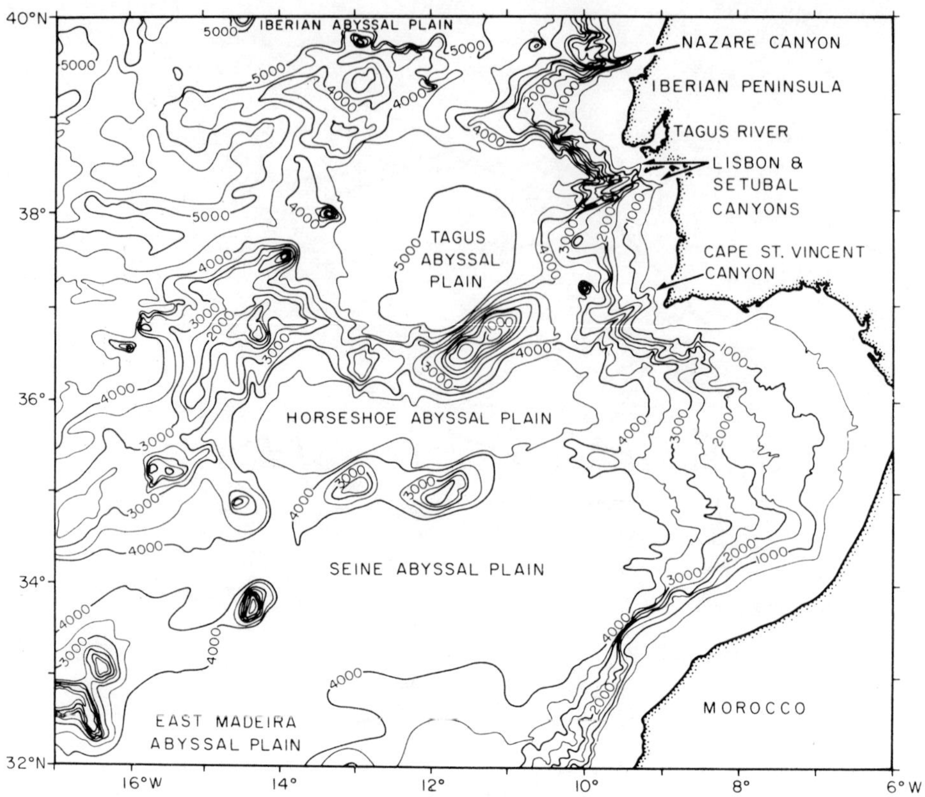

FIGURE 4.3

Bathymetric map. The contours have been redrawn in meters from unpublished bathymetric maps furnished by the British National Institute of Oceanography (courtesy of Dr. A. Laughton). The track density of the original charts is much greater than that of Fig. 4.2.

The five topographic character charts should be examined concurrently. Transparencies may be made of the figures and then overlaid in different combinations for best viewing of the bottom character of the sea floor. For example, Fig. 4.6 contains a qualitative description of the bottom, but it needs Figs. 4.7 and 4.8 to give quantitative meanings to the bottom features. A combined viewing of Figs. 4.9 and 4.10 would show the types of microtopography, as interpreted from the field records, which can ride on different bottom characters of the sea floor. Relationships between microtopography and physiographic provinces or between microtopography and bathymetry can also be sought.

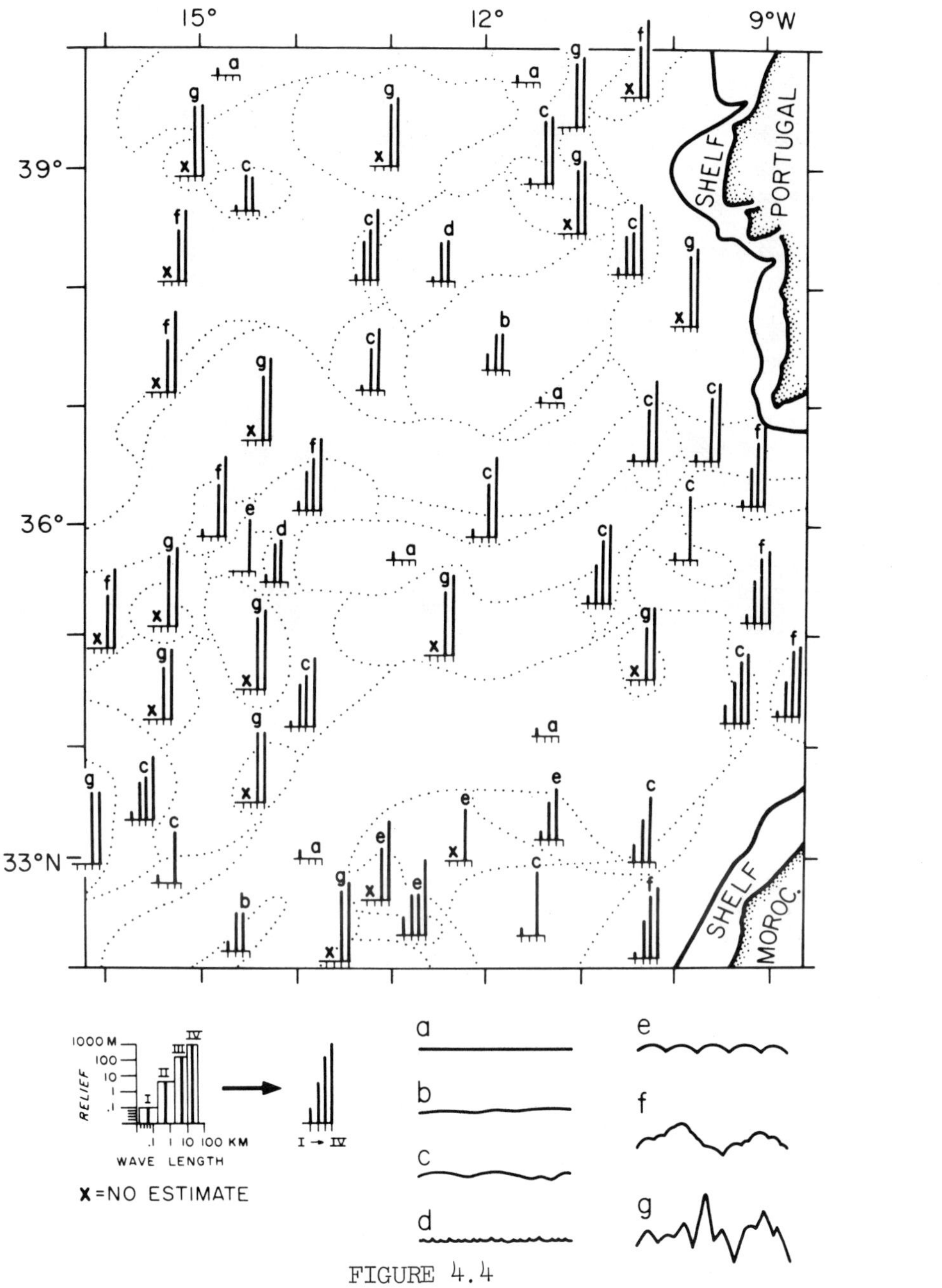

FIGURE 4.4

Spectral analysis. This chart has combined and simplified the information shown on Figs. 4.5 - 4.9. Each histogram shows the dominant or the roughest topography and microtopography within an area enclosed by dots. Explanations of keys a-g are given in Fig. 4.6 caption.

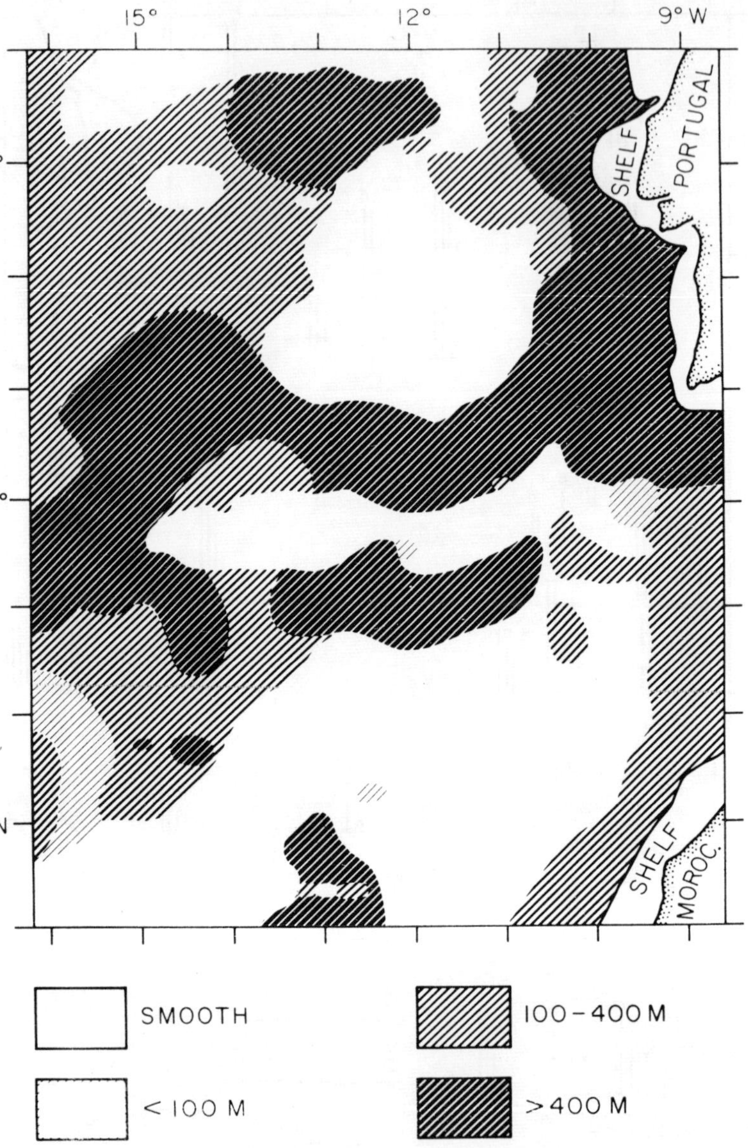

FIGURE 4.5

Large-scale relief. These features have wave lengths in the range of 12-60 km. The wave lengths were measured between peaks which stand higher than intervening peaks.

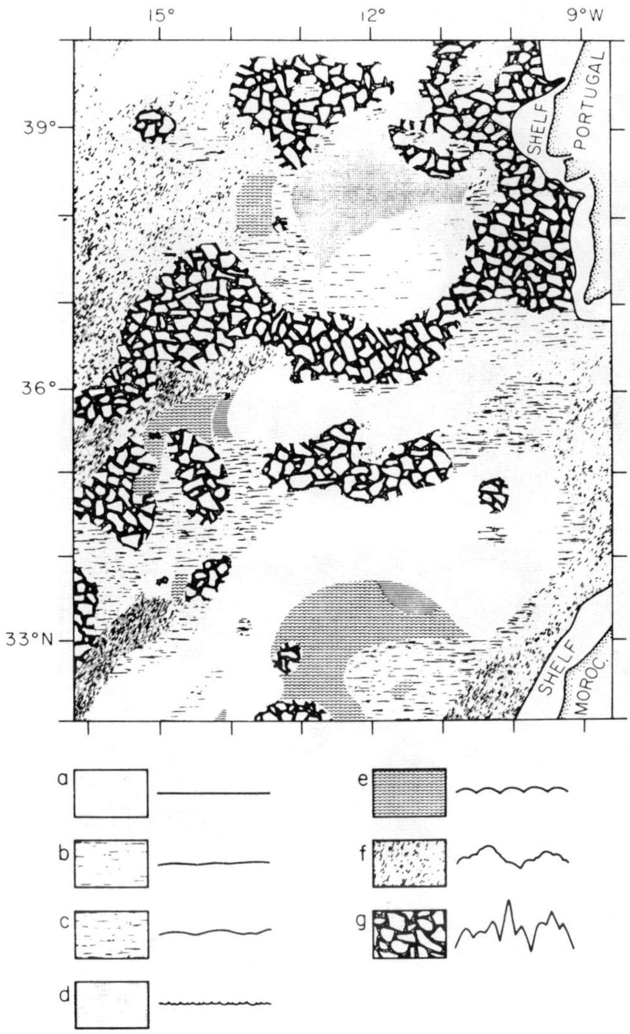

FIGURE 4.6

Small-scale topographic characters of the sea floor. These bottom characters give a visual impression of the shape of the small-scale features as seen on the echograms. (a) denotes very smooth surface; (b) represents a smooth surface but slightly wavy; (c) includes gently undulating to rolling surface and large benched surface; (d) shows a surface having many very small irregularities; (e) consists of a surface having small, regular wave-like features; (f) is a surface having irregular, wave-like features (hummocky); and (g) consists of large to intermediate, sharp to somewhat rounded hyperbolas.

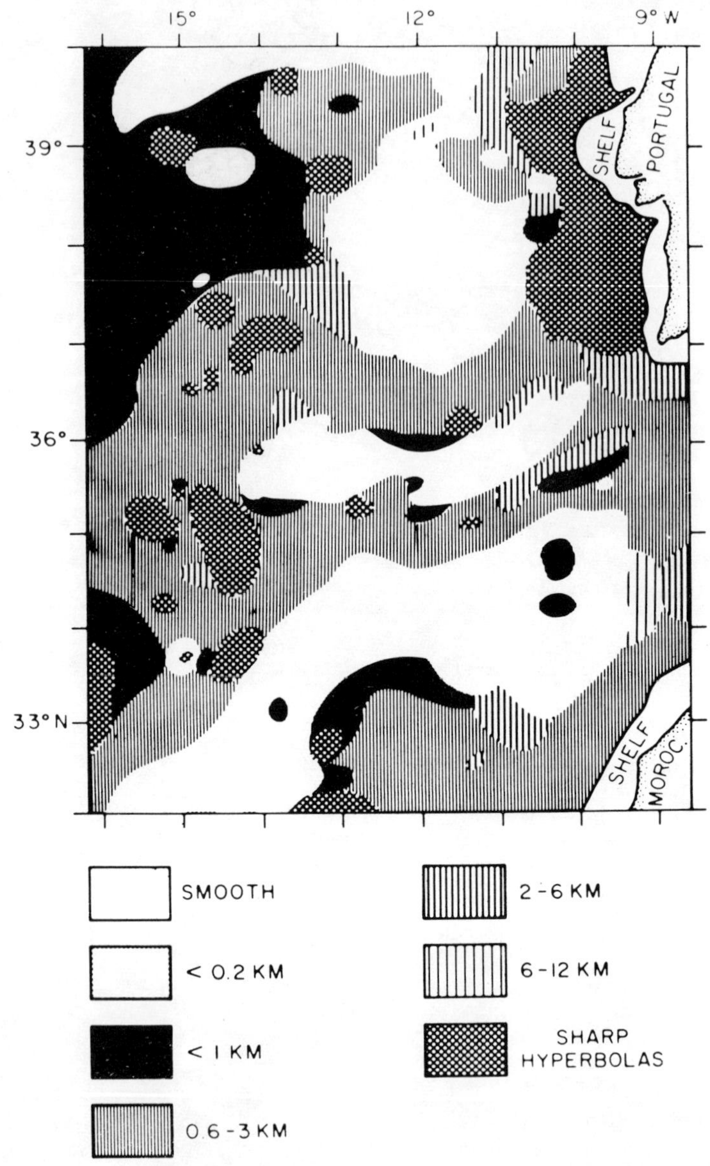

FIGURE 4.7

Topographic wavelengths for small-scale features.

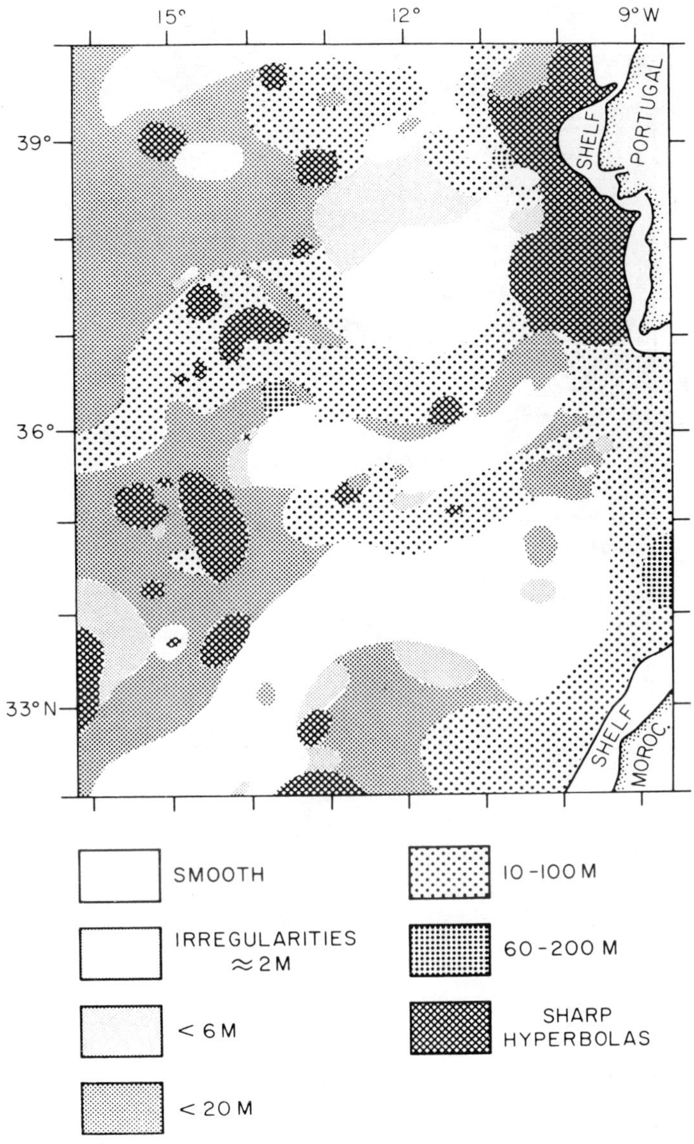

FIGURE 4.8

Relief for small-scale features.

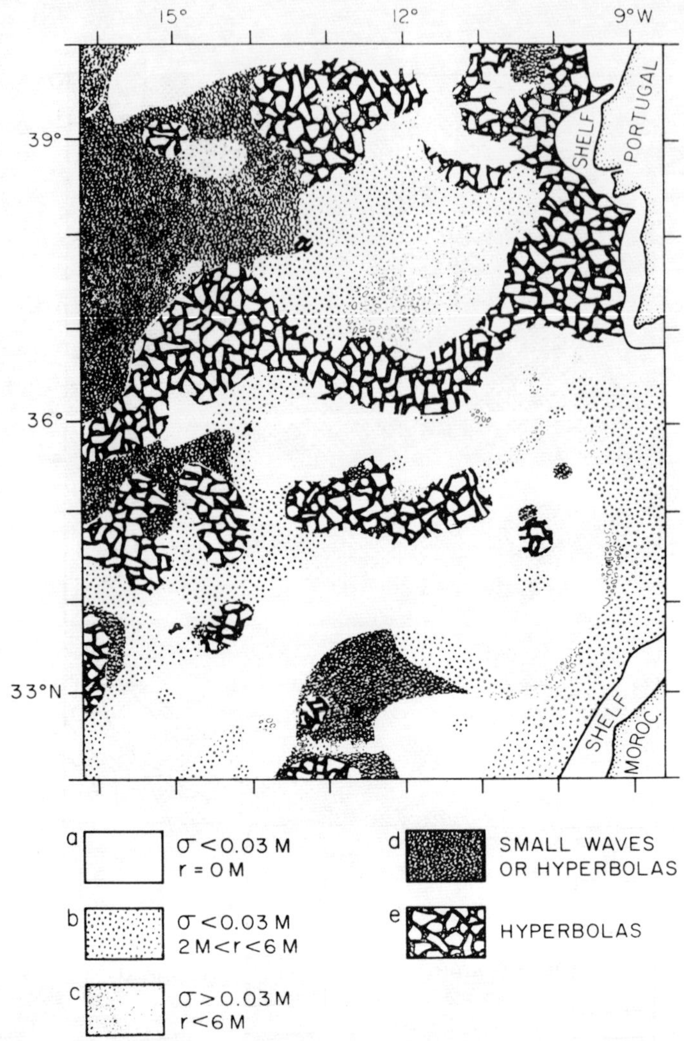

FIGURE 4.9

Microrelief. The presence and magnitude of microrelief were estimated from echo characters seen on the echograms and 3.5 kHz records. In (a), (b), and (c), σ was estimated from the phase coherency and prolongation of the echoes and r was measured between adjacent irregularities along a jagged echo trace; (d) includes small, regular waves and irregular waves and small hyperbolas, some of which may be tangential to the bottom echoes; and (e) embraces the large to intermediate, sharp to somewhat rounded hyperbolas.

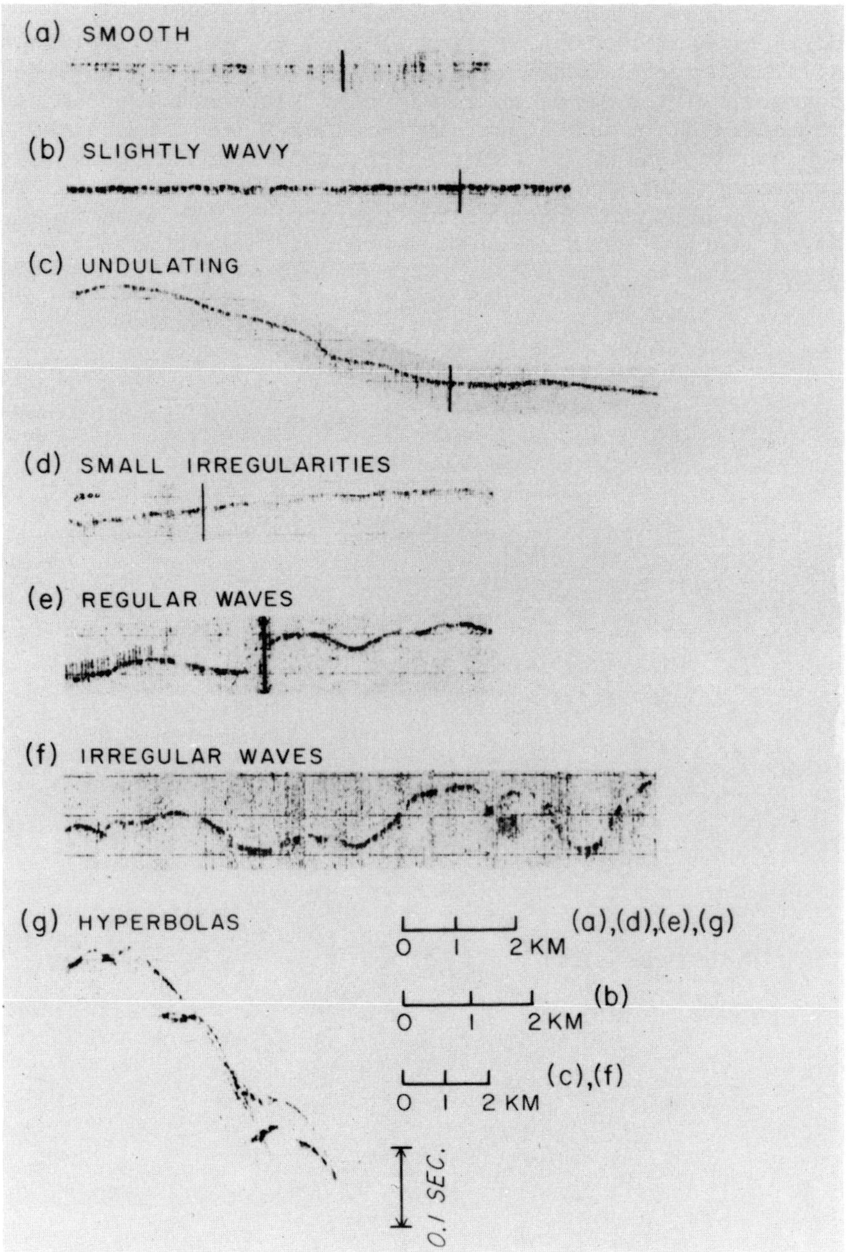

FIGURE 4.10

Echograms showing the small-scale topographic characters of the sea floor of Fig. 4.6 (see Fig. 4.6 caption for explanations) (U.S. Naval Oceanographic Office records).

At the subprovince level a threefold classification of topography, based on regional relief (Fig. 4.5), can be applied to the area. The Moroccan continental rise and abyssal plains have smooth topography; the Moroccan continental slope and the oceanic rise have moderately rough topography (relief < 400 m); and the Portuguese continental slope and the Horseshoe seamount group have rough topography (relief > 400 m). The following description of the small-scale topography and microtopography of the area is discussed in light of this large-scale relief background.

5. GEOLOGICAL RELATIONSHIP TO ROUGHNESS SPECTRA

The main emphasis of the following discussion is to relate the topography and roughness spectra of the abyssal plains (Tagus, Horseshoe, Seine, and East Madeira) and the Moroccan continental rise to the geological processes of the region. We hope these results can be used to predict the spectra of bottom roughness in geologically similar areas.

Throughout this study, care was taken to maintain a clear distinction between deductions made from field data and synthesis of these deductions. We have found that the occurrence of rough microtopography is not restricted to a particular topographic type nor to a particular physiographic province. In a regional study such as this, we can only speculate on the nature of the roughness features and the mechanics of the bottom roughening processes. Conclusive evidence on their existence must await more bottom photographs and current measurements. This study, however, has provided a body of preliminary information and ideas for future inquiry about the geology of the area.

Both tectonic and sedimentary processes can cause a sea floor to have roughness of widely ranging scales related to folds and faults, to slump features, and to bed forms sculptured by deep sea currents (turbidity currents and contour following currents). Modern marine sediments exhibit quasi-liquid behavior and can form beds of different geometry depending mainly on current velocity and particle size and specific gravity of the sediment (excellent references about turbidity currents and their deposits are given in Ewing et al., 1970).

It is well for us to bear in mind that conclusions drawn from remote sensing records, such as the sonic profiles, need direct measurements or observations to substantiate their validity. However, inferential evidence such as the lithological properties of a sediment can give clues to the hydrodynamic conditions at the time of sedimentation. To this end, Fig. 5.1 was constructed to show the distribution of the deep sea sediments. In preparing this figure, we tended to emphasize the coarse-fraction of a sediment sample because gross structures formed by a tractive mode of transportation almost always involved coarse-grained sediments. Turbidities were recorded on Fig. 5.1 wherever turbidity units as defined by Bouma (1962) were present in a core and the units were not necessarily at the top of the core column (assumed to be the present sea floor).

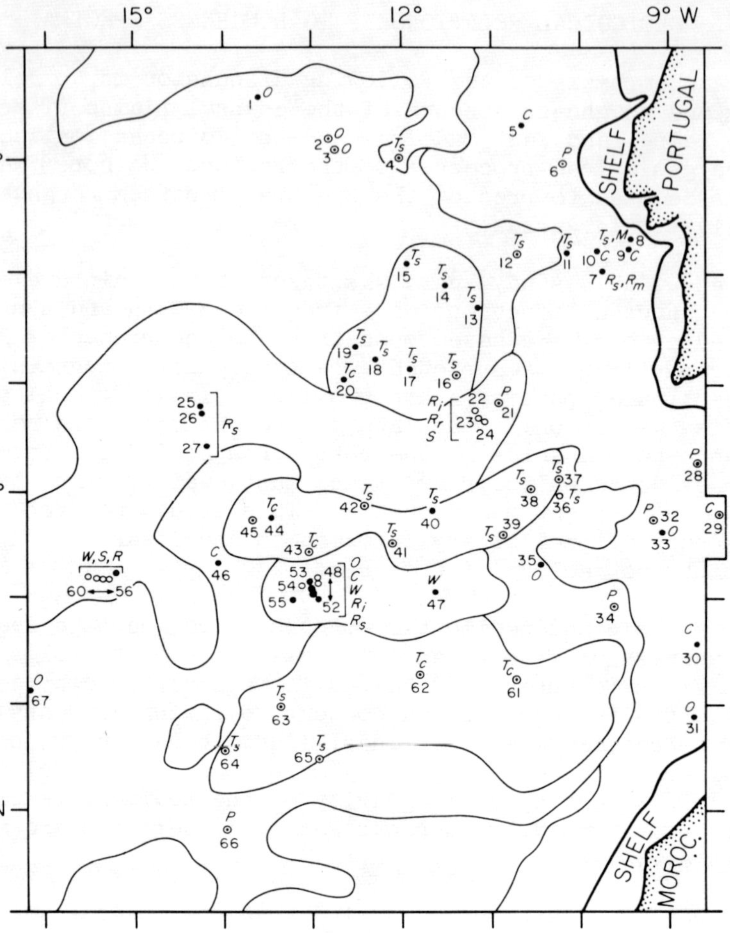

FIGURE 5.1

Location map showing core, dredge, bottom photograph, and nepheloid stations (shown by a dot and number) and the major bottom sediment types for each station. Symbols used: a solid dot, •, indicates station without photographs; ⊙, photographs showing a tranquil, smooth bottom having animal mounds and tracks; ○, photographs showing a rough bottom having ripple and scour marks, windrows, rock fragments and outcrops, and small vertical faults; O, oozes, foraminifera and coccoliths; C, clays; P, normal pelagic sediments of oozes and clays, as interpreted from photographs; W, winnowed sediments, coarse and moderately to well sorted; M, slump debris, coarse and unsorted; T_s, turbidites containing sand and silt units which are moderately developed; T_c, turbidites containing mostly fine-grained sediments, clays; S, mineral sands; R_s, rock fragments, sedimentary, limestones and shales; R_r, reef debris, corals; R_i, igneous, basic porphyry; and R_m, metamorphic, marble (see Appendix for source of data; the classification schemes for deep sea sediments are those of Ericson et al., (1961) and L-DGO Core Laboratory unpublished manuals).

Moroccan Continental Rise

The most striking topographic expressions of the Moroccan continental rise are the small, regular wave-like features. Most of the small-scale features have relief less than 20 m and wave lengths between 0.6 and 3 km. These features may be related, directly or indirectly, to tectonism. The occurrence of large earthquakes can set off a chain of geological processes which can roughen a sea floor, beginning with slumping and warping of the sediments to the rushing of turbidity currents onto the abyssal plains.

Down-slope movements of sediments, in the form of slumps and slides, and faulting may have deformed the lower rise into the present wave-like topography. A correlation of the fault traces (Fig. 5.2) with the small, regular wave-like features at B and E is possible. The sediment block at B can be attributed to down-slope decollement. It appears to have been moved from D as a coherent body.

Figure 5.2 shows rough microtopography at A and C. We suggest that the microrelief consists of sediment and rock debris eroded from adjacent inclines.

The gently undulating topography of the upper Moroccan continental rise appears not to have been affected by large-scale movements of sediments or by gravity faulting.

Tagus Abyssal Plain

The Tagus abyssal plain has a rough bottom in intervals ΛI and ΛII, as interpreted from the texturally rough bottom echo traces (Fig. 4.1c). Prolongation of echo time is not pronounced.

We suggest that fast flowing turbidity currents were responsible for the rough microtopography. We further suggest that sediments needed to generate high velocity turbidity currents were furnished by the Tagus river alluvium and the continental terrace deposits (Fig. 5.3). Duplaix et al. (1965) have correlated the coarse-fraction of the Tagus turbidities with the sands of the Tagus river. The Lisbon and Setubal canyons provided the direct route of sediment transport.

Horseshoe and Seine Abyssal Plains

The Horseshoe and Seine abyssal plains have acoustically smooth bottoms. These abyssal plains reflect a coherent bottom echo trace and a shallow (10 ms, 7.5 m) subbottom echo trace (Figs. 4.1a, 4.10a, and 5.4).

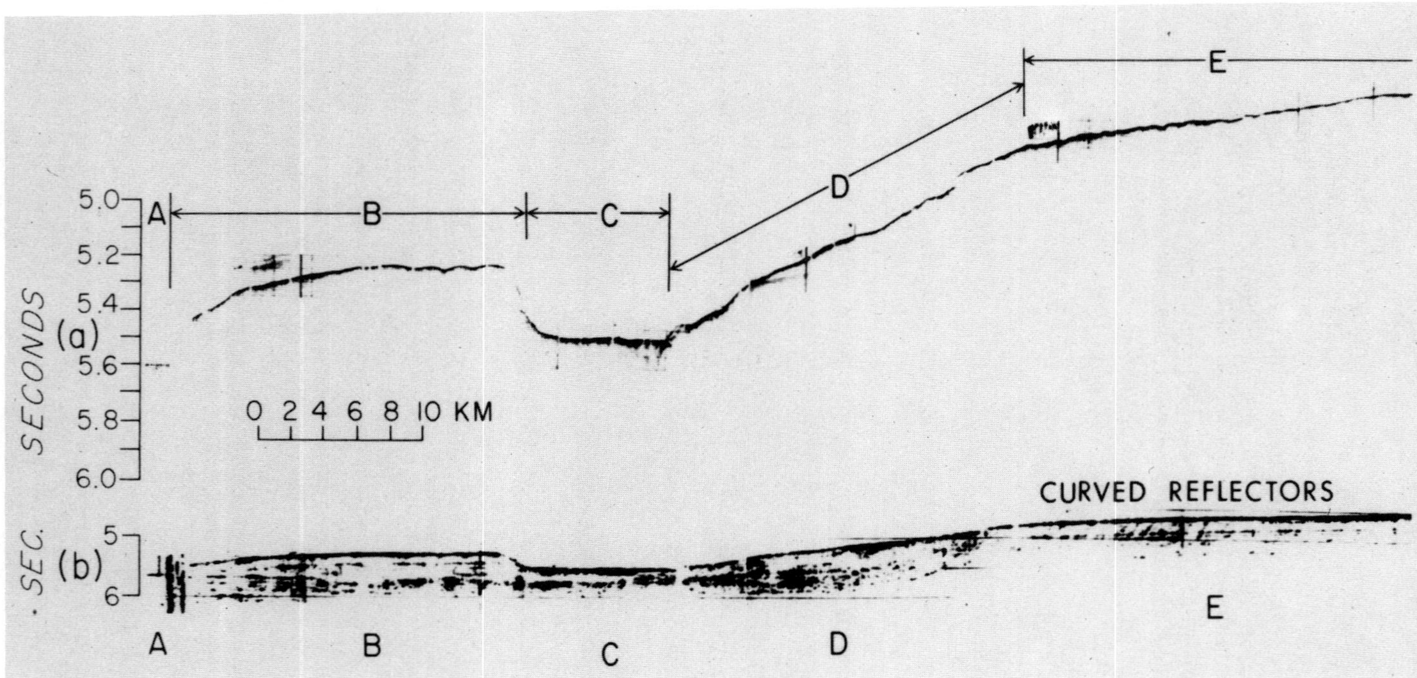

FIGURE 5.2

Records showing the structural-topographic features of the Moroccan continental rise. Echogram (a) and reflection profile (b) were taken from the same track and centered at about 32.3°N, 12.5°W. The wave-like bottom features shown on (a) can be interpreted as a result of slumping and small-scale faulting of the sediments. On record (b) the curved reflectors at E may be the result of gravity faulting. Sediment block at B also contains curved reflectors (not obvious at this scale of reproduction). (U.S. Naval Oceanographic Office records).

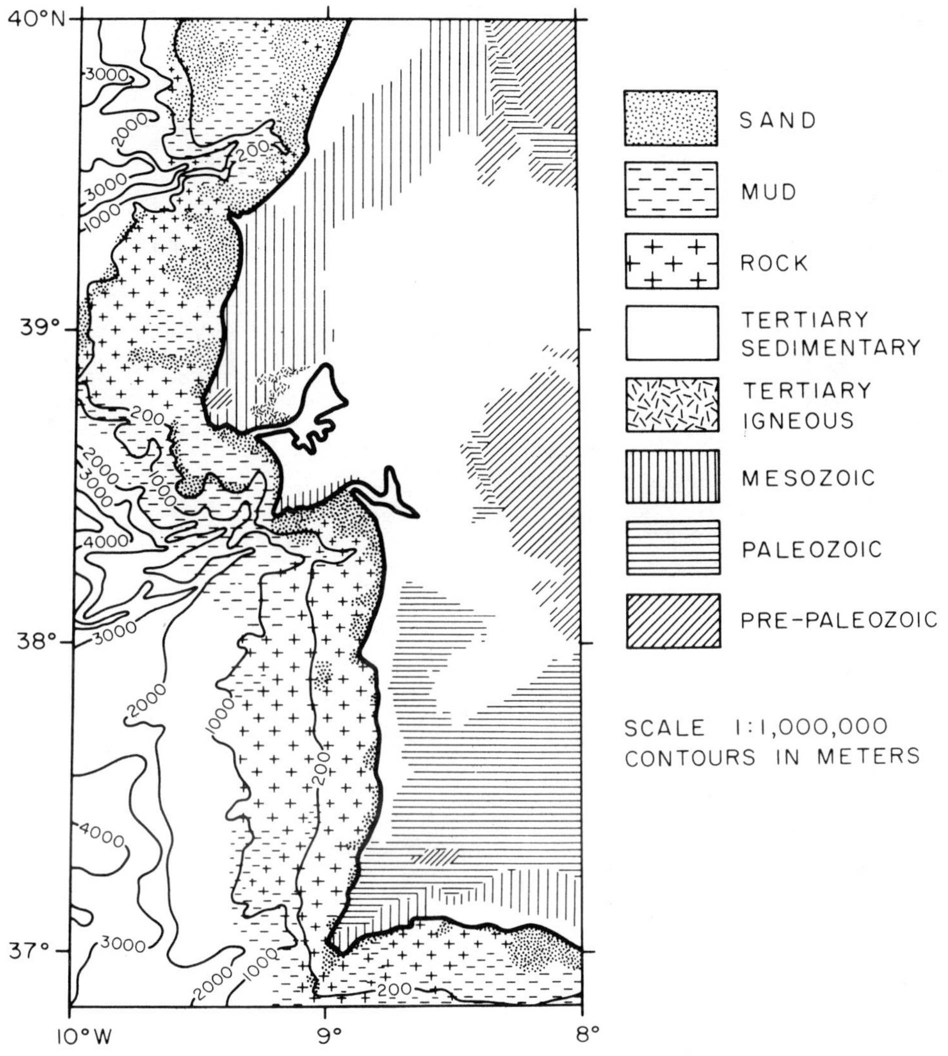

FIGURE 5.3

A broad outline of the geology and offshore bottom sediments of the western Iberia. Depths are given in meters. This outline was simplified after the Geological Map of Portugal, 1952, and Geological Map of Portugal, 1969; both maps were published by the Geological Survey of Portugal.

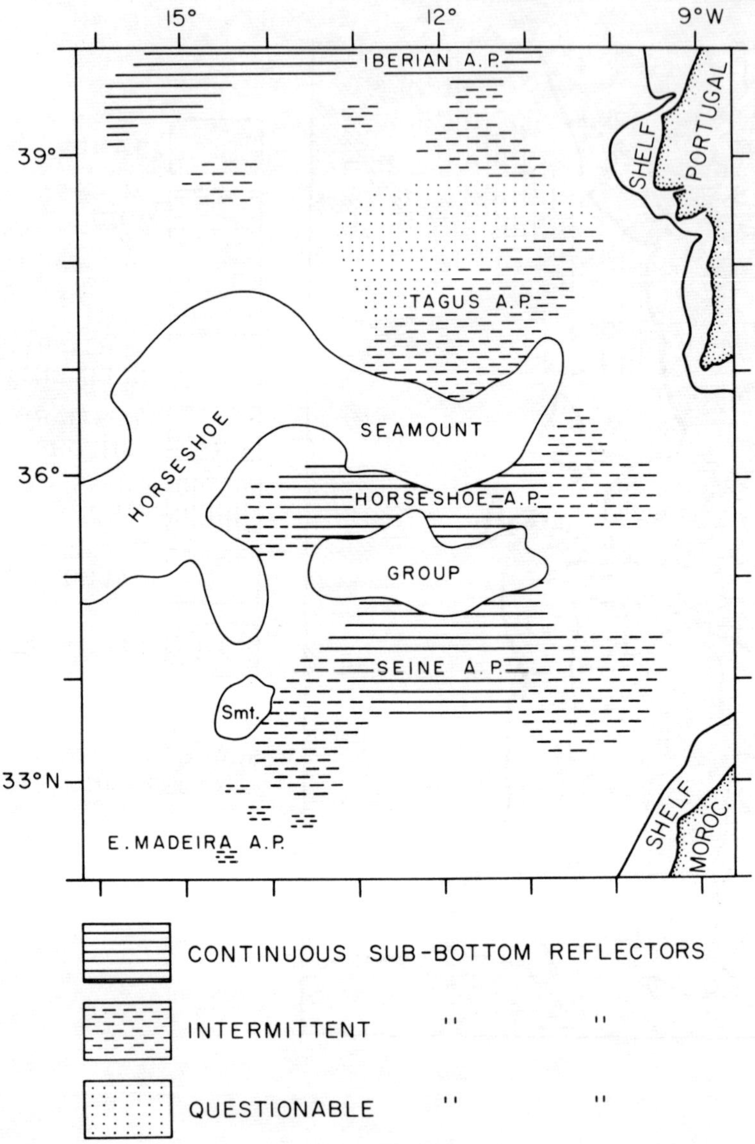

FIGURE 5.4

Map showing the distribution of shallow (< 10 ms) subbottom reflectors. Well defined subbottom reflections are associated with smooth bottom.

Although the Horseshoe abyssal plain is smooth, its eastern end is bordered by few patches of rough microtopography. Bottom photographs taken near the southeastern margin of the plain by LePichon et al. (1971, Figs. 5.3 and 5.4) show a small fault having a 1.2 m vertical offset along a fresh fault scarp. This suggests that the bordering microrelief may have a fault origin. The western end of the plain is also bordered by rough microtopography. There are small hyperbolas whose crests are tangent to the bottom echoes (Vema 27, 3.5 kHz profiles, L-DGO). Beneath them is another set of hyperbolas. These tangential hyperbolas probably are acoustic images of small vertical faults rather than sediment waves.

The Horseshoe abyssal plain, like the Tagus abyssal plain, is fed by a canyon, filled with turbidites, and bordered by seamounts. But, unlike the Tagus abyssal plain, it is a very smooth plain. Why? The answer may lie in the type and the amount of source sediment deposited at the upper continental slope where the sediment supply is poor (Fig. 5.3). Turbidity currents generated in this region probably are smaller and weaker and thus less competent to roughen the sea floor than those of the Tagus basin.

East Madeira Abyssal Plain

Rough microtopography ($r > 2m$, Fig. 5.4) occurs in two small areas on the East Madeira abyssal plain. A tectonic origin may be invoked to account for the rough microtopography at the eastern edge of the plain, which is bordered by rugged topography. The rough microtopography on the central portion of the plain is depicted on the echograms as a subdued, wavy surface. Laughton (1962) suggested that there were sediment waves (interpreted from the characters of the echoes) on the nearby Madeira abyssal plain, which is southwest of the East Madeira abyssal plain. It is possible that the central strip of microrelief of the East Madeira abyssal plain may also consist of small sediment waves.

Additional Research

The Tagus abyssal plain is microtopographically rough, as interpreted from the echo character it exhibits. Large-scale turbidity currents have been suggested as a plausible bottom roughening mechanism. Future investigations should concentrate on revealing the nature of the roughness features and their distribution relative to the Lisbon and Setubal canyons.

We have also suggested that tectonism plays an important role in forming the present small-scale and micro-scale topographics. This suggestion needs further testing. The lower Moroccan continental rise would be a choice site for a closely spaced, high resolution profiling and sediment sampling survey. Several geological points of interest are posted below for future investigation:

1. The extent of slump and slide masses and the prograding nature of the rise at the expense of the Seine abyssal plain.
2. The possibility of the rise being a temporary sediment reservoir for the turbidities of the Seine abyssal plain, which is canyonless.
3. The importance of tectonism, for example, gravity faults, on the formation of the small-scale and micro-scale topographies of the bottom and subbottom strata.
4. The existence of channelized deep sea currents.

These items are a natural outgrowth of the present study, although they may have less than a direct bearing on the formation of a rough sea floor topography.

6. SYNTHESIS OF MARINE GEOPHYSICS, GEOLOGY, AND SOUND SCATTERING THEORY

The sonar system designer may wish to estimate the bottom loss for his system in an area. The geophysicist may wish to use quantitative echo soundings to improve his estimates of the bottom. We have discussed the roughness of the bottom, how it affects the coherence of reflected signals, and the use of the coherence to estimate the roughness. Estimation of the scattered sound is more difficult because it depends upon more parameters. To illustrate how the bottom parameters and the sonar system interact, we will use the geological data and scattering theory to estimate the bottom reflectivity and "bottom loss." Since we don't have enough information to do it very well, we regard this as a numerical exercise. We have three reasons for doing it: first, an example of how to put geology and sonar together is needed; second, the attempt will display the kind of information which is needed; and third, to suggest that by reversing the steps, the scattering functions S and the bottom losses can be used to estimate the sea floor parameters.

Numerous assumptions will be needed as we proceed. To avoid confusing the reader and giving the impression that we know more than we do, assumptions, extrapolations, and pure guesses will be set apart and identified by '. . .'.

The bottom loss is measured in the following experiment: A sonar system is used to measure the mean square signal $\langle p^2 \rangle$ for a reflection path at a particular location, frequency, and beamwidth. Second, this result is compared to the theoretical signal that would be obtained if the bottom were a perfect reflector p_o^2. Thus, bottom loss is

$$\text{bottom loss} \equiv -10 \log (\langle p^2 \rangle / p_o^2) \tag{6.1}$$

$\langle p^2 \rangle$ is, Eq. A28,

$$\langle p^2 \rangle = \frac{P_o^2 R_o^2}{R_1^2} \frac{A}{R_2^2} S \tag{6.2}$$

$$A = 4XY \tag{6.3}$$

The theoretical signal p_o^2 is

$$\langle p_o^2 \rangle = P_o^2 R_o^2 / (R_1 + R_2)^2 \tag{6.4}$$

$$\langle p^2 \rangle / p_o^{\,2} = \frac{A(R_1+R_2)^2}{R_1^{\,2} R_2^{\,2}} \, S \tag{6.5}$$

To proceed, we assume the operating characteristics of the sonar equipment.

Assumption A. Sonar system

'3.5 kHz echo sounders are routinely used in many oceanographic surveys and a 3.5 kHz carrier frequency is used in this section. The theory in Appendix I gives $\langle p^2 \rangle$. This can be measured by transmitting a long ping and then using the middle portion for the average when all of the scattered signals are present. Alternatively, the integral of p^2 can be measured for short ping transmissions.

The sonar illuminates an area on the bottom and the amplitude of the incident pressure signal on the bottom is, Eq. A6,

$$D_o = \exp[-(x^2/X^2) - (y^2/Y^2)] \tag{6.6}$$

where x and y are measured from the maximum of the sonar beam. To convert X and Y to beamwidths, $\Delta\phi$ and $\Delta\chi$, we let

$$\begin{aligned} (X/R_1) \cos \theta_1 &= \sin \Delta\phi \\ (Y/R_1) &= \sin \Delta\chi \end{aligned} \tag{6.7}$$

The sonar is assumed to have the same beamwidths in the x and y directions, and $\sin \Delta\phi = \sin \Delta\chi$.'

Assumption B. The Reflection Path

'The bottom loss measurement is made at vertical incidence

$$\begin{aligned} R_1 &= R_2 \\ \theta_1 &= \theta_2 = 0 \end{aligned} \tag{6.8}$$

At 3.5 kHz, subbottom reflections are received. In this area, the analysis is applied to the first few meters depth of the sea floor. The assumptions about the sea floor sediments are given later.'

The substitution of the assumptions into Eq. 6.5 yields

$$\langle p^2\rangle/p_o^{\,2} = 16 \sin^2 \Delta\phi \; S \qquad (6.9)$$

Most of the Appendix is devoted to the calculation of S. Knowledge of the bottom and its topography are needed to continue.

Assumption C. The Typography of the Sea Floor

'We restrict ourselves to the non-mountainous physiographic provinces in the area. To decide the portion of the roughness spectrum which affects the measurement, we use the dimension of the Fresnel zone x_f, Eq. 3.2. At 4 km depth, x_f is about 17 m. Features having wave lengths of the order of 30 m (i.e. $\sim 2x_f$) and less are important in the computation of S. From Section 4, these features are in the wave length interval ΛI and shown on Figs. 4.9 and 4.10.

In the smooth areas (on Fig. 4.9, a and b) we assume that small ripples and other irregularities may be on the bottom (Laughton, 1963). We assume σ is about 2 cm in the smooth areas. The relief is larger in the c class of microrelief and here σ is estimated to be 10 cm. Areas having small waves or hyperbolas (d) are more difficult because there may be a large number of 30 m dimensioned patches which are large enough to be locally smooth. Without experimental data to verify it, we assume that σ is 2 cm in these areas.

In addition to σ, the correlation function, ψ, and the correlation distance are needed. Heezen and Hollister, in "Face of the Deep" (Fig. 9.13, 1971), have plotted the relief of features such as ripples and sand waves versus the distance between crests. They do not show the correlation function or spectral analysis of the data. By extrapolating Luyendyk's data, Fig. 3.4, to a different ocean, we assume that ψ, Eq. 3.6, is approximately a linear function. For ridge to ridge spacing Λ, we assume that the first null, L_1, is roughly $\Lambda/4$. The relief of a sinusoidal feature r and the rms σ are proportional as follows:

$$r \cong 8^{1/2}\,\sigma \qquad (6.10)$$

It is presumptious to use these approximations, but since we are making only crude estimates, we proceed. From Heezen and Hollister's Fig. 9.13 (1971), a straight line passed through their data yields the following equation for σ and L_1:

$$\sigma \cong L_1^{\,0.8}/15 \text{ m} \qquad (6.11)$$

$$L_1 \cong 30\sigma^{1.25} \text{ m} \qquad (6.12)$$

Sand waves have longer ridges than the crest to crest distances and the correlation function along the ridge direction would be longer than L_1. Even so, we assume L_1 represents an average value and let ψ be the same in both directions.

$$\begin{aligned} \psi &= 1 - \xi/L_1 - \eta/L_1 \\ &= 0 \quad \xi > L_1,\ \eta > L_1 \end{aligned} \tag{6.13}$$

Because it is important, we reiterate, we <u>do not have direct measurements of</u> σ, L_1, <u>or</u> ψ <u>in this area.</u>'

On the basis of Assumption C, we use Eqs. 3.9 to 3.11 to calculate S for the linear correlation function. The substitution of $\Delta\phi$ for X/R gives

$$s = 8^{1/2} k\, \sigma^2 (L_1 \sin \Delta\phi)^{-1} \tag{6.14}$$

$$u \leqq 2\pi^{-1/2} \left[s + (s^2 + 4/\pi)^{1/2} \right]^{-1} \tag{6.15}$$

$$S \cong \frac{\mathcal{R}^2}{16} \frac{u^2}{(\sin \Delta\phi)^2} \tag{6.16}$$

and the bounds of S are

$$\left.\frac{\mathcal{R}^2 L_1^2}{12\ 8\pi\ k^2 \sigma^4}\right|_{s \gg 1} < S < \left.\frac{\mathcal{R}^2}{16\ (\sin \Delta\phi)^2}\right|_{s \to 0} \tag{6.17}$$

In addition to k, σ, and L_1, it is obvious that S depends upon the beamwidth of the sonar. Computations of S are shown on Fig. 6.1. The reflection coefficient is one for simplicity. Equation 6.12 is used, so that the choice of σ also gives L_1. For reference, the limiting case for an incident plane wave is shown. The dependence of S upon the beamwidth is substantial and it is obvious that scattering measurements must be corrected for beamwidth if they are to be applied to another system.

The calculation of $\langle p^2 \rangle / p_o^2$ follows directly from the calculation of S and is

$$\langle p^2 \rangle / p_o^2 = \mathcal{R}^2 u^2 \tag{6.18}$$

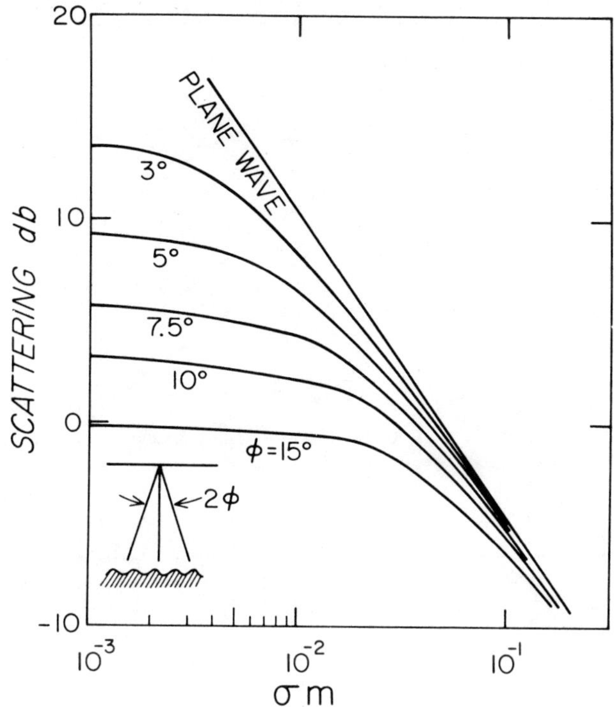

FIGURE 6.1

Scattering function. The signal frequency is 3.5 kHz. The scattering function is shown as a function of σ for ψ having a linear correlation function, and $L_1 = 30\ \sigma^{1.25}$. The reflection coefficient at the bottom is unity for this calculation. The water depth is 3000 m.

The limits for large s and small s are

$$\frac{\mathcal{R}^2}{\pi}\ \frac{L_1^{\ 2}\ (\sin \Delta\phi)^2}{8k^2\sigma^4}\Bigg|_{s\gg1}\ \ \langle\ \ \langle p^2\rangle/p_o^{\ 2}\ \langle \mathcal{R}^2\Bigg|_{s=o} \tag{6.19}$$

The apparent reflectivity or "bottom loss" of the bottom depends upon the beamwidth of the measuring system in addition to the roughness of the bottom. Graphs of $\langle p^2\rangle/p_o^2$ are shown on Fig. 6.2. To display the effects of roughness, we have let the reflection coefficient be unity.

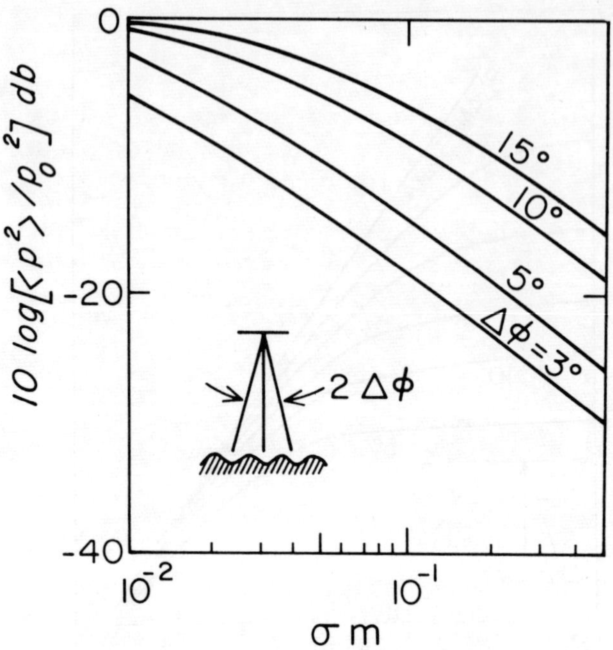

FIGURE 6.2

Apparent reflectivity at a rough bottom. $10 \log [\langle p^2\rangle/p_o 2]$ is shown as a function of rms roughness σ. The signal frequency is 3.5 kHz. The correlation function ψ is linear and $L_1 = 30\ \sigma^{1.25}$. The reflection coefficient is unity. The water depth is 3000 m.

Assumption D. The Reflection Coefficient

'The reflection coefficient depends upon the density, sound speed, and subsurface layers of the sediments. In the abyssal plains the first meter of sediment is predominantly silty clay. The remaining basinal sediment of the area is mostly clay. Cores taken in the abyssal plains have thin layers of sandy silt (or different % of sand and silt) interbedded in marl. We use Lamont-Doherty Geological Observatory sediment information and Hamilton's (1971) methods to estimate ρ_1 and c_1 at the water sediment interface. In this, the top layer is assumed to be lithologically homogeneous. Estimates of the reflection at the water-sediment interface are shown on Fig. 6.3.

The sandy silt layers within the marl are known to be good reflectors and are included. The effective reflection of a thin layer, medium (2), which has medium (1) above and medium (3)

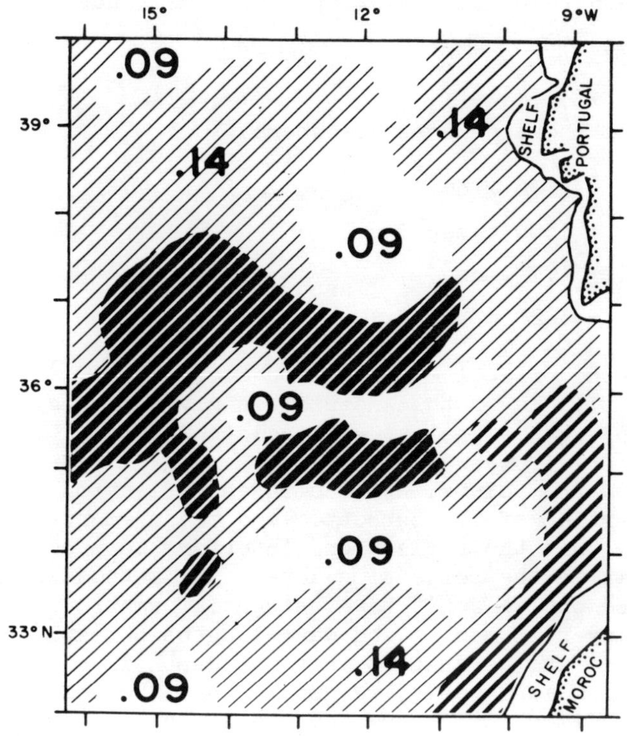

FIGURE 6.3

Estimates of reflection coefficient, R. Top of the water sediment interface where the top layer (surficial) is treated as an average homogeneous layer.

below, is, for vertical incidence,

$$\mathcal{R}_{23} = \frac{R_{23} + R_{34}\, e^{2i\gamma_3 h_3}}{1 + R_{23}\, R_{34}\, e^{2i\gamma_3 h_3}} \qquad (6.20)$$

$$R_{23} = \frac{\rho_3 c_3 - \rho_2 c_2}{\rho_3 c_3 + \rho_2 c_2} \qquad (6.21)$$

where R is the reflection coefficient at an interface shown by the subscripts, γ_3 is the wave number in medium 3, and h is the thickness of the layer of interest. We assume that the properties of the marl above and below the layer are the same; thus

$$R_{23} = -R_{34} \quad . \tag{6.22}$$

The properties of the sediments are where

$$\begin{aligned} \rho_2 &= 1.26 \text{ g cm}^{-3} \\ c_2 &= 1500 \text{ m sec}^{-1} \end{aligned} \quad \text{marl}$$

$$\begin{aligned} \rho_3 &= 1.7 \text{ g cm}^{-3} \\ c_3 &= 1620 \text{ m sec}^{-1} \end{aligned} \quad \text{sandy silt layer}$$

A curve showing the effective reflection coefficient, $\mathcal{R}_{23}$, for different sandy silt layer thicknesses, is given in Fig. 6.4. The curve is for a 3.5 kHz signal at vertical incidence.

Elsewhere in the volume, Akal compares reflection coefficient based upon properties of the subbottom and measurements. These comparisons show that the reflections interfere to produce sharp maxima and minima in the reflection data. The method is the proper one to use but we don't have enough information to use it. The sandy silt layer is many sound wave lengths beneath the water sediment interface and we suspect that the depth is quite variable. To make the problem manageable, we assume that the reflections from the water sediment interface and the sandy silt layer are incoherent and add as the sum of the squares. The effective reflection loss RL is

$$RL \cong -10 \log \left(R_{12}^{\;2} + \mathcal{R}_{23}^{\;2}\right) \tag{6.23}$$

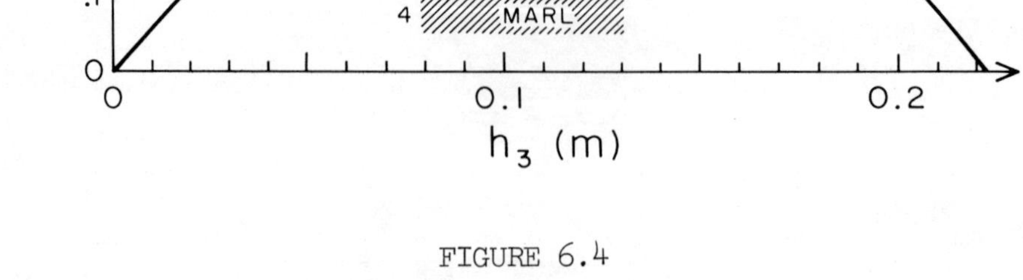

FIGURE 6.4

Reflection coefficient of a sandy silt layer within marl.

Figure 6.5 is a chart of the effective reflection loss in dB for the area, assuming an average of 10 cm thick sandy silt layer to lie within the top few meters of the Tagus abyssal plain sediment and an average of 5 cm thick sandy silt layer beneath the remaining abyssal plains of the area. The reflectivity of the abyssal plains is greatly influenced by the presence and thickness of a shallow subbottom sandy silt layer.'

The total bottom loss, BL in dB, is the sum of the reflection loss, RL, the bottom, Fig. 6.5, and the scattering loss, $-10 \log (\langle p^2 \rangle / p_o^2)$ for $\mathcal{R} = 1$.

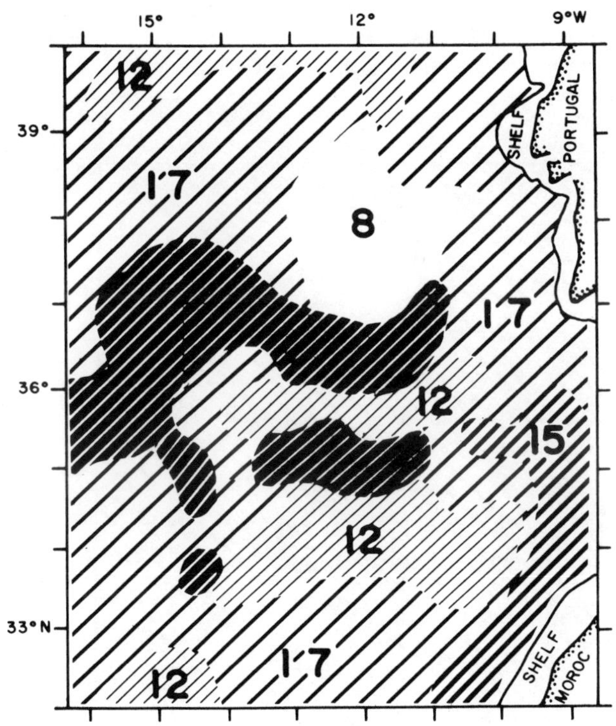

FIGURE 6.5

Average reflection loss in dB for the top few meters of sediment which includes a thin sandy silt layer interbedded in marl or silty clay sediments. An average of 10 cm thick sandy silt layer is estimated for the Tagus abyssal plain and a 5 cm thick sandy silt layer for the remaining abyssal plains of the area.

Assumption E. Sonar Transducer Beamwidth

'The beamwidth $\Delta\phi$ is 5°.'

The bottom loss chart for a vertical incidence 3.5 kHz sonar was constructed as follows: At 36.5°N and 12°W, the bottom is type b on Fig. 4.9. In Assumption C, σ is estimated to be 10 cm here. On Fig. 6.2, the scattering loss is about 15 dB at 5° beamwidth and $\sigma = 0.1$ m. The reflection loss, Fig. 6.5, is 8 dB. Thus, the bottom loss is 23 dB. The procedure is repeated to complete Fig. 6.6. From Fig. 6.2, the apparent reflectivity is larger for larger beamwidths and a chart constructed for 15° sonar would have smaller bottom losses than those shown on Fig. 6.6.

We had to make many assumptions about the nature of the sea floor to obtain a bottom loss chart. Actually, bottom losses are only one kind of acoustic measurement and are in a real sense unprocessed data about the sea floor. To develop a description of

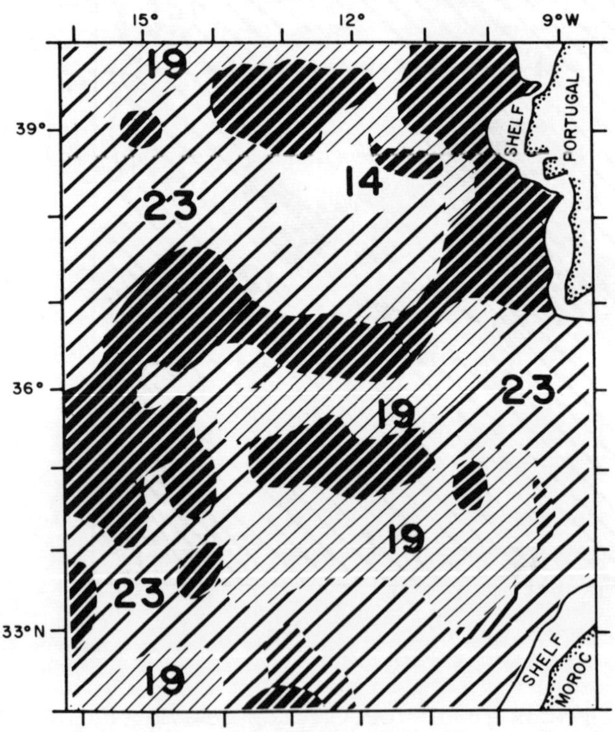

FIGURE 6.6

Total bottom loss in dB. Estimates are based on a 5° beamwidth, 3.5 kHz signal frequency at vertical incidence (see Fig. 6.1).

the sea floor, more information is needed, for example, the kind of spatial correlation function, the roughness, the reflection coefficient, and whether the local area is statistically homogeneous. This information can come from seismic profiles, wide-angle reflections, side-scanning sonar, and long cores.

Epilogue

I believe that we are dealing with problems far more important than making a bottom loss calculation, or for that matter, using bottom losses to estimate the roughness of the sea floor. It has to do with looking ahead. It has to do with predicting the future. It has to do with man's needs. Asimov has said to predict the future, predict the obvious. Thus, man will exploit the minerals on the sea floor. Man will seek patches of manganese nodules on the great abyssal plains. Man will seek rich minerals in brine-filled hot spots at spreading centers.

To find the resources on the large and complex sea floor we have some choices. Some would bet on the port holes of a submarine. Some would bet on deep diving man. I am betting on very high resolution sonars. Perhaps it will be on a cable, or on a mechanical beast, or on a ship. However the sonar is used, we will need all of the physics of sound in the sea for our quest.

Acknowledgments

We wish to thank the U.S. Naval Oceanographic Office and the Lamont-Doherty Geological Observatory (L-DGO) for data support. Individually, we wish to thank Dr. B. Olson for the loan of the Marine Geophysical Survey echograms and reflection profiles; Drs. J. Ewing, G. Bryan, and D. Hayes for permission to examine the Vema 27 cruise reflection, 3.5 kHz, and echo sounding records at L-DGO; Mr. R. Capo for the core description; and Mr. J. Brakl for the loan of the high resolution pulse-compression echo sounding (PCES) records. The illustrations were created by Mr. Paul Dombrowski. We also wish to thank the Office of Naval Research and the Geophysical and Polar Research Center, University of Wisconsin, for their generous financial support.

REFERENCES

Bouma, A. H., Sedimentology of Some Flysch Deposits, Elsevier, Amsterdam, p. 168, 1962.

Brakl, J., Two Magnetic and Bathymetric Profiles across the Atlantic Ocean and the Mid-Atlantic Ridge, Technical Report No. 141, Hudson Laboratories of Columbia University, p. 11, 1968.

Clay, C. S., Notes on Ocean Acoustics, University of Wisconsin, Geophysical and Polar Research Center Research Report, 71-2, August, 1971.

Clay, C. S., Coherent and Incoherent Components of Sound Scattered at a Time Dependent Rough Surface, Naval Postgraduate School, December, 1972.

Clay, C. S., and G. A. Sandness, Effect of Beamwidth on Acoustic Signals Scattered at a Rough Surface, XVII AGARD Symposium, Colorado Springs, Colorado, 1971.

Duplaix, S., W. D. Nesteroff, and B. C. Heezen, Mineralogie Comparée des Sediments du Tage (Portugal) et de Quelques Sables Profonds de la Plaine Abyssal Correspondants, Deep-Sea Res., 12, p. 211-217, 1965.

Ericson, D. B., M. Ewing, G. Wollin, and B. C. Heezen, Atlantic Deep-Sea Sediment Cores, Geol. Soc. Amer. Bull., 72, p. 193-286, 1961.

Ewing, M. W., B. F. Ryan, H. D. Needham, and B. C. Schreiber, Turbidity Currents on the Ocean Bottom, U.S. Naval Civil Engineering Laboratory, Report No. CR70.018, p. 149, 1970.

Hamilton, E. L., Prediction of In-situ Acoustic and Elastic Properties of Marine Sediments, Geophysics, V. 36, p. 266-284, 1971.

Hayes, D. E., A. C. Pimm et al., Initial Reports of the Deep Sea Drilling Project, Vol. 14, p. 947, Washington (U.S. Government Printing Office), 1972.

Heezen, B. C., and C. D. Hollister, The Face of the Deep, Oxford University Press, New York, p. 659, 1971.

Lamont-Doherty Geological Observatory unpublished manuals: (1) a practical classification of pelagic and hemipelagic sediments by Ansis Kaneps, and (2) the identification and classification of rocks and minerals in deep sea sediments: A general review for core describers by Maury Morgenstein.

Laughton, A. S., Discrete Hyperbolic Echoes from an Otherwise Smooth Deep-Sea Floor, Deep-Sea Res., 9, p. 218, 1962.

Laughton, A. S., Microtopography, in The Sea, Vol. 3, edited by M. N. Hill, p. 437-473, Interscience, New York, 1963.

Leong, W. K., T. K. Kan, and C. S. Clay, Use of Acoustic Scattering Theory to Interpret Marine Geophysical Data, University of Wisconsin Geophysical and Polar Research Center Research Report, 71-7, April, 1971.

Leong, W. K., Ph.D. Thesis, University of Wisconsin, 1973.

LePichon, X., J. M. Auzende, G. Pautot, and J. Francheteau, Deep Sea Photography of an Active Seismic Fault Zone Near Gibraltar Strait, Nature, 230, p. 110-111, 1971.

Lonsdale, Peter, and F. N. Spiess, Use of 3.5 to 4 kHz Echo Sounder Operating Close to the Deep Sea Floor to Examine Fine Scale Layering and Use of a Side-Looking Sonar to Examine Variations in Backscattering, Marine Physical Laboratory, Scripps Institution of Oceanography, University of California, 1973.

Luyendyk, B. P., Ph.D. Thesis, University of California at Lajolla, 1969.

U.S. Naval Oceanographic Office, Geology and Geophysics, SP95-5-5, Vol. 5, 1967a.

U.S. Naval Oceanographic Office, Oceanographic Stations and Velocity Profiles, SP95-5-6, Vol. 6, 1967b.

U.S. Naval Oceanographic Office, unpublished Precision Depth Recorder (PDR) and Seismic Reflection records taken by research vessels Atlantic Seal and Arctic Seal, 1965-66. These records are part of the Marine Geophysical Survey (MGS) Area 5 data file.

APPENDIX I

SCATTERING OF ACOUSTIC SIGNALS AT A ROUGH SURFACE

To analyze signals that have been reflected and scattered at the sea floor, an analytical theory needs to match the measurement conditions. Briefly, measurements are made with transducers having beamwidths ranging from a few degrees to omnidirectional. Water depths are about 3 km. The signal frequencies range from about 10 Hz to 25 kHz. The theory must also be adaptable to a wide range of bottom types.

The sea floor can be described by a reflection coefficient and probability density functions (PDF). The purpose of this development is to relate the statistical descriptions of the sea floor to the scattered sound signals. The first moment of the scattered signal is simply related to the PDF of the surface elevations. The second moment or mean square scattered sound has a two dimensional Fourier transformation relationship to the bivariate characteristic function of the surface elevations.

The scattered sound depends upon the geometry of the measurement, the illuminated area, and the rough surface. The analysis shows that the conventional interpretations of measurements to determine a scattering cross section yield (in general) functions that still depend upon the system parameters and geometry of the measurements. Interpretations based upon incident plane waves and a small illuminated area can be very misleading in the context of marine geophysical measurements.

The derivation of the scattered sound used the Helmholtz theorem and is adapted from Eckart (1953). The Fresnel approximation is used for the spherically diverging wave front. The necessity of including x^2 and y^2 or the non plane wave terms is well established [Clay (1960), Melton and Horton (1970), Clay and Medwin (1970), and Clay and Sandness (1971)]. The use of a spherically diverging wave front is also an approximation. Since an array of sources is used to form a directional "beam" or illumination function, the resulting wave front is not spherically symmetric. In the main lobe and at "large range," the wave front is approximately spherical.

This development is based upon two technical reports (Clay, 1971 and 1972).

Scattered Sound

On assuming local plane reflections, the "reflected" pressure signal, p', is related to the incident signal p_1 as follows:

$$p' = \mathcal{R} p_1 \tag{1}$$

$$\frac{\partial p'}{\partial n} = - \mathcal{R} \frac{\partial p_1}{\partial n} \tag{2}$$

where $\mathcal{R}$ is the reflection coefficient and $\partial/\partial n$ is the normal derivative.

Eckart (1953) used a small slope approximation in which he replaced the normal derivative by $\partial/\partial z$. The slopes can be included by performing an integration by parts [Tolstoy and Clay, p. 196-199 (1966)]. The result is a geometrical factor which is the same as obtained with $\partial/\partial n \simeq \partial/\partial z$ in the specular direction.

With the aid of the Helmholtz theorem [Born and Wolf, p. 375 (1965)] the scattered sound pressure is

$$p = \frac{1}{4\pi} \int_S \left(p' \Big|_S \frac{\partial U}{\partial n} - \frac{\partial p'}{\partial n} \Big|_S \right) ds \tag{3}$$

where

$$U = e^{i(kR' - \omega t_2')} / R' \tag{4}$$

$$p_1 = B\, D_o e^{i(kR - \omega t_1)} / R \quad \text{for } kR \gg 1 \tag{5}$$

$$B^2 = \Pi \rho c (2\pi)^{-1} \qquad D_o = \exp\left[- \frac{x^2}{X^2} - \frac{y^2}{Y^2} \right] \tag{6}$$

Π = source power

ρ is the density, c is sound speed, and the normal is drawn toward the receiver. R and r' are the source and receiver distances to ds (as x, y, z). In applying the Helmholtz theorem, the surface is closed by a hemisphere at infinity. There are no other sources and all waves are outgoing from our source. Hence, the contribution to the integral for the surface at infinity can be ignored and the integral over the illuminated interface is used. The substitution of (1) and (2) into (3) gives

$$p = \frac{1}{4\pi} \int_S \mathcal{R} \frac{\partial}{\partial n} (p_1 U) \Big|_S ds \tag{7}$$

In vector notation and with the aid of the geometry in Fig. (1), $\tilde{R}$ and $\tilde{R}'$ are

$$|\tilde{R}| = |\tilde{R}_1 + \tilde{i}x + \tilde{j}y + \tilde{k}z|$$
$$|\tilde{R}'| = |\tilde{R}_2 - \tilde{i}x - \tilde{j}y - \tilde{k}z| \qquad (8)$$

where $\tilde{i}$, $\tilde{j}$, and $\tilde{k}$ are the unit vectors along with x, y, and z. We assume the source is sufficiently directive and that the wave number along R can be replaced by $\tilde{k}_1$, the wave number along $\tilde{R}_1$. Similarly, $\tilde{k}_2$ is approximately the wave number along $\tilde{R}'$.

$$kR \simeq \tilde{k}_1 \cdot (\tilde{R}_1 + \tilde{i}x + \tilde{j}y + \tilde{k}z), \text{ etc.}$$

The normal derivative needs to be calculated. For small slopes, $\partial/\partial n$ is approximately $\partial/\partial z$ and we make this approximation and then use a boundary factor F which includes the slopes, Tolstoy and Clay (1966). The integrand becomes (ignoring time)

$$BD_o \mathcal{R} \frac{\partial}{\partial z}\left(\frac{1}{RR'} \exp i\left[\tilde{k}_1(\tilde{R}_1 + \tilde{i}x + \tilde{j}y + \tilde{k}z) + \tilde{k}z_2 \cdot (\tilde{R}_2 - \tilde{i}x - \tilde{j}y - \tilde{k}z)\right]\right)\Big|_S ds \qquad (9)$$

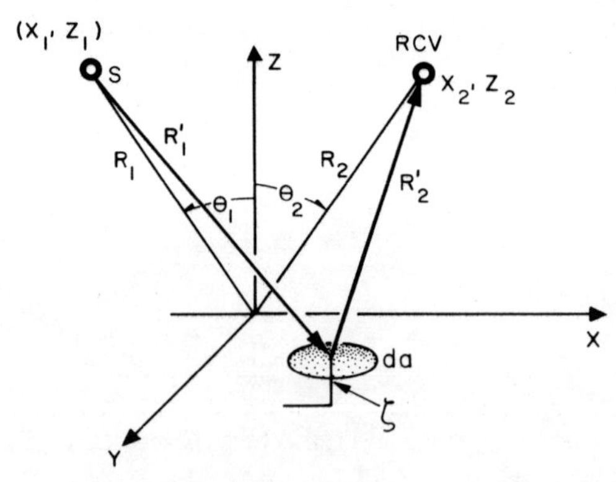

FIGURE 1

Geometry. ds is at the position x, y, ζ relative to the origin. θ_3 is the angle between the planes containing R_2 and z and the x-z plane. R_1 is in the x-z plane. θ_3 is not shown for clarity.

The main dependences on x and y are in the phase function in the exponential and the illumination factor D_o. The range of z is limited to very small values, relative to R_1, R_2, x, and y. Hence, the dependence of R^{-1} and R'^{-1} upon z can be ignored. R^{-1} and R'^{-1} can be replaced by R_1^{-1} and R_2^{-1}. The evaluation of Eq. (9) on the rough surface $\zeta(x,y)$ is indicated by letting z become ζ. On taking the differentials in Eq. (9) we obtain

$$BD\mathcal{R}R_1^{-1} R_2^{-1} (\underset{\sim}{k}_1-\underset{\sim}{k}_2)\Big|_{z\ comp} \exp i[k(R+R')] \tag{10}$$

We use k(R+R') in the exponential for convenience.

$$p \simeq -\frac{ik}{2\pi} B\mathcal{R}F \frac{e^{-i\omega t}}{R_1R_2} \int_S D_o \exp i[k(R+R')]ds$$
$$t = t_1 + t_2' \tag{11}$$

$$F = \begin{cases} (\cos\theta_1 + \cos\theta_2)/2, & \text{for } \partial/\partial n \simeq \partial/\partial z \\ \dfrac{1+\cos\theta_1\cos\theta_2 - \sin\theta_1\sin\theta_2\cos\theta_3}{\cos\theta_1+\cos\theta_2}, & \text{slope factor} \end{cases} \tag{12}$$

$$\text{note:}\quad \underset{\sim}{k}_1 - \underset{\sim}{k}_2\Big|_{z\ comp.} = -k(\cos\theta_1+\cos\theta_2)$$

where, for a moderately directive source $\mathcal{R}$, R_1 and R_2 are nearly constant over the illuminated area and have been removed from the integral.

We now expand k(R+R') for a moderately directive source. From Fig. 1 R and R' are

$$R^2 = (R_1\sin\theta_1 + x)^2 + y^2 + (R_1\cos\theta_1 - \zeta)^2$$

$$R'^2 = (R_2\sin\theta_2\cos\theta_3 - x)^2 + (R_2\sin\theta_2\sin\theta_3 - y)^2 + (R_2\cos\theta_2 - \zeta)^2 \tag{13}$$

On using the binomial expansion of $(1+a)^{1/2}$, (R+R') can be written to first order in ζ and second order in x and y as follows.

$$k(R+R') \simeq k(R_1+R_2) + 2(\alpha x+\beta y+\gamma\zeta) + x_f^{-2}x^2 + y_f^{-2}y^2$$

$$x_{-f}^{-2} \equiv \frac{k}{2}\left(\frac{\cos^2\theta_1}{R_1} + \frac{1-\sin^2\theta_2\cos^2\theta_3}{R_2}\right), \tag{14}$$

$$y_{-f}^{-2} \equiv \frac{k}{2}\left(\frac{1}{R_2} + \frac{1-\sin^2\theta_2\sin^2\theta_3}{R_2}\right)$$

$$\begin{aligned} 2\alpha &= k(\sin\theta_1 - \sin\theta_2\cos\theta_3) \\ 2\beta &= -k(\sin\theta_2\sin\theta_3) \\ 2\gamma &= -k(\cos\theta_1 + \cos\theta_2) \end{aligned} \tag{15}$$

The notation x_f and y_f was used because x_f and y_f are the half widths of the first Fresnel. α, β, and γ follow the notation of Tolstoy and Clay (1966).

The substitution of Eq. (15) in Eq. (14) yields:

$$p \simeq \frac{ik\ B\mathcal{R}F}{2\pi\ R_1R_2}\exp i[k(R_1+R_2) - \omega t]$$

$$\iint_S D_o \exp\left[i(x^2/x_f^2 + y^2/y_f^2 + 2i(\alpha x+\beta y+\gamma\zeta)\right]dydx \tag{16}$$

Equation (16) is the expression for a continuous wave signal. For a transient signal, one can use the Fourier integral theorem and integrate it over the spectrum of the signal. Actually, pings or gated sine waves have a very narrow spectrum and we can evaluate the equation at the frequency of the ping and use it as an approximation to the integration over the signal spectrum.

Coherently Reflected Signals and Determination of the PDF of the Surface

The signals are presumed to have random characteristics, so we do an ensemble average on the signals to determine the coherent component. The ensemble average is made by averaging the repeated transmissions of many transmissions. Between each transmission the rough surface is changed. The statistical descriptions of each section of the surface are the same. The random character of the surface enters only in ζ. Hence, the averaging operation $<p>$ yields the product of the reflection from a smooth surface, p, and the effect of the randomness as follows:

$$<p> = p_o \int_{-\infty}^{\infty} W(\zeta)\ \exp(2i\gamma\zeta) d\zeta \tag{17}$$

where $W(\zeta)$ is the probability density function, PDF, of the rough surface. It has the form of a Fourier transformation and one can write the inverse transformation as follows:

$$W(\zeta) = \frac{1}{\pi} \int_{-\infty}^{\infty} (<p>/p_o)\ \exp(-2i\gamma\zeta) d\gamma \tag{18}$$

At small grazing angles it is necessary to include the shadowing of the surface. Wagner (1966) used a ray-optics approximation to derive a shadowing function which can be used to transform W to an apparent probability density, W_a. With the aid of Wagner's Eq. 21, 22, 40, and 41 (slight notational changes),

$$W_a = \frac{2B[-B\mathrm{erfc}(\zeta\sigma^{-1}2^{-1/2})]}{1 - \exp(-2B)}\ W \tag{18a}$$

where B is the shadowing parameter. For specular reflection, Wagner gives

$$B = (2v)^{-1}\pi^{-1/2}\ (e^{-v^2} - v\pi^{1/2}\ \mathrm{erfc}\ v)$$
$$v \equiv \cot\theta\ (2\sigma^2\ |\psi''(o)|)^{-1/2} \tag{18b}$$

and $\psi''(o)$ is the second spatial derivative of the correlation function at $\xi = 0$. The effect of the shadowing is to reduce the relative values of W_a at positive ζ and to increase values of W_a at negative ζ. The width of W_a tends to be narrower than W. Partial shadowing of the surface increases the amount of coherent component.

The laboratory measurements of sound scattered at a wind blown surface demonstrate this effect (Clay, Medwin, and Wright, 1973). The measured W of the surface and a Gaussian fit to the measurements are shown in Fig. 2(a). Experimental and theoretical coherent components of the scattered sound are shown in Fig. 2(b). The frequency of the sound signal was varied to change the roughness parameter g, where

$$g \equiv 4\gamma^2\sigma^2$$

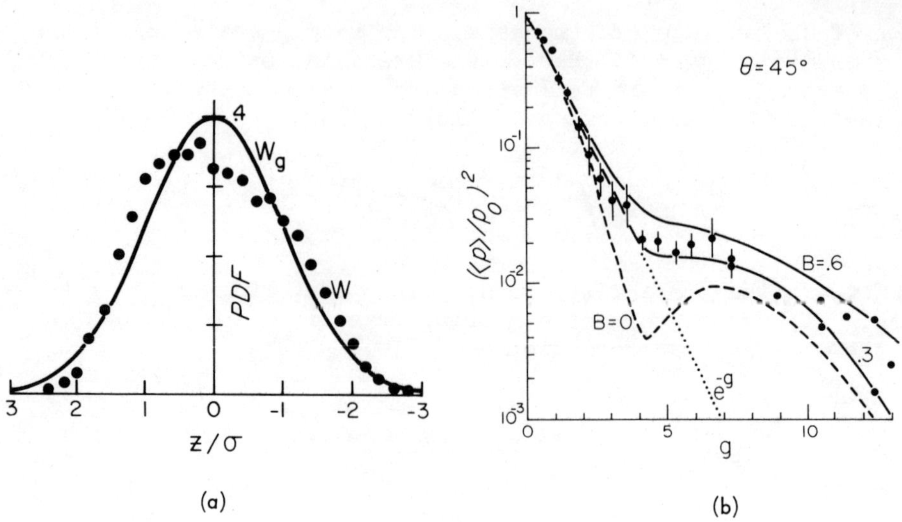

FIGURE 2

(a) Probability density function of the "model sea surface," W. The rms roughness $\sigma = 0.45$ cm. Five fans were used to generate this surface. The surface depression z is positive downward. The wave staff measurements are circles and the solid line, W_G, is a Gaussian PDF for comparison.

(b) Coherently scattered signal at 45° incidence angle. The experimental data points are the circles. The bars indicate the extent of fluctuations. Theoretical values of $(\langle p\rangle/p_o)^2$ calculated from the measured PDF and for shadowing parameter B = 0.3 and 0.6 are solid lines. No shadowing, B = 0 is the dashed line. For comparison e^{-g} for Gaussian PDF and no shadowing is the dotted line. $\sigma = 0.45$ cm, $g = 4\gamma^2\sigma^2$. [From Clay, Medwin, and Wright (1973)]

Theoretical curves are shown for different values of the shadowing function. It wasn't possible to use the measured values of ψ to estimate $\psi''(o)$ and Eq. (18b) to calculate B. The curve for a Gaussian W_g is indicated as exp(-g).

For a surface having a Gaussian PDF and an rms roughness, σ, $\langle p\rangle$ is

$$\langle p\rangle = p_o \exp(-2\gamma^2\sigma^2) \tag{19}$$

The validity of the expression has been demonstrated for Gaussian PDF by Proud, Beyer, and Tamarkin (1960) and in Tolstoy and Clay (1966). If the PDF of the surface is not known, Eq. (19) can be used to estimate σ.

Over an ideal plane interface, the reflected signal is $\mathcal{R}$ times the incident signal so long as the angle of incidence is not too near the critical angle. The effect of surface roughness can be included in the reflection coefficient by defining a coherent reflection coefficient $\langle\mathcal{R}\rangle$

$$\langle\mathcal{R}\rangle = \langle p\rangle/p_o \tag{19a}$$

where p_o is the signal which would be received for $\mathcal{R} = 1$. We believe that the term "reflection" should be applied to the coherent component and, correspondingly, is associated with the first moment of the ensemble of signals.

Mean Square Signal

The ensemble average of the mean square signal is estimated as follows:

$$\left\langle p^2\right\rangle = \left\langle \frac{1}{T}\int_{T_o}^{T_o+T} pp^*dt \right\rangle \quad p^* = \text{complex conjugate of } p \tag{20}$$

where T is the duration of the pulse. The substitution of Eq. (16) in the preceding expression yields a multiple integral within the ensemble average. Since the randomness enters in the rough surface, the averaging operation is moved to within the integrals and performed on the function of $\zeta(x,y)$ and $\zeta(x+\xi, y+\eta)$. Having described the procedure, we do it. Starting with Eq. (16), we have

$$\left\langle p^2\right\rangle \simeq \frac{k^2B^2\mathcal{R}^2F^2}{4\pi^2R_1^2R_2^2}\iiiint_{\infty}^{\infty} D_oD_o' \exp i\left\{\left[(x^2-x'^2/x_f^2) + (y^2-y'^2)/y_f^2 + 2\alpha(x-x') + 2\beta(y-y')\right]\right\} \left\langle \exp[i2\gamma(\zeta-\zeta')]\right\rangle dydsdy'dx \tag{21}$$

where D_o' and ζ' are functions of x' and y'. The average is taken over by the bivariate PDF. The averaging operation yields the corresponding characteristic function C_2.

$$C_2 = \left\langle \exp[i2\gamma(\zeta-\zeta')] \right\rangle \qquad (22)$$

If the PDF is not Gaussian, measurements of the bivariate PDF or C_2 can be used.

On <u>assuming</u> the surface roughness has a Gaussian distribution over the illuminated area, the average within the integral can be evaluated for a bivariate Gaussian function. The result is

$$C_2 = \exp[-4\gamma^2\sigma^2(1-\psi)]$$

$$\psi = \frac{1}{\sigma^2 A}\int_A \zeta(x,y)\ \zeta(x+\xi,y+\eta)\,dy\,dx \qquad (23)$$

ψ is a function of ξ and η, or $(x-x')$ and $(y-y')$. To evaluate Eq. (21), we need to choose an illumination function. I prefer the Gaussian type because of their similarity to the main beam of a sonar transducer, lack of side lobes, and, most importantly, ease of integration. D_o is

$$D_o = \exp\left[-\left(\frac{x^2}{X^2} + \frac{y^2}{Y^2}\right)\right] \qquad (24)$$

The following change of variables will permit a double integration over x" and y":

$$x = x'' + \xi/2 \qquad y = y'' + \eta/2$$

$$x' = x'' - \xi/2 \qquad y' = y'' - \eta/2 \qquad (25)$$

The product of the Jacobians is one. $\langle p^2 \rangle$ becomes, after integration over x" and y", the following:

$$\left\langle p^2 \right\rangle \simeq \frac{k^2B^2\mathcal{R}^2F^2XY}{8\pi R_1^2 R_2^2}\iint_{-\infty}^{\infty} D\exp(2i[\alpha\xi+\beta\eta])\ \exp[-4\gamma^2\sigma^2(1-\psi)]\,d\xi\,d\eta \qquad (26)$$

$$D \equiv \exp\left\{-\left(\frac{X^{-2}}{2x_f^4}\ 1 + \frac{x_f^4}{X^4}\right)\xi^2 - \frac{Y^2}{2y_f^4}\left(1 + \frac{y_f^4}{Y^4}\right)\eta^2\right\} \qquad (27)$$

The integral over ξ and η cannot be done until ψ is known. As mentioned earlier, we are interested in the inverse problem and the determination of ψ. One could measure the experimental scattering function and solve the integral equation for ψ (Proud, Beyer, and Tamarkin, 1960). We will guess the form of typical correlation functions and compute scattering functions for comparison to experimental data. Before proceeding, we cast Eq. (26) into the form of the sonar equation. Quantities like source strength and positions are separated from the scattering integral as follows:

$$\langle p^2 \rangle = \frac{P_o^2 R_o^2}{R_1^2} \frac{A}{R_2^2} S \tag{28}$$

$$A \equiv 4XY$$

$$P_{so}^2 = B^2/R_o^2, \; R_o \text{ is a unit distance}$$

$$S \equiv \frac{k^2 \mathcal{R}^2 F^2}{32\pi} \iint_{-\infty}^{\infty} D \exp[2i(\alpha\xi+\beta\eta)] \exp[-4\gamma^2\sigma^2(1-\psi)]d\xi d\eta \tag{29}$$

S is the scattering function. Many calculations have been made with the assumption that S is only a function of the roughness of the surface, σ and ψ. Unfortunately, this assumption is not valid because D is a function of the illumination, X and Y, the geometry, and the frequency in x_f and y_f.

Acoustic Measurement of the Characteristic Function

In a previous section, the Fourier transformation was used to determine the PDF of the rough interface. The scattering function, Eq. (29), also has the form of a Fourier transformation. A similar operation can be performed on <ppx> or S. On using C_2 (Eq. 22) instead of making the Gaussian assumption, S is

$$S = \frac{k^2 \mathcal{R}^2 F^2}{32\pi} \iint_{-\infty}^{\infty} DC_2 \exp[2i(\alpha\xi+\beta\eta)]d\xi d\eta \tag{30}$$

For moderately narrow beamwidths, D is a slowly varying function of ξ and η. Thus, the two dimensional Fourier transformation yields

$$C_2 = \frac{32}{\pi k^2 \mathcal{R}^2 F^2} D^{-1} \iint_{-\infty}^{\infty} S \exp 2i[(\alpha\xi+\beta\eta)]d\alpha d\beta \tag{31}$$

Obviously, one's ability to do the inverse transformation depends upon having accurate values of S or $<pp^*>$. I assume the user would do the transforms numerically. Although the integral has infinite limits, actual measurements of S give both upper and lower limits. There is another consideration. Since D operates as a spatial high pass filter in Eq. (30), I doubt if C_2 would have accuracy for ξ and η much larger than $2^{\frac{1}{2}}x_f^2/X$ and $2^{\frac{1}{2}}y_f^2/Y$.

If the surface has been determined to have a Gaussian PDF by measuring the coherent component and Eq. (18), then the Gaussian bivariate characteristic function, Eq. (23), can be used to estimate ψ.

Evaluation of the Scattering Function

Our method of calculating S is based on evaluating the scattering integral over sub areas in ξ and η space. Let us start with the assumption that ψ is a set of measured points and we are free to make any reasonable analytic fit to the points. We choose to make a sub-area fit. The two dimensional ψ is approximated by means of second order polynomials in each sub-area. Hence, the integral is transformable to a Gaussian function. Since ψ is symmetric, ξ and η have absolute value signs. Examples of ξ dependence are sketched on Fig. 3. For the sub-areas shown on Fig. 4, ψ can be expressed as follows:

$$\psi_{ij} = c_{ij} - b_{ij}\,\xi - a_{ij}\,\xi^2 - f_{ij}\,\eta - e_{ij}\eta^2 \tag{32}$$

$$\text{for } \xi_{i-1} < \xi \le \xi_i \text{ and } {}_{j-1} < \eta \le \eta_j$$

$$\psi = 0 \text{ for } \xi > \xi_n \quad \eta > \eta_m \quad .$$

For the first sub-area, the coefficients can be expressed in terms of the previously used notation

$$c_{11} = 1,\ b_{11} = L_1^{-1},\ a_{11} = L_2^{-2},\ f_{11} = W_1^{-1},\ e_{11} = W_2^{-2} \tag{33}$$

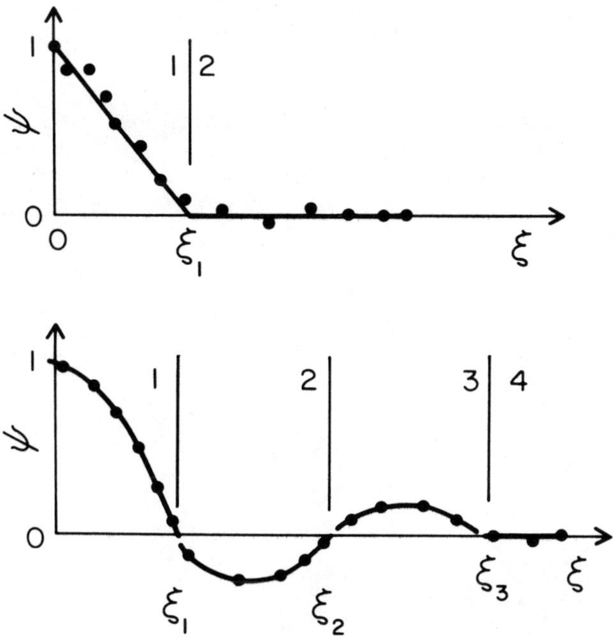

FIGURE 3

Analytic approximation to correlation function. The correlation measurements are indicated as the dots. The piece-wise fits are solid curves between 0, 1, 2, etc. The solid curves are first and second order functions of ξ.

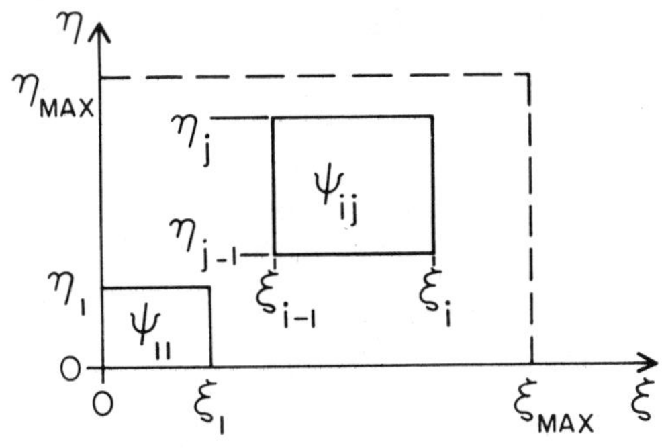

FIGURE 4

Sub-areas of the correlation function ψ. The function is symmetric and one quadrant is shown.

In many applications ψ can be approximated by two kinds of functions, ψ_{11} (ξ, η) near the origin and zero for the rest of the area.

The manipulation of the scattering integral into a form which is convenient for evaluation consists of straightforward algebra. To establish a sense of direction, we begin with the scattering integral Eq. (29) and subdivide the integration as follows:

$$S = \sum_{i}^{n} \sum_{j}^{m} S_{ij} \tag{34}$$

$$S_{ij} = \frac{k^2 \mathcal{R}^2 F^2}{32\pi} \int_{\xi_{i-1}}^{\xi_i} \int_{\eta_{j-1}}^{\eta_j} D \exp[2i(\alpha\xi + \beta\eta)] \exp\left[-4\gamma^2\sigma^2(1-\psi_{ij})\right] d\xi d\eta \tag{35}$$

In the calculation of S, we have already integrated over the illuminated area so the constants contained in α, β, γ, and D are the same for all of the sub-areas. As the calculation has been developed, the double integration can be separated into the product of two integrals provided ψ is separable. This lets us do the formal manipulation on a single integral and apply it to the double integral.

The scattering integral will be changed to a general form of the error integral. Integrals of complex error functions are given in the National Bureau of Standards Handbook [Abramowitz and Stegun (1965)]. From the Handbook, pp. 297-328, the functions are (7.1.1 and 7.1.2)

$$\mathrm{erf}(z) \equiv \frac{2}{\pi^{1/2}} \int_0^z e^{-t^2} dt \tag{36}$$

$$\mathrm{erfc}(z) \equiv \frac{2}{\pi^{1/2}} \int_z^\infty e^{-t^2} dt = 1 - \mathrm{erf}(z) \tag{37}$$

erf(z) and erfc(z) are tabulated for real arguments. They define and tabulate the function w(z) for complex arguments (7.1.3).

$$w(z) \equiv e^{-z^2}\left[1 - \frac{2i}{\pi^{1/2}}\int_0^z e^{t^2}\,dt\right] \tag{38}$$

$$w(z) = e^{-z^2}\,\mathrm{erfc}(-iz)$$

$$w(ix) = e^{x^2}\,\mathrm{erfc}(x) \text{ for real } x \tag{39}$$

$$w(z^*) = w^*(-z) \quad .$$

The introduction of the absolute value signs in ψ requires the following alteration of the infinite integral:

$$\int_{-\infty}^{\infty} f(x)\, e^{2i\alpha x - a|x|}\,dx = \int_{0}^{\infty} f(x)\, e^{-ax}\left(e^{2i\alpha x}+e^{-2i\alpha x}\right)dx \tag{40}$$

where all even dependence upon x has been collected into f(x) and the linear dependence on x is expressed in the exponential.

Having established the form of the tabulated functions, our next task is to cast the scattering function into that form. With the aid of C_{ij}, U_{ij}, and V_{ij}, to be defined later,

$$S_{ij} = \frac{k^2\mathcal{R}^2F^2}{128}\,C_{ij}\,U_{ij}\,V_{ij} \tag{41}$$

U_{ij} and V_{ij} are error functions for the integrals of ξ and η respectively. C_{ij} contains the constants and with the aid of Eqs. (32) and (35) is

$$C_{ij} = \exp\left[-4\gamma^2\sigma^2(1-c_{ij})\right] \quad . \tag{42}$$

The ξ and η integrals have the same form, so we use an integral on x and let it stand for either ξ or η. For the ijth sub-area, we define

$$U_{ij} = \frac{2}{\pi^{1/2}}\left(\int_{x_{i-1}}^{x_i} \exp-\left[\left(\frac{x}{A_{xij}}\right)^2 + B_{xij}\,x\right]dx + \int_{x_{i-1}}^{x_i} \exp-\left[\left(\frac{x}{A_{xij}}\right)^2 + B^*_{xij}x\right]dx\right) \tag{43}$$

and for the y-dependence, the exponentials have the following forms:

$$\left(\frac{y}{A_{yij}}\right)^2 + B_{yij}y, \quad \left(\frac{y}{A_{xij}}\right)^2 + B^*_{yij}y \tag{44}$$

where the use of B_{xij} and B^*_{xij} result from the application of Eq. (40). Ordinarily, A_{xij} is real and B_{xij} may be complex. These integrals are transformed to the standard error integral by completing the square and changing variables as follows:

$$t \equiv \frac{x}{A_{xij}} + s_{ij} \quad \text{or} \quad t \equiv \frac{y}{A_{yij}} + t_{ij}$$

$$s_{ij} \equiv A_{xij}\, B_{xij}/2 \tag{45}$$

$$s^*_{ij} \equiv A_{xij}\, B^*_{xij}/2$$

Using q to designate the transformed limits of integration, the lower and upper limits are

$$q_{ij}(i-1) \equiv \frac{\xi_{i-1}}{A_{xij}} + s_{ij}$$

$$q_{ij}(i) \equiv \frac{\xi_i}{A_{xij}} + s_{ij} \tag{46}$$

We define the y-dependence as follows:

$$r_{ij}(j-1) \equiv \frac{\eta_{j-1}}{A_{yij}} + t_{ij}$$

$$r_{ij}(j) \equiv \frac{\eta_j}{A_{yij}} + t_{ij} \tag{47}$$

$$t_{ij} \equiv A_{yij} B_{yij}/2, \text{ etc.}$$

with the aid of Eq. (37) and the changes of variable, Eqs. (45) and (42) become

$$U_{ij} = 2\, A_{xij}\, u_{ij} \tag{48}$$

where the expression u_{ij} is defined as follows:

$$u_{ij} \equiv \frac{1}{2}\left\{e^{s_{ij}^{2}} \operatorname{erfc}\left[q_{ij}(i-1)\right] + e^{s_{ij}^{*2}} \operatorname{erfc}\left[q_{ij}^{*}(i-1)\right]\right\}$$
$$- \frac{1}{2}\left\{e^{s_{ij}^{2}} \operatorname{erfc}\left[q_{ij}(i)\right] + e^{s_{ij}^{*2}} \operatorname{erfc}\left[q_{ij}^{*}(i)\right]\right\} \quad . \tag{49}$$

Without going through the details, the y-dependent function is

$$V_{ij} = 2\, A_{yij}\, v_{ij} \tag{50}$$

where v_{ij} has the same form as u_{ij}, t_{ij} replaces s_{ij}, etc.

These functions can be expressed as a combination of w(z) type functions. Choosing the first term Eq. (49), one can write it as follows:

$$e^{s_{ij}^{2}} \operatorname{erfc}\left[q_{ij}(i-1)\right] = \exp\left[s_{ij}^{2} - q_{ij}^{2}(i-1)\right]$$
$$e^{q_{ij}^{2}(i-1)} \operatorname{erfc}\left[q_{ij}(i-1)\right] \tag{51}$$

Recalling Eq. (39) and letting $z = i\, q_{ij}(i-1)$, Eq. (49) becomes

$$e^{s_{ij}^{2}} \operatorname{erfc}\left[q_{ij}(i-1)\right] = \exp\left\{-\left(\xi_{i-1}/A_{xij}\right)^{2}\right\}$$
$$\exp\left\{-\left(\xi_{i-1}B_{xij}\right)\right\} w\left[iq_{ij}(i-1)\right] \tag{52}$$

With the aid of the symmetry relation, Eq. (39), the complex conjugate $w(iq^{*})$ becomes $w^{*}(iq)$. On applying the latter and Eq. (52) to Eq. (50), one obtains the following.

$$u_{ij} = \frac{1}{2} \exp\left[-\left(\xi_{i-1}/A_{xij}\right)^2\right]\left\{\exp\left[-\left(\xi_{i-1}B_{xij}\right)\right] w\left[iq_{ij}(i-1)\right] + \exp\left[-\left(\xi_{i-1}B_{xij}^{*}\right)\right] w^{*}\left[iq_{ij}(i-1)\right]\right\} - \frac{1}{2} \exp\left[-\left(\xi_i/A_{xij}\right)^2\right]\left\{\exp\left[-\left(\xi_i B_{xij}\right)\right] w\left[iq_{ij}(i)\right] + \exp\left[-\left(\xi_i B^{*}_{xij}\right)\right] w^{*}\left[iq_{ij}(i)\right]\right\} \tag{53}$$

where v_{ij} = similar expression. The identification of the constants and gathering of the pieces remains. On combining Eqs. (41), (48), and (49), one obtains

$$S_{ij} = \frac{k^2 \mathcal{R}^2 F^2}{32} \exp\left\{-\left[4\gamma^2\sigma^2(1-c_{ij})\right]\right\} A_{xij}\, A_{yij}\, u_{ij}\, v_{ij} \tag{54}$$

Comparison of the scattering integral in Eqs. (35), (27), and (43) gives the coefficients of the quadratic and linear terms (after substitution for D and ψ_{ij}). For the ijth area, the coefficients are

$$A_{xij} \equiv \left[\frac{X^2}{2x_f^{\,4}}\left(1 + \frac{x_f^{\,4}}{X^4}\right) + 4\gamma^2\sigma^2 a_{ij}\right]^{1/2} \tag{55}$$

$$B_{xij} \equiv 4\gamma^2\sigma^2 b_{ij} \pm 2i\alpha \tag{56}$$

$$A_{yij} \equiv \left[\frac{Y^2}{2y_f^{\,4}}\left(1 + \frac{y_f^{\,4}}{Y^4}\right) + 4\gamma^2\sigma^2 e_{ij}\right]^{-1/2} \tag{57}$$

$$B_{yij} \equiv 4\gamma^2\sigma^2 f_{ij} \pm 2i\beta \tag{58}$$

The arguments of u_{ij} and v_{ij} are given by Eqs. (45) through (47). Many calculations will reduce to the sum of complex conjugates having the following form:

$$u = [w(iq) + w^{*}(iq)]/2 \tag{59}$$

where w(iq) is the tabulated error function. The tables do not give values for the larger arguments. We have used the approximations of Abramowitz and Stegun (1965), (7.1.13) and (7.1.23), for the following, where $q = x + iy$:

$$\frac{2\pi^{-1/2}}{x+(x^2+2)^{1/2}} < w(ix) \leq \frac{2\pi^{-1/2}}{x+(x^2+4/\pi)^{1/2}} \quad x > 0 \tag{60}$$

and

$$u \simeq \frac{1}{2\pi^{1/2}} \left[\frac{2x}{x^2+y^2} + \frac{y^2-x^2}{(x^2+y^2)^2} \right] \tag{61}$$

$$x > 1$$

$$y > 2$$

In the specular direction $\theta_1 = \theta_2$, $\theta_3 = 0$, and the parameters simplifiy as follows:

$$\alpha = \beta = 0$$

$$F = \cos\,\theta_1$$

$$x_f^{-2} = k\,\cos^2\,\theta,\ (R_1^{-1}+R_2^{-1})/2 \tag{62}$$

$$y_f^{-2} = k(R_1^{-1}+R_2^{-1})/2$$

From Eq. (56) and Eq. (58), B_{xij} and B_{yij} are real; thus q and r are real. For real q Eq. (49) simplifies to

$$\begin{aligned} u_{ij} = {} & \exp\left\{-\left[\left(\frac{\xi_{i-1}}{A_{xij}}\right)^2 + \xi_{i-1}\,B_{xij}\right]\right\} w\,[iq_{ij}(i-1)] \\ & -\exp\left\{-\left[\left(\frac{\xi_i}{A_{xij}}\right)^2 + \xi_i\,B_{xij}\right]\right\} w\,[iq_{ij}(i)] \end{aligned} \tag{63}$$

where $w(iq)=e^{q^2}\,\mathrm{erfc}(q)$, and q is real. In many underwater sound problems, E (Eq. (60)) can be used to approximate w(iq) and the resulting u_{ij} are easy to evaluate numerically. The somewhat arbitrary definitions of u_{ij} and v_{ij} may seem more reasonable to the reader if he evaluates them for the specular reflection at a smooth surface. The region of integration becomes zero to infinity and the functions are the following:

for

$$\alpha = \beta = \sigma = B_{x11} = B_{y11} = 0 \tag{64}$$

$$w(0) = 1$$

$$w(\infty) = 0$$

$$u_{11} = v_{11} = 1 \tag{65}$$

$$S = \frac{k^2 \mathcal{R}^2 F^2}{32} A_{x11} A_{y11} \tag{66}$$

The far field signal is

$$\left\langle p^2 \right\rangle = 4 \frac{P_{so}^{\ 2} R_0^{\ 2}}{R_1^{\ 2} R_2^{\ 2}} \quad XYS = \frac{P_0^{\ 2} R_0^{\ 2}}{(R_1+R_2)^2} \mathcal{R}^2 \tag{67}$$

The image solution is the expected result.

REFERENCES

Abramowitz, M., and I. A. Stegun, Handbook of Mathematical Functions, N.B.S. Applied Mathematics Series 55, U.S. Government Printing Office, Washington, D.C. (1964).

Born, M., and E. Wolf, Principles of Optics, Pergamon Press, Oxford, England (1965).

Clay, C. S., Coherent and Incoherent Components of Sound Scattered at a Time Dependent Rough Surface, Naval Postgraduate School, December (1972).

Clay, C. S., J. Acoust. Soc. Am. 32, p. 1547-1551 (1960).

Clay, C. S., Notes on Ocean Acoustics, Univ. of Wis. Geophysical and Polar Research Center Research Report, No. 71-2, August (1971).

Clay, C. S., and H. Medwin, J. Acoust. Soc. Am. 47, p. 1412-1418 (1970).

Clay, C. S., H. Medwin, and W. M. Wright, J. Acoust. Soc. Am. 53, p. 1677 (1973).

Eckart, C., J. Acoust. Soc. Am. 25, p. 195 (1953).

Horton, C. W., Sr., J. Acoust. Soc. Am. 51, p. 1049-1061 (1972).

Melton, D. R., and C. W. Horton, Sr., J. Acoust. Soc. Am. 47, p. 290-298 (1970).

Proud, J. M., R. T. Beyer, and P. T. Markin, J. Appl. Phys. 31, p. 543 (1960).

Tolstoy, I., and C. S. Clay, Ocean Acoustics: Theory and Experiment in Underwater Sound, McGraw-Hill Book Co., New York (1966).

Wagner, R. J., J. Acoust. Soc. Am. 41, p. 138 (1966).

APPENDIX II
IDENTIFICATION OF CORES AND PHOTOGRAPHS

The station numbers refer to the Fig. 5.2 for locations. The particular cores and sources of bottom photographs are identified. V refers to cores taken by the R/V Vema and RC to cores taken by R/V Robert Conrad.

APPENDIX II

STA NO.	PISTON CORE ID (DREDGE ID)	PHOTO ID	REFERENCES
1	V27-144	---	Lamont-Doherty Geological Observatory (L-DGO) data files
2	RC9-210	RC9-186	L-DGO
3	RC9-209	RC9-185	L-DGO
4	V27-145	V27-K86	L-DGO
5	(V27-20)	---	L-DGO
6	---	Photo	Heezen and Hollister, 1971, Plate 4
7	(V27-21)	---	L-DGO
8	V4-40	---	L-DGO
9	V4-39	---	L-DGO
10	V4-41	---	L-DGO
11	V4-42	---	L-DGO
12	V27-146	V27-K87	L-DGO
13	V4-44	---	L-DGO; Ewing et al., 1970
14	V4-45	---	L-DGO; Ewing et al., 1970
15	V4-46	---	L-DGO; Ewing et al., 1970
16	V27-147	V27-K88	L-DGO; Ewing et al., 1970
17	T1-7	---	Ewing et al., 1970
18	V4-47	---	L-DGO; Ewing et al., 1970
19	AS5-4	---	U.S. Naval Oceanographic (NAVOCEANO), 1967b
20	V4-48	---	L-DGO; Ewing et al., 1970
21	---	Photo	Heezen and Hollister, 1971, Plate 3
22	(RC9-6)	RC9-183	L-DGO
23	RC9-208 (RC9-7)	RC9-184	L-DGO
24	RC9-206	RC9-182	L-DGO
25	(V27-15)	---	L-DGO
26	(V27-16)	---	L-DGO
27	(V27-14)	---	L-DGO
28	---	Photo	Heezen and Hollister, 1971, Fig. 9.55L
29	RC9-205	RC9-181	L-DGO
30	V4-33	---	L-DGO
31	V10-79	---	L-DGO
32	---	Photo	NAVOCEANO, 1967b, Fig. 2
33	AS5-2	---	NAVOCEANO, 1967b
34	---	Photo	Heezen and Johnson, 1967, Fig. 15AA
35	J(JOIDES)135	---	Hayes, Pimm et al., 1972

APPENDIX II (cont'd)

STA NO.	PISTON CORE ID (DREDGE ID)	PHOTO ID	REFERENCES
36	---	Photo	LePichon et al., 1971
37	V27-156	V27-K95	L-DGO
38	V27-155	V27-K94	L-DGO
39	V27-154	V27-K93	L-DGO
40	V27-153	---	L-DGO
41	V27-152	V27-K92	L-DGO
42	V27-151	V27-K91	L-DGO
43	V27-150	V27-K90	L-DGO
44	V27-148	---	L-DGO
45	---	V27-K89	L-DGO
46	V27-149	---	L-DGO
47	V4-32	---	L-DGO
48	---	Photo	Heezen and Hollister, 1971, Fig. 12.18U,L
49	V4-27	Photo	L-DGO; Heezen and Hollister, 1971, Fig. 2.9UL, UR
50	V4-25	---	L-DGO
51	V4-23	---	L-DGO
52	V4-24	---	L-DGO
53	V4-22	---	L-DGO
54	V4-20	Photo	L-DGO; Heezen and Hollister, 1971, Figs. 9.18L, 9.19L, 12.15 U,L
55	R5-50	---	Ericson et al., 1961
56	V4-19	---	L-DGO
57	---	Photo	Heezen et al., 1959, Plate 15-1,2,3; Laughton, 1963, Figs. 3, 22, 29
58	V4-15	Photo	L-DGO; Laughton, 1963, Fig. 30
59	---	Photo	Heezen and Hollister, 1971, Fig. 4.2UL; Heezen et al., 1959, Plate 15-6
60	---	Photo	Heezen and Hollister, 1971, Fig. 2.79UR; Laughton, 1963, Fig. 16
61	V27-157	V27-K96	L-DGO
62	V27-158	V27-K97	L-DGO
63	V27-159	V27-K98	L-DGO
64	V27-161	V27-K100	L-DGO
65	V27-160	V27-K99	L-DGO
66	---	Photo	NAVOCEANO, 1967b, Fig. 1
67	J136	---	Hayes, Pimm et al., 1972

ACOUSTICAL CHARACTERISTICS OF THE SEA FLOOR: EXPERIMENTAL TECHNIQUES AND SOME EXAMPLES FROM THE MEDITERRANEAN SEA

TUNCAY AKAL

NATO SACLANT ASW Research Centre

La Spezia, Italy

ABSTRACT

The acoustical properties of the sea floor are primarily determined by the physical properties of the sediments, layering in the sediments, and the roughness of the bottom and sub-bottom. Over the last decade many cores and bottom photographs have been taken and acoustical measurements made to improve our understanding of the mechanism governing the physics of sound in marine sediments. The results of such studies made by NATO SACLANT ASW Research Centre in different regions of the Mediterranean Sea are presented. Methods of observation and data analysis are briefly discussed. The relationships among those properties of the sediments that affect the acoustical characteristics of the bottom are given and results are compared with the acoustical measurements.

INTRODUCTION

As the physical properties of the sea floor affect the propagation of acoustic energy in this medium, knowledge of these properties and their relation to acoustical parameters is essential for the understanding of the physics of sound in marine sediments.

Research on the physical and acoustical properties of the Mediterranean sea floor has received considerable attention in the SACLANT ASW Research Centre and a number of theoretical and experimental studies have been made since the early 1960's. In the studies conducted the main emphasis has been:

i) To determine the magnitude of bottom reflection losses with respect to frequency and the angle of incidence, the distortion of the reflected waves, and scattering.

ii) To relate these acoustic parameters to the environmental characteristics of the sea floor, such as physical properties of the sediments, layering in the sediments, and the roughness of the bottom.

The Mediterranean Sea is one of the major closed basins of the world and is composed of a western and an eastern basin separated by a shallow sill across the Strait of Sicily (Fig. 1)[Akal 1972a]. These basins have many physiographic and sedimentary features that are characteristic of oceanic basins, such as well-developed continental shelves (comparatively narrow in the Mediterranean), continental slopes (steep and generally cut by submarine canyons), continental rises and abyssal plains (extensive areas where the depths are about 3000 meters in the western basin, the greatest depth of 5081 meters being in the eastern basin) where sediment deposition mostly occurred by turbidity currents. These two basins can be subdivided further into a complex of small basins with quite arbitrary boundaries. The western basin contains the Alboran-Balearic zone, which includes an extensive abyssal plain, and the Tyrrhenian basin, with a central abyssal plain and its characteristic volcanic seamounts. The eastern basin, which is remarkably different from the western basin, mainly comprises (i) a small abyssal plain in the Ionian basin, (ii) the Mediterranean ridge extending from the Ionian basin and covering the central part of the Levantine basin, (iii) small plains located around the Mediterranean ridge, and (iv) the Nile cone.

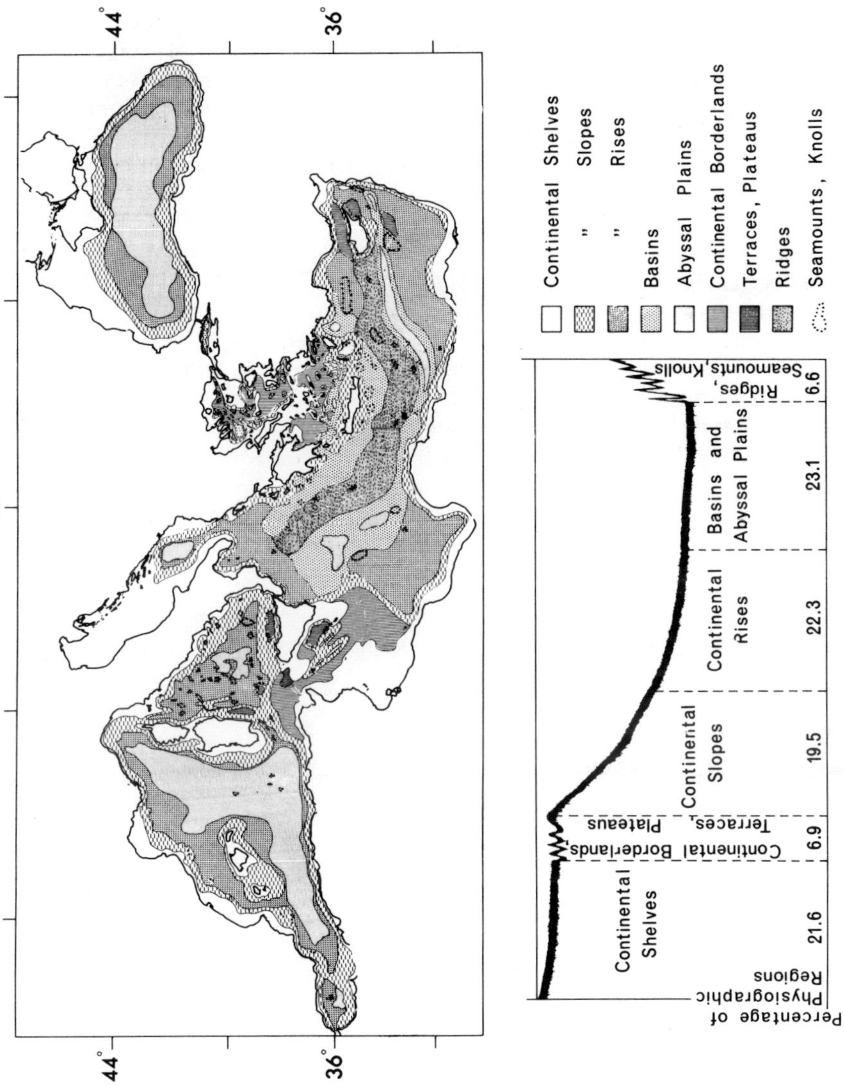

FIGURE 1

MAJOR PHYSIOGRAPHIC REGIONS OF THE MEDITERRANEAN AND BLACK SEAS. [AFTER AKAL (1972a)].

METHODS OF DATA ACQUISITION AND ANALYSIS

Acoustical Parameters

Most of our present knowledge of the sea floor has been obtained through geophysical methods using acoustics as a tool. Seismic continuous profiling, reflection, and refraction measurements are three important and "direct" uses of acoustics in sea floor studies. Generally, continuous profiling methods give the gross features of the configuration of bottom and sub-bottom, and reflection/refraction measurements give detailed information of the velocities in the layers, losses, signal distortion and impulse responses.

Reflection methods have been used at SACLANTCEN to investigate the acoustical properties of the bottom, with emphasis on the intensity of reflection, the distortion of the reflected wave, and scattering. Since all these parameters are frequency and angle dependent, explosive sound sources with firing-ship and receiving-ship combinations have been used to cover a large frequency band and different angles of incidence.

There are mainly two techniques used during reflectivity measurements, as seen in Fig. 2. Although the technique of operating

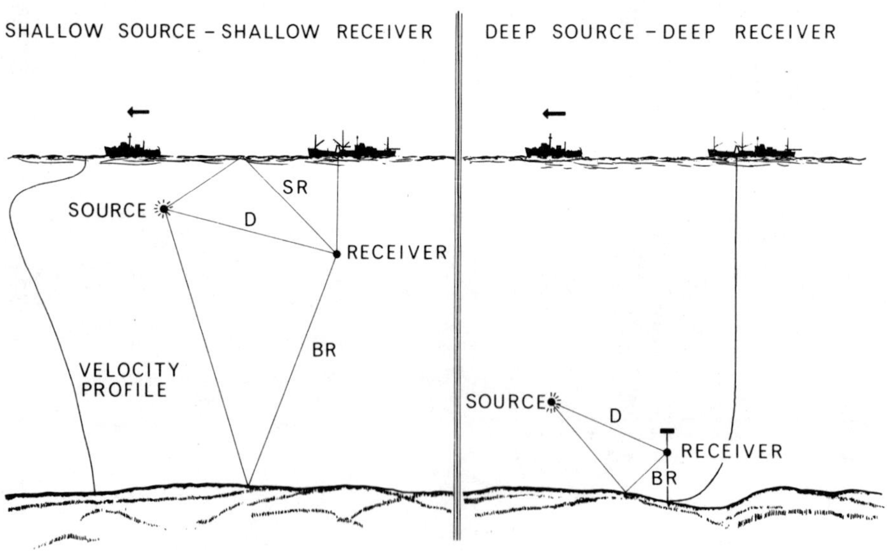

FIGURE 2

BOTTOM REFLECTIVITY MEASURING TECHNIQUES.

with shallow source and receiver is very rapid, the deep source and receiver technique has the advantages that it is self-calibrating, no surface reflections are involved, rays pass through a rather constant environment, and small areas are insonified.

The signals are generally recorded on both ships in digital form. Figure 3 shows the block diagram of the equipment most often used. A typical example of a recorded signal is shown in Fig. 4, where direct and bottom reflected signals are apparent.

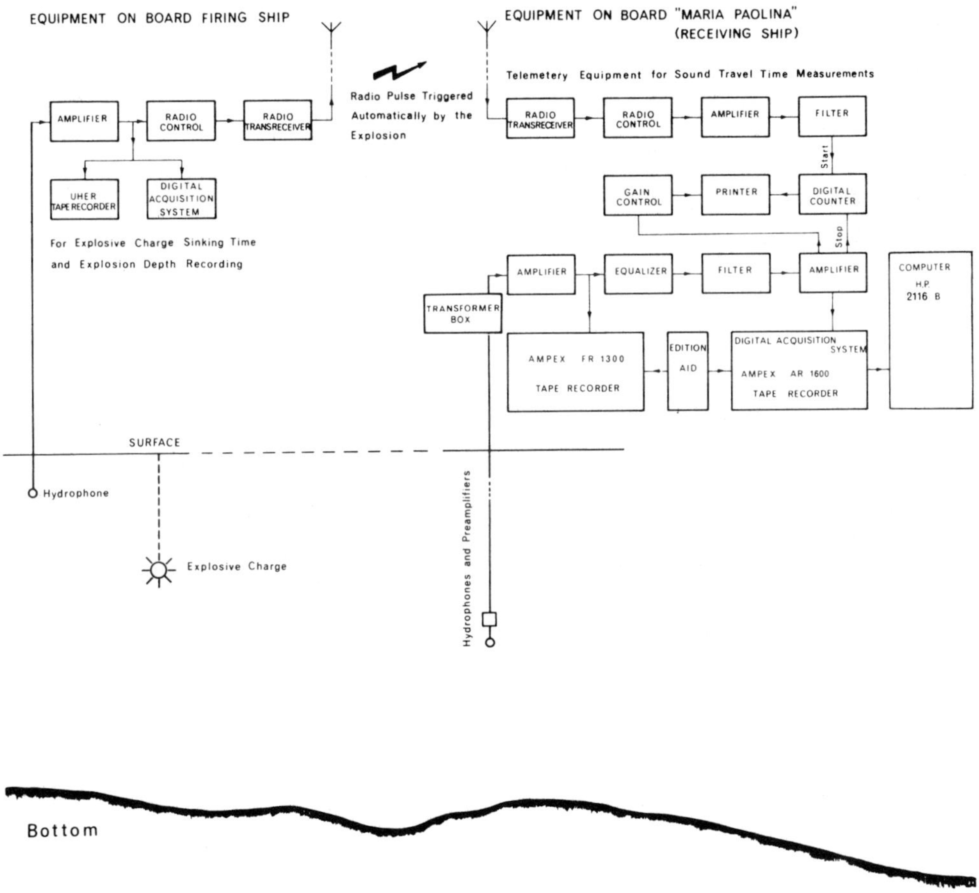

FIGURE 3

BLOCK DIAGRAM OF THE EQUIPMENT.

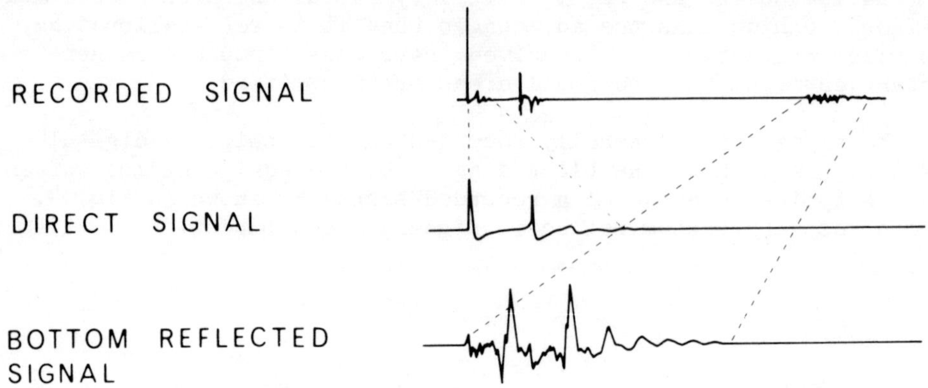

FIGURE 4

AN EXAMPLE OF A RECORDED SIGNAL AND ITS COMPOSITION.

Many different techniques exist to compute reflection losses from experimental results. The methods used at SACLANTCEN were developed by O. Hastrup, and references [Hastrup 1966a,b,c., 1967a,b., 1968, 1969a,b.] give details of these techniques. These methods are briefly discussed below.

1. For water/sediment interface. When the first bottom layer is sufficiently thick, the reflected explosive source signals from the water-sediment interface can be resolved in time from the deeper reflections. Thus it is possible in a simple way to calculate the frequency-independent reflection loss as the ratio between the amplitude of the shock pulse and the peak of the first reflection (after correction for phase shift, absorption, and differences in spreading loss). Recorded signals and reflection losses as a function of grazing angle are shown together with the physical properties of the cores in Fig. 5 [Michelozzi 1973, Akal et al. 1972].

2. For layered bottom. Since sea floors are usually layered, the reflection losses for the whole bottom cannot be obtained by the peak amplitudes method because of the frequency dependence. In this case the analysis (Fig. 6) is based on considering the bottom to be a linear system and computing its transfer function (or the reflection coefficient) from the deconvolution of the direct signal with the reflected signal, using Fourier transforms. Examples of the phase shift and reflection loss as a function of frequency are also shown in Fig. 6.

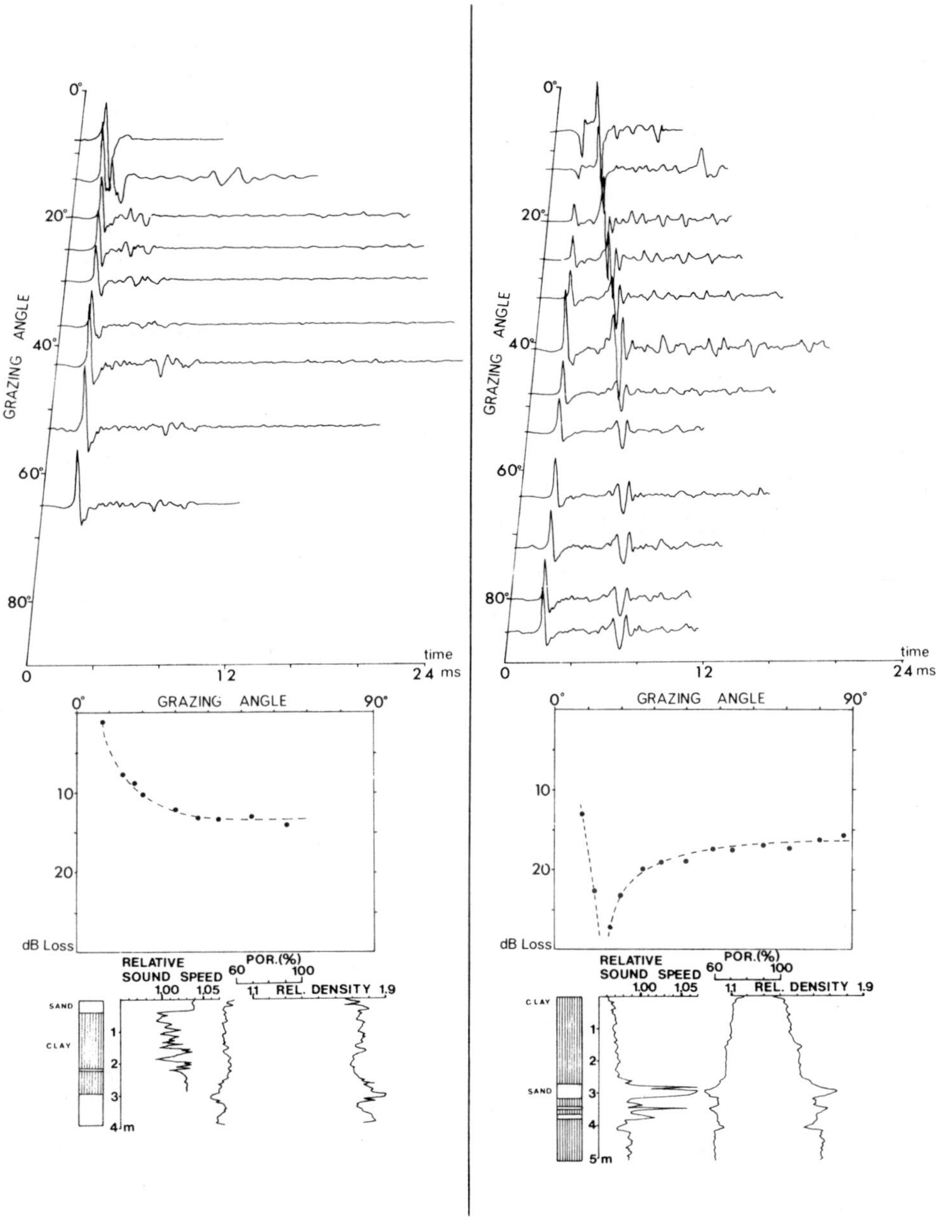

FIGURE 5

RECORDED SIGNALS AND REFLECTION LOSS AS A FUNCTION OF GRAZING ANGLE AND CORE PROPERTIES. [AFTER AKAL (1972a) AND MICHELOZZI (1973)].

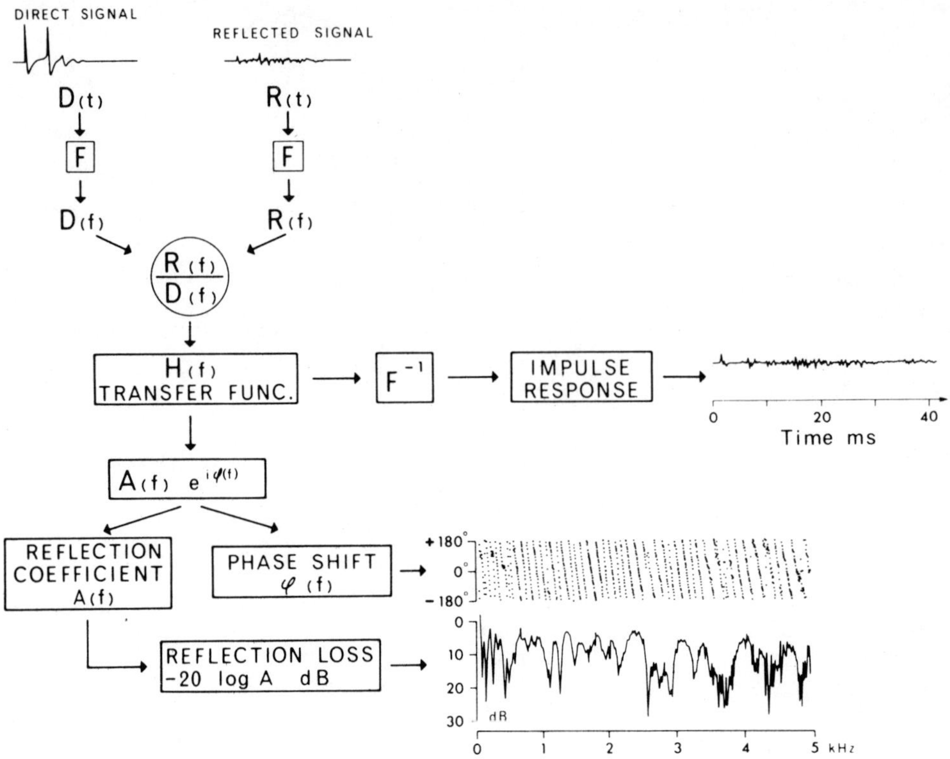

FIGURE 6

DATA ANALYZING TECHNIQUE FOR LAYERED BOTTOM.

The reflectivity can also be described in the time domain by the impulse response, which is the inverse Fourier transform of the transfer function, also shown in Fig. 6. Here the effect of the bubble pulse has disappeared due to the use of a deconvolution process (Fig. 7) developed by J. Hovem [1969]. Figure 8 shows some of the results obtained with these analysis techniques during a reflectivity measurement.

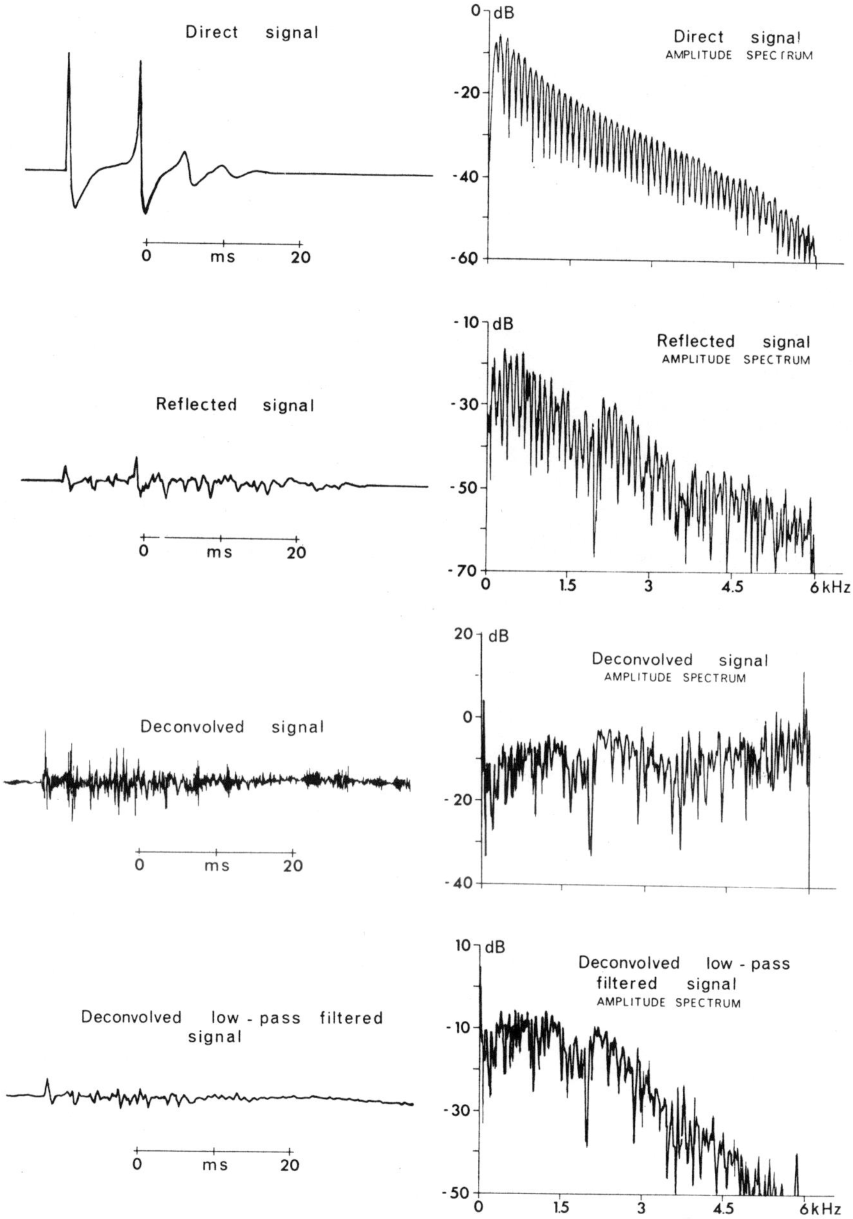

FIGURE 7

PROCESSING TECHNIQUE TO REMOVE THE EFFECT OF THE BUBBLE PULSE. [AFTER HOVEM (1969)].

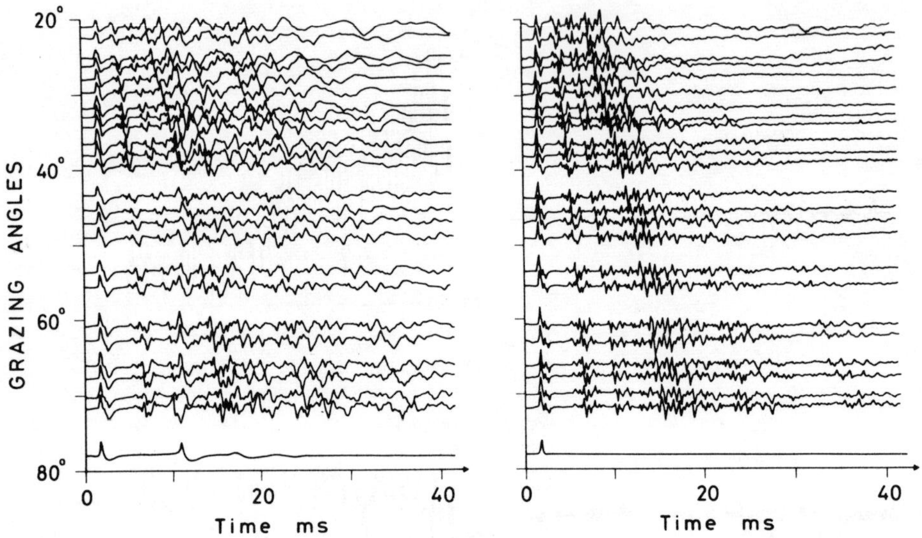

FIGURE 8

REFLECTED SIGNALS AS A FUNCTION OF GRAZING ANGLE BEFORE AND AFTER DECONVOLUTION PROCESSING. [AFTER HOVEM (1969)].

ENVIRONMENTAL PARAMETERS

Physical Properties of Sediments

One of the important objectives in acoustic studies of the sea floor is to know to what extent bottom reflectivity can be predicted from a knowledge of the physical properties of the sediments. Most of our knowledge of the physical properties of sediments is acquired through core sampling. Cores are obtained with the 12 m long and 12 cm diameter SACLANTCEN sphincter corer [Kermabon et al. 1966] and analyzed at 2 or 5 cm intervals along the core to obtain the parameters shown in Fig. 9. Sound speeds are measured by a pulse technique, where the accuracy is approximately ±0.1% in clay and ±0.4% in sand layers. Samples on which physical measurements are made are removed from the same levels at which sound speed measurements were taken. Physical measurements are made of porosity, wet and dry density, water content, void ratio, resistivity, and grain size, with estimated accuracies of about ±1% for porosity, ±0.2% for water content, ±1% for wet density, ±6% for dry density, ±0.2% for void ratio [Kermabon 1967]. After being cut in half the cores are photographed and lithological descriptions are annotated.

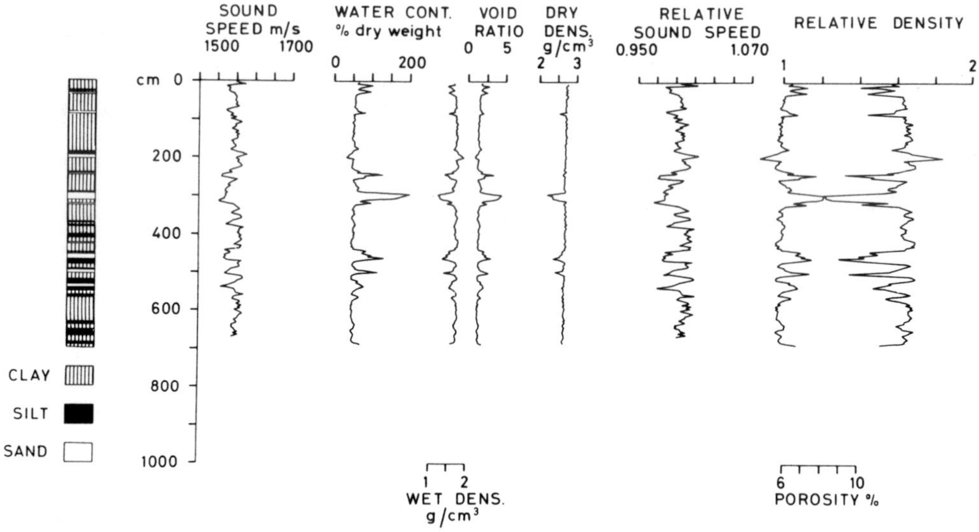

FIGURE 9

PARAMETERS MEASURED AND COMPUTED ALONG A CORE.

To establish a range of normal incidence reflection loss for any particular zone [Akal 1972a], curves can be constructed giving relative porosity from the measured data on all cores taken in the zone. From these the normal incidence reflection coefficient can be computed to give the reflection loss versus porosity relationship (Fig. 10).

It is well known that the reflectivity of the bottom can be expressed as a function of the compressional and shear velocities, the attenuation of compressional and shear waves, and the density. Since our analysis does not include a determination of attenuation, nor of shear velocity, these quantities are deduced from our physical measurements using the results obtained by Hamilton [1971a,b and 1972]. These data are then used as inputs to a computer model to predict bottom losses and impulse responses. The measured and computed results are shown in Fig. 11, where it can be seen that the agreement is remarkably good [Hastrup 1969a].

Layering in the Sediments

The layering of the upper 10 m has a significant effect on the reflection process. An inspection of deep water cores taken in the Mediterranean shows that generally they are of two distinct types. As seen in Fig. 12, core No. 137 consists of many layers (turbidity sediments), whereas core No. 138 is very homogeneous (pelagic sediments). They were both taken from the Ionian basin about 220 km apart. The changes of sound speed, density, and porosity as a function of depth are shown in the same figure. The changes in the properties are also evident from plots of sound speed and density versus porosity. Changes in these parameters naturally cause the reflection characteristics of the bottom to change. The signals

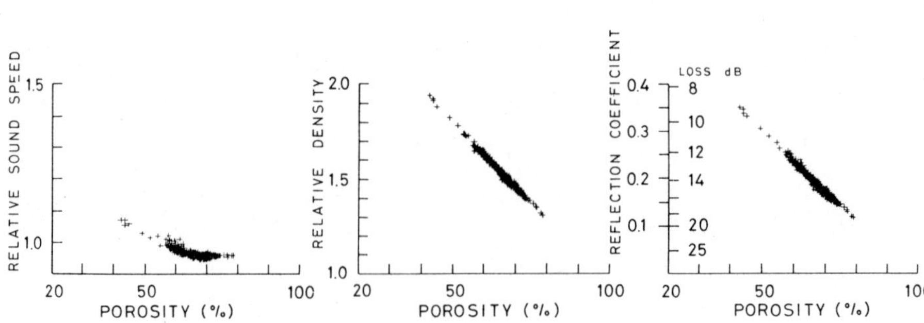

FIGURE 10

RELATIONSHIP BETWEEN SOME OF THE MEASURED PARAMETERS AND COMPUTED NORMAL INCIDENCE BOTTOM LOSS FOR ONE AREA.

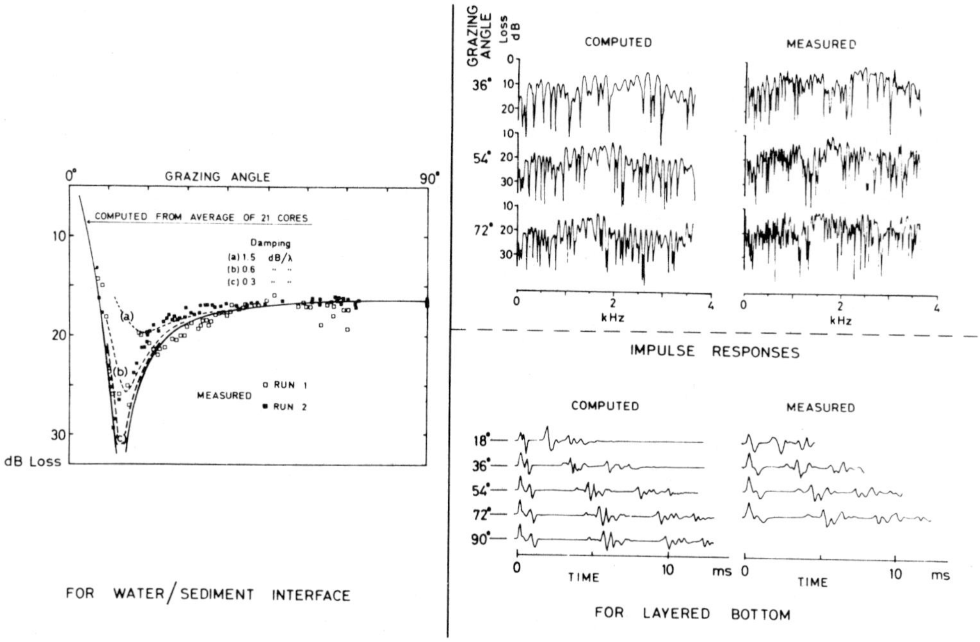

FIGURE 11

MEASURED AND COMPUTED RESULTS FOR WATER/SEDIMENT INTERFACE AND LAYERED BOTTOM. [AFTER HASTRUP (1969a)].

received from a layered bottom are characterized by multiple arrivals corresponding to reflections from the sub-bottom layers (core No. 137), whereas the homogeneous bottom, by contrast, shows a strong pulse from the water/sediment interface followed by a decaying part (core No. 138) (Fig. 12).

Hastrup [1969b] established that very low losses result from a system of regular layering when the thickness of double layers equals half the acoustic wavelength. Such periodicity exists in the areas where turbidity currents play an important role in the sedimentation process. To summarize the information concerning layering in the sediment, we may examine the spectra of the various physical parameters measured in the cores. Figure 13 gives the spectrum of the density versus depth plots obtained from three cores taken from the same area.

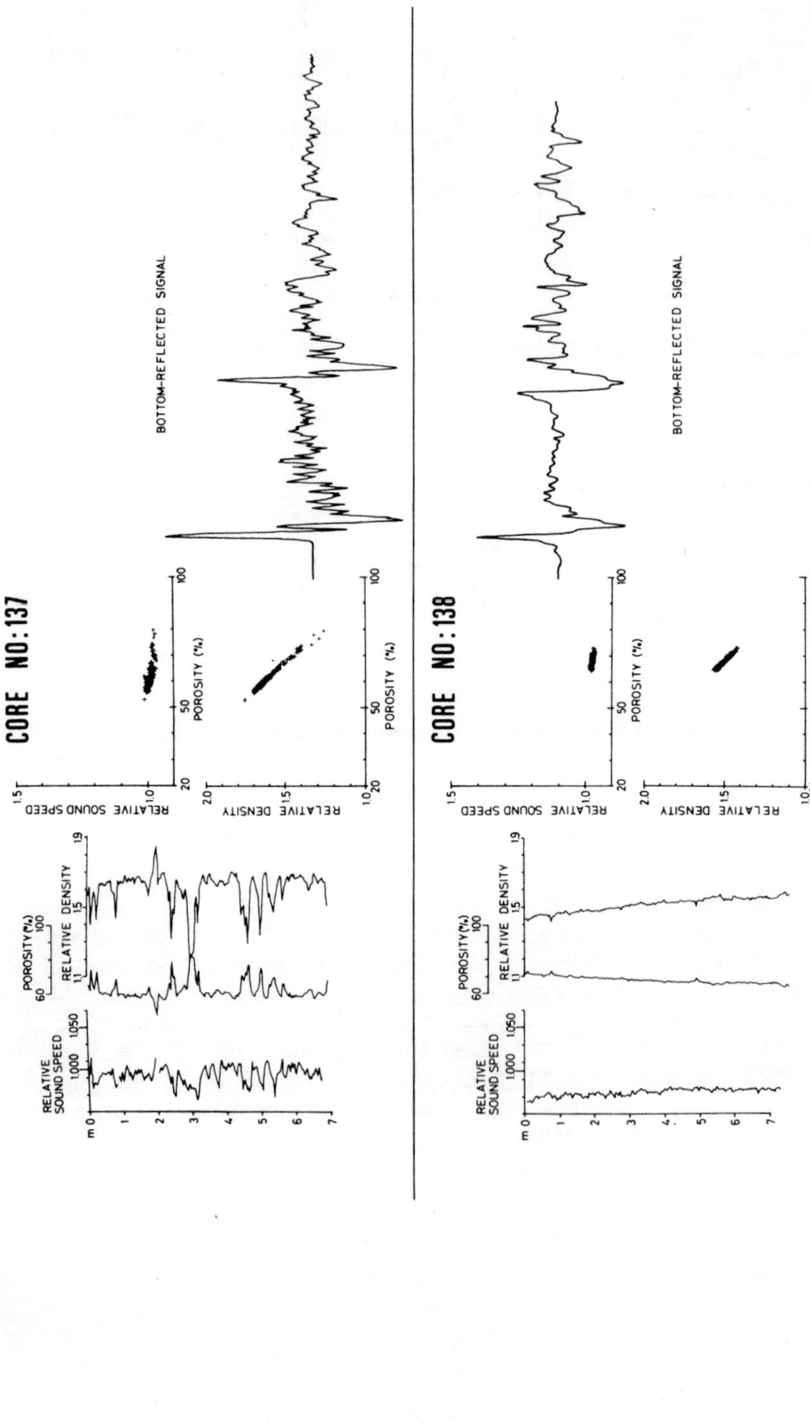

FIGURE 12

MEASURED PARAMETERS FROM CORES NO. 137 AND 138 AND SAMPLES OF BOTTOM REFLECTED SIGNALS FROM THE SAME ZONE.

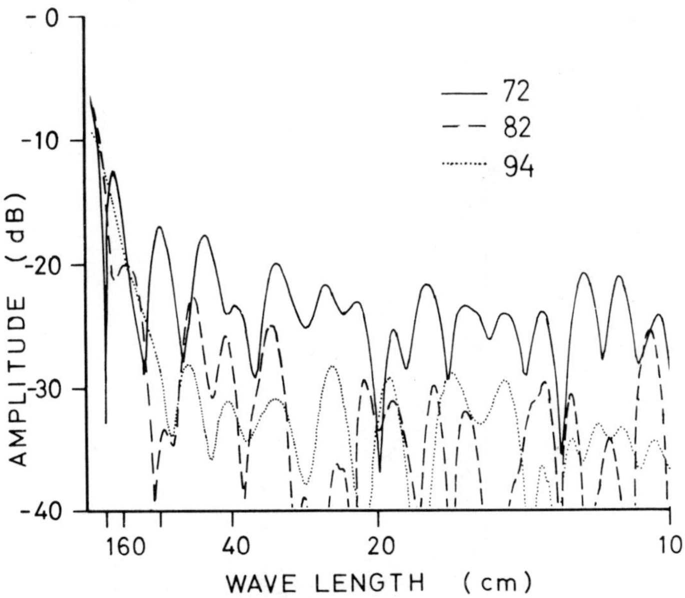

FIGURE 13

DENSITY SPECTRA OF THREE CORES TAKEN FROM THE SAME AREA. THE ORDINATE IS NORMALIZED SPECTRAL AMPLITUDE IN dB. WAVELENGTH IS DISTANCE UNITS ALONG THE CORE.

Sea Floor Roughness

Another important environmental parameter to be considered in reflectivity studies is the roughness of the bottom. The sea floor contains a wide spectrum of topographic roughness, from features of the order of tens to hundreds of kilometers, to those of the order of centimeters. This very broad spectrum can be divided into three parts, as shown in Fig. 14:

a. Gross features, of the order of tens to hundreds of kilometers, are mainly physiographic features such as ridges, abyssal plains, trenches, etc.

b. Intermediate features, of the order of hundreds or thousands of meters, are usually part of the gross features (banks, valleys, hills, etc.).

c. Small features, of the order of centimeters to tens of meters, are the most important ones in reflection processes because they are of the same order as the acoustic wavelengths (ripples, boulders, mounds, rock outcrops, etc.).

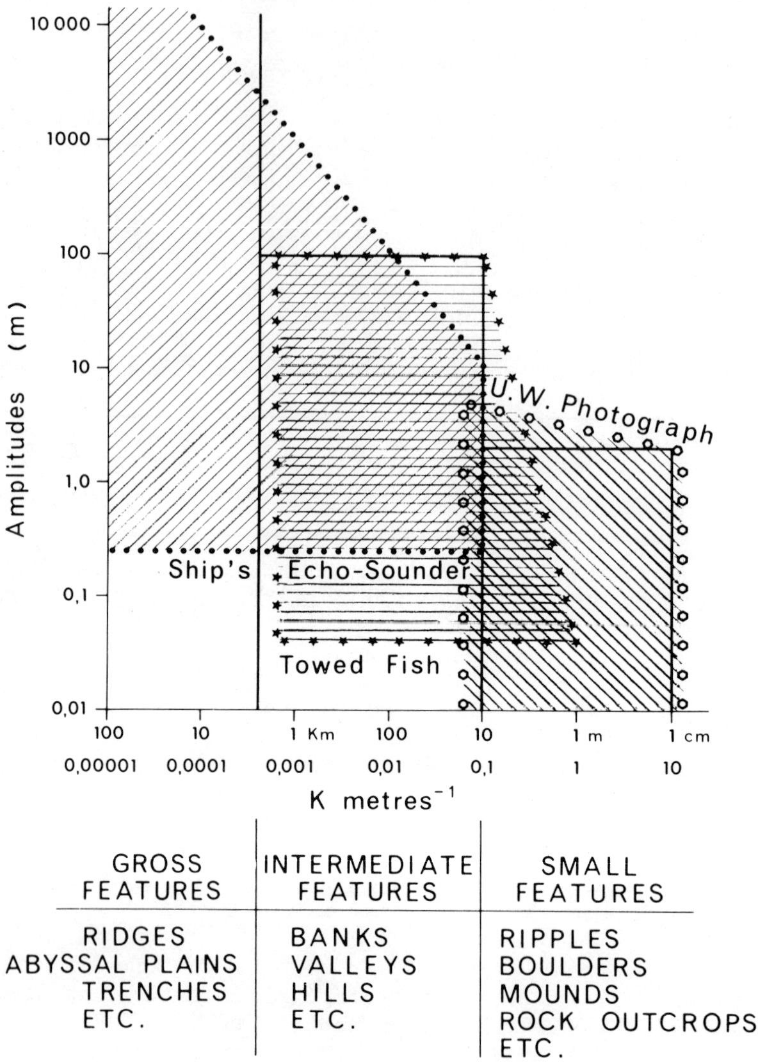

FIGURE 14

SPECTRUM RANGE OF THE SEA FLOOR ROUGHNESS AND RESOLUTION CAPABILITIES OF THE MEASUREMENT TECHNIQUES.

The gross scale features of the sea floor are reasonably well known due to the convenience of such measuring techniques as the use of echo sounders. On the other hand there is little known about small scale roughness due to the difficulties of measurement. Figure 14 also compares the resolution capabilities of the equipment used to study the roughness of the sea floor.

Small scale roughness is measured at SACLANTCEN by means of a stereo underwater camera and a 50 kHz narrowbeam echo-sounder towed close to the bottom (5-10 m). It is a very recent development at SACLANTCEN and we have not yet collected many records for analysis, but the technique works well in water depths less than 200 m. Figure 15 shows examples of the recordings obtained with different sampling techniques and the spectra obtained from these recordings.

Whereas the narrowbeam echo-sounder provides information along individual tracks, the stereo cameras can provide a mosaic over small areas and photogrammetric techniques can be subsequently employed to provide very fine-scale contour charts (of the order of 5 mm resolution) of the sea floor. We have developed a very simple method of digitizing these contours to obtain, through the computer, two-dimensional power spectra and autocorrelation functions of the surfaces, as shown in Fig. 16. The technique has the advantage of revealing any fine roughness directionality of the sea floor.

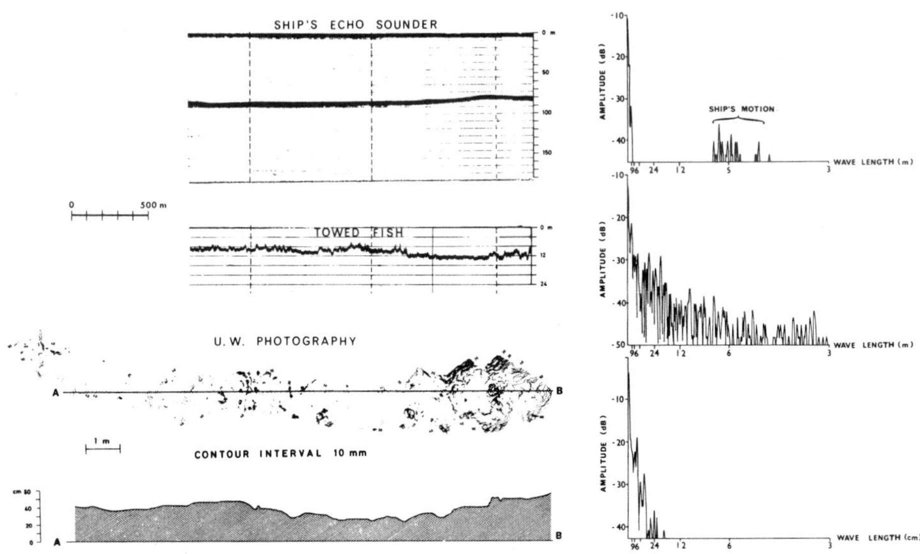

FIGURE 15

EXAMPLES OF RECORDINGS OBTAINED WITH DIFFERENT TECHNIQUES AND THE SPECTRA OBTAINED FROM THESE RECORDINGS.

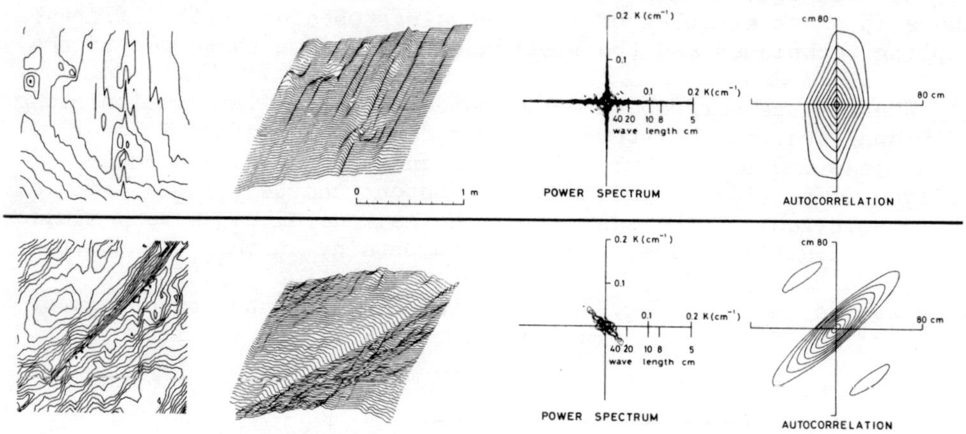

FIGURE 16

CONTOUR PLOTS OBTAINED FROM STEREO PHOTOGRAPHS, PERSPECTIVE VIEWS OF THE SURFACES, TWO-DIMENSIONAL POWER SPECTRA, AND AUTOCORRELATION FUNCTIONS OF THE SURFACES.

THE PHYSICAL PROPERTIES OF THE SEDIMENTS IN SOME OF THE MEDITERRANEAN PHYSIOGRAPHIC REGIONS

Figure 17 shows the areas from which information concerning the physical properties of the sediments was obtained. Figure 18 shows the positions of recent cores taken from different regions of the Mediterranean together with bathymetric profiles recorded in their vicinity. For each area a figure (Figs. 19-27) has been prepared to summarize the data. These show:

a. In the upper part, porosity versus relative sound speed, relative density, and computed normal incidence reflection coefficient.

b. At the lower left, relative sound speed, porosity, and relative density measured along a typical core.

c At the lower right, the spectrum of the relative density measured along some of the cores taken in the area.

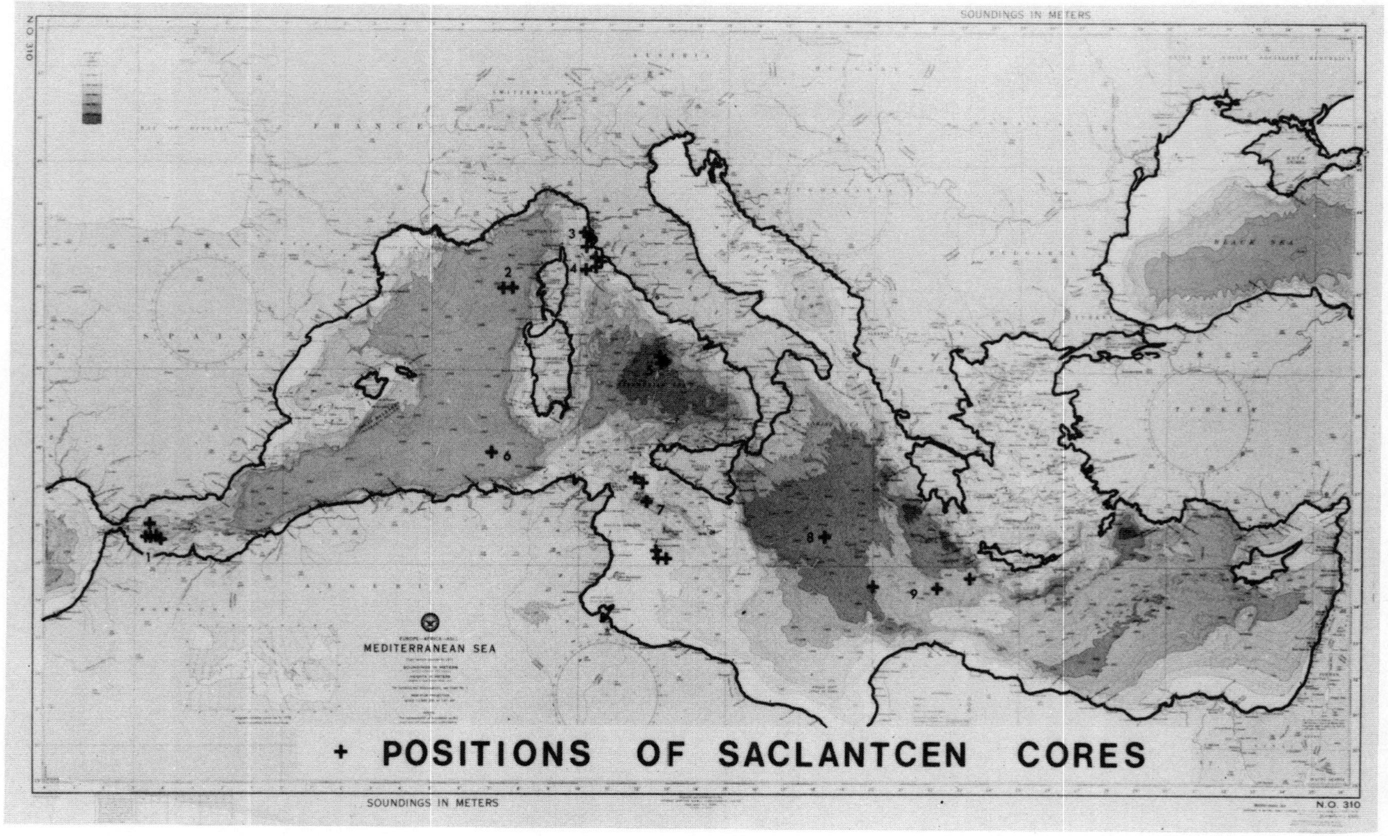

FIGURE 17

POSITIONS OF THE SACLANTCEN CORES.

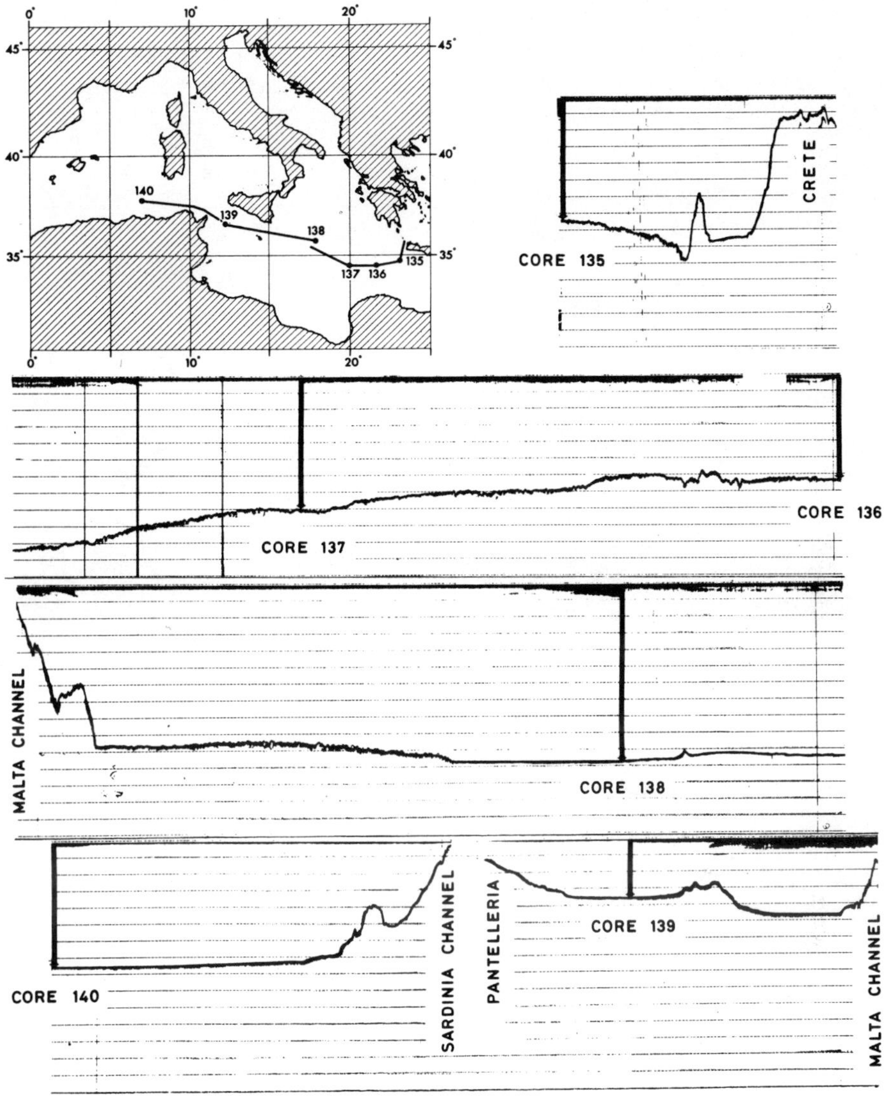

FIGURE 18

ECHO-SOUNDING RECORDS AND POSITIONS OF RECENT SACLANTCEN CORES.

1. <u>(Fig. 19) Western Alboran Basin</u>: Sedimentological investigation of the western basin has been made by extensive coring [Gehin et al. 1971]. The cores taken in this area show homogeneous material with some strips of sand layers believed to be caused by turbidity currents. As can be seen from the figure homogeneous material has more than 60% porosity, with most of the data falling between 60% to 80%, giving sound speeds of less than that in water. The spectra of relative density indicate marked layers approximately 12, 30, 50, 160, and 300 cm thick.

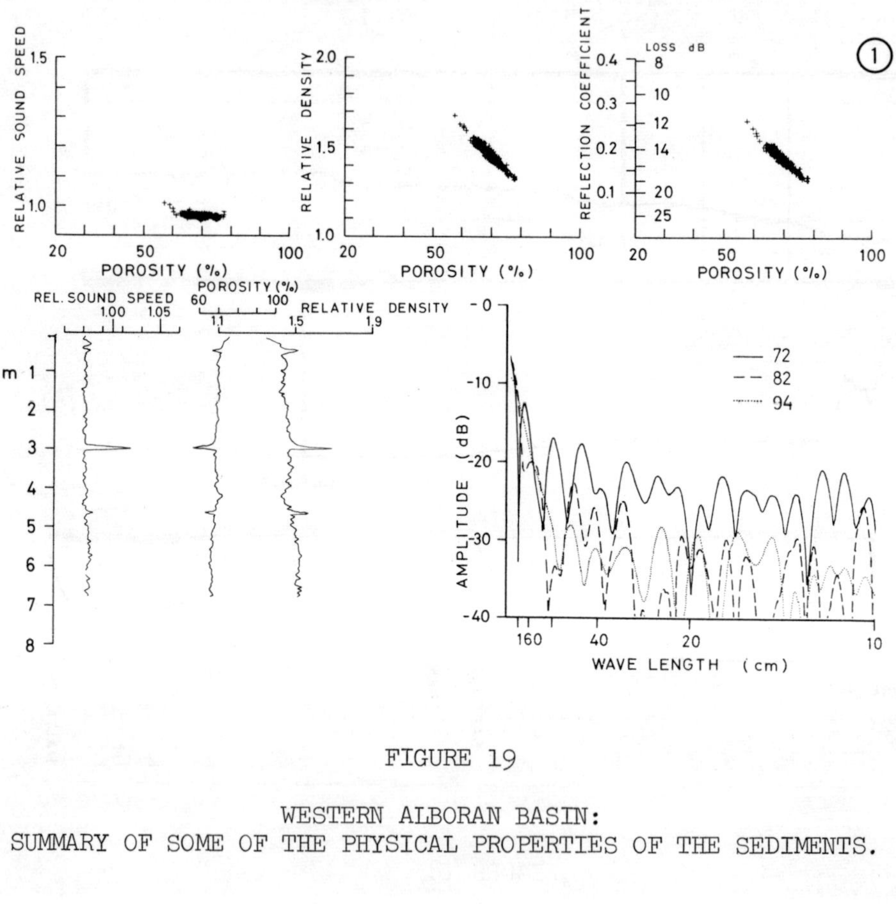

FIGURE 19

WESTERN ALBORAN BASIN:
SUMMARY OF SOME OF THE PHYSICAL PROPERTIES OF THE SEDIMENTS.

2. (Fig. 20) Northern Balearic Basin: The cores taken in this area contain alternating layers of sand, silt, and clay that appear as equally distributed data points of 45% to 75% porosity. There are also marked layers whose thicknesses vary from 10 to 250 cm.

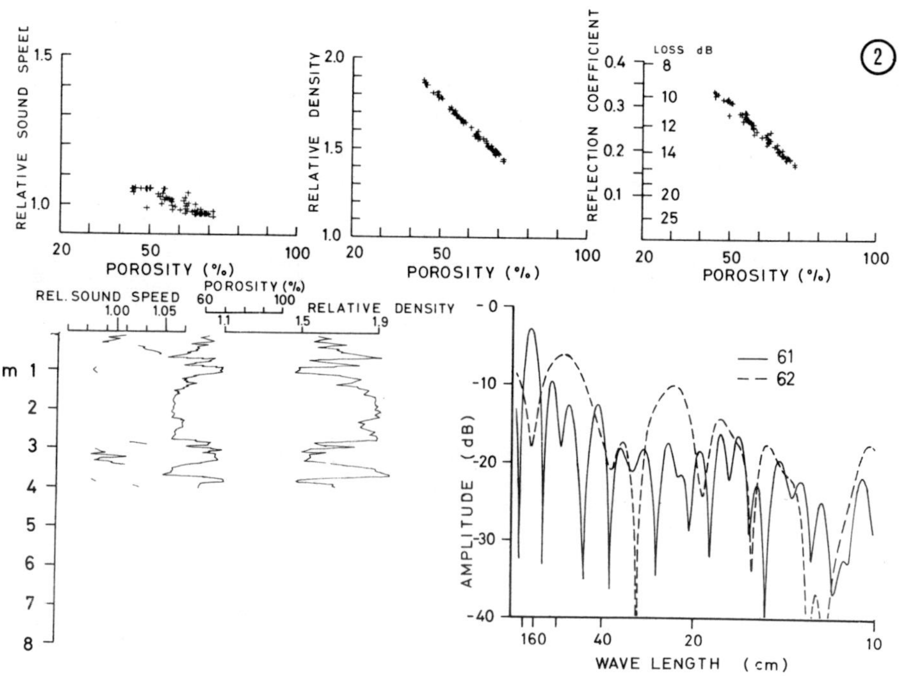

FIGURE 20

NORTHERN BALEARIC BASIN:
SUMMARY OF SOME OF THE PHYSICAL PROPERTIES OF THE SEDIMENTS.

3. <u>(Fig. 21) Northern Elba Shelf</u>: The cores taken in this zone also show mixed layers of clay, silt, and sand, and the porosities measured vary between 40% and 80%.

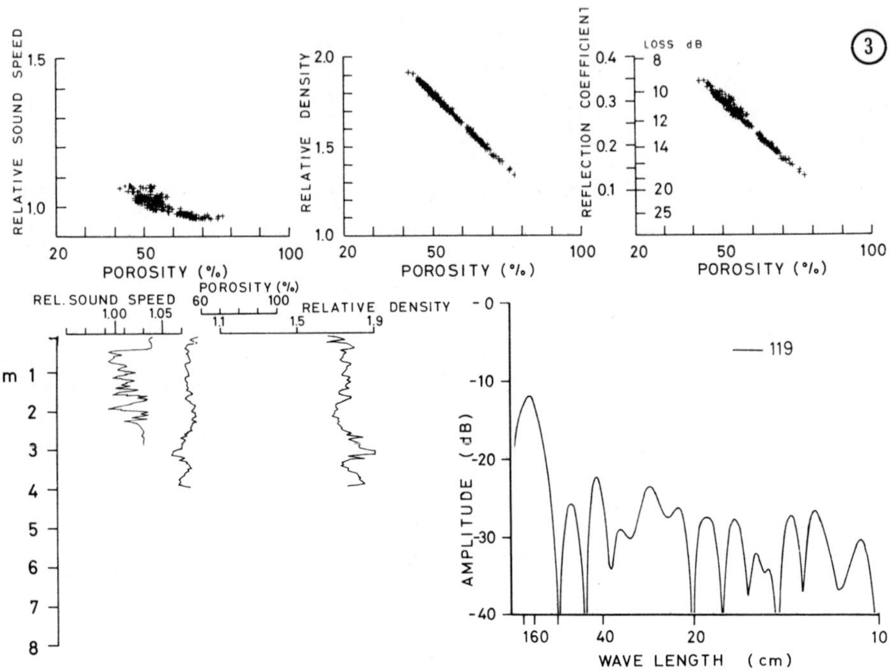

FIGURE 21

NORTHERN ELBA SHELF:
SUMMARY OF SOME OF THE PHYSICAL PROPERTIES OF THE SEDIMENTS.

4. <u>(Fig. 22) Southeastern Elba Shelf</u>: This area shows completely different characteristics from those in the northern zone. The cores contain homogeneous clay material with some layers of sand. Data points on the plots are mostly in the 55% to 80% porosity range except for a few points with lower porosity that correspond to the sand layers. A layer approximately 3 m thick is a common feature of most cores taken in this area.

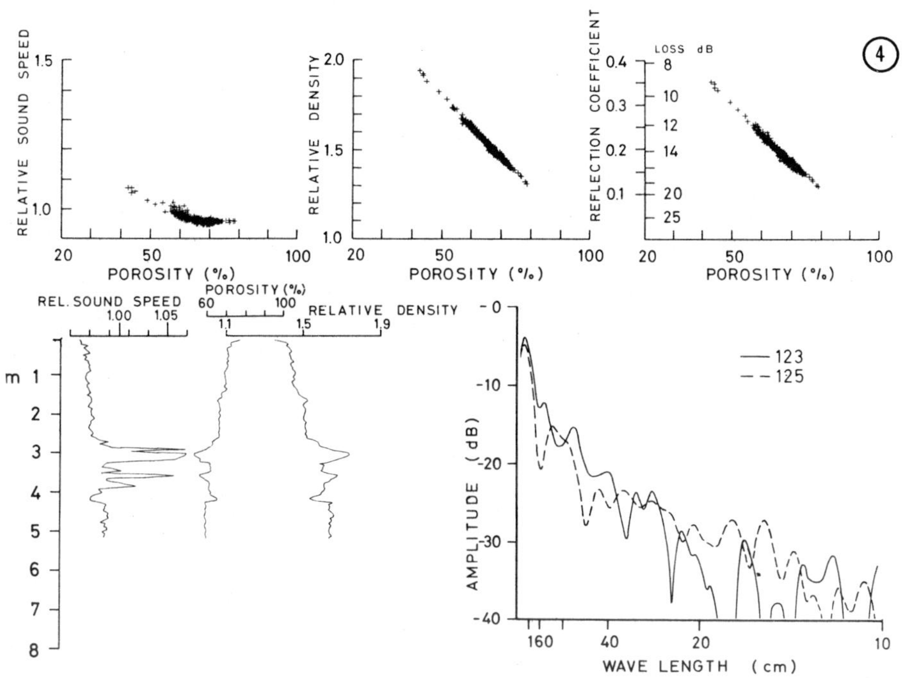

FIGURE 22

SOUTHEASTERN ELBA SHELF:
SUMMARY OF SOME OF THE PHYSICAL PROPERTIES OF THE SEDIMENTS.

5. (Fig. 23) Tyrrhenian Abyssal Plain: Analyses of the cores taken in this plain [Kermabon et al. 1968] indicate clay, silt, and sand layers. The thicknesses of the clay layers vary from a few millimeters to 3 m, the silt and sand layers from 1 cm to 1 m. Coarse material, which makes a sharp contrast with the clay layers, was most probably caused by turbidities or volcanic ash depositions. The spectrum of the relative densities measured along the cores shows good correlation from one core to another.

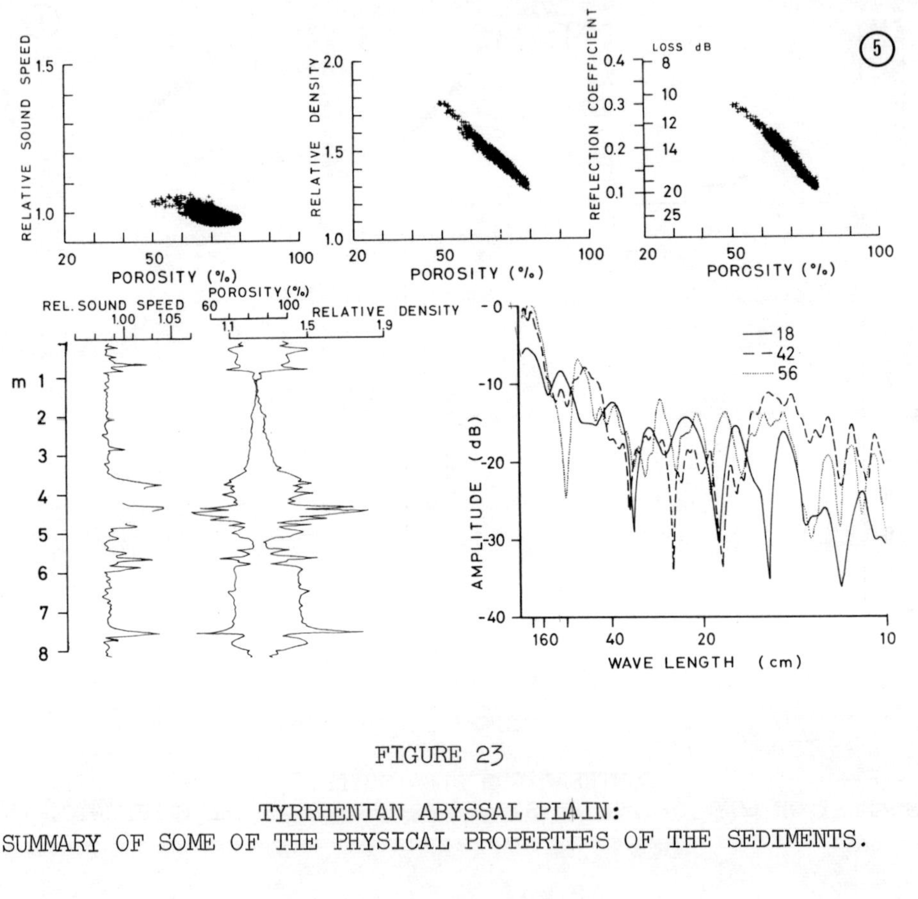

FIGURE 23

TYRRHENIAN ABYSSAL PLAIN:
SUMMARY OF SOME OF THE PHYSICAL PROPERTIES OF THE SEDIMENTS.

6. (Fig. 24) Southeastern Balearic Basin: This core (140) contains alternating layers of clay and sand. The data points fall in two clusters, with the clay material having porosities of more than 60%. There is a high density material at a depth of about 3 m.

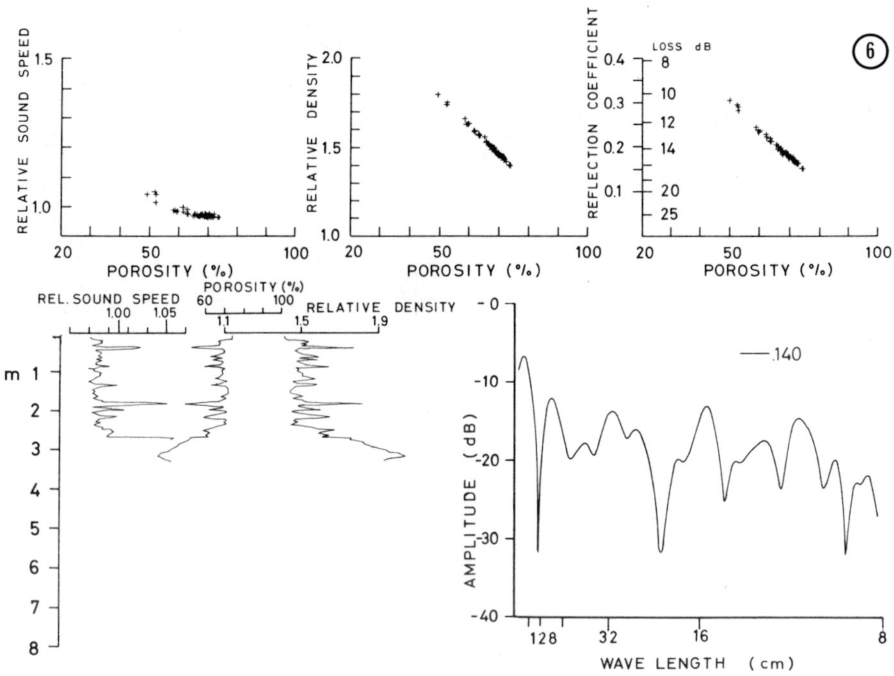

FIGURE 24

SOUTHEASTERN BALEARIC BASIN:
SUMMARY OF SOME OF THE PHYSICAL PROPERTIES OF THE SEDIMENTS.

7. <u>(Fig. 25) Pantelleria Basin</u>: The core (139) taken from this small basin (southeast of the island of Pantelleria) is very homogeneous. As shown by the plots of porosity relationships, the relative sound speeds are less than one and the porosities vary between 65% and 80% [Akal 1972b].

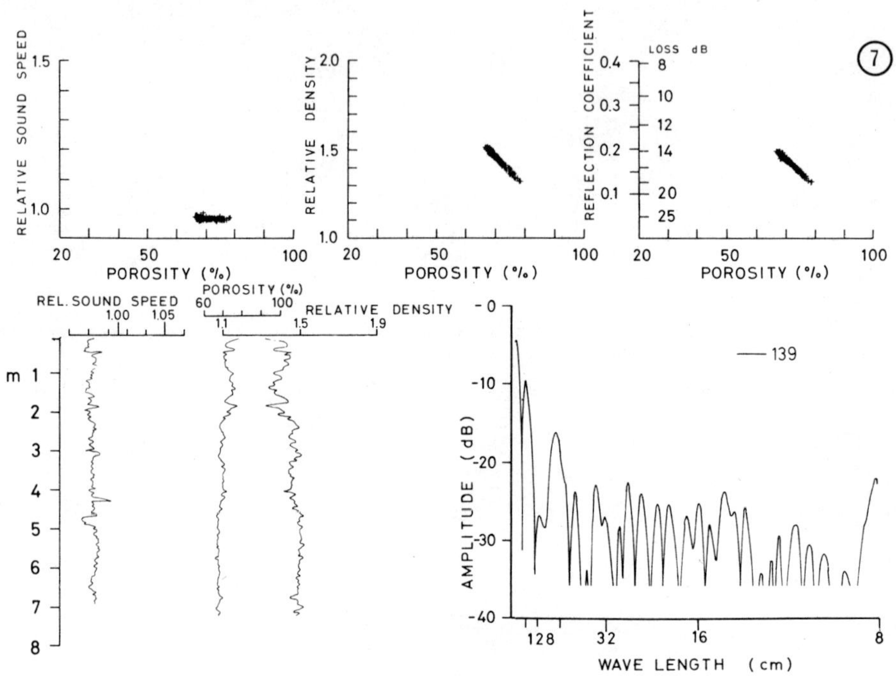

FIGURE 25

PANTELLERIA BASIN:
SUMMARY OF SOME OF THE PHYSICAL PROPERTIES OF THE SEDIMENTS.

8. <u>(Fig. 26) Messina Abyssal Plain</u>: This core (138) is also very homogeneous and the porosity of the clay material varies between 65% and 75%.

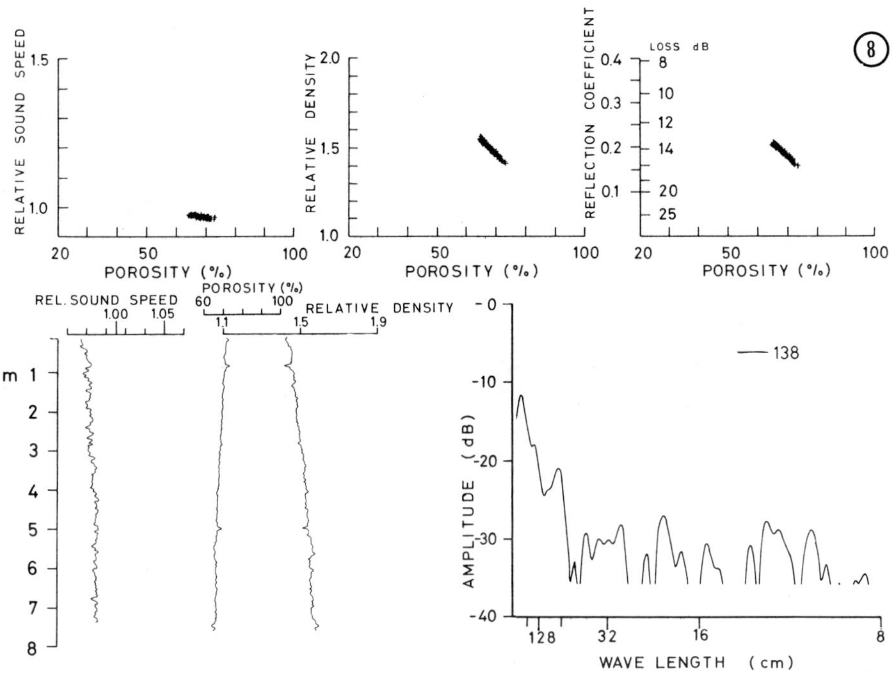

FIGURE 26

MESSINA ABYSSAL PLAIN:
SUMMARY OF SOME OF THE PHYSICAL PROPERTIES OF THE SEDIMENTS.

9. <u>(Fig. 27) Mediterranean Ridge</u>: The cores taken in this area show completely different structure from those taken from the Messina Abyssal Plain. They contain layers of clay, silt, and sand with different thicknesses and colors. The porosities of the material vary between 50% and 80%, with most of the data points falling between 50% and 70%. There are many layers whose thicknesses correlate very well from one core to another.

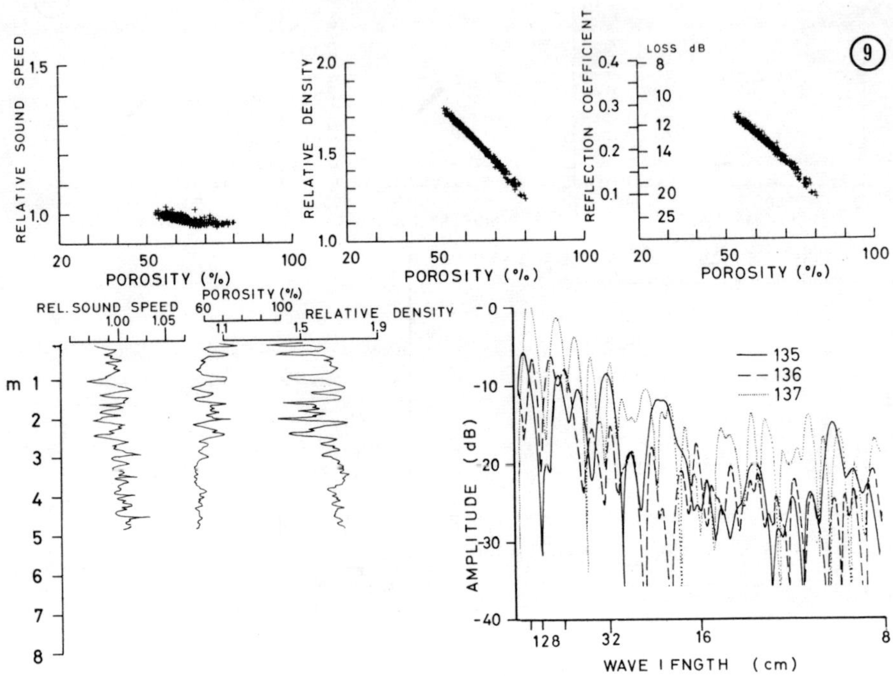

FIGURE 27

MEDITERRANEAN RIDGE:
SUMMARY OF SOME OF THE PHYSICAL PROPERTIES OF THE SEDIMENTS.

CONCLUSIONS

1. Acoustical characteristics of the sea floor can be predicted from a knowledge of the physical properties of the sediments.

2. Displays of relative sound speed, relative density, and computed normal incidence reflection coefficient versus porosity give a general idea of the acoustical characteristics of an area.

3. The porosity of the marine sediments stands out as the most important parameter causing variations in compressional sound speed and density.

4. The relationships between relative sound speed, relative density, and computed normal incidence reflection coefficient versus porosity for different Mediterranean regions agree well with the statistical relationships previously calculated [Akal 1972a] from an analysis of world data (Fig. 28).

5. Layering information can be summarized by obtaining the spectra of the various physical parameters along the core.

6. The sea floor contains a wide spectrum of topographic roughness that can be divided into three parts and can be resolved by different sampling techniques. Univariate and/or bivariate spectra of the bottom roughness give statistical information for the roughness of the sea floor.

7. The cores taken from different regions of the Mediterranean are, in general, of two distinct types.

   a. Homogeneous cores [Pelagic sediments], generally consisting of clay whose porosity varies between 65% and 80%.

   b. Layered cores (turbidity sediments), consisting of clay, silt, and sand layers where porosities vary between 40% and 80%.

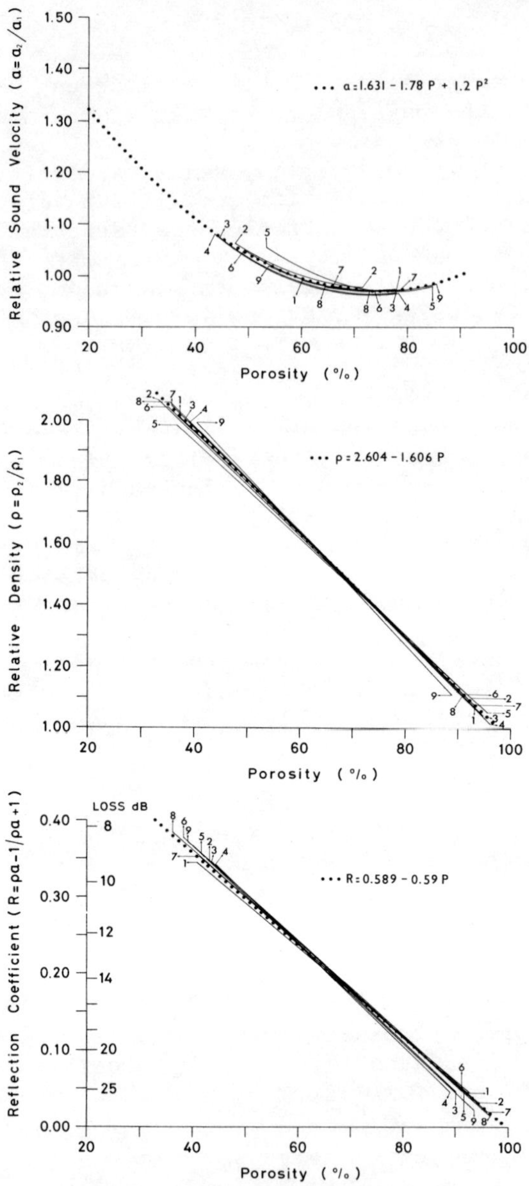

FIGURE 28

COMPARISON OF POROSITY RELATIONSHIPS FROM DIFFERENT MEDITERRANEAN REGIONS (NUMBERED) WITH THOSE CALCULATED (AKAL 1972a) FROM WORLD DATA (DOTTED). 1. WESTERN ALBORAN BASIN, 2. NORTHERN BALEARIC BASIN, 3. NORTHERN ELBA SHELF, 4. SOUTHEASTERN ELBA SHELF, 5. TYRRHENIAN ABYSSAL PLAIN, 6. SOUTHEASTERN BALEARIC BASIN, 7. PANTELLERIA BASIN, 8. MESSINA ABYSSAL PLAIN, 9. MEDITERRANEAN RIDGE.

REFERENCES

Akal, T., "Bathymetry and bottom structure of zones near the Island of Elba used for acoustical trials in shallow water," SACLANT ASW Res. Cen., Tech. Memo 162, La Spezia, Italy, 24 pp., 1970, (AD 879 590).

Akal, T., "The relationship between the physical properties of underwater sediments that affect bottom reflection," Marine Geology, 13, 251-266, 1972a.

Akal, T., The general geophysics and geology of the Strait of Sicily, in Oceanography of the Strait of Sicily, (T. D. Allan, T. Akal, R. Molcard, Eds.), Proceedings of a Conference held at SACLANTCEN, La Spezia, Italy, 11-12 April 1972, SACLANT ASW Res. Cen., Conf. Proc. 7, La Spezia, Italy, pp. 177-192, 1972b, (AD 750 753).

Akal, T., C. Gehin, B. Matteucci, and B. Tonarelli, "Measured and computed physical properties of sediment cores, Island of Elba zone," SACLANT ASW Res. Cen., Spec. Rpt. M-82, La Spezia, Italy, 69 pp., 1972.

Gehin, C., C. Bartolini, D. J. Stanley, P. Blavier, and B. Tonarelli, "Morphology and late quaternary fill of the western Alboran basin, Mediterranean Sea," SACLANT ASW Res. Cen., Tech. Rpt. 201, La Spezia, Italy, 78 pp., 1971, (AD 742 464).

Hamilton, E. L., "Elastic properties of marine sediments," J. Geophys. Res., 76, 579-604, 1971a.

Hamilton, E. L., "Prediction of in-situ acoustic and elastic properties of marine sediments," Geophysics, 36, 266-284, 1971b.

Hamilton, E. L., "Compressional-wave attenuation in marine sediments," Geophysics, 37, 620-646, 1972.

Hastrup, O. F., "Reflection of plane waves from a solid multilayered damping bottom," SACLANT ASW Res. Cen., Tech. Rpt. 50, La Spezia, Italy, 52 pp., 1966a, (AD 479 437).

Hastrup, O. F., "Distortion of bottom reflected pulses," SACLANT ASW Res. Cen., Tech. Rpt. 51, La Spezia, Italy, 34 pp., 1966b, (AD 482 740).

Hastrup, O. F., "The influence of a low velocity layer on the reflectivity from the sea floor," SACLANT ASW Res. Cen., Tech. Rpt. 61, La Spezia, Italy, 42 pp., 1966c, (AD 483 680).

Hastrup, O. F., "Impulse reponse of a layered bottom," SACLANT ASW Res. Cen., Tech. Rpt. 85, La Spezia, Italy, 30 pp., 1967a, (AD 811 104).

Hastrup, O. F., and E. Schunk, "Numerical determination of experimental bottom reflection losses," SACLANT ASW Res. Cen., Tech. Rpt. 104, La Spezia, Italy, 21 pp., 1967b, (AD 824 826).

Hastrup, O. F., "The reflectivity of the top layers in the Naples and Ajaccio abyssal plain," SACLANT ASW Res. Cen., Tech Rpt. 118, La Spezia, Italy, 42 pp., 1968, (AD 839 816).

Hastrup, O. F., "A detailed analysis of acoustic reflectivity in the Tyrrhenian abyssal plain," SACLANT ASW Res. Cen., Tech. Rpt. 145, La Spezia, Italy, 61 pp., 1969a, (AD 854 842).

Hastrup, O. F., "The effect of periodic bottom layering on acoustic reflectivity," SACLANT ASW Res. Cen., Tech. Rpt. 149, La Spezia, Italy, 30 pp., 1969b, (AD 856 029).

Hastrup, O. F., "Digital analysis of acoustic reflectivity in the Tyrrhenian abyssal plain," J. Acoust. Soc. Am., 47, 181-190, 1970.

Hovem, M. J., "Removing the effect of the bubble pulses when using explosive charges in underwater acoustic experiments," SACLANT ASW Res. Cen., Tech. Rpt. 140, La Spezia, Italy, 25 pp., 1969, (AD 849 890).

Kermabon, A., P. Blavier, U. Cortis, and H. Delanze, "The sphincter corer: a wide-diameter corer with watertight core catcher," Marine Geology 4, 149-162, 1966.

Kermabon, A., and P. Blavier, "Principles and methods of core analysis at the SACLANT ASW Research Centre," SACLANT ASW Res. Cen., Tech. Rpt. 71, La Spezia, Italy, 44 pp., 1967, (AD 818 754).

Kermabon, A., C. Gehin, P. Blavier, and B. Tonarelli, "Acoustic and other physical properties of deep sea sediments in the Tyrrhenian abyssal plain," SACLANT ASW Res. Cen., Tech. Rpt. 113, La Spezia, Italy, 47 pp., 1968, (AD 834 802).

Michelozzi, E., "Studio sui primi strati dei sedimenti marini recenti dell'Arcipelago Toscano," Unpublished thesis, University of Pisa, Inst. of Geology, 1973.

STATISTICAL CORRELATION OF PHYSICAL PROPERTIES AND SOUND VELOCITY IN SEDIMENTS

ROCKNE S. ANDERSON

Acoustical Oceanography Division

U. S. Naval Oceanographic Office, Washington, DC 20373

ABSTRACT

A systematic stepwise regression analysis is used to determine the statistical correlation between sound velocity and physical properties measured in the laboratory on sediments from 82 cores taken in the Atlantic, Pacific, and Indian Oceans, and the Mediterranean Sea. Carbonate content and shear strength show no significant correlation with sediment sound velocity. Mean grain size in phi units and porosity are the two parameters showing the highest significant correlation with sound velocity. The inclusion of void ratio, wet density, specific gravity, or the higher moments of the mean grain size statistics does not significantly improve the prediction equation for velocity. The data is grouped by different water depths, oceans, or physiographic provinces to determine the effect of these parameters on the velocity-porosity and velocity-grain size relationships. Results show that the prediction of velocity as a function of grain size is improved by using these groupings of data. In particular, for a given grain size, the velocity is higher in the Atlantic Ocean than in the Pacific Ocean. For a given porosity, the velocity is higher in the oceanic rise provinces than elsewhere. Overall, for all other provinces the data is well represented by one equation relating sound velocity V in meters per second to porosity P in percent. The equation is $V = -22.9P + 0.1500P^2 + 2367$ with a standard error of 35 m/sec. One application is made using this equation and another equation which relates density to porosity. For each data point, a bottom loss is determined from the value of the sediment velocity and density predicted from a measured porosity, and then compared to a bottom loss determined from the actual measured velocity and density. The difference between these two determinations is less than 3 dB for each data point having a specific gravity within the broad range of 2.18 to 3.43 gm/cm^3.

INTRODUCTION

In recent years, in order to achieve a better understanding of reflectivity of sound from the sea floor and the relationships between the acoustic and physical properties of the ocean bottom, numerous studies relating the physical properties of sediments to the velocity of sound through the sediments have been conducted. Breslau [1965] related actual normal incidence bottom loss measurements at 12 kHz with the measured porosity and grain size of the ocean bottom samples. Several authors have attempted to predict a reflection coefficient at normal incidence to the ocean bottom from a measured sediment density or porosity [Akal, 1972; Faas, 1969; Hamilton, 1970a]. The significance of this relationship is diminished by the fact that the actual reflectivity of the ocean bottom may be quite different from the predicted reflection coefficient, especially at lower frequencies. The implication of a high correlation coefficient between the predicted reflection coefficient and porosity, for example, is misleading [Blackman, 1971]. Factors other than physical properties which affect reflectivity are: layering and the persistence of layers, attenuation, and bottom roughness. Bottom photography, bathymetric and seismic profiles, and core data can provide information on these other factors. Nevertheless, the ability to predict sound reflectivity can be improved by knowing to what degree each physical property of the sediment influences sound velocity. In an area where measurements are lacking, these physical properties may be estimated from knowledge of the submarine geology.

The data used in this paper are compiled from the reports of the Marine Geophysical Survey (MGS), which was conducted from 1965 to 1967 under the direction of the U. S. Naval Oceanographic Office. Alpine Geophysical Associates, Inc., and Texas Instruments, Inc., (TI) conducted a number of surveys in different areas of the world oceans. In the present study, data from a total of 82 cores were used from Alpine areas 6 (near Hawaii), 15 (South China Sea), and ST (eastern South Atlantic and Indian Ocean), and TI areas 1 (Norwegian Sea), 2 and 3 (eastern North Atlantic), 7/4 (western North Atlantic), and 6 (Mediterranean Sea) [U. S. Naval Oceanographic Office, 1968c, 1969b, 1968d, 1967, 1968a, 1968b, 1969a; Horn et al., 1967]. The core locations are shown in Fig. 1, except for the Indian Ocean cores, and listed in Table 1. Other MGS data have been ignored in this study since the results are in graphical form only rather than in tabular form and are impractical to use for exact determinations of actual values.

The purpose of the present study is two-fold: (1) to determine by the use of stepwise regression analysis whether the prediction of sediment sound velocity is improved by the use of more than one physical property, and (2) to determine what factors, other than physical properties of the sediment, enhance the prediction of sound velocity. With regards to the latter, the factors that are considered in this paper as to their affect on the relationship between sediment sound velocity and physical properties are: general ocean area, physiographic province, and water depth.

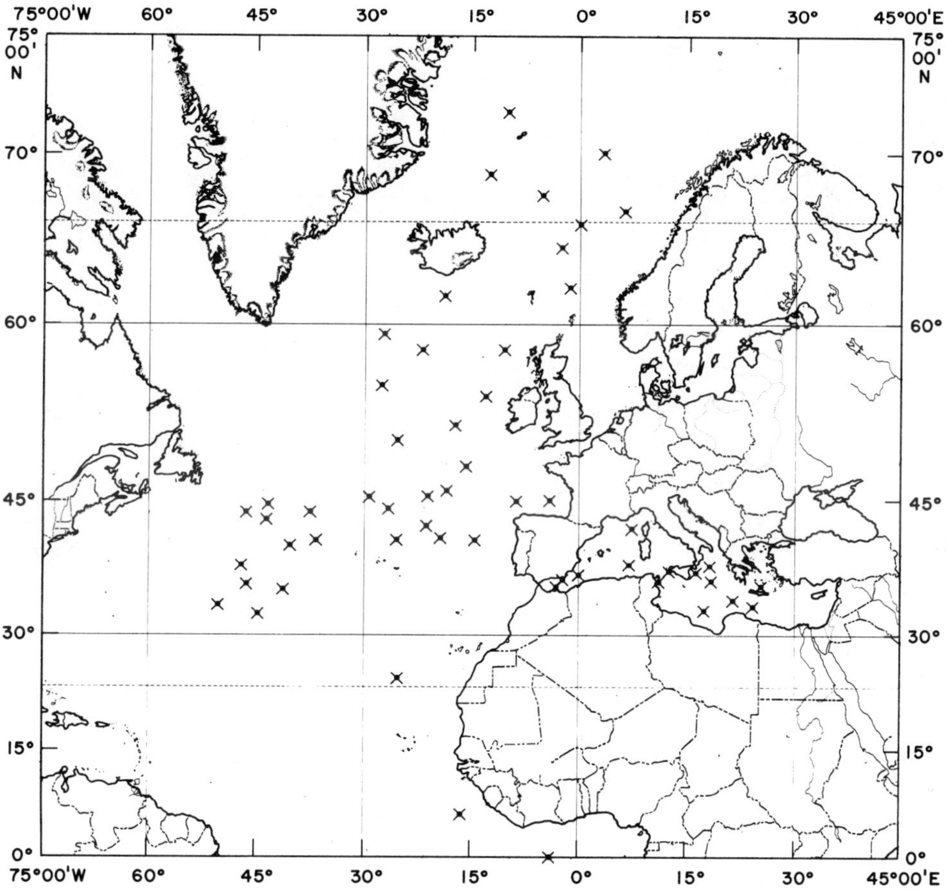

FIGURE 1A

LOCATIONS OF 41 MGS CORES IN THE ATLANTIC OCEAN (INCLUDING THE NORWEGIAN SEA) AND 14 MGS CORES IN THE MEDITERRANEAN SEA

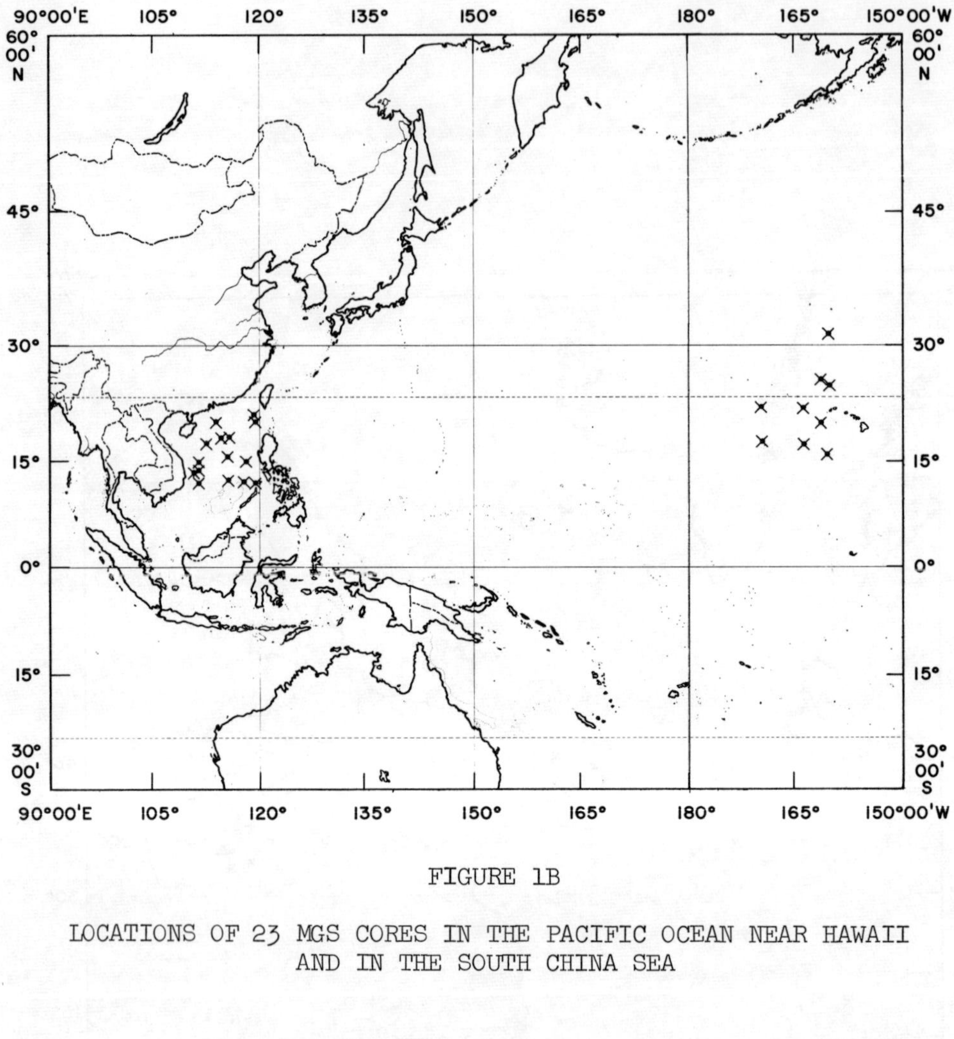

FIGURE 1B

LOCATIONS OF 23 MGS CORES IN THE PACIFIC OCEAN NEAR HAWAII AND IN THE SOUTH CHINA SEA

Core	Latitude	Longitude	Depth (meters)	Physiographic Province	Core Length (cm)	No. of data pairs Velocity, Grain Size	Velocity, Porosity
BA-6-11-1	15°51'N	160°27'W	5550	Abyssal Hills	440	9	10
BA-6-11-2	17°14'N	163°40'W	5515	Abyssal Hills	600	9	9
BA-6-11-3	17°42'N	169°39'W	5266	Marcus Necker Rise	365	3	3
BA-6-16-4	20°10'N	161°28'W	4640	Abyssal Hills	1045	16	16
BA-6-16-5	22°11'N	163°50'W	4720	Seamount Group S. flank Hawaiian Ridge	625	14	14
BA-6-16-6	22°06'N	169°28'W	4420	Abyssal Hills and Rise	542	11	11
BA-6-17-7	25°06'N	160°30'W	4775	Hawaiian Arch	30	2	2
BA-6-17-8	25°53'N	161°33'W	4960	Hawaiian Arch	32	4	4
BA-6-17-9	31°28'N	160°14'W	5600	Abyssal Hills	400	4	12
RA-15-3-1	20°22'N	113°51'E	102	Continental Shelf	263	5	5
RA-15-3-2	13°25'N	110°37'E	2730	Plateau off South Vietnam	155	4	4
RA-15-3-3	13°43'N	111°08'E	2801	Plateau off South Vietnam	510	17	17
RA-15-3-4	11°55'N	111°23'E	3100	Scarp on Continental Slope	395	6	8
RA-15-3-5	12°16'N	115°30'E	4365	Continental Rise	1195	45	48
RA-15-3-6	12°01'N	117°40'E	3340	Continental Rise	475	19	19
RA-15-4-1	14°42'N	118°14'E	4130	Outer Ridge of Manila Trench	575	23	28
RA-15-4-2	20°20'N	119°43'E	3350	Continental Rise west of Taiwan Ridge	580	14	14
RA-15-4-3	21°13'N	19°24'E	3130	Continental Rise west of Taiwan Ridge	575	17	16
RA-15-4-4	18°03'N	114°29'E	3350	Continental Rise north of Macclesfield Rise	475	10	10
RA-15-4-5	17°28'N	112°29'E	1385	Fracture Marginal Plateau	450	7	8
RA-15-4-6	18°11'N	115°54'E	3810	Continental Rise	590	16	17
RA-15-5-1	14°52'N	111°18'E	1375	Marginal Plateau	555	15	15
RA-15-5-2	15°39'N	115°41'E	4200	Abyssal Plain east of Macclesfield Bank	1195	27	33
RA-ST-1B-1	24°27'N	25°24'W	5240	Canary Island Rise	1064	55	57
RA-ST-1B-2	06°10'N	16°15'W	4930	Sierra Leone Abyssal Plain	1120	26	27
RA-ST-1B-3	00°00'N	03°55'W	5110	Guinea Basin Abyssal Plain	1125	22	23
RA-ST-1C-1	29°58'S	32°31'E	4945	Mozambique Abyssal Plain	530	11	11
RA-ST-1C-2	20°00'S	51°00'E	4820	Continental Rise near Madagascar	250	7	8
RA-ST-1D-1	06°40'S	62°00'E	3980	Madingley Abyssal Plain	485	10	10
RA-ST-1D-3	04°37'N	86°28'E	4040	Ganges Cone	125	7	7

TABLE 1

CORE DESIGNATIONS, LOCATIONS, WATER DEPTH, CORE LENGTH, AND NUMBER OF DATA PAIRS.

Core	Latitude	Longitude	Depth (meters)	Physiographic Province	Core Length (cm)	No. of data pairs Velocity, Grain Size	Velocity, Porosity
AS1-2	70°04'N	3°14'E	3191	Norwegian Basin-Jan Mayen Abyssal Plain	850	19	45
AS1-3	71°59'N	10°24'W	2414	Norwegian Basin Ridge	1010	21	61
AS1-4	69°00'N	12°46'W	1832	Norwegian Basin	845	12	32
AS1-5	67°53'N	5°13'W	3475	Icelandic Ridge	775	23	42
AS1-6	62°36'N	1°24'W	1609	Upper Continental Rise	975	8	22
AS1-7	64°00'N	2°35'W	2853	Lower Continental Rise	775	9	21
AS1-8	67°04'N	6°04'E	1317	Helgeland Plateau	1000	10	31
AS1-9	66°21'N	00°18'E	3017	Lower Continental Rise	330	3	11
AS2-1	48°17'N	15°44'W	4721	Biscay Abyssal Plain	920	13	25
AS2-2	54°28'N	13°15'W	2891	Abyssal Plain	818	16	33
AS2-3	58°13'N	10°36'W	1885	Rockall Rise	868	9	18
AS2-4	62°00'N	18°55'W	1894	Icelandic Rise	940	8	23
AS2-5	58°10'N	22°00'W	2928	Abyssal Plain	650	11	28
AS2-6	51°58'N	17°35'W	4310	Continental Rise	1070	13	22
AS2-7	59°15'N	27°18'W	2128	Middle Step (Reykjavik Ridge)	1056	13	21
AS2-8	55°32'N	27°34'W	2855	Abyssal Hills	877	9	19
AS2-10	50°36'N	25°29'W	3989	Upper Step (Middle Atlantic Ridge)	639	11	23
AS3-2	45°00'N	4°03'W	4209	Continental Rise	1055	13	36
AS3-3	41°02'N	14°42'W	5216	Iberian Abyssal Plain	971	12	32
AS3-4	46°06'N	18°28'W	3989	Lower Step (Middle Atlantic Ridge)	906	12	18
AS3-5	45°28'N	21°19'W	3935	Middle Step (Middle Atlantic Ridge)	1186	14	24
AS3-6	41°10'N	19°24'W	4593	Lower Step (Middle Atlantic Ridge)	764	12	22
AS3-7	42°24'N	21°29'W	3865	Seamount Group	559	10	17
AS3-8	40°55'N	25°26'W	3459	Upper Step	800	11	21
AS3-10	44°09'N	26°38'W	2494	High Fractured Plateau	290	7	11
AS6-1	33°28'N	24°13'E	1902	Ridge	813	20	28
AS6-2	35°47'N	25°23'E	1737	Abyssal Plain	669	15	16
AS6-3	37°34'N	16°17'E	2121	Continental Rise	940	19	21
AS6-4	37°50'N	18°15'E	3164	Continental Rise	734	21	29
AS6-5	34°13'N	21°30'E	2195	Ridge	559	13	21
AS6-6	33°10'N	17°35'E	2158	Continental Rise	827	17	27
AS6-7	37°20'N	12°40'E	168	Continental Shelf	770	16	16
AS6-8	36°22'N	11°13'E	100	Continental Shelf	558	9	17
AS6-9	36°20'N	18°35'E	3840	Abyssal Plain	897	15	16
AS6-10	42°03'N	07°03'E	2570	Abyssal Plain	505	20	30
AS6-11	38°05'N	06°58'E	2743	Abyssal Plain	846	15	27

TABLE 1 (cont'd)

Core	Latitude	Longitude	Depth (meters)	Physiographic Province	Core Length (cm)	No. of data pairs Velocity, Grain Size	Velocity, Porosity
AS6-12	37°03'N	00°09'E	2606	Abyssal Plain	468	12	19
AS6-13	36°32'N	02°20'E	494	Continental Slope	983	12	17
AS6-14	35°38'N	03°05'W	1015	Ridge	908	11	7
AS7-1	32°37'N	44°45'W	3850	Middle Atlantic Ridge	325	5	14
AS7-2	34°27'N	40°33'W	3700	Middle Atlantic Ridge	587	7	22
AS7-3	35°23'N	41°05'W	3694	Middle Atlantic Ridge	680	6	27
AS7-4	35°59'N	46°17'W	4938	Sohm Abyssal Plain	1192	0	37
AS7-5	33°38'N	50°24'W	4462	Seamount Group	549	8	21
AS7-6	40°22'N	40°11'W	4517	Newfoundland Basin	960	11	31
AS7-8	38°12'N	47°02'W	5121	Sohm Abyssal Plain	330	5	10
AS4-1	45°19'N	29°17'W	2853	Ridge Slope	981	15	33
AS4-2	43°48'N	37°29'W	3859	Middle Atlantic Ridge	506	7	17
AS4-3	43°01'N	43°31'W	4664	Newfoundland Basin	945	12	33
AS4-4	40°53'N	36°46'W	4481	Middle Atlantic Ridge	1199	17	36
AS4-5	44°33'N	43°27'W	4660	Newfoundland Basin	365	9	16
AS4-6	43°50'N	46°24'W	3877	Continental Slope	1194	19	36

TABLE 1 (cont'd)

METHOD

All sediment sound velocity and physical property data were punched onto computer cards for subsequent use in correlation analyses. The data for the TI cores are listed in the reports in separate tables for depth versus each measured property: sound velocity, shear strength, carbonate content, porosity with related properties, and grain size parameters. After the data in these tables were punched onto cards, a tape file was created listing all properties measured within two centimeters of a sound velocity measurement in the core.

A stepwise regression analysis computer program (BMD02R) [Dixon, 1965] was used throughout this study. It computes a sequence of multiple linear regression equations in a stepwise manner. At each step, a variable is added to the regression equation which makes the greatest reduction in the error sum of squares. Equivalently, it is the variable which, if it were added, would have the highest F value. By use of control cards, arithmetic operations on the input variables can be used to generate new variables. Thus, one variable may be a higher order of an existing variable or a crossterm of two existing variables. This flexibility also allows the program to be used for multiple polynomial regression analysis.

The program determines which variables are most meaningful in defining the dependent variable (sound velocity) for a given set of data with several independent variables. Each variable added gives a new equation with a correlation coefficient, r. Morgan [1969] has used the F-test in determining significant variables to be added to the regression equation in his study of the dependence of sound speed on physical properties in Lake Erie. Lund [1971] used the stepwise and stagewise regression procedures in another area of research. More details on the stepwise method can be found in Draper and Smith [1966, ch. 6].

The data for a given physical property were partitioned into groups within a given category. The categories in this study are: water depth, physiographic province, and ocean. After a regression curve has been determined for each group of the partitioned data, the F-test is used to determine if partitioning the data is justified. The F-test is discussed in detail later in this paper.

CORE ANALYSES

In addition to the MGS reports, details of the core analyses are also reported by Schreiber [1968] for the cores analyzed at Alpine Geophysical Associates, Inc., and by Horn et al. [1968a] for the TI cores analyzed at the Lamont-Doherty Geological Observatory.

The compressional wave sound velocities were measured across the diameter of a core with a 400 kHz signal sediment velocimeter described by Winokur and Chanesman [1966]. The sound velocity values in the reports have been corrected to laboratory conditions of 1 atmosphere and 20°C by using an additive term equivalent to that required to adjust the sound velocity of sea water according to the tables published by the U. S. Naval Oceanographic Office [1962].

The shear strength was measured with a Fall Cone Penetrometer meter described by Hansbo [1957]. The Beckman Air Comparison Pycnometer (Model 930) was used to obtain data for moisture content, porosity, void ratio, wet density, and specific gravity. Calcium carbonate content [Hulsemann, 1966] was also determined for most cores.

The sediment grain size distribution in phi units was determined using procedures outlined by Folk [1964]. The phi unit φ is the negative logarithm to the base 2 of the diameter in millimeters. The division between sand and silt is $\varphi = 4.0$ and between silt and clay is $\varphi = 8.0$. The statistical parameters defined by Folk and Ward [1957] are given in the reports. These are: mean grain size, inclusive graphic standard deviation, inclusive graphic skewness, and transformed kurtosis.

RESULTS

Relationship between Velocity and Physical Properties

Plots of velocity versus the various individual measured physical properties are in each MGS report and in Horn [1968a] for TI areas 1 and 6. Schreiber [1968] shows relationships from the western North Atlantic MGS core data not included in this study (Alpine areas 1 and 2). General conclusions drawn from these figures are summarized as follows:

a. Velocity shows the greatest dependence on porosity as a single factor.
b. Velocity versus grain size also has a well defined relationship.
c. Velocity versus standard deviation of the grain size distribution has a defined relationship. A lesser relationship exists with skewness and kurtosis.

Horn et al. [1968b, 1969a, 1969b] using MGS cores not included here plus other cores from the western North Atlantic and North Pacific found a strong dependence of sound velocity on the mean grain size and wet density.

One of the purposes of the present study is to determine whether the inclusion of more than one physical property significantly reduces the standard error of estimate of the predicted velocity. Similar studies have been made by Sutton et al. [1957] on 26 cores with 37 data points from the eastern North Atlantic, Shumway [1960] with 111 data points from predominately shallow cores on the Pacific coast off California, and Morgan [1969] from 68 cores taken in Lake Erie.

In the present study, the analysis is first applied to each of four groups of core data grouped according to water depth. The depth division was chosen in order to separate different sedimentary regimes in an arbitrary but quantitative way. The observation by Hamilton [1971] that the sound velocity dependence on porosity and grain size varies with three types of provinces suggest that a division by depth would be helpful as a first approximation to separating the data into groups according to different depositional processes and types of sediments. The data are separated by water depths of 1500 m, 3000 m, and 4200 m. If a physical property has a strong influence on the velocity of sound despite a variation in sediment type, then it should appear as a significant independent variable in the regression equation for each group.

Following is a list of the physical properties with their respective symbols as used as independent variables in the regression equations to predict sound velocity V in meters per second:

1. Shear strength, γ (t/m^2)
2. Carbonate content, C (%)
3. Porosity and related parameters:
   a. Porosity, P (%)
   b. Wet density, ρ (gm/cm^3)
   c. Moisture content, M (%)
   d. Void ratio, Vr
   e. Specific gravity, G (gm/cm^3)
4. Grain size distribution in phi units:
   a. Mean grain size, φ
   b. Standard deviation of the grain size, σ
   c. Skewness, Sk
   d. Kurtosis, K

In addition, the square of porosity (P^2) and mean grain size (φ^2) were included as independent variables. Not all properties were necessarily measured at the same location in the core nor with the same frequency.

The equations derived in the stepwise regression analysis for the group for each depth range are given in Table 2, which lists the corresponding standard error of estimate, SE, the multiple

Group	SE	r	Increase in r^2	Regression equation
1. < 1500m n = 41	25.9	.85	.72	$V = -21.8\phi + 1697$
	18.5	.93	.14	$V = -25.7\phi - 1.51C + 1772$
	17.1	.94	.02	$V = -20.3\phi - 1.25C + 52.0\rho + 1640$
	15.1	.96	.03	$V = -16.2\phi - 0.80C + 71.5\rho + 133.2K + 1498$
	13.5	.97	.02	$V = -34.5\phi - 0.97C + 67.1\rho + 132.8K + 134\phi^2 + 1566$
2. 1500-3000m, n = 125	37.4	.81	.66	$V = -25.2\phi + 1738$
	28.0	.90	.15	$V = -16.9\phi - 2.84P + 1851$
	26.3	.91	.02	$V = -15.7\phi - 12.11P + 0.0765P^2 + 2113$
	25.2	.92	.02	$V = -50.1\phi - 11.13P + 0.0692P^2 + 2.42\phi^2 + 2193$
3. 3000-4200m, n = 172	33.1	.68	.46	$V = 148.5\rho + 1291$
	32.7	.69	.02	$V = 128.9\rho - 3.86\phi + 1352$
	32.3	.70	.02	$V = 141.1\rho - 6.96\phi - 35.6Sk + 1366$
	31.9	.71	.01	$V = 190.0\rho - 7.12\phi - 33.0Sk + 9.2Vr + 1267$
	31.0	.73	.03	$V = 12.3\rho - 7.00\phi - 36.3Sk + 37.9Vr - 5.69M + 1729$
4. > 4200m n = 125	55.0	.95	.90	$V = 352.2\rho + 1002$
	48.5	.96	.02	$V = 324.1\rho + 52.9K + 1004$
	46.9	.96	.01	$V = 326.1\rho + 51.3K + 8.0Vr + 984$
	37.1	.98	.03	$V = 367.9\rho + 28.5K + 29.4Vr - 15.1\phi + 1002$

TABLE 2

STEPWISE REGRESSION EQUATIONS FOR DATA GROUPED BY WATER DEPTH WHERE n = NUMBER OF DATA POINTS, SE = STANDARD ERROR OF ESTIMATE, r = CORRELATION COEFFICIENT. SOUND VELOCITY V AS A FUNCTION OF: CARBONATE CONTENT C, SPECIFIC GRAVITY G, KURTOSIS K, MOISTURE CONTENT M, POROSITY P, SKEWNESS Sk, VOID RATIO Vr, MEAN GRAIN SIZE φ, WET DENSITY ρ, AND SHEAR STRENGTH γ.

correlation coefficient, r, and the increase in r^2. The r^2 value is meaningful in that it measures the proportion of total variation about the mean sound velocity explained by the regression [Draper and Smith, 1966, p. 26]. The table lists the equations derived with the stepwise method up to where the addition of another variable does not increase r^2 by at least 0.02 (i.e., the additional variable does not explain at least 2% of the variation). This 2% limit is completely arbitrary. A more rigorous test is the F-ratio of the variable entered. In each case this ratio is significant at the 99% level. The number of data points available to derive the equations for each depth grouping are indicated in the left hand column in Table 2. There are a total of 463 points with measurements of all the above properties.

The absence of the shear strength term γ in Table 2 indicates that there is no linear relationship between shear strength and velocity for any of the four data sets. Hamilton [1970b] summarizes the relationship between shear strength and velocity which is dependent on sediment type. However, he states the lack of a usable relationship.

The mean grain size φ appears in the final equation for each group although its related parameter kurtosis K has a dominate effect in the deepest water depths. Both porosity P and wet density ρ do not appear in the same equation because of the close relationship of wet density to porosity. However, the presence of one of these terms in each equation shows the strong influence of porosity on velocity.

The carbonate content significantly affects the predictability of sound velocity only in the shallow-water core samples (group 1). This variable does not appear in the equations for other groups. Schreiber [1968] discusses the relationships of carbonate content with grain size, porosity, and sound velocity. The relationship to velocity is complex, and although the relationship exists in the shallow cores of this study, the low number of cores in this group indicates that this may not be representative of shallow cores in general. The average value for percent carbonate content in each of the respective groupings is 28.3%, 27.8%, 26.4%, and 22.3%. Since the average value in the shallowest cores is not much higher than the others, the influence of carbonate content on velocity in group 1 cannot be the result merely of high carbonate content in the shallower cores. Therefore, carbonate content is not considered here as an important variable in predicting sound velocity in general.

The first analysis of the data (Table 2) shows that carbonate content and shear strength need not be considered as independent variables in the regression analysis. Since fewer measurements of carbonate content were taken on the cores than any other property, there are many more points available for a second analysis if this physical property is omitted. A total of 858 data points are available with coincident measurements of sound velocity, porosity and related parameters, and grain size distribution data. The results of the second analysis is given in Table 3. The additional data points in each group has caused a change in the order of importance of the significant variables. The wet density ρ in Table 2 gives way to the more definitive first and second order porosity terms. A noticeable change due to the additional data is the high correlation of mean grain size with sound velocity for the deepest cores (group 4) which was lacking in Table 2. The increase in the standard error for group 1 in Table 3 as compared to Table 2 is caused by additional scatter due to the nine added data points rather than to the deletion of the carbonate content factor. In both tables, the first equation for group 1 has only mean grain size as a variable. Even with this increase in the standard error, the correlation coefficient, r, has improved slightly as a result of the additional data points which not only adds to the spread of the data but also increases the slope of the regression line.

One general conclusion that can be observed from Table 3 is that the mean grain size, porosity, and their square terms appear as the most significant independent variables. Even though there is a close relationship between porosity and wet density, the analysis shows that porosity is a better parameter than wet density.

A third analysis was performed constraining the second variable to be the square of the first and using only mean grain size or porosity. Thus the requirement for both porosity and grain size

Group	SE	r	Increase in r^2	Regression equation
1. < 1500m, n = 50	33.4	.87	.75	$V = -27.7\phi + 1754$
	27.4	.92	.09	$V = -17.1\phi - 2.15P + 1810$
	20.9	.95	.07	$V = 8.9\phi - 21.41P + 0.1468P^2 + 2350$
	17.4	.97	.03	$V = -5.6\phi - 30.30P + 0.2332P^2 - 30.3Vr + 2576$
2. 1500-3000m, n = 227	37.3	.79	.62	$V = -4.75P + 1834$
	28.6	.88	.16	$V = 3.09P - 15.1\phi + 1851$
	27.5	.89	.02	$V = -11.21P - 13.5\phi + 0.0668P^2 + 2078$
3. 3000-4200m, n = 263	33.3	.68	.47	$V = -2.83P + 1714$
	31.3	.73	.06	$V = -13.25P + 0.0808P^2 + 2038$
	30.7	.74	.02	$V = -13.23P + 0.0825P^2 + 18.7K + 2018$
4. > 4200m, n = 318	54.0	.84	.70	$V = -34.3\phi + 1825$
	46.0	.89	.08	$V = -76.9\phi + 3.27\phi^2 + 1936$
	40.8	.91	.05	$V = -59.5\phi + 2.79\phi^2 - 2.38P + 1993$
	37.4	.93	.03	$V = -38.8\phi + 1.63\phi^2 - 12.36P + 0.0749P^2 + 2224$

TABLE 3

STEPWISE REGRESSION EQUATIONS AS IN TABLE 2 WITH ADDITIONAL DATA AVAILABLE DUE TO THE ELIMINATION OF CARBONATE CONTENT AND SHEAR STRENGTH PARAMETERS.

data on each sample is eliminated and more points are available. The number of data points used in this analysis are listed for each core in the right hand column of Table 1. The regression equations are listed in Table 4 with the respective curves plotted in Figs. 2 and 3. There is a noticeable increase in the standard errors listed in Table 4 corresponding to the same depth groupings in Table 3 except for group 3 within the porosity set. This is due not only to the constraint of using one measured variable but is also due to the use of additional data points. From a comparison of the standard errors for the two sets of equations in Table 4, it appears that, overall, porosity is a better predictor of sound velocity than is grain size. The same conclusion holds true for a similar set of equations (not shown) using just the data points of Table 3. However, since grain size is commonly used as a velocity predictor, both grain size and porosity were studied in a parallel manner in attempting to find another factor which could reduce the standard error about the regression line.

The similarity in the velocity-porosity equations for the four groups shows that the porosity is a good independent variable to be used in predicting sound velocity in that it is relatively unaffected by water depth. In a later section the curves of Figure 3 are shown to be not significantly different. Grain size, on the other hand, does not have the same consistency in predicting sound velocity.

In developing the model for the velocity equation, the third and fourth degree terms of the independent variable were considered. However, in several of the regression equations generated using these terms, the F-ratio was not sufficiently increased to warrant adding these higher terms. Therefore, only the first and second degree terms have been used. Combining all 1080 data points for

Group	n	SE	r	Regression equation
1. < 1500m	69	30.7	.87	$V = -34.7\phi - 0.59\phi^2 + 1772$
2. 1500-3000m	306	42.1	.72	$V = -57.6\phi - 2.51\phi^2 + 1828$
3. 3000-4200m	305	40.0	.58	$V = 8.9\phi - 1.73\phi^2 + 1572$
4. > 4200m	400	46.1	.88	$V = -80.0\phi + 3.48\phi^2 + 1946$
1. < 1500m	116	20.5	.93	$V = -27.58P + 0.1868P^2 + 2506$
2. 1500-3000m	532	35.1	.78	$V = -17.07P + 0.1072P^2 + 2171$
3. 3000-4200m	491	25.0	.83	$V = -19.29P + 0.1253P^2 + 2238$
4. > 4200m	609	40.3	.90	$V = -24.13P + 0.1545P^2 + 2433$

TABLE 4

REGRESSION EQUATIONS FOR VELOCITY V AS A FUNCTION OF MEAN GRAIN SIZE φ (UPPER SET OF EQUATIONS) AND AS A FUNCTION OF POROSITY (LOWER SET) FOR DATA GROUPED BY WATER DEPTH.

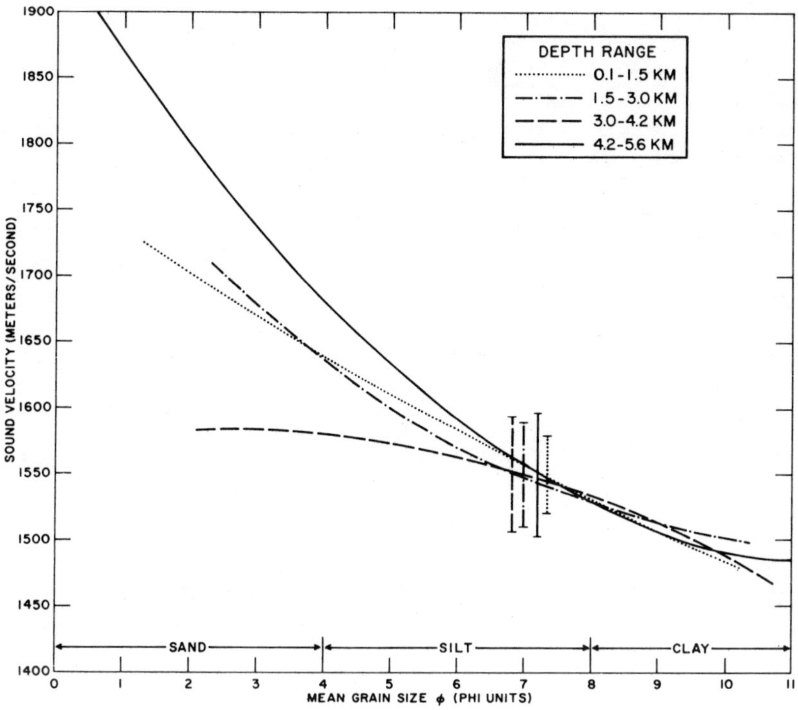

FIGURE 2

REGRESSION CURVES FOR SOUND VELOCITY AS A FUNCTION OF MEAN GRAIN SIZE FOR FOUR SETS OF DATA GROUPED BY WATER DEPTH. VERTICAL BARS INDICATE PLUS OR MINUS ONE STANDARD DEVIATION. CORRESPONDING EQUATIONS ARE GIVEN IN TABLE 4.

all depths where the mean grain size φ and the sound velocity V were measured, the equation relating grain size to velocity is:

$$V = -68.156\varphi + 3.053\varphi^2 + 1874.4 \text{ m/sec} \quad , \qquad (1\text{-a})$$

with a standard error of 45.7 m/sec. Using 1748 data points for all depths where both percent porosity P and velocity were measured, the equation relating porosity to velocity is:

$$V = -22.906P + 0.14997P^2 + 2367.1 \text{ m/sec} \quad , \qquad (1\text{-b})$$

with a standard error of 34.9 m/sec. Equations 1-a and 1-b are referred to henceforth as Model A for either the velocity-grain size or velocity-porosity relationship.

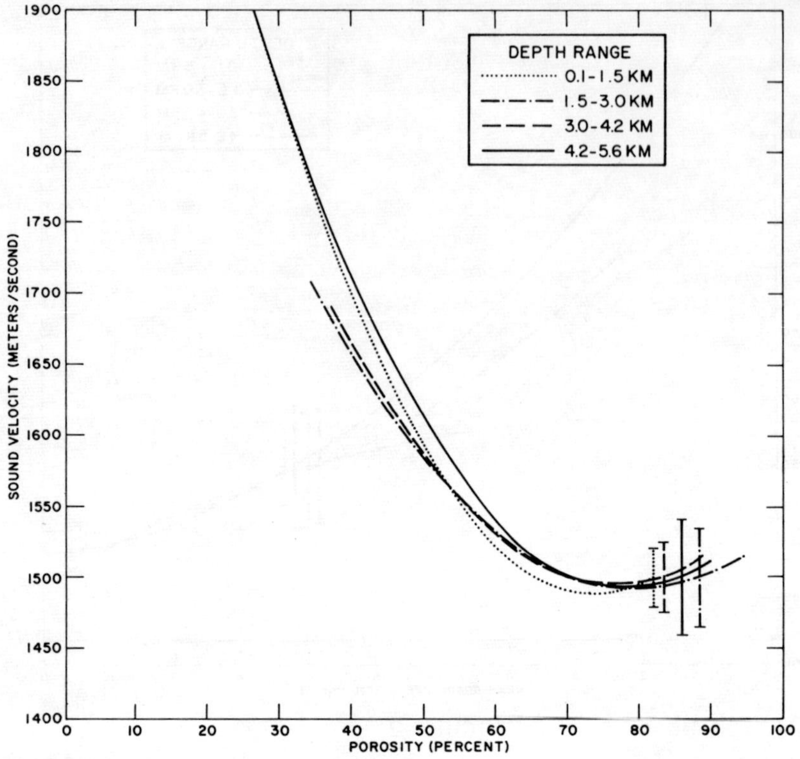

FIGURE 3

REGRESSION CURVES FOR SOUND VELOCITY AS A FUNCTION OF POROSITY FOR FOUR SETS OF DATA GROUPED BY WATER DEPTH. VERTICAL BARS INDICATE PLUS OR MINUS ONE STANDARD DEVIATION. CORRESPONDING EQUATIONS ARE GIVEN IN TABLE 4.

Morgan [1969] has used a similar approach in determining those physical properties having most influence upon sound velocity. Although his cores represent only 5 feet of penetration into the sediments in Lake Erie, the relationship between porosity and velocity is very close to that shown by Eq. 1-b. His equation is:

$$V = -21.97P + 0.1333P^2 + 2380 \text{ m/sec}$$

Factors Affecting the Velocity Relationships with Porosity or Grain Size

Effect of water depth. Since water depth appears to affect the relationship between velocity and grain size (see Fig. 2), a model for velocity as a function of mean grain size φ and water depth D was considered. Since it is apparent from Fig. 2 that there is definitely not a linear relationship between the curves and water depth, a single additive depth term would not be sufficient. Therefore, the equation of the form of Eq. 1, referred to as Model A, was extended to include a third order depth dependence. The new equation is referred to as Model B. The development of the model was limited to a third degree equation, since each additional degree requires an additional three terms and the equation becomes extremely cumbersome. The equation can be written in the form of a power series expansion:

$$V = \sum_{i=o}^{2} \sum_{j=o}^{3} C_{ij} D^j \varphi^i$$

Ignoring the fourth degree terms and above, and using a single index constant k in place of C, the Model B equation is:

$$V = k_1 + k_2 D + k_3 D^2 + k_4 D^3 + k_5 \varphi + k_6 \varphi D + k_7 \varphi D^2 + k_8 \varphi^2 + k_9 \varphi^2 D \quad (2)$$

The stepwise regression analysis was used to determine the coefficients for the Model B equation by progressively adding the next most significant term in the same manner as used in determining the equations for Tables 2 and 3. The coefficients of the final Model B equation have standard errors no greater than 10%.

The Model B equation for mean grain size as a variable is plotted in Fig. 4. The curves in Fig. 4 represent the Model B equation defined in the figure using values of water depth indicated for each curve. The values chosen for display correspond to the depths for the division of data in Fig. 2. The Model B equation was also determined for the velocity-porosity relationship and is shown in Fig. 5. The residual sum of squares for the Model B equation is less than that for the combined residual sum of squares for the four depth equations of Model A, indicating that this is a meaningful expression of the depth dependence of the velocity-grain size or velocity-porosity relationship. The F-ratio was calculated to show the significance of adding the depth terms to the total data equation. The F-ratio for Model B over Model A is 25.4 for mean grain size and 78.1 for porosity. These are both highly significant values.

Effect of province. Sound velocity versus mean grain size was plotted for each of the 82 cores. The ranges of values of velocity for each core were compared with a 1%, 3%, and 5% deviation of velocity from the regression curve for Model A (Eq. 1-a). A number determined by the greatest deviation was assigned to each

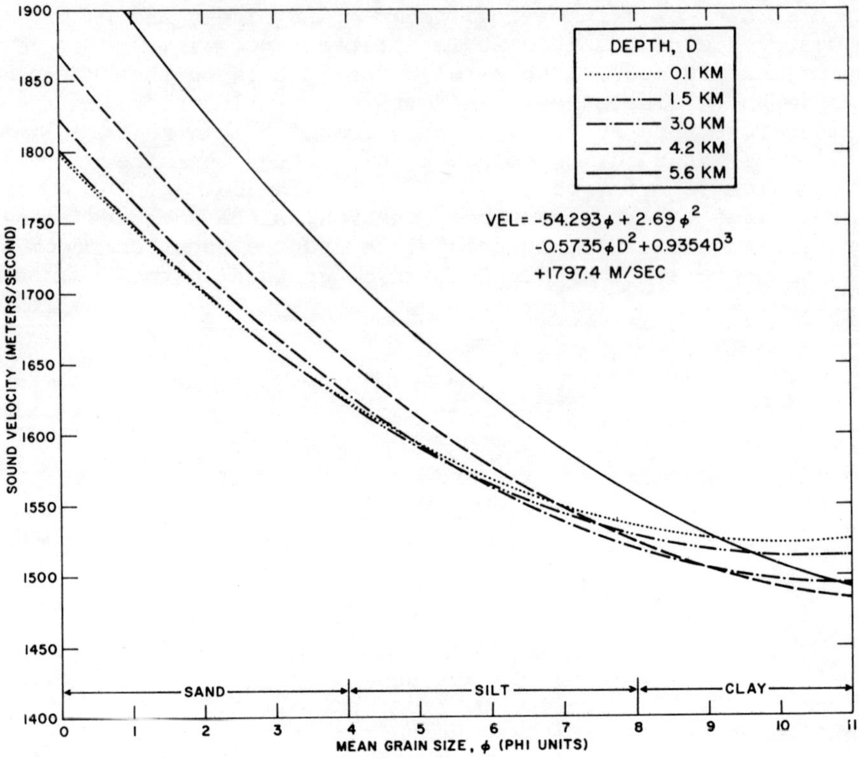

FIGURE 4

MODEL B EQUATION FOR SOUND VELOCITY AS A FUNCTION OF MEAN GRAIN SIZE φ. EACH CURVE IS THE EVALUATED SOUND VELOCITY AS A FUNCTION OF φ FOR A FIXED VALUE OF DEPTH D.

core. This number indicates not only the spread of velocities at a given grain size, but also depends upon the difference in the shape of the curve from the Model A regression curve. The number is larger for cores having low values of phi (large grain size) because the spread in velocities is much greater for large particle sizes than for clays. An index computer card for each core was punched with this number on it, as well as general area, physiographic province, core length, water depth, and range of mean grain size. The cards were then sorted by each of these factors to determine which were possibly related to the spread. A division by physiographic provinces became apparent using this indicator. A province name corresponding to the location of each core was obtained from the MGS reports. These province names are used in this study (Table 1) and no further division or redefinition of province has been made. In some cases the province names are somewhat misleading. A basin in the Norwegian Sea and an abyssal plain

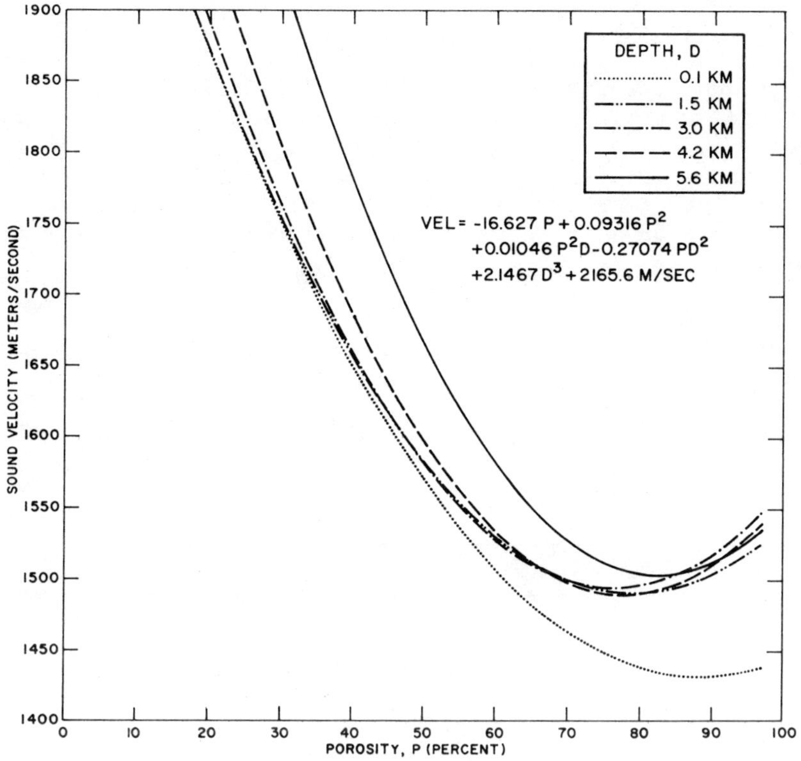

FIGURE 5

MODEL B EQUATION FOR SOUND VELOCITY AS A FUNCTION OF POROSITY P. EACH CURVE IS THE EVALUATED SOUND VELOCITY AS A FUNCTION OF P FOR A FIXED VALUE OF DEPTH D.

in the Mediterranean Sea, both shallower than 1900 m, have been included in the 'abyssal plains and basins' group along with the deep ocean abyssal plains. Also, the cores combined under the one label 'seamount group and arches' are in water depths greater than 3800 m. The seamount group consists of cores at abyssal depths within a region with seamounts. The two cores from the Hawaiian Arch were included with the three cores from the seamount group initially only because of the similar water depths although the few data points do not coincide for the two groups.

Abyssal hills have a very limited range of grain size and velocity, consisting mainly of red clays. The spread of velocities for a given grain size is the least for this province. Abyssal plains and basins containing coarse sands have the largest spread; mid-oceanic ridges and continental rises are in between. Separate regression curves were determined for each province using velocity versus mean grain size and velocity versus porosity. The data from the provinces having very similar curves were combined into one group to produce one curve. This resulted in six distinct curves for the velocity-grain size and three for the velocity-porosity relationship. The regression curves for these provinces are given in Table 5 and illustrated in Figs. 6 and 7. In these figures, the Model A curve is shown as a dashed line. The curves representing only the data for each individual province group are shown as solid lines.

The following conclusions about the average velocities for a province can be drawn from the curves for velocity versus mean grain size:

a. Abyssal hills have considerably lower velocities with no high velocity throughout the whole range of mean grain size ($5<\varphi<11$).

		Independent Variable							
		Mean grain size ϕ				Percent porosity P			
Province	N	n	SE	r	Regression equation	n	SE	r	Regression equation
Continental rise	15	231	39.5	.73	$V = -29.3\phi + 0.57\phi^2 + 1722$				
Continental slope	4								
Continental shelf	3								
Plateaus	6	430	45.5	.86	$V = -101.5\phi + 5.18\phi^2 + 1997$	1182	31.2	.91	$V = -24.09P + 0.1594P^2 + 2399$
Abyssal plains and basins	21								
Abyssal hills	6	58	14.7	.44	$V = 22.5\phi - 1.72\phi^2 + 1428$				
Seamount group and arches	5	38	26.5	.54	$V = -194.6\phi + 12.44\phi^2 + 2250$				
Oceanic rises	4	82	47.2	.86	$V = -146.1\phi + 8.48\phi^2 + 2151$	101	54.6	.79	$V = -49.13P + 0.3360P^2 + 3312$
Ridge	18	241	35.9	.41	$V = -3.7\phi - 0.55\phi^2 + 1594$	465	27.4	.66	$V = -12.89P + 0.0792P^2 + 2019$
Combined (Model A)	82	1080	45.7	.78	$V = -68.2\phi + 3.05\phi^2 + 1874$	1748	34.9	.86	$V = -22.91P + 0.1500P^2 + 2367$

TABLE 5

REGRESSION EQUATIONS FOR VELOCITY AS A FUNCTION OF MEAN GRAIN SIZE φ OR POROSITY P BASED ON DATA GROUPED BY INDIVIDUAL PROVINCES OR COMBINATION OF PROVINCES.

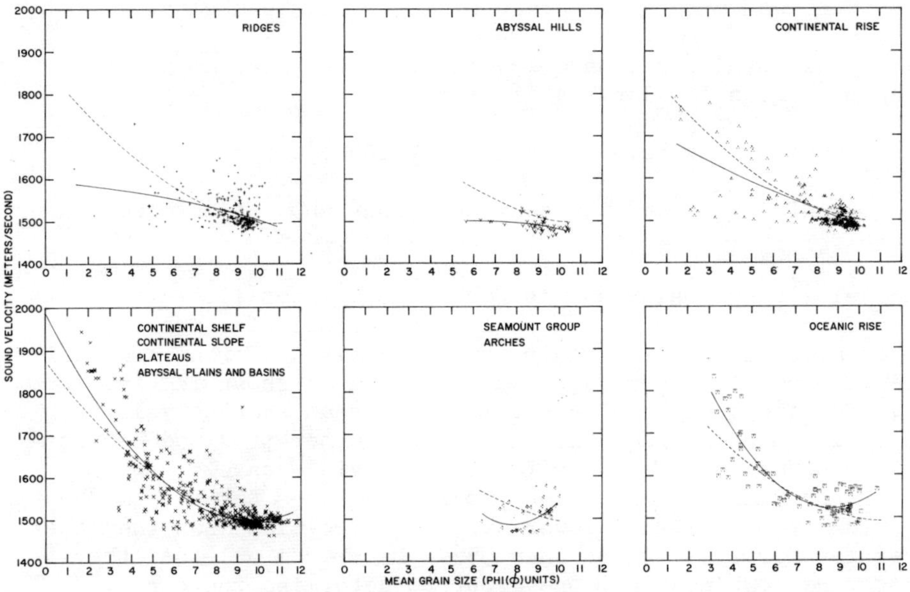

FIGURE 6

SOUND VELOCITY AS A FUNCTION OF MEAN GRAIN SIZE IN PHI UNITS FOR DIFFERENT PROVINCES. THE REGRESSION CURVE FOR EACH SET OF DATA IS SHOWN BY A SOLID LINE; THE REGRESSION CURVE DEVELOPED BY COMBINING DATA FROM ALL PROVINCES (MODEL A) IS INDICATED BY THE DASHED LINE AND IS COMMON TO ALL PLOTS.

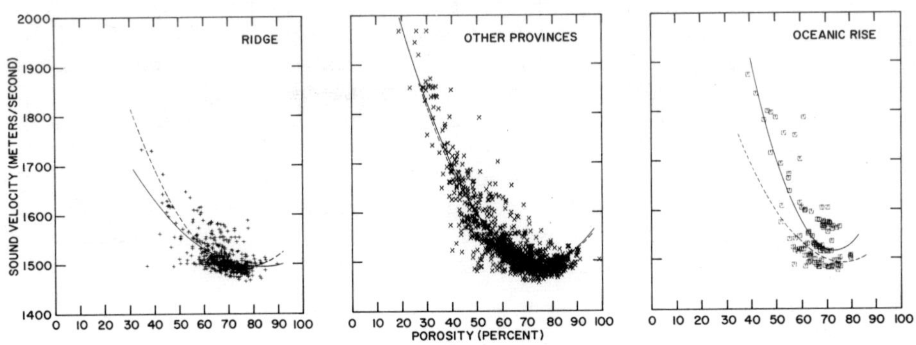

FIGURE 7

SOUND VELOCITY AS A FUNCTION OF POROSITY FOR THREE PROVINCES. THE REGRESSION CURVE FOR EACH SET OF DATA IS SHOWN BY A SOLID LINE; THE REGRESSION CURVE DEVELOPED BY COMBINING DATA FROM ALL PROVINCES (MODEL A) IS INDICATED BY THE DASHED LINE AND IS COMMON TO ALL PLOTS. EQUATIONS FOR SOLID LINES IN BOTH FIGURES ARE GIVEN IN TABLE 5.

b. Ridges generally have velocities less than 1600 m/sec throughout the whole range of grain size for silt and clay.

c. Continental rises have lower velocities than do the remaining provinces for large grain size.

d. Oceanic rises have the highest velocities.

e. Oceanic rises and the seamount-group show an inverse relationship with velocities increasing with $\varphi > 8$.

The velocity-porosity curves for the provinces listed in Fig. 6 are very close together. Only the oceanic rise and the ridge provinces have relationships sufficiently different from all the others to permit a separation of the data as shown in Fig. 7. The curve for the oceanic rise province shows an average velocity from 20 to 100 m/sec higher than any of the other province curves for a given porosity. The data for this province have a higher standard error than for any other province. Cores taken from the oceanic ridge province show lower average velocities than the other provinces for porosities less than 55%. Data from all the other provinces were combined into one group to determine one curve.

Effect of different oceans. The data were separated into four groups according to the major body of water from which each core was taken: the Pacific Ocean, the Atlantic Ocean, the Indian Ocean, and the Mediterranean Sea. The regression equation for velocity versus grain size and velocity versus porosity was determined for each group. The corresponding regression curves are plotted in Figs. 8 and 9. In Fig. 8 there is little separation between the Atlantic and Mediterranean curve for silt and clay size sediments. For a given grain size, the Pacific Ocean sediment sound velocities are lower than those in the Atlantic Ocean. However, it should be kept in mind that the Pacific Ocean is represented only by cores near Hawaii and from the South China Sea. The large standard error and the few data points (only 4 cores) from the Indian Ocean make it impossible to draw meaningful conclusions for this ocean from the set of curves. The velocity-porosity curves shown in Fig. 9 for the Atlantic and Pacific Oceans diverge for the high porosity values which in general represent clayey sediments.

Comparison with Other Studies

A comparison of the empirical relationship between sound velocity and porosity from this study and other studies is presented in Fig. 10. All curves have been adjusted to a measurement temperature of 20°C. Also shown is the theoretical equation by Wood [1941] as applied by Urick [1947] to zero-rigidity suspensions. In calculating the Wood equation, the suspended material is assumed to be

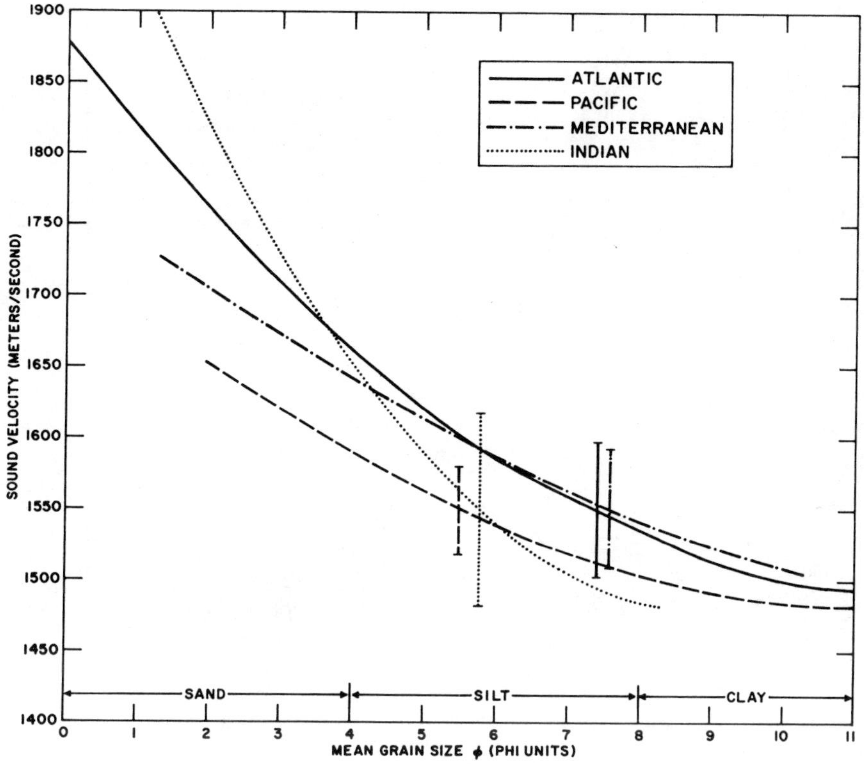

FIGURE 8

REGRESSION CURVES FOR SOUND VELOCITY AS A FUNCTION OF MEAN GRAIN SIZE FOR CORE DATA FROM THE ATLANTIC, PACIFIC, AND THE INDIAN OCEANS, AND THE MEDITERRANEAN SEA. VERTICAL BARS INDICATE PLUS OR MINUS ONE STANDARD DEVIATION.

a varying proportion of quartz and clay. Below 35% porosity the compressibility of quartz is used (2.7×10^{-12} $cm^2/dyne$) and above 70% that of clay is used (1.0×10^{-12} $cm^2/dyne$). In the 35% to 70% porosity range the value of compressibility is linearly interpolated between these two extremes. The compressibility of water is assumed to be 41.5×10^{-12} $cm^2/dyne$; the density of water, 1.03 gm/cm^3; and the density of the suspended material, 2.7 gm/cm^3.

The Model A curve and those from Morgan [1969], Shumway [1960], and Akal [1972] lie close to the Wood curve, whereas the curve of Nafe-Drake [1957] using n = 5 has considerably higher values for velocities at the lower porosities and has much lower velocities for porosities between 60% and 80%. Morgan's data base is Lake Erie and Shumway's is the Pacific shelf, whereas Akal's extensive data are from various oceans and seas including most of the velocity-

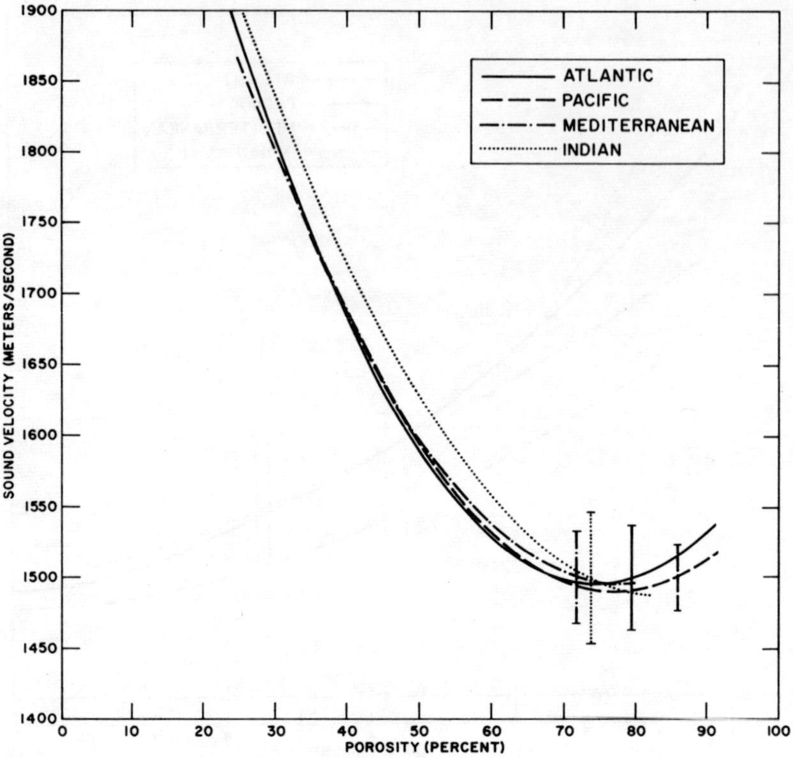

FIGURE 9

REGRESSION CURVES FOR SOUND VELOCITY AS A FUNCTION OF POROSITY FOR CORE DATA FROM THE ATLANTIC, PACIFIC, AND INDIAN OCEANS, AND THE MEDITERRANEAN SEA. VERTICAL BARS INDICATE PLUS OR MINUS ONE STANDARD DEVIATION.

porosity data used in the present study. In considering the actual data rather than the average curves it is obvious in the studies mentioned here that there are many values of velocities at low porosities which fall below or on the Wood equation velocity. In general, though, most of the points are higher showing the effect of rigidity, which is neglected in the Wood curve. Cernock's [1970] data in the Gulf of Mexico tends to have a rather high velocity for a given porosity. Hamilton's [1970b] measurements were taken only from the upper 30 cm of the ocean floor. In his abyssal plains data (indicated by curve AP in Fig. 10) and abyssal hills data (AH), high velocity sediments are missing. On the other hand, he has many high velocity points on the continental terraces (T) whereas a very limited amount of continental terrace data is contained in this study. In general, his velocities are considerably higher than those reported here.

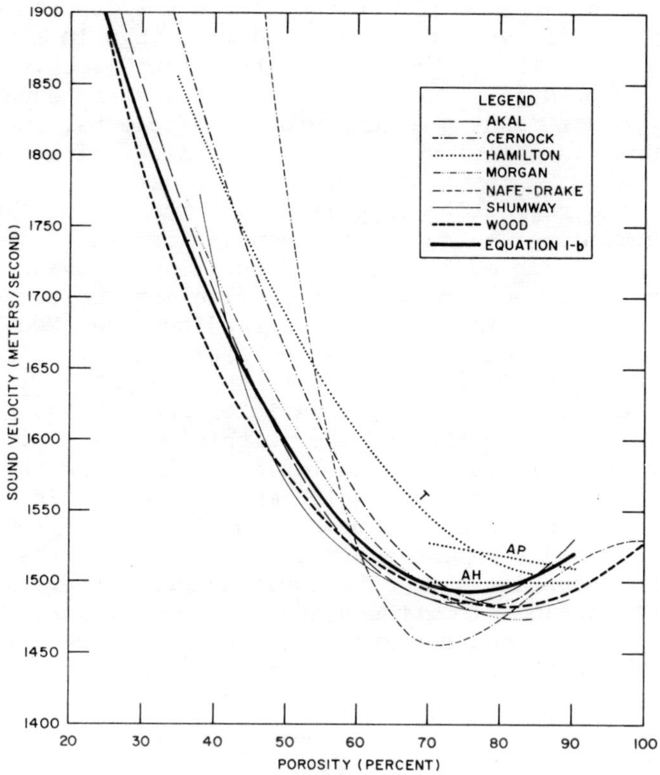

FIGURE 10

SOUND VELOCITY AS A FUNCTION OF POROSITY FROM OTHER STUDIES AND EQ. 1-b (MODEL A). HAMILTON'S CURVES ARE LABELED ACCORDING TO HIS SEPARATION OF DATA BY PROVINCE: CONTINENTAL TERRACE (T), ABYSSAL PLAINS (AP), AND ABYSSAL HILLS (AH).

The curves in Fig. 10 representing a variety of areas are nearly coincident at 70% porosity with the exception of Hamilton's terraces and abyssal plains curves and the Nafe-Drake curve.

Evaluating Partitioned Data

This section deals with the problem of determining whether or not the regression equations representing the partitioned data are more meaningful than the Model A equation. In other words, are the reduced number of data points used in producing each of a number of equations truly representative of different populations? The method used to answer this question is the analysis of variance. This method is valid for multiple (or polynomial) regression equations in an analogous manner to the common application of determining the significance of subdivided groupings in a sample (see Draper and Smith [1966], ch. 9; Kullback [1959], ch. 10; and Griffiths [1967], p. 453). The degrees of freedom are equal to the number of data points minus the number of coefficients, including the constant, used in defining the regression line. Instead of using variances about a mean, the sum of the squares of the residuals about the regression line is used. The 'within' variation is defined by combining the sum of the squares for the individual regression equations for each of the groups. The 'between' variation is the difference between this value and the sum of squares of residuals about the common regression line (Model A) in this application. The mean square is the residual sum of squares divided by the degrees of freedom. As in the analysis of variance, the F-test can be used by taking the ratio of the mean square of the 'between' variation to that of the 'within' variation. An example of this analysis is shown in Table 6. In using the F-ratio, certain prerequisites are assumed which may not be met in the data, such as constant variances and a normal distribution. From visual inspection of the data (Figs. 6 and 7), it is apparent that the variances are not constant along the curve and that there may not be a normal distribution along the y axis for each value of x. Attempts to meet these requirements could be made by weighting the data, but this may still not fully satisfy these requirements. In order to circumvent this difficulty, the table values of significance

Source of variation	Deg. of freedom	Residual sum of squares	Mean square
Model A regression equation	1745	2,121,656	
Individual equations for each group			
Mid oceanic rise province	98	292,428	
Mid oceanic ridge province	462	346,373	
All other provinces	1179	1,147,077	
'Within' variation	1739	1,785,878	1027
'Between' variation	6	335,777	55,963

$$\text{F-ratio} = \frac{\text{mean square of the 'between' variation}}{\text{mean square of the 'within' variation}} = \frac{55{,}963}{1{,}027} = 54.5$$

TABLE 6

CALCULATION OF THE F-RATIO TO DETERMINE WHETHER GROUPING THE DATA BY PROVINCE IS MORE SIGNIFICANT THAN USING THE MODEL A EQUATION FOR PREDICTING VELOCITY.

levels of the F-distribution have been ignored and the F-ratios are instead compared to those calculated from an analyses of the velocity-grain size data grouped by random sampling.

In order to produce a random division of the data, all the data points for a core were kept together as the cores were assigned positions in the tape file according to a nonrepeating list of random numbers. Using this one sequence of data, division was made into three parts and again into four parts of varying sizes. The computed F-ratio for each case was 6.0 and 2.8, respectively. The value of 6.0 is higher than the 1% F-distribution, but comparatively lower than the value for F-tests on the non-random data. No single test could definitely determine what F value could be expected for the data if the division were not a significant improvement. Nevertheless, it appears that the values above 15 are most likely highly significant. The F-ratio offers a criterion to estimate whether or not the division of data is due to a meaningful factor as opposed to a division by an unrelated factor which would result in a random division. However, in order to actually 'prove' the significance of the choice, it would have to first be proved that the data distribution satisfies the F statistic.

Five methods of predicting sediment sound velocity from either mean grain size or porosity have been discussed in this paper. They are: (1) division of data by province, (2) division by ocean, (3) division by water depth, (4) incorporation of water-depth terms in a single equation (Model B), and (5) using one second-degree equation (Model A). Table 7 summarizes the results of the F-ratios used to test the significance of using each of the first four methods in place of the Model A equation. The lower part of the table lists the degrees of freedom, residual sum of squares and the mean square used in determining the F-ratio in the upper part of the table. For the province, ocean, and depth categories, these values are derived as shown in Table 6. The average standard error is given in order to present a more conceptual comparison between the different equations or sets of equations. It is analogous to the standard deviation about a mean but is applicable only near the mean of the velocity and independent variable.

In considering mean grain size as the independent variable, the F-test indicates that all of the first four methods are significant improvements over the use of just one second-degree equation. A disadvantage in using the division of data by province as mentioned earlier is the possible ambiguity in defining the provinces. The Model B equation leaves no ambiguity because water depth is a known quantity. The division by oceans on the other hand provides the simplest improvement on the general velocity-grain size equation. Since the Atlantic and Mediterranean curves almost coincide, there are essentially only the Pacific and Atlantic curves to consider.

Method of grouping	F-ratios		Number of groups	
	Mean Grain size ϕ	Porosity P	Mean Grain size ϕ	Porosity P
Province	19.7	54.5	6	3
Ocean	22.2	4.80	4	4
Depth	21.8	12.8	4	4
Model B	25.4	78.1	1	1
Random	6.0		3	
Random	2.8		4	

Category	Mean grain size equations				Porosity equations			
	Deg. of freedom	Residual sum of sq.	Mean square	SE[a]	Deg. of freedom	Residual sum of sq.	Mean square	SE[a]
Province	1062	1,758,773	1656.1	40.7	1739	1,785,879	1027.0	32.0
Oceans	1068	1,893,269	1772.1	42.1	1736	2,070,119	1192.5	34.5
Depth	1069	1,932,408	1807.7	42.5	1736	1,989,571	1146.1	33.9
Model B	1075	2,062,734	1918.8	43.8	1742	1,870,205	1073.5	32.8
Model A	1077	2,247,561	2086.9	45.7	1745	2,121,656	1215.8	34.9

[a] average standard error SE = $(\text{mean square})^{1/2}$

TABLE 7

THE UPPER PART OF TABLE IS A SUMMARY OF THE F-RATIOS CALCULATED TO DETERMINE SIGNIFICANCE OF USING VARIOUS METHODS OF DATA GROUPING AS OPPOSED TO THE MODEL A EQUATION. THE LOWER PART OF THE TABLE LISTS THE INFORMATION USED TO CALCULATE THE F-RATIOS.

(The Indian Ocean curve is too poorly represented to be considered.)

In considering porosity as the independent variable, the high F-ratio for the Model B equation indicates that the water depth from which a core is taken does affect the velocity-porosity relationship. The F-ratio value of 12.8 indicates that grouping by depth may not be significant. Also, the average standard error is less for the Model B equation than for the depth grouping indicating that, overall, the Model B fits the data better. The 0.1-1.5 km curve in Fig. 3 does not coincide with the band bounded by the D = 0.1 and D = 1.5 km curves in Fig. 5 outside the 55 to 75% porosity range. This is because the data is more heavily weighted by deeper cores. Nevertheless, the complexity of the Model B equation diminishes the usefulness of this relationship. The F-test also indicates that the separation in curves for oceans (Fig. 9) is not significant. The high F-ratio and low average standard error indicates that grouping by province is a significant improvement over using just Model A.

The average standard error of the province grouping of data is 32 m/sec for the velocity-porosity relationship compared to values greater than 40 m/sec for any of the velocity-grain size relationships presented. In addition, there are only three curves for the porosity data with two-thirds of the data points represented by one curve. In Fig. 7 it is apparent that the ocean rise province has higher velocities than any of the other provinces. The reduction of only about 3 m/sec in the average standard error despite the wide digression of the ocean rise province curve from the total curve is a result of the few (6%) data points from this province. It is also apparent that the curve for the ridge province diverges from Model A due to a few points which causes the curve to drop down at lower porosities. Thus this data could also be combined with all but the oceanic rise data. Since all but 6% of the points can be combined into one curve, the Model A curve can be used very conveniently with a high degree of confidence if the exception for the oceanic rise province is kept in mind. The F-ratio indicates that the separation of the ridge and oceanic province data from the rest of the data is significant, but from Fig. 7 it appears that the separation of the oceanic rise province is what makes this division of data significant.

Although the different factors studied show significant influences on the prediction of sound velocity, the single porosity curve provides the method with the greatest confidence. This conclusion is based on: (1) the largest number of data points, (2) the minimum standard error, (3) the greatest confidence in coefficients, and (4) the simplicity in application.

PREDICTION OF BOTTOM LOSS

For a homogeneous ocean bottom with no attenuation and a flat smooth water-sediment interface, the theoretical bottom loss in decibels is given by:

$$BL = -20 \log[(\rho V - \rho_w c_w)/(\rho V + \rho_w c_w)], \tag{3}$$

where the density ρ_w and velocity c_w of sea water is 1.03 gm/cm^3 and 1522 m/sec, respectively, at 20°C, 1 atmosphere, and a salinity of 35‰. No corrections need to be applied for in situ conditions because the difference in the computed bottom loss at laboratory and the actual ocean bottom temperature is only a few tenths of a decibel [Hamilton, 1971]. Using this equation a bottom loss for each of the 1748 data points has been calculated using the measured density ρ and measured sediment sound velocity V. For the purposes of this paper, this is defined as the calculated bottom loss, BL_c. Bottom loss computed from the predicted velocity, based on Eq. 1b, and a predicted density is defined here as the predicted bottom loss. The predicted density is determined from the regression of the measured density on the measured porosity for the 1748 data points. The regression line shown by a dashed line on the scatter diagram in Fig. 11 is:

$$\rho = 2.68 - 0.0166P \tag{4}$$

In predicting sound velocity from the equations developed in this study, the scatter or standard error about the predicted value places limits on the usefulness of the prediction. The predicted bottom loss is plotted against the calculated bottom loss in Fig. 12. These data points center about the 45° line ($BL_\rho = BL_c$) because the equations used to predict both density and sound speed are derived from regression equations on the same data. However, the scatter on this type of plot is important. Lines representing an arbitrary criterion of ±3 dB difference between the predicted bottom loss and a calculated bottom loss are indicated. The points outside these lines correspond to the diamond symbols in Fig. 11. All of these data points have densities significantly different from that expected for a given porosity and their calculated particle dry density values fall outside of the range 2.18 to 3.43 gm/cm^3. It appears that at least some of these anomalous points are erroneous in the initial reports. The points within the ±3 dB band comprise 98% of the data.

If the anomalous points indicated by diamonds in Fig. 11 are discounted, then all the predicted bottom loss values are within 3 dB of the calculated bottom loss. The equation determined from the regression of the calculated bottom loss BL_c on the value of

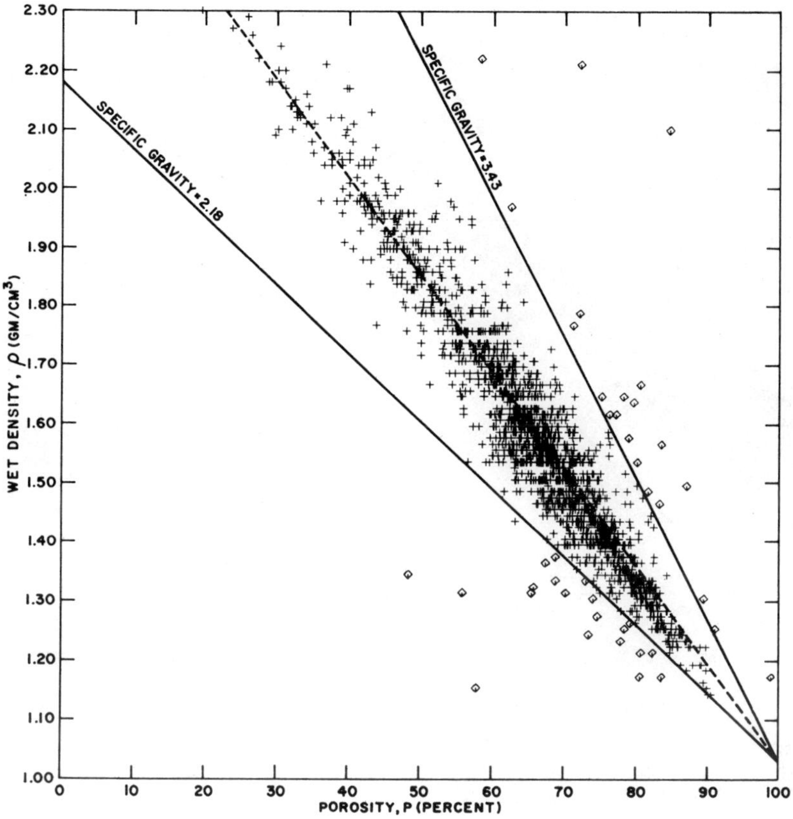

FIGURE 11

WET DENSITY VERSUS POROSITY. THE SOLID LINES CORRESPONDING TO SPECIFIC GRAVITIES OF 2.18 AND 3.43 gm/cm^3 ARE BASED ON A WATER DENSITY OF 1.03 gm/cm^3.

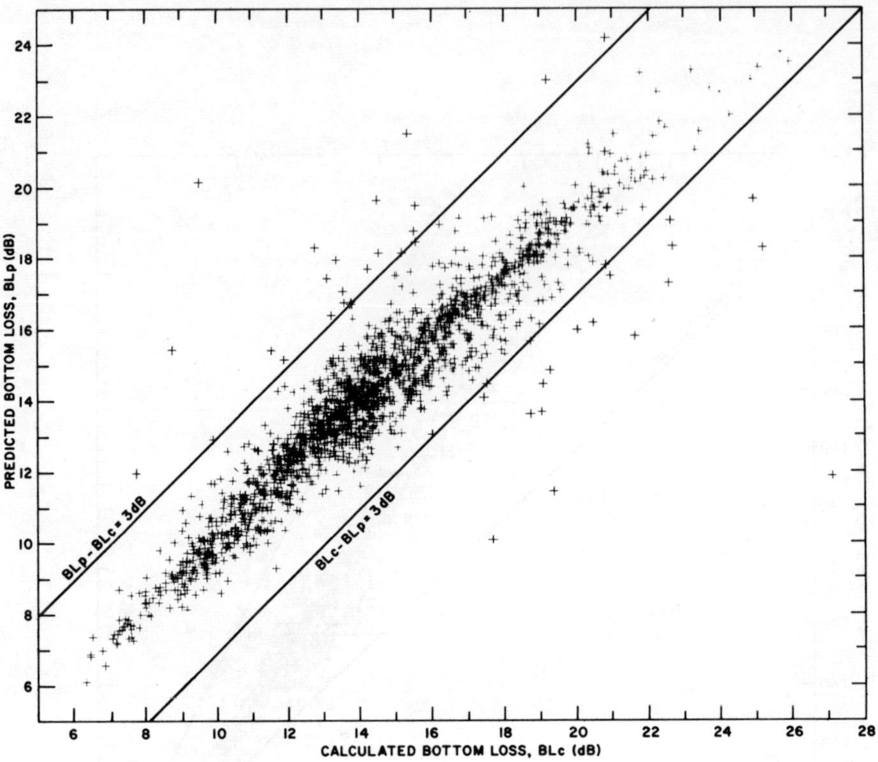

FIGURE 12

PREDICTED BOTTOM LOSS BL_P BASED ON A PREDICTED VALUE OF VELOCITY AND DENSITY FOR A MEASURED POROSITY VERSUS CALCULATED BOTTOM LOSS BL_C BASED ON THE MEASURED VALUES OF VELOCITY AND DENSITY. SOLID LINES REPRESENT THE ±3 dB DIFFERENCES IN THE TWO BOTTOM LOSS VALUES.

porosity for each respective data point is:

$$BL_c = 0.1458P - 0.0032P^2 - 9.426 \tag{5}$$

This equation is graphically displayed in Fig. 13. The curve is about 1 dB greater bottom loss than the curve from Fig. 7 of Hamilton [1970a]. The predicted bottom loss BL calculated by Eqs. 1b, 3, and 4 is also presented in Fig. 13. The two curves cannot exactly coincide since the latter is of a much more complicated form than that of Eq. 5.

The implications of these predicted equations must be restricted as emphasized by Blackman [1971]. No empirical

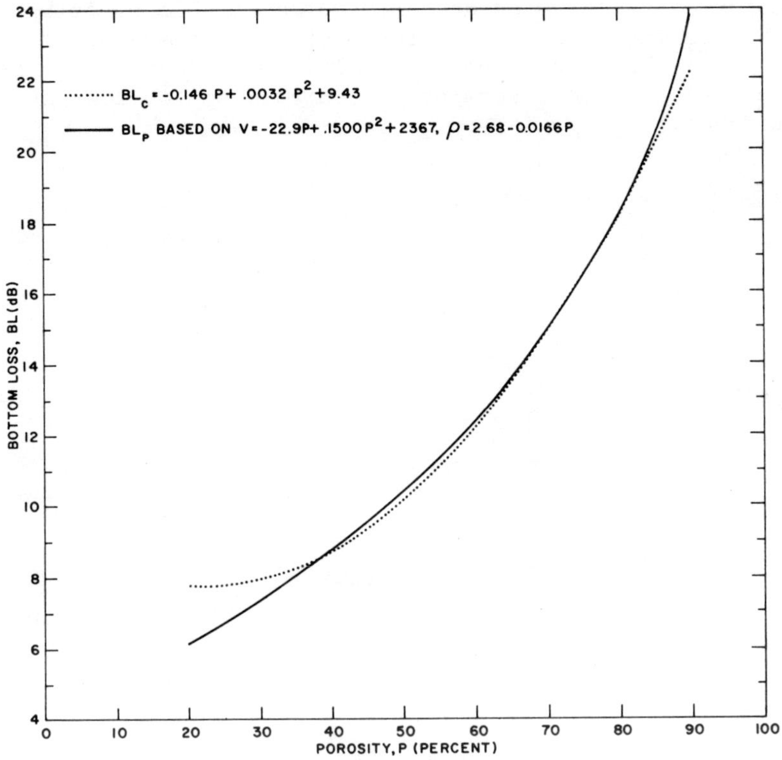

FIGURE 13

BOTTOM LOSS AS A FUNCTION OF POROSITY. THE SOLID LINE IS THE PREDICTED BOTTOM LOSS BL_P; THE DOTTED LINE IS THE REGRESSION CURVE OF THE CALCULATED BOTTOM LOSS BL_C.

relationship is implied between bottom loss and porosity, since actual bottom loss values are not involved in deriving the equations. Breslau [1965] made comparisons showing a difference up to 6 dB between the actual and measured bottom loss at normal incidence and a theoretical value based on actual measured porosity of the ocean bottom sediment. Bottom loss was determined from peak pressure measurements of a 12 kHz signal on the continental shelf. The high frequency somewhat precludes the influence of layering within the sediment on measured bottom loss due to a high attenuation, therefore enhancing the ability to predict bottom loss in this experiment.

The predicted bottom loss in the present study is useful only when considering the simplest type of layer below the water interface. More useful models for determining bottom loss involve the whole range of incident angles and are complicated by the incorporation of multilayers, attenuation, and possible shear mode conversion.

The prediction of velocity or density from available porosity measurements are important to the formulation of these models. Therefore, using porosity to predict velocity or density per se is much more useful than to determine a predicted bottom loss at normal incidence.

SUMMARY

The following are conclusions for predicting sediment sound velocity based on a detailed statistical analysis of measured sediment properties and compressional wave velocities for a variety of sediment types sampled during the MGS program.

(1) The incorporation of more than one physical property in the prediction model does not significantly improve the ability to predict velocity.

(2) Porosity is a better predictor for velocity than the mean grain size. These are the only two primary factors influencing the velocity when compared with the other grain size distribution parameters, shear strength, carbonate content, specific gravity, moisture content, or wet density.

(3) The grouping of data by provinces or by oceans improves the ability to predict velocity based on either porosity or mean grain size. However, all of the data except those from the oceanic rise province are well represented by one second degree equation with porosity as the independent variable for predicting velocity. Porosity measurements from the oceanic rise province tend to have higher corresponding velocities.

(4) The grouping of data by water depth or the incorporation of a water depth factor into an equation with mean grain size or porosity slightly improves the ability to predict velocity.

(5) By using the derived equation containing only the porosity factors for predicting sound speed, bottom loss can be predicted within 3 dB of that calculated by using the measured density and measured sound speed.

Acknowledgments

I want to thank Robert Winokur and Nicholas DiPiazza for their critical review of the manuscript and helpful suggestions. Joseph Kullback, Maurice Owens, and Leonard Howes were very helpful in the application of statistics. I also wish to express my appreciation for the most generous advice and encouragement provided by the late Dr. Abner Blackman.

REFERENCES

Akal, T., The relationship between the physical properties of underwater sediments that affect bottom reflection, Marine Geology, 13, 251, 1972.

Blackman, Abner, Discussion on 'Analysis of the relationship between acoustic reflectivity and sediment porosity' by R. W. Faas, Geophysics, 36, 613, 1971.

Breslau, L., Classification of sea-floor sediments with a shipborne acoustical system, Le Petrole et al. Mer No. 132, 11 pp., 1965.

Cernock, P. J., Sound velocities in Gulf of Mexico sediments as related to physical properties and overburden pressures, Ph.D. thesis, Texas A&M University, College Station, 1970.

Dixon, W. J. (ed.), BMD Biomedical Computer Programs, School of Medicine, University of California, Los Angeles, Rev. Ed. Sept., 1965.

Draper, N., and H. Smith, Applied Regression Analysis, 407 pp., John Wiley and Sons, Inc., New York, 1966.

Faas, R. W., Analysis of the relationship between acoustic reflectivity and sediment porosity, Geophysics, 34, 546, 1969.

Folk, R. L., Petrology of Sedimentary Rocks, Hemphill's, Austin, 1964.

Folk, R. L., and W. C. Ward, Brazos River Bar: A study in the significance of grain size parameters, J. Sediment., Petrol., 27, 3, 1957.

Griffiths, John C., Scientific Method in Analysis of Sediments, 508 pp., McGraw-Hill, Inc., New York, 1967.

Hamilton, E. L., Reflection coefficients and bottom losses at normal incidence computed from Pacific sediment properties, Geophysics, 35, 995, 1970a.

Hamilton, E. L., Sound velocity and related properties of marine sediments, North Pacific, J. Geophys. Res., 75(23), 4423, 1970b.

Hamilton, E. L., Prediction of in situ acoustic and elastic properties of marine sediments, Geophysics, 36, 266, 1971.

Hansbo, S., A new approach to the determination of the shear strength of clay by the fall-cone test, Proc. Roy. Swedish Geotech. Inst., 14, 1957.

Horn, D. R., B. M. Horn, and M. N. Delach, Correlations between acoustical and other physical properties of Mediterranean deep-sea cores, Lamont Geological Observatory Tech. Rep. No. 2, Palisades, New York, 1967.

Horn, D. R., B. M. Horn, and M. N. Delach, Correlation between acoustical and other physical properties of deep-sea cores, J. Geophys. Res., 73(6), 1939, 1968a.

Horn, D. R., B. M. Horn, and M. N. Delach, Sonic properties of deep-sea cores from the North Pacific Basin and their bearing on the acoustic provinces of the North Pacific, Lamont Geological Observatory Tech. Rep. No. 10, Palisades, New York, 1968b.

Horn, D. R., M. Ewing, B. M. Horn, and M. N. Delach, A prediction of sonic properties of deep-sea cores, Hatteras Abyssal Plain and environs, Lamont Geological Observatory Tech. Rep. No. 1, Palisades, New York, 1969a.

Horn, D. R., M. Ewing, M. N. Delach, and B. M. Horn, A prediction of sonic properties of deep-sea cores, Sohn Abyssal Plain and environs, Lamont Geological Observatory Tech. Rep. No. 2, Palisades, New York, 1969b.

Hulsemann, J., On the routine analysis of carbonates in unconsolidated sediments, J. Sediment, Petrol., 36, 622, 1966.

Kullback, Solomon, Information Theory and Statistics, 395 pp., John Wiley and Sons, Inc., New York, 1959.

Lund, Iver A., An application of stagewise and stepwise regression procedures to a problem of estimating precipitation in California, J. Appl. Meteor., 10, 892, 1971.

Morgan, Nabil A., Physical properties of marine sediments as related to seismic velocities, Geophysics, 34, 529, 1969.

Nafe, John E., and Charles L. Drake, Variation with depth in shallow and deep water marine sediments of porosity, density, and the velocities of compressional and shear waves, Geophysics, 22, 523, 1957.

Schreiber, B. Charlotte, Sound velocity in deep sea sediments, J. Geophys. Res., 73(4), 1259, 1968.

Shumway, G., Sound speed and absorption studies of marine sediments by a resonance method, Part 1, Geophysics, 25, 451, 1960.

Sutton, G. H., H. Berckhemer, and J. E. Nafe, Physical analysis of deep sea sediments, Geophysics, 22, 779, 1957.

U. S. Naval Oceanographic Office, Tables of sound speed in sea water, Special Publication SP-58, 1962.

U. S. Naval Oceanographic Office, Marine Geophysical Survey Program 1965-1967; North Atlantic Ocean, Norwegian Sea, and Mediterranean Sea - Oceanographic stations and velocity profiles, prepared by Alpine Geophysical Associates, Inc., Special Publications: SP 95-1-6, 1967; SP 95-2-6, 1968a; SP 95-3-6, 1968b; SP 95-7/4-6, 1968a.

U. S. Naval Oceanographic Office, Marine Geophysical Program 1965-1967; Western North Pacific Ocean and South China Sea - Summary report of geophysical information, prepared by Texas Instruments, Inc., Special Publications: SP 96-6-8, 1968c; SP 97-15-5, 1969b; SP 97-ST-2, 1968d.

Urick, R. J., A sound velocity method for determining the compressibility of finely divided substances, J. Appl. Phys., 18, 983, 1947.

Winokur, R. S., and S. Chanesman, A pulse method for sound speed measurements in cored ocean bottom sediments, U. S. Naval Oceanographic Office, I. M. 66-5, 10 pp., 1966.

Wood, A. B., A Textbook of Sound, 578 pp., Bell, London, 1941.

PRELIMINARY STUDY OF GLOBAL SEDIMENT SOUND VELOCITIES FROM SONOBUOY DATA

ROBERT E. HOUTZ

Lamont-Doherty Geological Observatory of
Columbia University
Palisades, New York 10964

ABSTRACT

Sonobuoy records from Lamont-Doherty's collection are routinely reduced and analyzed for velocity information. Computed solutions are stored on disk where they are easily available for global studies. Velocity gradients (valid for 5 to 10 Hz waves) have been computed statistically from 16 geographical regions, based on sonobuoy interval velocity solutions from all over the world.

Distinct curvature in the sonobuoy travel-time data from certain continental shelf sediments and from deep-sea basement (Layer 2), show that typical airgun frequencies (15 to 20 Hz) can respond to the detailed velocity structure as if the structure is defined by a continuous increase of velocity with depth.

Some revised techniques for the routine analysis of sonobuoy records, and estimates of the accuracy of sound-velocity solutions from sonobuoy data, are discussed. The need for converting sediment isopach data, now in units of reflection time, into units of thickness, and the feasibility of doing so, is also discussed.

INTRODUCTION

The continuing program of sonobuoy deployment, reduction, and analysis at Lamont-Doherty has resulted in a growing accumulation of data. The data consist of velocities and thicknesses from sedimentary and crustal layers. This work has led to numerous publications of general interest as well as applications to regional structure, particularly on the continental shelves and margins. Sonobuoy solutions used for these papers and unpublished solutions are routinely stored in card files and on disk where they are instantly available for additional research. Now that 7,000 such solutions are on disk at Lamont, it has become feasible, by means of statistical techniques, to identify and characterize the sound velocity structure of the world's major sediment bodies. It is therefore appropriate now to undertake lines of research that were not possible a few years ago when there were insufficient data for statistical control. Locations of all sonobuoy stations, as of January 1973, appear in Fig. 1.

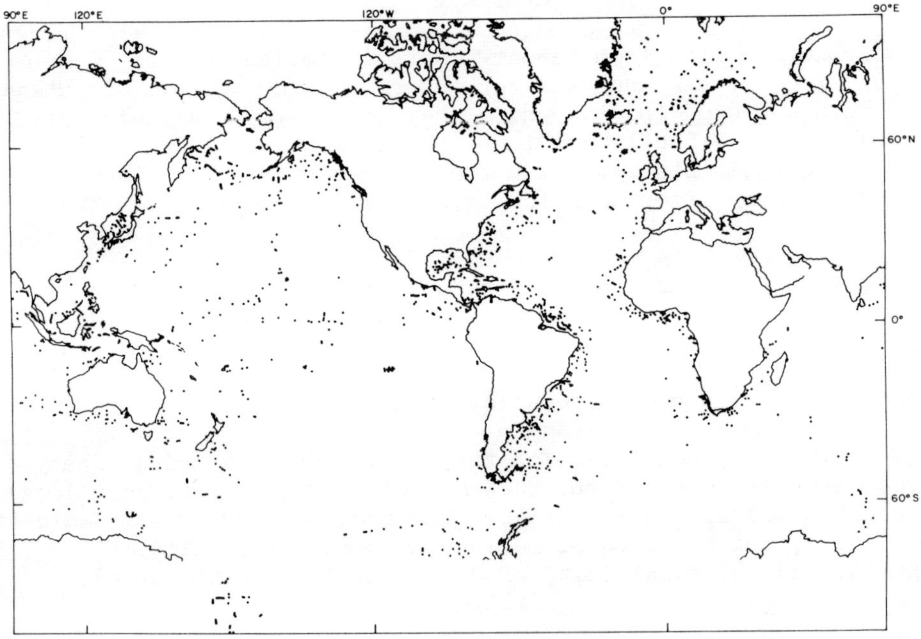

FIGURE 1

LOCATION MAP OF LAMONT-DOHERTY SONOBUOY STATIONS.
JANUARY 1973

This paper is an attempt to show some preliminary results of generalizing part of our world-wide data. Most of the results were completed several years ago, and will be up-dated and extended in a more complete paper, now in preparation. Certain techniques for the reduction and analysis of sonobuoy data have changed during the past few years, since the original publication by Le Pichon et al. (1968). These changes in data reduction and analysis are discussed, as well as some of the world-wide statistical data. Other topics discussed include the accuracy of the technique, observational evidence of velocity gradients on continental shelves (and possibly also in the oceanic basement), and the conversion of sediment isopach data in units of reflection time into units of thickness.

Reduction of Data

The technique used at Lamont-Doherty for computing interval velocities from variable-angle reflection data assumes random dips of plane interfaces between 'constant velocity layers', as described by Le Pichon et al. (1968). The original computer program developed by Le Pichon et al. has remained in use with no more than minor changes. The Lamont-Doherty airgun sound source for the sonobuoys is free-firing, so that timing corrections are required to reconcile the apparent water depth of the sonobuoy recording with the true water depth. These corrections arise because the recording cycle is triggered by the signal reaching the towed hydrophone array, which is far enough astern to produce a significant delay in the data. This and other corrections for dip, first onset, and changes of ship's speed are entered before computation, all of which is described by Le Pichon et al. Refraction lines and intercepts are also picked and catalogued, but are not solved for true velocity and layer thickness unless the data are to be published. This is because a good deal of geophysical judgment and background data are required before refraction sections can be computed with confidence.

Interval velocity solutions are double-checked in several ways before being finally accepted. T^2/X^2 plots of digitized data are plotted by machine in order to edit bad data before interval velocities are computed. Symptoms of bad data in the T^2/X^2 plots include excessive scatter, curvature (usually due to undetected dip), abrupt changes of slope (due to undetected changes in the ship's speed), and distortions near $X^2=0$ (often caused by inaccurate marking of launch time). After computation, the solutions are examined to make sure they lie within the known range of values for sound velocity in sediments, and for their consistency with solutions from deeper and/or shallower layers (sediment sound velocities normally increase with depth of burial). The computer program outputs the standard deviation of the velocity, so that solutions with standard deviations greater than about 20% of the computed value can be rejected. The program also lists some rather technical but useful diagnostic information for judging quality, such as rejected points (caused by irregularities in the reflection curves, probably from hidden sub-bottom topography), and the amount by which the reduced T^2/X^2 data miss the origin when projected along a straight line. The mean vertical water velocity is an input which is used to compute the surface sound channel velocity from the D-wave and the seafloor reflection curve. The computed sound channel velocity is checked against the known value (if the sea surface temperature is known) or a predicted value from hydrographic tables. Deviation from the true value is a measure of how much the ship's apparent speed deviated from the true speed, as a result of a poorly recorded D-wave. Refraction velocities measured from sonobuoy records are corrected if the deviation in the ship's speed exceeds the accuracy of the refraction measurement. Interval velocities are not affected by incorrect observations of ship's speed, as long as the ship's speed is constant.

Some Revisions in Data Reduction Techniques

Repeated tests have shown that interval velocities from variable-angle sonobuoy data become unreliable when the water depth is greater than 10 to 15 times the layer thickness. When the layer is too thin compared to water depth, the solutions tend to give velocities that are unrealistically high and with large standard deviations. These errors occur because the real ray path in the water does not conform to the constant-velocity assumption. Strictly speaking, the relative ray lengths and angles of incidence, rather than relative layer thickness, lead to the errors. If elaborate corrections for the true water velocity structure are made to the ray paths in the water, the interval velocity and its standard deviation are appreciably reduced. This technique was used only to locate the source of error, and is too inefficient and lengthy a program to apply routinely; accordingly, we no longer routinely pick reflection horizons thinner than about 1/12th water depth.

This finding suggested that interval velocities could be picked very accurately in shallow water. Although critical refractions are normally the primary source of data at shallow water stations, we found that many of these stations provided adequate sub-bottom reflection curves. These reflection data yielded accurate interval velocities that were superior to the refraction picks because the standard deviation provides a measure of reliability, and the ambiguity that exists between genuine refractions and reflection crossovers (Houtz and Ewing, 1963) is avoided. It is now routine procedure to pick all sub-bottom reflection curves in shallow-water refraction stations. In the final interpretation the interval velocities are given more weight than the refraction velocities.

Occasionally an interval velocity from multi-layered sonobuoy data will be unacceptable. It can be seen in the schematic T^2/X^2 plot of Fig. 2 that one poorly picked reflection curve results in two bad interval velocities, V_2 and V_3, because the two layers share an interface. Velocities are determined essentially (but not exactly) by the difference between the slopes of the adjacent T^2/X^2 lines; an erroneously low value for V_2 produces an erroneously high value for V_3 and vice versa. Clearly, the correct solutions can be obtained by eliminating the third reflector in Fig. 2 and recomputing as a three-layer problem. This procedure consumes a lot of computer time, assumes a constant velocity within the layer (an assumption that becomes less realistic as the layer is made thicker), and requires that all the dips in layers below V_1 be changed. The latter is a frequent source of error by omission.

For these reasons, we choose to compute the combined velocity, V_{12}, according to Eq. (1),

$$V_{12} = \frac{\Delta_1 V_1 + \Delta_2 V_2}{\Delta_1 + \Delta_2} \qquad (1)$$

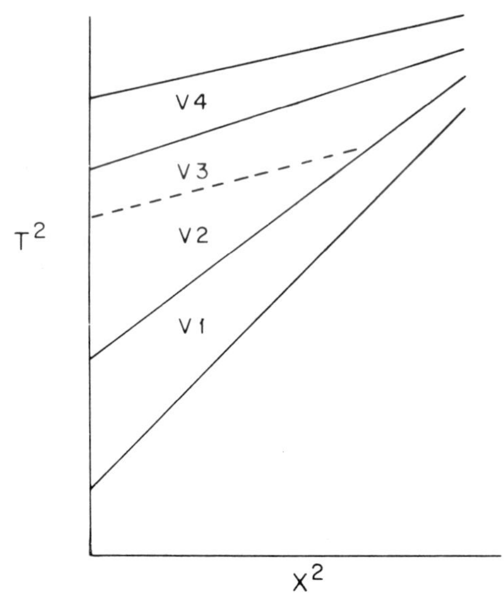

FIGURE 2

SCHEMATIC T^2/X^2 PLOT, SHOWING HOW A POORLY PICKED INTERFACE (BROKEN LINE) LEADS TO TWO BAD VELOCITY SOLUTIONS.

where V_1, V_2, and Δ_1, Δ_2 are the upper and lower layer velocities and interval times, respectively. To compute the standard deviation, σ_{v12}, of the combined velocity, we could use Eq. (2),

$$\sigma_{v12} = \left[\left(\frac{\partial V_{12}}{\partial V_1} \right)^2 \sigma_{v1}{}^2 + \left(\frac{\partial V_{12}}{\partial V_2} \right)^2 \sigma_{v2}{}^2 + 2r \left(\frac{\partial V_{12}}{\partial V_1} \right) \left(\frac{\partial V_{12}}{\partial V_2} \right) \sigma_{v1}\ \sigma_{v2} \right]^{1/2} \tag{2}$$

where σ_{v1} and σ_{v2} are the computed standard deviations of the original velocities, and r is the correlation coefficient between them. Since the errors tend to be self-compensating, r is negative; hence a pessimistic estimate of the error in the combined velocity can be found by assuming r=0, as expressed in Eq. (3).

$$S_{v12} = \frac{(\Delta_1 \sigma_{v1}) + (\Delta_2 \sigma_{v2})^{1/2}}{\Delta_1 + \Delta_2} \tag{3}$$

We realize that V_{12} is not necessarily that velocity with the lowest possible standard deviation, because we cannot conveniently find a weighting function. A smaller value for S_{v12} (and probably closer to the real value) can be obtained by letting r become negative in Eq. (2).

Distributions of Velocity and of Reflection Depth

Sonobuoys were originally deployed from Lamont ships mainly to accumulate data for velocity functions that could be used to convert profiler data into thickness profiles. To accomplish this we plot interval velocities against one-way travel time to the mid-point (in time) of each layer, which yields an approximation of the instantaneous velocity. The method is identical to that outlined by Houtz et al. (1968). We have tried repeatedly to fit polynomials of order greater than one (up to five) to velocity versus reflection depth data, but have never succeeded in demonstrating (by means of variance ratio tests) that the standard deviation is significantly reduced. Even though there is no statistical reason for making more than a linear fit to the data, there may be geological reasons for doing so, or even for using a different kind of function. Houtz et al. (1970) chose to use second-order polynomials, and constrained the curves to pass through the intercept at 1.52 km/sec.

Regression constants of instantaneous velocity versus reflection depth (from 16 regions of the world) are listed in Table 1. The standard deviation of the velocity, the depth to

TABLE 1

$$V = V_o + kT$$

Area	V_o	k	σ_v	r	N	Max depth (sec)
Coral Sea	1.74	1.96	0.20	0.85	17	0.55
Western North Atlantic Rise	1.58	1.24	0.20	0.91	87	1.65
Western North Atlantic Abyssal Plains	1.67	0.97	0.17	0.82	33	0.80
Guyana Basin-Amazon Cone	1.77	0.90	0.19	0.83	52	1.15
Brazil Rise	1.56	1.87	0.16	0.90	24	0.70
Argentine Basin and Rise	1.62	1.17	0.24	0.85	45	1.35
Bering Sea	1.34	1.65	0.22	0.95	84	1.65
Equatorial Pacific	1.46	3.90	0.17	0.90	29	0.35
Alexander Arch	1.42	2.28	0.22	0.92	32	0.90
Aleutian Trench	1.75	1.51	0.16	0.92	10	0.90
Gulf of Mexico	1.74	0.97	0.18	0.91	92	2.15
Bay of Bengal	1.57	1.68	0.20	0.97	40	2.15
Bellingshausen Sea	1.63	1.43	0.20	0.80	45	0.75
Chile Trench	1.41	2.71	0.23	0.93	19	0.80
Guinea-Sierra Leone Basins	1.49	1.57	0.17	0.92	50	1.00
Ross Sea, Antarctica	1.50	2.02	0.20	0.81	21	1.10

which the regressions are valid, the number of datum points, and the correlation coefficient are included in the table. The regression constants are of the form shown in Eq. (4), where v_o is the initial velocity, k is the acceleration factor, and T is one-half the reflection time.

$$v = v_o + kT \tag{4}$$

The value k varies from 0.9 to 3.9 km/sec^2 in the various parts of the world we have investigated. At least some of this diversity is caused by simple changes in the thickness of the unconsolidated overburden on deeper layers whose sound speeds are independent of depth and usually in the 3 to 4 km/sec range. For example, as the overburden thins and the high-speed material comes nearer the seafloor, the velocity will seem to increase more rapidly with depth. This is shown in Table 2 where the acceleration factor in the Caribbean increases dramatically from 1.1 km/sec^2 in the West Caribbean to 2.6 in the mid-Caribbean to 4.91 in the East Caribbean as the high-speed layers progressively rise towards the seafloor. Clearly, the acceleration factor by itself need not have any relation to the physical properties of the sediment. Similarly, we can attach little significance to the intercept value v_o, which is simply an artifact of the general distribution of velocities at greater depths. However, frequencies in the 5-10 Hz range, corresponding to wavelengths about equal to the thickness of typical resolved layers, will react with satisfactory accuracy to these gross regional velocity models.

TABLE 2

$$V = V_o + kT$$

Area	V_o	k	σ_v	r	N	Max depth (sec)
West Caribbean	1.64	1.13	0.28		38	
Central Caribbean	1.25	2.58	0.37		50	
East Caribbean	1.11	4.91	0.28		60	

It is sometimes possible to screen out the effects of depth-independent high-speed layers, by inspection of the velocity versus reflection depth plots. This type of screening is facilitated if the unconsolidated cover is quite variable in thickness, which results in identifiable scatter. Note also that the standard deviations of the Caribbean data are much higher than the average in Table 1; this may be an indicator of a variable cover of unconsolidated sediment on high-speed layers. Table 1 shows that there is a persistent minimum value of $k \approx 1.0$ km/sec^2 in many parts of the world. We suspect that this value would occur most frequently in basins with high sedimentation rates (see Gulf of Mexico, Guyana Basin, West Caribbean, in Tables 1 and 2) if our data were always controlled well enough to eliminate velocities from the depth-independent layers.

By simple integration of Eq. (4), the depth as a function of one-way travel-time is given by Eq. (5).

$$h = v_o T + \frac{kT^2}{2} \qquad (5)$$

This expression does not apply in areas where deep erosion has locally removed the overburden. It is worth noting that the intercept value, v_o, should ideally be about 1.52 km/sec in deep water (it has been measured accurately by other means). Hence Eq. (5) will not be very reliable in the first layer if it is quite thin and v_o is very different from 1.52 km/sec. In general, it would be best to assume that Eq. (5) is not valid within layers whose thickness is less than 1/12th the water depth. A new value for k in Eq. (4) could be estimated by connecting 1.52 km/sec to the computed value of v at 1/12th water depth; at greater depths the regular value should be used.

Some of the regressions in Table 1 are so similar that they can be treated as coming from the same type of sediment. One could therefore make a combined estimate with two (or more) regressions to attain improved accuracy. However, the depth to which the data can be applied may vary among the component regressions, so that the combined regression cannot be used in a particular region at depths greater than the regional limit shown in Table 1.

Accuracy of the Sonobuoy Technique

Lamont-Doherty sonobuoy surveys in various parts of the world are usually carried out by launching sonobuoys end-to-end to form a continuous section. Frequently the top layer is quite uniform, so that velocity solutions from this layer can be considered as repeated measurements on the same layer. The standard deviation of the mean of these observations is a measure of the reliability of the sonobuoy technique if we can assume that the layer is uniform. We therefore have independent estimates of the reliability from various parts of the world, but the mean values from each survey are not necessarily the same because of interregional differences in the sediment. We have more faith in an estimate of the reliability from a large sample than we do from a small one. By weighting the standard deviations from each of eight regions according to the number of measurements, we find a weighted average standard deviation of 102 m/sec. The 90% confidence level for this average standard deviation is ±50 m/sec. The value of 102 m/sec is pessimistic because it includes some of the scatter that must exist in each of the layers, which were assumed to be uniform.

A comparison between predicted sound velocities from Table 1 and observed sound velocities from DSDP data (drill string length divided by one-half the reflection time) appears in Fig. 3. The histogram is scaled in deviation units of mis-match between predicted and observed sound velocities; the standard deviations were taken from the appropriate regional curve in Table 1. The drill hole measurements are also subject to considerable error in places where the seafloor is quite soft. The data were taken from DSDP Legs 1 through 11. Sediments (Layer β) from the Horizon 'A' outcrop area have sound speeds of nearly 3.0 km/sec at the seafloor (Houtz et al., 1968), and do not bear comparison with predicted values for uneroded sediments. Their mis-match values appear as solid squares in Fig. 3. Likewise, the cross-hatched values are

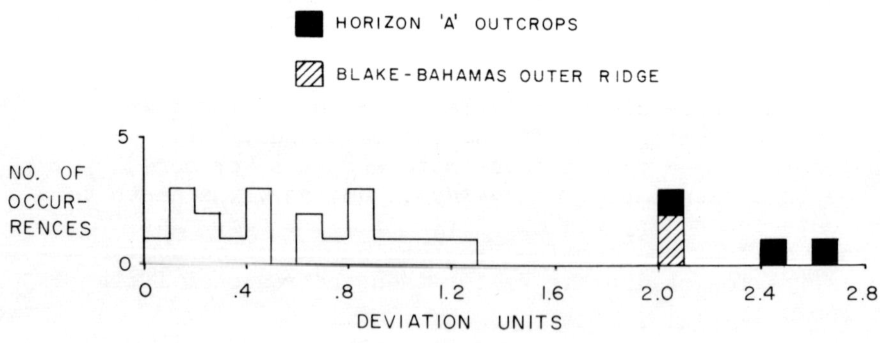

FIGURE 3

HISTOGRAM OF MIS-MATCH BETWEEN OBSERVED DSDP MEAN SEDIMENT SOUND VELOCITY AND PREDICTED VALUES FROM TABLE 1. MIS-MATCH PLOTTED IN DEVIATION UNITS.

from the Blake-Bahamas Outer Ridge, an area of anomalous sound velocities, possibly due to gas hydrates (Lancelot and Ewing, 1972). The balance of the DSDP measurements are less than 1.3 deviation units from the predicted value.

Continuous Velocity Variations with Depth

Some typical sonobuoy records with pronounced curvature in the travel-time data appear in Fig. 4. The curvature is evidence for a continuous increase of velocity with depth. These records were obtained from sonobuoys launched on continental shelves where the geometry of curved ray paths permits the curvature of the primary refraction data to be observed. In deep water the sediments are rarely thick enough nor the velocity gradient steep enough to accommodate curved rays whose arrivals appear on the travel-time graph earlier than the seafloor reflections. The evidence for velocity gradients in deep water sediments is convincing (Ewing and Nafe, 1963) if extra data reduction and analysis are undertaken, but the evidence is not directly observable as it is in shallow-water records. Many, perhaps most, sediments on the shelves are better represented by straight refraction lines. However, the application of artificial straight lines to curved travel-time data leads to serious errors. Houtz and Davey (1973) found (in data from the Ross Sea, Antarctica) that straight line approximations lead to thicknesses that on the average are 25% less than thicknesses obtained with velocity gradients. This error decreases as the number of straight-line segments is increased. Hence, important structures under a variable cover of sediment will be severely distorted by forcing a few straight lines through the travel-time data, a common practice. The identification of ray curvature, which implies a steady accumulation of rather uniform sediment if the velocity increases continuously, provides an insight into the sedimentation process.

Eldholm and Ewing (1972) used the Weichert-Herglotz equation on digitized sonobuoy data to find velocity and depth in the Barents Sea. Houtz and Cavey assumed a linear increase of velocity with depth ($v=v_0+ah$) in the Ross Sea where most of the sonobuoy records show curved travel times. The assumption of a linear increase of velocity reduces the problem to computing a, since v_0 is observed. The parameter a can be found by trial and error (probably the most accurate method) or by measuring derivatives of the graph to compute a. Testing individual sonobuoy data in the Ross Sea, Houtz and Davey showed that a was invariant over various depth ranges, which justified the assumption of a constant gradient, at least to within the accuracy of their measurements.

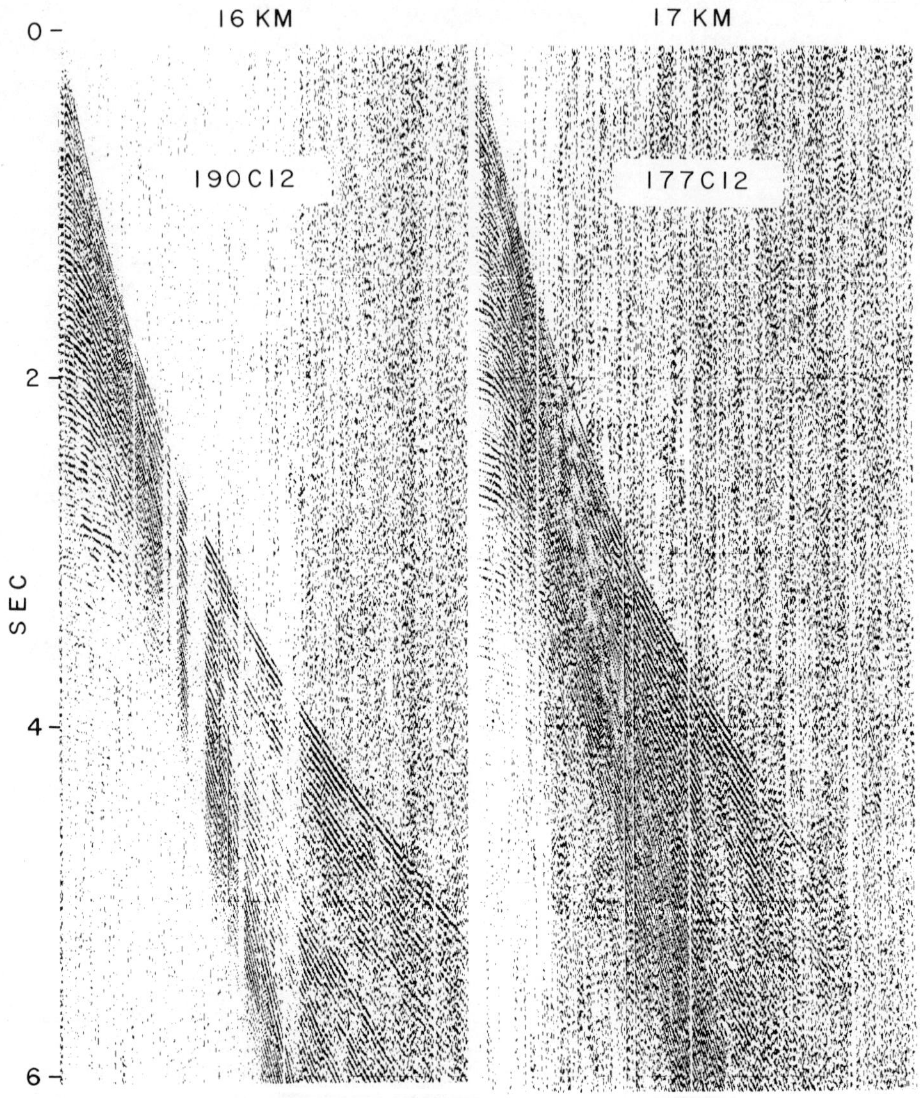

FIGURE 4

CONTINENTAL SHELF SONOBUOY RECORDS SHOWING CURVED TRAVEL-TIME DATA.

A continuous increase of velocity with depth may also occur in the basement. The sonobuoy records in Fig. 5 show a strong tendency for the second arrival basement refraction data to be curved at the shorter ranges. Approximate lines fitted to the data reveal a velocity increase from about 4.4 to 5.2 km/sec. Schreiber et al. (1972) have reported similar values in the laboratory under comparable overburden pressures using DSDP core samples. Houtz et al. (1970) reported the frequent occurrence of basement refraction lines that pass below the basement reflection curve. In Fig. 5 the mis-match between the basement refraction and the reflection curve can be attributed to the velocity gradient in the basement. However, many sonobuoy records show a large mis-match and have unquestionably straight refraction lines.

Sediment Isopach Maps in Units of Thickness

One of the major reasons for compiling velocity data from sonobuoys has been to convert reflection times observed in the profiler data to thicknesses. The need for doing so is illustrated in Fig. 6 where a section of the continental margin south of Adelaide, Australia (Houtz and Markl, 1972), appears in units of seconds of reflection time (upper section), and in units of kilometers (lower section). For the sake of comparison, the scale of the lower section has been adjusted to maintain the same apparent water layer 'thickness' as the upper section. It is clear that a first-order basement depression (the Otway rift) is concealed unless the sound velocities from sonobuoy data are incorporated in the reflection data. In fact, the publication of isopach maps based solely on subbottom reflection times can be misleading, particularly in those regions where the sediments are thickest, i.e., near the continental margins.

It is our intent to apply the wealth of sonobuoy data to the isopach maps to convert them into units of thickness without losing the accuracy and geological judgment that went into the original isopach maps. In closed basins, such as the Chile trench, where one velocity function (based on regressions such as that in Table 1), can be applied, the conversion is relatively easy (after simple integration of the velocity to a depth function).

In other basins, such as the Caribbean, more than one velocity function applies. This leads to artificial discontinuities in the thickness at the boundary between two regions where different velocity functions apply. To avoid this problem, some sort of smoothing process is required. This is doubly important because most of the world's sediment occurs in large bodies, such as the Atlantic, where 7 or 8 different velocity functions were required. An efficient smoothing process for such variable basins was

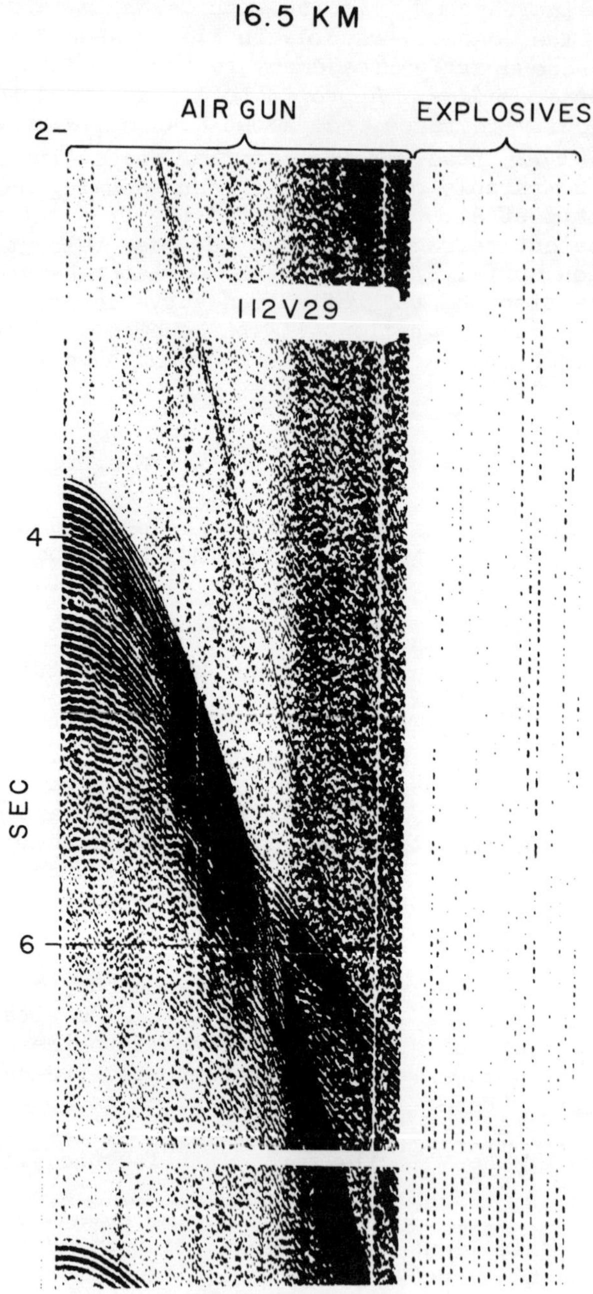

FIGURE 5

SONOBUOY RECORD FROM GREENLAND MARGIN SHOWING BASEMENT TRAVEL-TIME CURVATURE AND A SCALED LINE DRAWING.

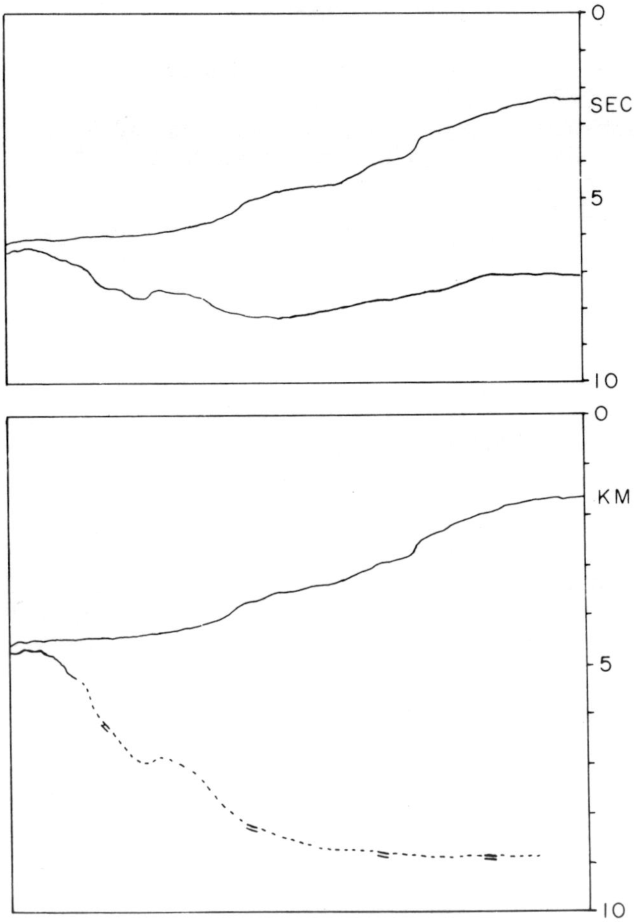

FIGURE 6

STRUCTURE SECTIONS SOUTH OF ADELAIDE, AUSTRALIA, BASED ON SONOBUOY DATA. UPPER SECTION IN UNITS OF REFLECTION TIME; LOWER IN KILOMETERS.

suggested by G. Bryan (personal communication). This method requires that the mean velocity of the entire sedimentary column be measured down to basement at each sonobuoy station where it is available. Careful contouring of the mean velocity data, either by hand or by machine-fitted trend surfaces, could then be carried out. The contours of equal mean velocity could then be overlaid on the original isopach contours. The data could be left in this form, or each intersection of a time and velocity contour could be converted to thickness and eventually recontoured.

ACKNOWLEDGMENTS

The sonobuoys and the support for this research were provided by the Office of Naval Research Contract No. N00014-67-A-0108-0004. The useful suggestions of the reviewers of this paper, John Ewing and George Bryan, are gratefully acknowledged.

REFERENCES

Eldholm, O., and J. Ewing, "Marine geophysical survey in the southwestern Barents Sea," J. Geophys. Res., 3832-3841, 1971.

Ewing, J., and J. Nafe, "The unconsolidated sediments," in The Sea, Vol. 3, John Wiley and Sons, New York, 1963, pp. 73-84.

Houtz, R., and F. Davey, "Seismic profiler and sonobuoy measurements in Ross Sea, Antarctica," J. Geophys. Res., 78, 3448-3467, 1973.

Houtz, R., and J. Ewing, "Detailed sedimentary velocities from seismic refraction profiles in the western North Atlantic," J. Geophys. Res., 68, 5233-5258, 1963.

Houtz, R., J. Ewing, and P. Buhl, "Seismic data from sonobuoy stations in the northern and equatorial Pacific," J. Geophys. Res., 75, 5093-5111, 1970.

Houtz, R., J. Ewing, and X. Le Pichon, "Velocity of deep-sea sediments from sonobuoy data," J. Geophys. Res., 73, 2615-2641, 1968.

Houtz, R., and R. Markl, "Seismic profiler data between Antarctica and Australia," in Antarctic Oceanology II: The Australian-New Zealand Sector, Antarctic Res. Ser., Vol. 19, D. E. Hayes (Ed.), A.G.U., Washington, D.C., 1972, pp. 147-164.

Lancelot, Y., and J. Ewing, "Correlation of natural gas zonation and carbonate diagenesis in Tertiary sediments from North-West Atlantic," in Hollister, C., et al., Initial Reports of the Deep Sea Drilling Project, Vol. XI, U.S. Government Printing Office, Washington, 1972.

Le Pichon, X., J. Ewing, and R. Houtz, "Deep-sea sediment velocity determination made while reflection profiling," J. Geophys. Res., 73, 2597-2614, 1968.

Schreiber, E., P. Fox, and J. Peterson, "Compressional sound velocities in semi-indurated sediments and basalts from D.S.D.P. Leg XI," in Hollister, C., et al., Initial Reports of the Deep Sea Drilling Project, Vol. XI, U.S. Government Printing Office, Washington, 1972, pp. 723-728.

MEASUREMENTS OF REFLECTIVITY BY EXPLOSIVE SIGNALS

RALPH R. GOODMAN AND A. Z. ROBINSON

Naval Research Laboratory, Washington, DC 20375
and
Naval Research Laboratory, Orlando, Florida 32806

ABSTRACT

Two types of reflectivity measurements were made in support of a long range acoustics propagation experiment. The objective was to obtain low frequency third octave reflection coefficients on a sloping bottom. The two types were made by (1) varying the depth of the shots at constant range and (2) varying the range of surface shots for the purpose of changing the angle of incidence. The direct and reflected arrivals were processed by a fast Fourier transform technique. The display of the arrivals and their transforms give insight to some of the reasons for the wide variability that is observed in the third octave reflection coefficients that were measured here and those reported by others. The variability in the data and the resulting uncertainty in reflection coefficients lead to the conclusion that both types of measurements yield unsatisfactory results for use in low frequency propagation models.

INTRODUCTION

In this paper we present some of our experiences with and opinions on the use of broadband acoustical sources for the determination of acoustical properties of the ocean bottom that are needed to determine long range transmission loss. Before beginning this presentation we wish to make a few remarks on the ultimate user requirements. The reason for these remarks is to give some idea of the requirements for the acoustic measurement accuracies that are needed in order to be of use to the Navy.

Generally we who concern ourselves with Naval applications have two reasons to study acoustic transmission through the ocean environment. These are: (1) to understand and predict the performance of current Navy systems that employ acoustic sensors, in order to deploy them as efficiently as possible; and (2) to develop new and improved systems through our knowledge of acoustic characteristics.

The first of these leads to much more specific studies than the second. This is obvious in that current systems require information concerning specific frequencies, geometries, sensors, and signal processing. The second reason naturally leads to a much broader and more liberal approach to studying the environment and is, therefore, much more pertinent to the nature of this Conference. There are, however, certain requirements for precision and accuracy that are set by the second category. New systems are expensive and their justification requires significant improvement in performance over current equipment. Thus, although early research goals may be broad, ultimately we must talk about decibels of improvement to an acoustic system, and it remains an important part of the scientist's and engineer's role to carry through what we learn from basic research to the hard requirements of systems comparison. In such comparisons even numbers as small as 3 dB improvement may be enough to have impact on systems decisions. Promising techniques that will allow the prediction of environmental affects on acoustic systems must therefore be pursued. Those that offer little or no promise must be rejected.

Currently our knowledge of the ocean floor is too incomplete to give either broad averages of acoustic scattering, transmission, or reflection, or to give clear pictures of fluctuations over time and space. From the wide spectrum of papers presented in these proceedings we can see progress in understanding both acoustics and the nature of sediments. But we can also see that we do not have the ability at this time to determine, on a routine basis, the acoustic properties of the ocean floor to a few decibels based on geological, mechanical, and even acoustical measurements. Thus we must determine what routes we can take that ultimately will give the design engineers and systems analysts the data and accuracies they require.

Currently the frequencies of interest in long range transmission in the sea (say for convenience on the order of 30 nm or greater) is in the kilohertz range and below. Our interest in low frequency long range sound transmission has led us to a need for measuring the reflection of sound from the sea floor.

The measurements that we report in this paper are nothing especially new in terms of techniques except in some of the details of analysis, which are not particularly of interest at this Conference. They offer a basis, however, of review of the method in which we can discuss how similar measurements may or may not satisfy requirements, and they serve to show geophysicists what some of the acoustician's needs are. We shall begin by giving a brief physical description of the experiment and then a rationale for the type of processing we chose. We then will present our results. We conclude by discussing the value of broadband results as deduced from our analysis.

EXPERIMENTAL METHOD

The Naval Research Laboratory has for some time been concerned with the theory and measurement of long range underwater sound. The experiments discussed here were in support of a long range propagation experiment. Their purpose was to provide numbers for reflection coefficients that could be used in long range propagation models. The long range sound transmission measurements were made using small explosives over ranges from tens to hundreds of miles. The processing of particular interest was third octave analysis centered at 50 to 100 hertz. Thus the requirement for reflectivity was to provide third octave reflectivity values for a range of angles over a region of the ocean floor that was known from ray tracing to be important to the calculation of transmission loss for the particular geometry of the experiment. Thus the reflectivity part is done more or less for the first reason given above; i.e., it is subservient to another requirement. We shall return later to a discussion of the requirement.

The basic techniques are well known. They employ explosive charges as signal sources. The receiver is a hydrophone suspended approximately 500 feet above the ocean floor. Figure 1 shows the configuration. The approximate depth is 1500 fathoms, so that individual returns from the surface are well separated. The hydrophone response is multiplexed and transmitted up a coaxial anchor cable. In the ship's laboratory the signal is demodulated passed through a 300 hertz low-pass filter, and FM recorded on magnetic tape. The details of the techniques are available, but are not given here.

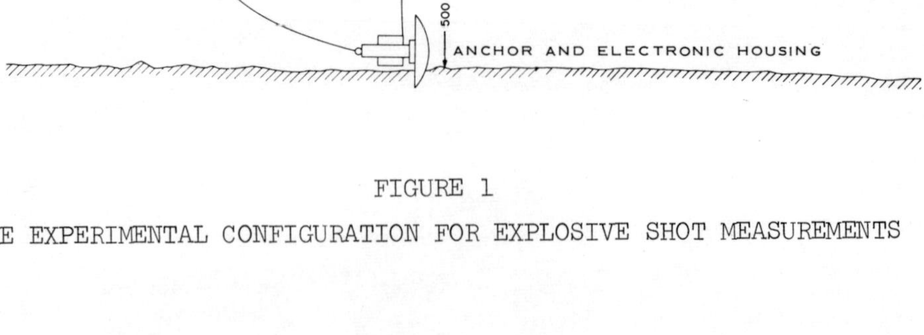

FIGURE 1

THE EXPERIMENTAL CONFIGURATION FOR EXPLOSIVE SHOT MEASUREMENTS

Two types of measurements were taken at each station. The first, "type 1", consisted of detonating 1/2 lb charges (with pressure-sensitive detonators) at depths ranging from 4000 to 11,000 feet (in 1000 ft increments) at a fixed horizontal range from the receiving hydrophone (0.8 to 2.5 nautical miles depending on the anchor scope of the receiving ship). The desired spread in the grazing angle was achieved by the variations in shot depth. The most serious problem anticipated with this experiment was the expected small energies likely to be available at frequencies below 200 Hz for the deeper charges. A reduction in acoustic energy at the lower frequency levels is expected at the greater depths due to an increased bubble oscillation frequency and to a decreased bubble size. This will be discussed later in the paper.

The second, or "type 2", measurement consisted of signals from 1/2 lb charges detonated just below the surface so that the gas bubble vented on its initial expansion. This required that changes in horizontal separation between the charge and the receiving hydrophone be used to achieve the desired range in the grazing angle. This was done by having a second ship launch the charges at 1/2 mile intervals as she traversed a straight course going directly over the receiving hydrophone.

These measurements were performed at four different stations several hundred miles apart. Thus there is no reason to expect close acoustic similarity in the results. Figure 2 shows the bathymetry at one of the stations which, for the purposes of this paper, can be said to be a typical geometry in terms of slope, contours, etc.

One can see immediately, from the nature of the contours in Figure 2 and a little knowledge of how contour charts are made, that angles of incidence will not be known to a great degree of accuracy. But this error is one that we must live with until bathymetric techniques become better (and presumably much more expensive). One expects the errors to be largest in regions of steeply sloping bottoms, and these are generally important in long range transmission since the regions where the bottom intrudes into the acoustic channel will generally be in areas that are not geometrically smooth or flat. This is an important point to remember in terms of acoustic requirements.

The experimental design concept was based upon the generation of a source signal of sufficiently short duration that the incident signal and the reflected signal at the suspended hydrophone could be separated in time. This time separation allows separate processing of the incident and reflected signals. The initial data that we examined were the deep-shot experiments. Several methods of data reduction were examined, but it became obvious that the superior method would involve translation of the magnetic

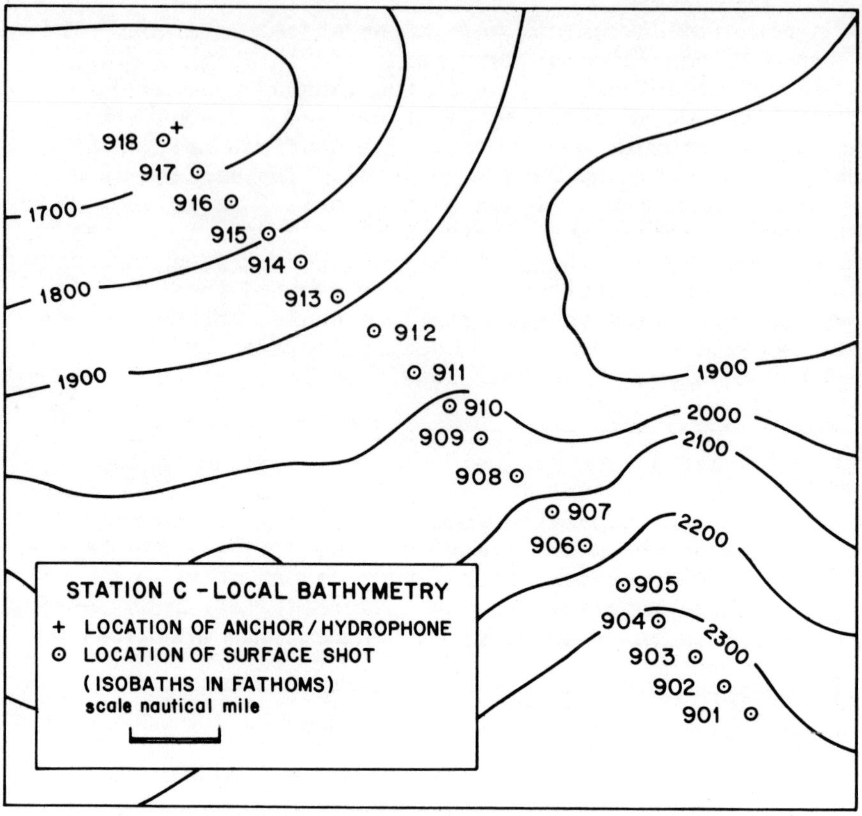

FIGURE 2

THE BATHYMETRIC CONTOUR CHART FOR ONE OF THE STATIONS

tape analog signals to a digital format for subsequent computer analysis. It was evident upon examination that the early data were contaminated by low-frequency radiated noise from the receiving ship which at times contained a discrete component of approximately 15 to 20 Hz. The data reduction process was then as follows:

1. The analog tapes were played back at four times real time. The recorder output was passed through a sharp cut-off high-pass filter with a cut-off frequency of 160 Hz (corresponding to 40 Hz in real time). This eliminated the low-frequency noise below 40 Hz without affecting the signal information above 40 Hz. This filter was used with all data except those from one series of shots.

2. The output of this filter with suitable amplification was fed into an A-D converter with a sampling rate of 15,574 Hz (3,938.5 Hz in real time). This was considerably above the Nyquist rate which thus allows for linear interpolation between sample points with negligible error. All of the recorded shots were translated to a digital form and again recorded. For each series of shots, digital samples of noise (taken between shots) were recorded. This noise analysis was intended to give a measure of the available signal-to-noise ratio in our data reduction process.

3. Each shot is called back from the tape and the pressure-time waveform of the shot plotted on a CRT display. This is done to allow for the positioning of time gates to separate the incident and reflected signals for future FFT analysis and to gain some insight into the physical processes of reflection.

In addition to the tape recordings made at each station, a log of ancillary information was maintained to allow for efficient reduction and interpretation of results. Blasting caps were also fired at the surface immediately after the anchor touched bottom at each station to give a measure of the actual height of the suspended hydrophone above the bottom. This height was 500 feet for three stations and 475 feet for another.

RESULTS

Since the purpose of this paper is to discuss the acoustic method of using broadband sources for measuring reflectivity, it is not necessary to go into detail on the determination of reflectivity curves for each location. Even if we did it is doubtful that they would be of any general value. Therefore we shall give some typical results and discuss them together with the method by which they were obtained.

From the analysis of the shot tapes, 50 and 100 hertz third octave reflection coefficients were found. From an analysis of arrival times and their differences it is relatively straightforward to compute the arrival angles. These computations, of course, assume a smooth slope and are therefore low order estimates. For the case of variable depth shots no sound speed gradient is assumed. This leads to insignificant errors for the grazing angles we achieved. The results given in this section will be separated into the two types of measurements: (1) variable depth, and (2) shallow sources.

Variable Depth

One of the virtues of the type of display of the data shown in Fig. 3 is that it is possible to observe and deduce some of the explanations for the variations. For this reason the results of several sets of measurements are shown in Figs. 3 to 8, with the left hand side showing the pressure-time history of the direct and reflected signals and the gating points used for data reduction. The direct signal, which, for all practical purposes, can be called the incident signal, is marked by D and the reflected signal gate is marked with an R. The right hand side of each figure shows the energy spectral densities as calculated by the FFT. The solid curves are for the incident signals and the dotted curves are the reflected signals.

The following characteristics should be noted.

1. The energy content of the signal is low at frequencies below 200 Hz.

2. The reflected arrival is complex in nature and shows as an oscillatory type of power spectrum.

The third octave energies that are processed for signals obviously also contain contaminating noise. To assess the noise contributions the digital noise samples were taken during each shot sequence and processed in exactly the same manner as the

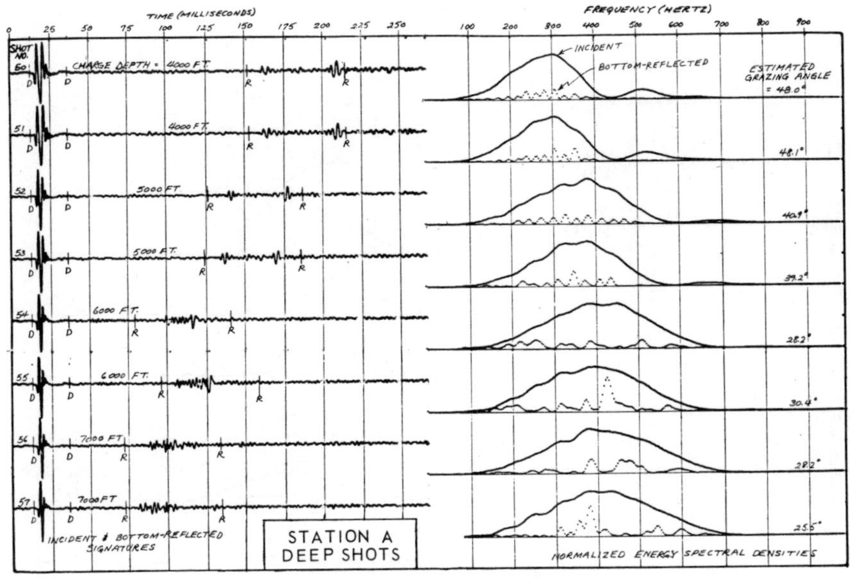

FIGURE 3

50 SERIES-SHOT ARRIVAL RECORDS AND THEIR POWER SPECTRA. EACH SHOT IS SEPARATELY NORMALIZED ("TYPE 1" MEASUREMENTS).

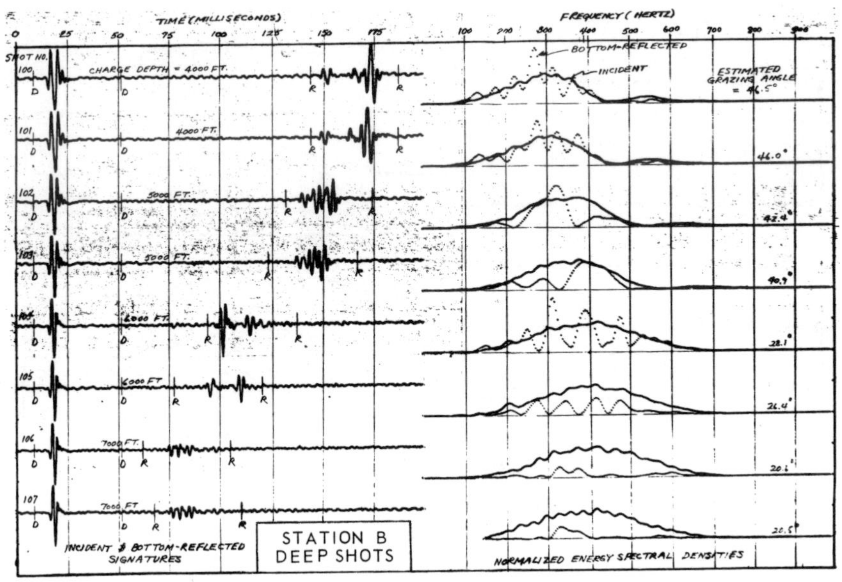

FIGURE 4

100 SERIES-SHOT ARRIVAL RECORDS AND THEIR POWER SPECTRA. EACH SHOT IS SEPARATELY NORMALIZED ("TYPE 1" MEASUREMENTS).

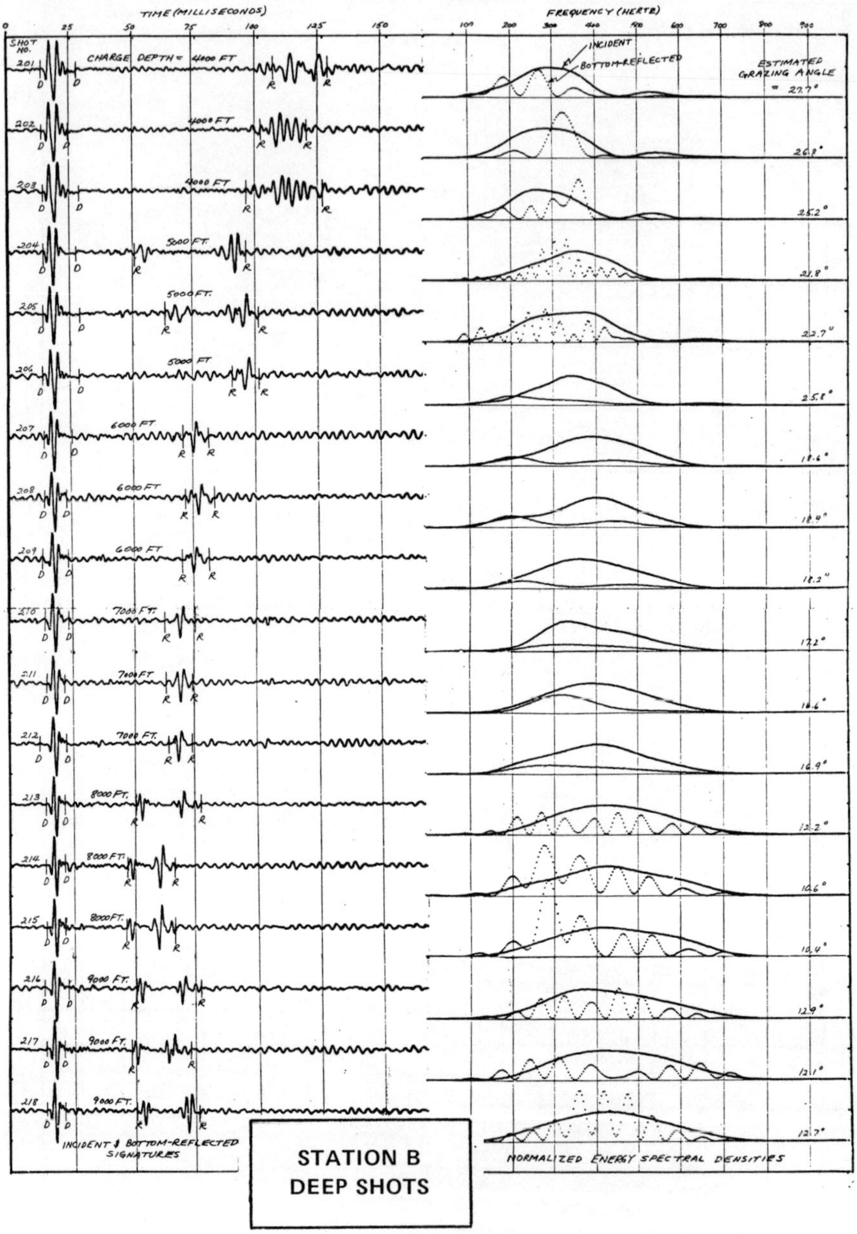

FIGURE 5

200 SERIES-SHOT ARRIVAL RECORDS AND THEIR POWER SPECTRA.
EACH SHOT IS SEPARATELY NORMALIZED ("TYPE 1" MEASUREMENTS).

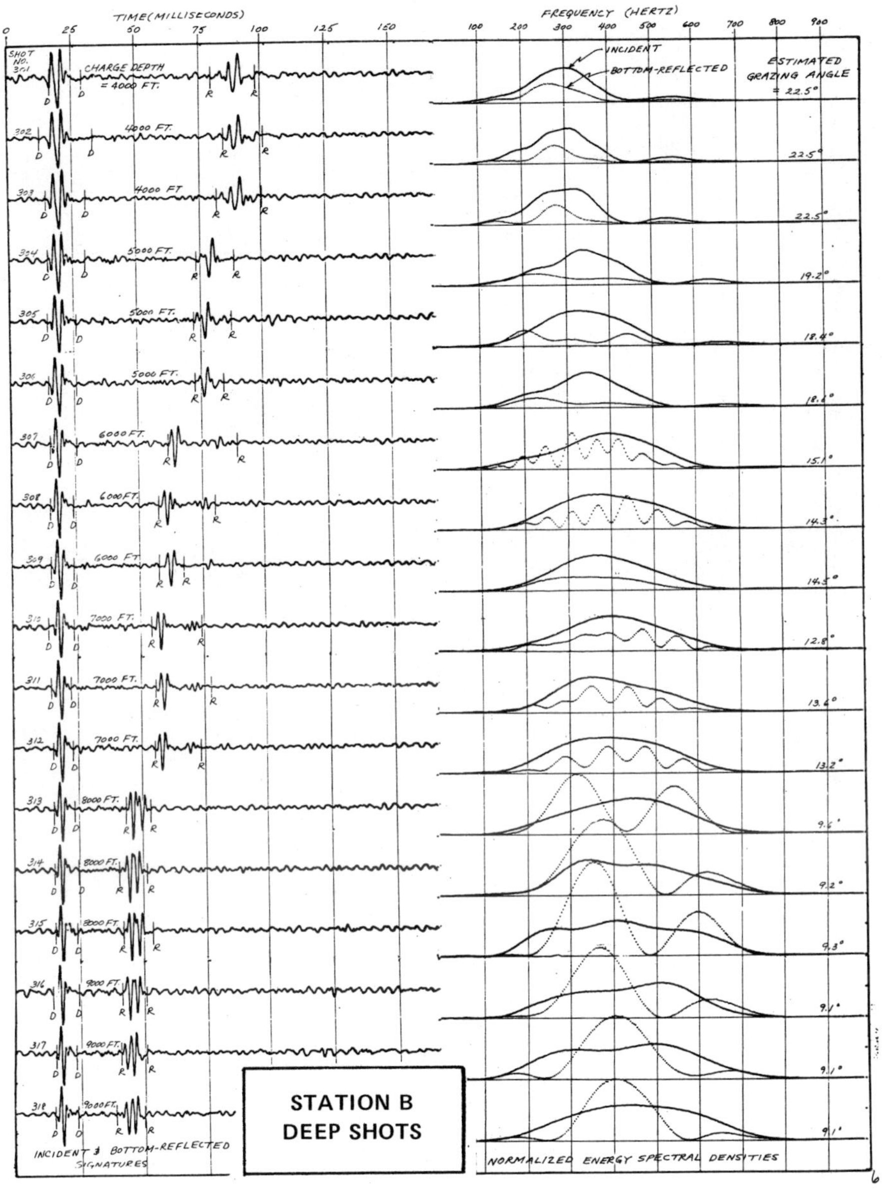

FIGURE 6

300 SERIES-SHOT ARRIVAL RECORDS AND THEIR POWER SPECTRA.
EACH SHOT IS SEPARATELY NORMALIZED ("TYPE 1" MEASUREMENTS).

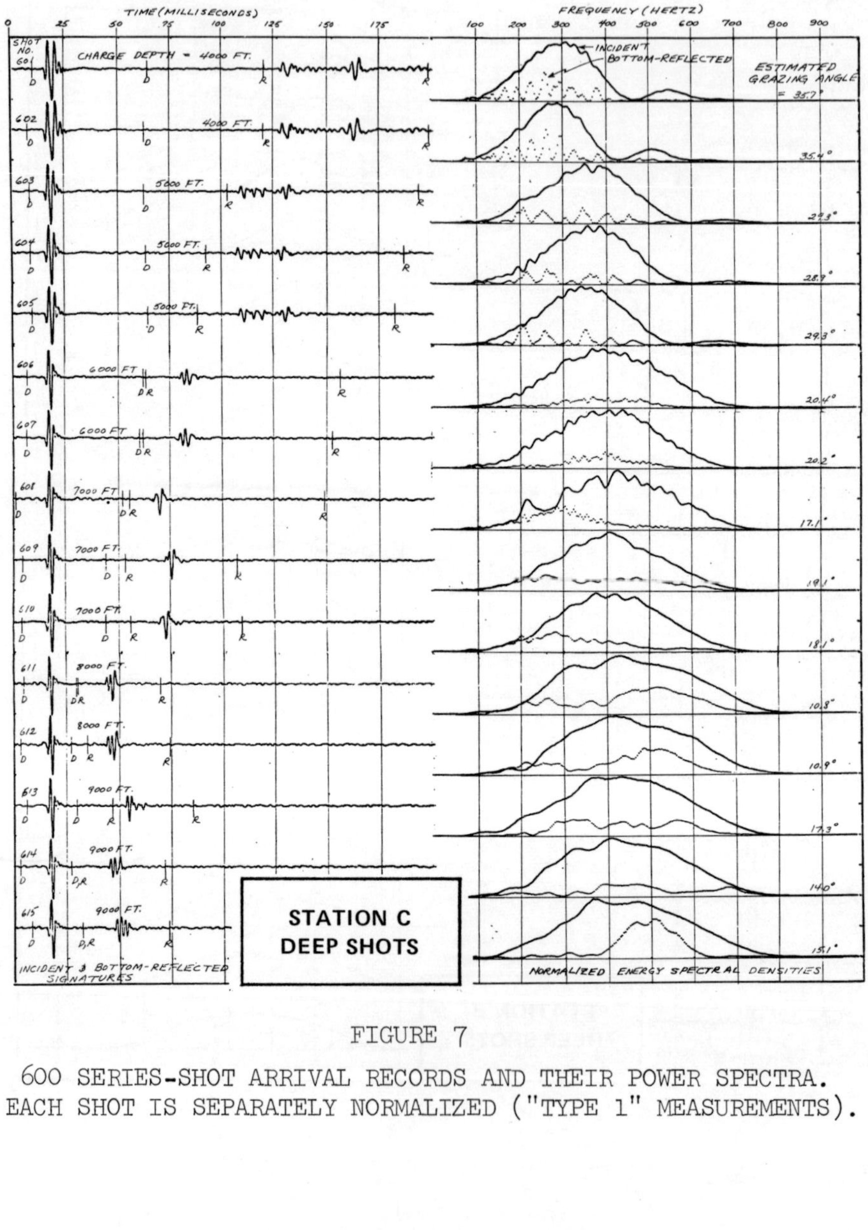

FIGURE 7

600 SERIES-SHOT ARRIVAL RECORDS AND THEIR POWER SPECTRA. EACH SHOT IS SEPARATELY NORMALIZED ("TYPE 1" MEASUREMENTS).

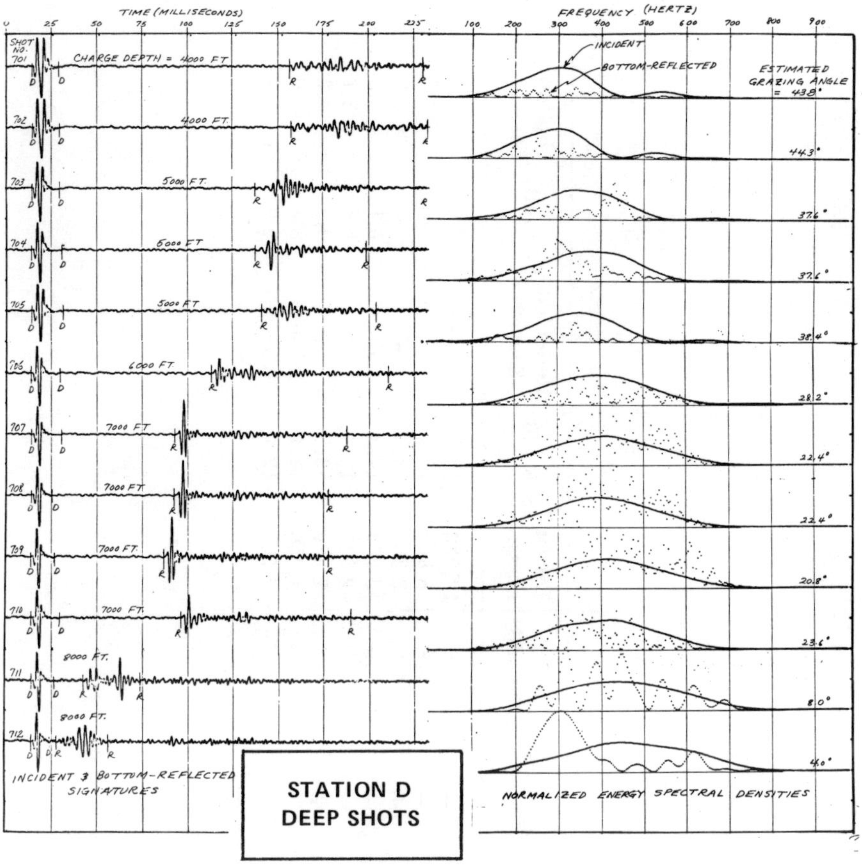

FIGURE 8

700 SERIES-SHOT ARRIVAL RECORDS AND THEIR POWER SPECTRA. EACH SHOT IS SEPARATELY NORMALIZED ("TYPE 1" MEASUREMENTS).

signal. This was done for twelve different times for each gate width and for each shot series and the average of these was used for each frequency to estimate the noise during the series. This noise energy is then subtracted from the third octave signals (+ noise) to give the best estimate of the signals alone, and thus the best estimate of the signal to noise energy ratios. The direct and reflected 1/3 octave energies then are used to determine reflectivities. These values, along with the appropriate signal to noise ratios and other important data, are tabulated in Tables I through V. In cases where the average energy of the noise was greater than the measured signal plus noise no value of signal energy or signal to noise is computed. Bottom loss is 10 log of the ratio of reflected signal to incident signal energy. All energies shown in the tables are arbitrary but consistent units. No values of bottom

TABLE I

BOTTOM-LOSS DATA
STATION A
SHOT SERIES 50

Shot			50 Hertz					100 Hertz				
No.	Estimated Angle	Depth (Ft)	Direct Energy	S/N (dB)	Reflected Energy	S/N (dB)	Bottom Loss (dB)	Direct Energy	S/N (dB)	Reflected Energy	S/N (dB)	Bottom Loss (dB)
50	45	4000	0.6736	11.4	—	—	>6.5	13.220	19.5	0.7392	3.0	12.5
51	45	4000	0.1838	5.8	—	—	>0.9	8.805	17.7	—	—	>13.8
52	37	5000	0.1983	6.1	—	—	>1.2	3.351	13.5	—	—	> 9.6
53	35	5000	0.1707	5.5	—	—	>0.6	3.979	14.3	1.490	6.1	4.3
54	24	6000	0.0470	-0.1	—	—	—	2.731	12.6	1.327	5.6	3.1
55	26	6000	0.0208	-3.7	0.0547	-4.4	—	2.986	13.0	1.148	4.9	4.2
56	20	7000	0.1345	4.4	0.0102	-11.7	—	1.915	11.1	2.390	8.1	-1.0
57	18	7000	0.0579	0.8	—	—	—	2.567	12.4	0.0954	-5.9	> 8.4

TABLE II

BOTTOM-LOSS DATA
STATION B
SHOT SERIES 100

Shot			50 Hertz					100 Hertz				
No.	Estimated Angle	Depth (Ft)	Direct Energy	S/N (dB)	Reflected Energy	S/N (dB)	Bottom Loss (dB)	Direct Energy	S/N (dB)	Reflected Energy	S/N (dB)	Bottom Loss (dB)
100	41	4000	—	—	—	—	—	9.080	18.0	20.340	21.5	-3.5
101	41	4000	—	—	—	—	—	6.850	16.8	15.860	20.4	-3.6
102	37	5000	0.384	3.1	0.120	-2.0	(GAIN ?)	5.030	15.4	4.970	15.4	0.1
103	36	5000	—	—	0.732	5.9	—	3.320	13.6	0.670	6.7	7.0
104	24	6000	—	—	—	—	—	0.834	7.6	0.921	8.1	-0.4
105	22	6000	0.026	-8.6	—	—	—	1.086	8.8	0.919	1.2	7.5
106	16	7000	0.033	-7.6	—	—	—	1.710	10.7	0.093	-1.9	>10.7
107	16	7000	0.187	-0.1	0.031	-7.8	—	0.595	6.2	0.147	0.1	6.1

TABLE III

BOTTOM-LOSS DATA
STATION B
100 HERTZ ONLY

Shot		Shot Series 200					Shot		Shot Series 300					Depth
No.	Est. Angle	Direct Energy	S/N (dB)	Reflected Energy	S/N (dB)	Bottom Loss (dB)	No.	Est. Angle	Direct Energy	S/N (dB)	Reflected Energy	S/N (dB)	Bottom Loss (dB)	(Ft)
201	25	6.070	10.8	25.310	12.8	-6.2	301	20	15.600	16.2	9.740	13.9	2.0	4000
202	24	0.903	1.7	0.890	0.1	—	302	20	8.150	13.3	12.600	15.0	-1.9	4000
203	22	9.890	12.9	2.330	2.5	6.3	303	20	6.330	12.2	8.320	13.2	-1.2	4000
204	11 (?)	2.150	2.1	7.750	5.9	—	304	17	3.010	9.0	3.390	8.3	-0.5	5000
205	14	4.440	9.4	15.700	9.0	-5.5	305	16	1.650	6.4	0.497	0.0	> 5.2	5000
206	23 (?)	4.950	9.7	14.500	8.7	-4.7	306	16	4.070	10.3	1.179	3.7	5.4	5000
207	16	3.890	9.3	—	—	> 2.9	307	13	1.971	7.2	2.226	7.5	-0.5	6000
208	17	2.530	7.0	—	—	> 1.1	308	12	1.636	6.4	0.029	-11.3	6.2	6000
209	16	1.310	4.4	0.983	0.7	—	309	13	0.459	0.8	0.280	-1.5	—	6000
210	15	2.590	7.3	2.570	4.7	0.0	310	11	0.120	-5.0	1.415	5.5	—	7000
211	14	1.780	5.8	0.740	-0.9	> 3.0	311	12	1.164	4.9	0.668	2.3	—	7000
212	14	0.723	2.0	2.160	3.7	—	312	11	2.211	7.7	1.083	4.4	3.1	7000
213	10	1.160	3.7	1.150	3.5	—	313	8	0.854	3.5	2.350	7.7	(GAIN)?	8000
214	9	—	—	3.330	5.2	—	314	8	0.596	2.0	1.600	6.1	(GAIN)?	8000
215	9	0.083	-7.6	4.950	6.5	—	315	8	—	—	—	—	—	8000
216	10	2.780	7.3	1.290	3.8	3.3	316	7	—	—	1.247	1.5	—	9000
217	9	—	—	3.560	6.0	—	317	7	1.107	4.7	1.064	0.9	—	9000
218	10	2.330	6.5	—	—	> 6.0	318	7	0.068	-7.4	2.047	3.7	—	9000

TABLE IV

BOTTOM-LOSS DATA
STATION C
SHOT SERIES 500 AND 600

Shot			50 Hertz					100 Hertz				
No.	Estimated Angle	Depth (Ft)	Direct Energy	S/N (dB)	Reflected Energy	S/N (dB)	Bottom Loss (dB)	Direct Energy	S/N (dB)	Reflected Energy	S/N (dB)	Bottom Loss (dB)
501	33	4000	0.2850	7.5	0.0740	-2.6	>3.3	8.720	18.7	5.740	12.6	1.8
502	33	4000	0.2620	7.1	0.1160	-0.6	>2.9	7.330	18.0	2.045	8.1	5.5
503	28	5000	—	—	—	—	—	3.100	14.2	1.955	7.9	2.0
504	26	5000	0.0470	-0.4	—	—	—	2.240	12.8	0.258	-0.9	>8.5
505	28	5000	—	—	—	—	—	2.340	13.0	1.215	5.9	2.8
506	19	6000	0.1340	4.2	—	—	—	1.073	9.6	2.335	8.7	-3.4
507	19	6000	—	—	—	—	—	1.093	9.7	1.255	6.0	-0.6
508	17	7000	0.0540	0.2	—	—	—	0.831	8.5	0.965	4.9	-0.6
509	14	7000	—	—	—	—	—	0.942	11.4	0.136	-4.1	>4.3
510	14	7000	—	—	—	—	—	0.523	8.9	0.043	-8.6	>2.2
511	6	8000	—	—	0.0840	2.2	—	0.516	8.8	0.095	-0.9	>6.4
512	8	8000	0.0240	-3.3	—	—	—	0.308	4.2	0.303	4.1	—
513	10	9000	0.0260	-2.9	—	—	—	0.595	7.1	0.265	3.6	3.5
514	8	9000	0.1030	3.1	—	—	—	0.422	5.6	0.346	4.7	0.9
515	9	9000	—	—	—	—	—	0.222	2.8	0.156	1.2	—

TABLE IV CONTINUED

STATION C
SHOT SERIES 500 AND 600

Shot			50 Hertz					100 Hertz				
No.	Estimated Angle	Depth (Ft)	Direct Energy	S/N (dB)	Reflected Energy	S/N (dB)	Bottom Loss (dB)	Direct Energy	S/N (dB)	Reflected Energy	S/N (dB)	Bottom Loss (dB)
601	33	4000	0.1715	4.9	—	—	—	6.070	14.7	1.770	7.0	5.4
602	33	4000	0.0842	1.8	0.0601	-3.1	—	10.660	17.1	7.950	13.5	1.3
603	27	5000	0.1870	5.3	0.1070	-0.5	>1.9	1.404	8.3	1.148	5.1	0.9
604	26	5000	0.0685	0.9	0.0138	-9.5	—	3.204	11.9	1.008	4.6	5.0
605	27	5000	0.0591	0.3	0.1820	1.8	—	2.774	13.4	0.569	2.1	6.9
606	18	6000	—	—	0.0022	-17.4	—	1.104	7.3	0.394	0.5	4.5
607	18	6000	0.0129	-6.3	0.0737	-2.2	—	2.344	10.5	0.768	3.4	4.8
608	14	7000	—	—	—	—	—	0.713	5.4	0.788	3.5	-0.4
609	16	7000	—	—	0.0389	-1.5	—	0.945	8.1	0.183	-0.5	>6.6
610	15	7000	0.0053	-8.4	—	—	—	1.183	9.0	0.251	0.8	6.7
611	8	8000	—	—	0.0828	+1.8	—	0.903	9.7	—	—	>7.9
612	8	8000	—	—	0.0089	-6.2	—	0.581	7.8	0.561	5.8	0.2
613	10	9000	—	—	0.0242	-1.8	—	1.794	12.7	—	—	>10.9
614	8	9000	0.0588	3.7	0.0725	+3.0	—	1.194	10.9	0.095	-1.9	>9.1
615	9	9000	0.0288	0.6	0.0519	+1.5	—	0.152	2.0	0.335	3.6	—

TABLE V

BOTTOM-LOSS DATA
STATION D
SHOT SERIES 700

Shot			50 Hertz					100 Hertz				
No.	Estimated Angle	Depth (Ft)	Direct Energy	S/N (dB)	Reflected Energy	S/N (dB)	Bottom Loss (dB)	Direct Energy	S/N (dB)	Reflected Energy	S/N (dB)	Bottom Loss (dB)
701	44	4000	0.3420	11.2	0.3120	4.6	0.4	9.940	17.2	4.015	7.9	3.9
702	44	4000	0.4160	12.1	0.1072	-0.1	> 5.8	9.080	16.8	2.725	6.3	5.2
703	38	5000	0.1043	6.1	0.0562	-2.9	> -0.2	6.880	15.6	2.625	6.1	4.2
704	38	5000	0.2030	9.0	0.0302	-5.6	> 2.7	2.580	11.3	5.385	9.2	-3.2
705	38	5000	0.0843	5.2	—	—	> -0.5	4.880	14.1	1.022	1.4	6.8
706	29	6000	0.2390	9.7	0.1140	-1.0	> 2.2	2.500	11.2	0.468	-3.4	> 3.9
707	22	7000	0.0166	-1.9	0.3100	3.3	—	2.110	10.4	1.968	2.8	0.3
708	22	7000	0.1340	7.2	0.0714	-3.1	> -0.3	1.109	7.6	1.758	2.4	-2.0
709	21	7000	0.0228	-0.5	0.2184	1.8	—	0.491	4.1	1.178	0.6	—
710	24	7000	0.0463	2.6	0.1630	0.5	—	1.127	7.7	2.468	3.8	-3.4
711	8	8000	0.0111	-3.0	—	—	—	0.591	6.3	0.474	-3.3	> -2.4
712	4	8000	0.1101	7.0	—	—	> -1.2	—	—	0.122	-9.2	—

loss are calculated when the estimated signal to noise ratio for the direct signal is less than 5 dB. In cases where the signal to noise value of the reflected signal is less than zero but the above criterion is met the bottom loss estimated is greater than the value obtained and tabulated, which is the 10 log of the noise to direct energy ratios. All of the data shown in these tables are for the variable depth shots.

Tables I through V give bottom loss as a function of grazing angle for each set of data. It is clear that it is not possible to make useful estimates for a reflection loss model. The 50 hertz data has, generally, too low signal to noise to be useful, and the 100 hertz has too wide a variation. But note: the values and their variation are consistent with others given at this conference and in the literature (Morris, 1972; Bucker et al., 1965) for low frequencies analyzed over third octave bands.

In order to understand better the reasons for the lack of precision it is useful to look at shot records in Figs. 3 to 8.

It is clear in many of these that two significant arrivals are present. This gives rise to a well known frequency interference effect which is mathematically demonstrated for a simple case in Appendix A. If the arrival time is of the order of the inverse bandwidth, the variation in the arrival times (τ) will vary the positions of the maxima and minima to a sufficient degree to give significant apparent "fluctuations" in the third octave band processed signal. This will be true even more when noise is present. Thus some of the problem is, in fact, due to the choice of third octave rather than to the actual phenomena occurring. As the signals at low frequency get weaker (as depth increases) noise becomes an important factor. Thus for this technique small charges become inadequate from a signal to noise point of view just at the low grazing angles that we normally associate with Rayleigh critical angles. This is obviously the case for our 50 Hz case and is still true for some of the 100 Hz data.

In this experiment the geological differences in the bottom are not significant quantitatively, and therefore are not discussed. It should be noted that the first arrivals vary over about one mile for the geometry of the experiment. On general slopes of a few degrees it should be considered rare to find a uniformly sloping ocean floor over a mile. Thus it shouldn't be surprising that large variations were observed. There is no way for us to determine the statistics of slope to a sufficient resolution on the slope to refine the estimate of angle of incidence beyond the lowest approximation.

Shallow Shots

Samples of the "type 2" measurements, with the surface shots, are shown in Figs. 9 to 11. Criteria for reflectivity are the same as for the "type 1" measurements. Note the similarities with the type 1, but note the much larger span of arrival time that is required to include all of the significant reflected energy. It is difficult to see any coherence in arrivals; the power spectra has a random character. This is of no great surprise since the signal interacts with a very large portion of the ocean floor purely due to the geometry of the measurement. It is clear from the geometry that the portion of the sea floor contributing to reflection is significantly greater than for the "type 1" measurements and that the region varies by much more as grazing angles change from 90° to 0°. It is also important to note that the direct arrival is not necessarily a single arrival and therefore, by our technique, also has an oscillatory power spectrum. A comparison of the power spectrum of the direct arrivals in Figs. 4, 5, and 6 with those of Figs. 9 and 10, all at the same station, shows this. Thus if both the direct and the reflected arrivals have interference structure it is not surprising to have the reflected arrival occasionally appear, in third octave analysis, to have more energy than the direct purely due to the banding in the power spectrum. This can give rise to the occasionally observed phenomena of reflectivities greater than unity. These can also occur when the reflected surface is concave and gives some degree of focusing and when the signal to noise ratio gives peculiar results.

Both types of measurements reported above show the enormous complexities that arise in a "simple" reflectivity experiment. Broadband results taken over such experiments clearly do not give slowly varying reflectivity numbers, for third octave low frequency values.

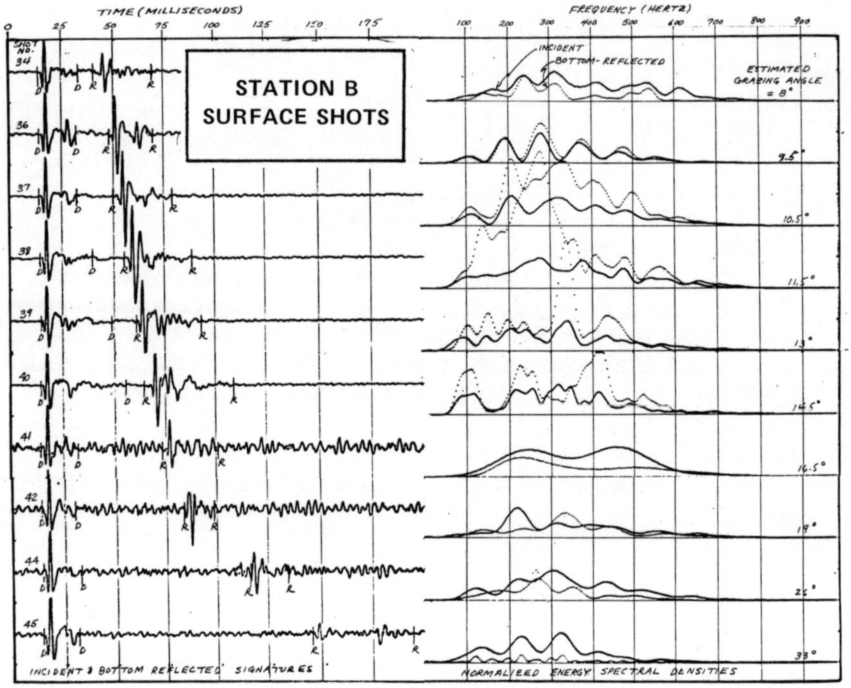

FIGURE 9

30 SERIES-SHOT ARRIVAL RECORDS AND THEIR POWER SPECTRA.
EACH SHOT IS SEPARATELY NORMALIZED ("TYPE 2" MEASUREMENTS).

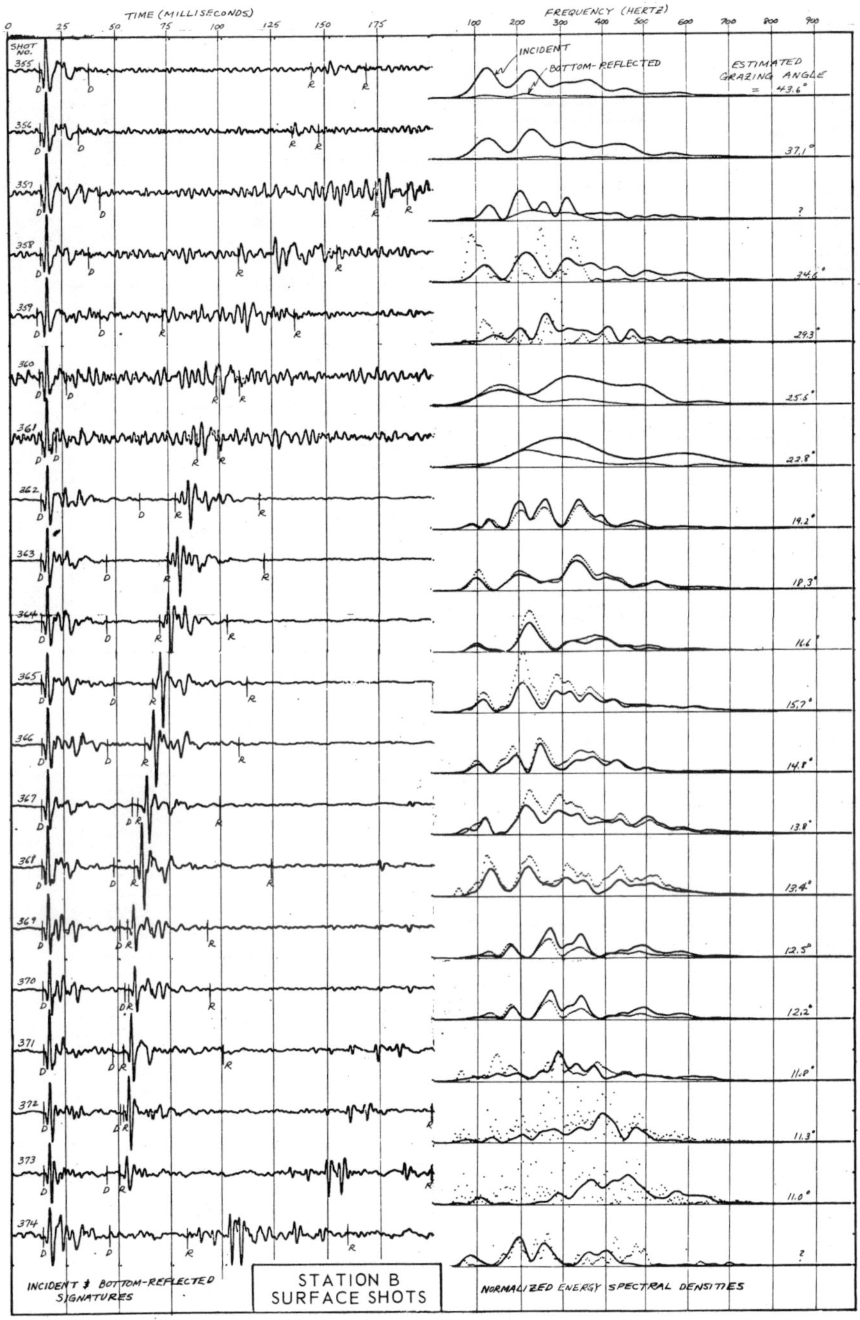

FIGURE 10

350 SERIES-SHOT ARRIVAL RECORDS AND THEIR POWER SPECTRA. EACH SHOT IS SEPARATELY NORMALIZED ("TYPE 2" MEASUREMENTS).

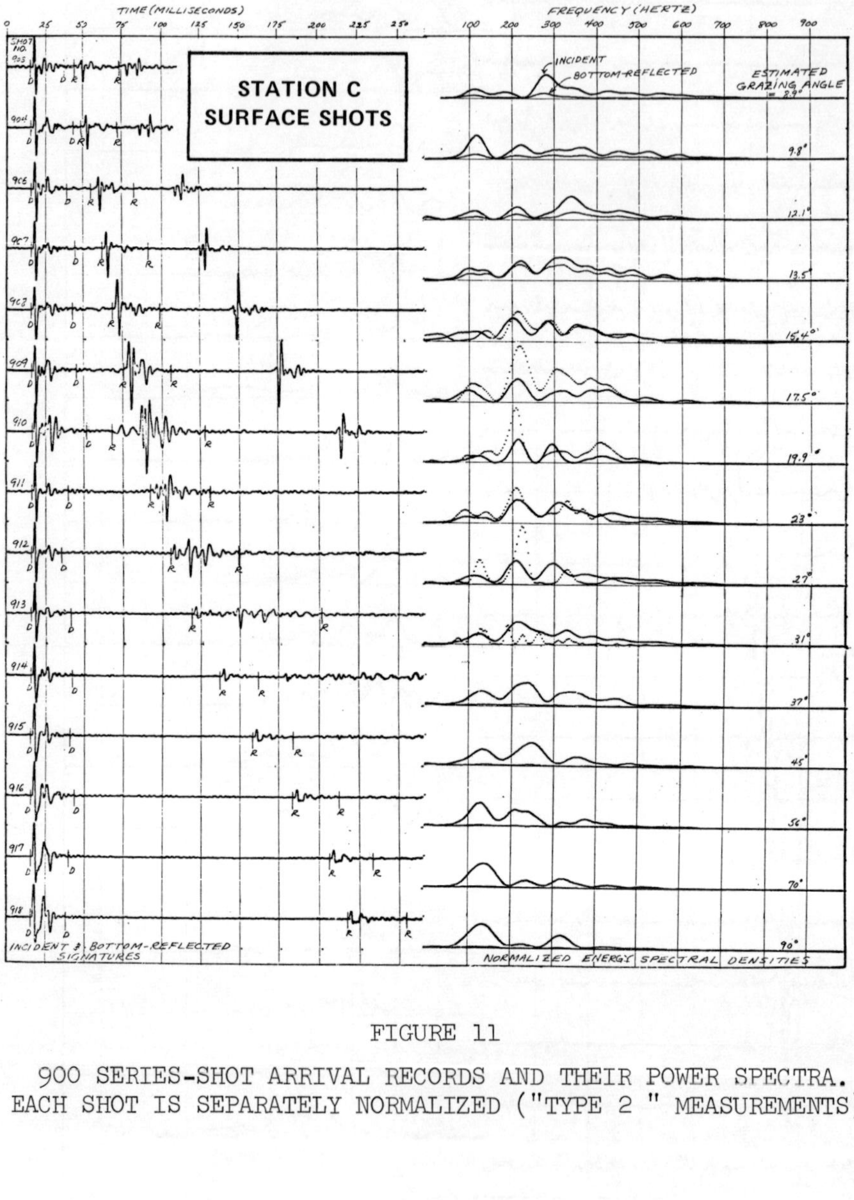

FIGURE 11

900 SERIES-SHOT ARRIVAL RECORDS AND THEIR POWER SPECTRA. EACH SHOT IS SEPARATELY NORMALIZED ("TYPE 2 " MEASUREMENTS).

CONCLUSIONS

Our broadband experiments have led us to a series of conclusions, many of which are obvious but have been ignored in some measurements and surveys in the past.

1. Deep shots are unsatisfactory unless they are of sufficient mass to give a substantial signal to noise ratio. Since the power increases as $W^{1/3}$, where W is the weight, large quantities of explosives are required, making the experiment cumbersome and expensive.

2. Surface shots also must be large, but in general measurements are easier to perform. The arrival structure is more complicated, however, and is difficult to interpret.

3. Due to the multi-arrival of layered media the use of indiscriminate broadband analysis can give rise to apparent fluctuations that are easily interpreted, incorrectly, as being due to geological changes.

4. The errors in estimation of angles of incidence on sloping ocean bottoms is impossible to assess without elaborate microbathymetric measurements heretofore not made.

5. Due to all of the large variations in arrivals and therefore in computed reflectivities, even in the cases where signal to noise is sufficient, it is currently impossible to discuss transmission loss as a single valued deterministic function of range when bottom interactions are important. This leads us to the conclusion that a stochastic approach to many systems must be used.

It must be remembered that our measurements were made on sloping ocean bottoms. Thus much of what we discuss is not the case for ideal flat ocean floor, the data for which are also difficult to interpret (Stickler 1972). However, for long range propagation over hundreds of miles the ocean bottom is generally important only when it intrudes into the sound channel, which means of course that it is sloping. It is important to do adequate ray tracing to know which regions need to be considered.

Our results conclusively show the need for narrowband processing of signals and a detailed comparison with the arrival structures to insure adequate understanding of the process of reflection. This vitiates the application of general survey techniques.

It is clear to us that the performance of any Navy system that utilizes low frequency bottom reflection is going to be difficult to predict with our current approaches and lack of good models of the ocean floor. We need to know enough to assess the small scale statistical character of the ocean floor as well as some of the deterministic and large scale properties that we have heard about at this conference.

APPENDIX A

The phenomena of frequency interference is easily understood by the following simple case. Consider two arrivals of a signal $f(t)$ differing by a time τ and differing in amplitude by a ratio A. The arrival is thus given by:

$$f(t) + A\, f(t+\tau)$$

The Fourier transform of this is given by

$$g(\omega) + A\, e^{i\omega\tau}\, g(\omega)$$

where $g(\omega)$ is the FT of $f(t)$. The power spectrum is

$$P(\omega) = g(\omega)\, g^*(\omega)\, (1+A\, e^{i\omega\tau})\, (1+A^*\, e^{-i\omega\tau})$$

$$= |g|^2 \left(1+A^*A + (A\, e^{i\omega\tau}+A^*\, e^{-i\omega\tau})\right)$$

or, if A is real,

$$P(\omega) = |g|^2\, (1+A^2+2A \cos \omega\tau) \quad .$$

Thus there is an oscillating term $2A \cos \omega\tau$. The coefficient of $|g|^2$ is an oscillating term varying between $(1+A)^2$ and $(1-A)^2$ in the radial frequency interval π/τ.

REFERENCES

H. E. Morris, "Comparison of Calculated and Experimental Bottom-Reflection Losses in the North Pacific," Naval Undersea Report NUC TP 327, December 1972.

H. P. Bucker et al., "Reflection of Low Frequency Sonar Signals from a Smooth Ocean Bottom," J. Acoust. Soc. Am. 37, 1037-1051, 1965.

D. Stickler, "The Principles Involved in the Measurement of Bottom Loss," Bell Telephone Laboratories Report 72R245, February 1972.

SUBJECT INDEX

Attendees of Symposium on
Physics of Sound in Marine Sediments
8, 9, 10 May 1973
Austin, Texas

1.	Malahoff, Alex	ONR
2.	Akal, Tuncay	SACLANT ASW Research Centre
3.	Muir, Tom	Applied Research Laboratories The University of Texas at Austin
4.	Celikkol, Barbaros	University of New Hampshire
5.	Inderbitzen, A. L.	University of Delaware
6.	Stevenson, Robert E.	ONR
7.	Lucas, Frank	ONR
8.	Hampton, Loyd D.	Applied Research Laboratories The University of Texas at Austin
9.	Goodman, Ralph R.	Naval Research Laboratory
10.	Lewis, Lloyd	NavFac
11.	Saunders, Don	Naval Intelligence Support Center
12.	Ahuja, Avtar S.	Rust College
13.	Anderson, Rockne S.	U. S. Naval Oceanographic Office
14.	Horn, David	Lamont-Doherty Geological Observatory of Columbia University
15.	Bunce, Elizabeth T.	Woods Hole Oceanographic Institution
16.	McKinney, Chester M.	Applied Research Laboratories The University of Texas at Austin
17.	Heinrichs, Donald	ONR
18.	Lewis, M. G.	Navy Electronics Systems Command
19.	Kibblewhite, A. C.	University of Auckland
20.	Gallagher, J. J.	Naval Underwater Systems Center New London Laboratory
21.	Bucker, Homer	Naval Undersea Center - San Diego
22.	Wilson, O. B.	Naval Postgraduate School
23.	Nacci, Vito A.	University of Rhode Island
24.	Lonsdale, Peter	Scripps Institution of Oceanography
25.	Stoll, R. D.	Columbia University of the City of New York
26.	Hamilton, Edwin L.	Naval Undersea Center - San Diego
27.	Maynard, G. L.	Georgia Institute of Technology
28.	Andrews, Robert W.	Naval Postgraduate School
29.	Clay, C. S.	University of Wisconsin
30.	Greenfield, R. J.	Applied Research Laboratory Penn State University
31.	Houtz, R. E.	Lamont-Doherty Geological Observatory of Columbia University
32.	Urick, R. J.	Naval Ordnance Laboratory
33.	Bell, Thad	Naval Underwater Systems Center New London Laboratory
34.	Sutton, George H.	University of Hawaii
35.	Anderson, A. L.	Applied Research Laboratories The University of Texas at Austin
36.	Shirley, Don J.	Applied Research Laboratories The University of Texas at Austin
37.	Thompson, Lewis	Applied Research Laboratories The University of Texas at Austin
38.	Bryan, George M.	Lamont-Doherty Geological Observatory of Columbia University

SYMPOSIUM ON PHYSICS OF SOUND IN MARINE SEDIMENTS
8, 9, 10 MAY 1973
AUSTIN, TEXAS